Advances in Experimental Medicine and Biology

Advances in Experimental Medicine and Biology presents multidisciplinary and dynamic findings in the broad fields of experimental medicine and biology. The wide variety in topics it presents offers readers multiple perspectives on a variety of disciplines including neuroscience, microbiology, immunology, biochemistry, biomedical engineering and cancer research. Advances in Experimental Medicine and Biology has been publishing exceptional works in the field for over 30 years and is indexed in Medline, Scopus, EMBASE, BIOSIS, Biological Abstracts, CSA, Biological Sciences and Living Resources (ASFA-1), and Biological Sciences. The series also provides scientists with up to date information on emerging topics and techniques.

More information about this series at http://www.springer.com/series/5584

Xiangdong Wang • Dragos Cretoiu
Editors

Telocytes

Connecting Cells

Springer

Editors
Xiangdong Wang
Zhongshan Hospital
Fudan University Medical School
Shanghai
China

Dragos Cretoiu
Department of Cellular and Molecular Med
Carol Davila Univ of Medicine&Pharmacy
Bucharest
Romania

ISSN 0065-2598 ISSN 2214-8019 (electronic)
Advances in Experimental Medicine and Biology
ISBN 978-981-10-1060-6 ISBN 978-981-10-1061-3 (eBook)
DOI 10.1007/978-981-10-1061-3

Library of Congress Control Number: 2016955050

Printed on acid-free paper

This Springer imprint is published by Springer Nature
The registered company is Springer Science+Business Media Singapore Pte Ltd.

In memory of Professor Laurentiu M. Popescu

Contents

Chapter 1
The History of Telocyte Discovery and Understanding

Jian Wang, Meiling Jin, Wen-huan Ma, Zhitu Zhu, and Xiangdong Wang

Abstract Telocytes (TCs) are identified as a peculiar cell type of interstitial cells in various organs. The typical features of TCs from the other cells are the extending cellular process as telopodes with alternation of podomeres and podoms. Before the year of 2010, TCs were considered as interstitial Cajal-like cells because of the similar morphology and immunohistochemical features with interstitial cells of Cajal which were found more than 100 years ago and considered to be pacemakers for gut motility. Subsequently, it demonstrated that TCs were not Cajal-like cells, and thus the new name "telocyte" was proposed in 2010. With the help of different techniques, e.g., transmission electron microscopy, immunohistochemistry, or omics science, TCs have been detected in various tissues and organs from different species. The pathological role of TCs in different diseases was also studied. According to observation in situ or in vitro, TCs played a vital role in mechanical support, signaling transduction, tissue renewal or repair, immune surveillance, and mechanical sensor via establishing homo- or heterogenous junctions with neighboring cells to form 3D network or release extracellular vesicles to form juxtacrine and paracrine. This review will introduce the origin, distribution, morphology, functions, omics science, methods, and interaction of TCs with other cells and provide a better understanding of the new cell type.

J. Wang • M. Jin • X. Wang (✉)
Zhongshan Hospital, Shanghai Institute of Clinical Bioinformatics,
Fudan University Center for Clinical Bioinformatics, Clinical Science Institute of Fudan University Zhongshan Hospital, Shanghai, China
e-mail: Xiangdong.wang@clintransmed.org

W.-h. Ma
Zhabei District Hospital of Traditional Chinese Medicine,
Yanchang Middle Road No. 288, Jingan District, Shanghai, China

Z. Zhu (✉)
Jinzhou Hospital of Liaoning Medical College, Jinzhou, China
e-mail: zhuzhitu@163.com

X. Wang, D. Cretoiu (eds.), *Telocytes*, Advances in Experimental Medicine and Biology, 913, DOI 10.1007/978-981-10-1061-3_1

1.1 Introduction

Telocytes (TCs) are described as a new type of interstitial cell with extremely long and thin telopodes (Tps). Since Popescu et al. [1] pioneered to name this special interstitial cell as "telocyte" to distinguish from other interstitial cell types (e.g., fibroblasts, fibrocytes, fibroblast-like cells, mesenchymal cells), increasing attention has been paid from lots of teams who have identified the existence of TCs in various cavitary and non-cavitary organs from human and laboratory mammals, respectively. Although flow cytometry and immunohistochemistry are introduced to study TCs, transmission electron microscopy (TEM) is still the "gold standard" to identify TCs in different tissues and organs so far. TCs are revealed to participate in intercellular signaling transduction, regeneration, or immunoregulation, but more unknown functions of TCs need to be explored.

There are a total of 132 records about "telocyte" between 2010 and 2015 with 2,829 citations in the Web of Science. The average citation of every article reaches up to 21.4. Importantly, the amount of publication and citation about TCs presents an ascending trend yearly (Fig. 1.1). This review will introduce the origin, distribution, morphology, functions, omics science, methods, and interaction of TCs with other cells to lift the veil of TCs.

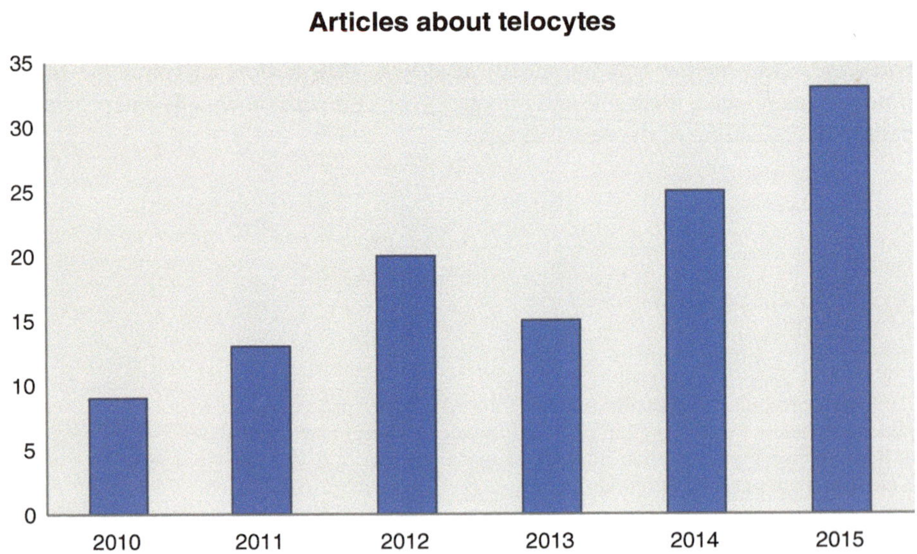

Fig. 1.1 The amount of publication about telocytes in between 2010 and 2015 based on the Web of Science

1.2 Origin

In 1889, the great Spanish neuroanatomist Santiago Ramo´n y Cajal was the first to describe a spindle-shaped or triangular cell with network-like appearance in the muscle coat of the gut with the Golgi stain and name as the "interstitial neurons" [2]. The particular cells were depicted to exist in the interstitium between nerve endings and smooth muscle cells (SMC) to intermingle with SMC and nerve fibers. However, the existence, nature, and function of the cells attracted a great of controversy in different scientific cliques in the subsequent seven to eight decades. By the 1960s, Taxi et al. [3] firstly described the interstitial cells of Cajal (ICC) as a new cell type which was distinguished from neurons, Schwann cells, smooth muscle cells, fibroblasts, and macrophages with ultrastructural evidences from TEM techniques. Further, Thuneberg, Faussone-Pellegrini, and Rumessen separately proved that ICC was not neuron and played a role as pacemakers for gastrointestinal movement and neurotransmission [4, 5].

Another development of knowledge in this field was that ICC was thought as the origin of gastrointestinal stromal tumors (GIST) and might be associated with several gastrointestinal disorders by pathologists [6]. ICC has been the fascinating subject to researchers and was reported by plentiful relevant publications every year. The positive-expressed c-kit/CD117 receptor in ICC was found to help investigators understand the cell better at the light microscopy level by immunohistochemical techniques, although TEM was the main choice to define ICC at present [7].

The interstitial cells of Cajal were continually detected outside the gastrointestinal musculature after the year 2000. As for the similar morphology and immunohistochemical features, these cells were renamed as "Interstitial Cajal-like cells (ICLC)." The presence of ICLC has been reported in the interstitium of several organs [1, 8]. Faussone-Pellegrini MS et al. [9] focused on gastrointestinal muscle coat of humans once again and proved that ICC and ICLC coexisted in the muscle coat as two unequivocally different cell types. They also listed the different ultrastructural and immunohistochemical characteristics of the ICC, ICLC, fibrocytes, and fibroblasts of the human GI tract.

Popescu et al. [10] and Ciontea et al. [11] confirmed that the ICLC was a distinct cell type from the classic ICC in the myocardium and uterus, respectively. It was necessary to give a new name for ICLC to avoid confusion from ICC because ICLC is a peculiar and independent cell type. In 2009, Popescu et al. [1] proposed to name ICLC as "telocyte" for the first time. The term "telocyte" means the cell bearing long prolongations on the basis of the peculiar feature that are extremely long (tens to up to hundreds of micrometers) and thin (mostly below 0.2 mm) and with a moniliform aspect under TEM [12]. From then on, a new scientific door was open to "telocyte."

1.3 Distribution and Related Diseases

TCs, formerly known as ICLC, are distributed widely in the connective tissue and have been identified in various organs from different species (Table 1.1). As reported, TCs have been found in cardiac system (e.g., epicardium [13], endocardium [14], myocardium [15, 16], or heart valves [17]), reproductive system (e.g., prostate [18, 19], penis [20], uterus [21, 22], placenta [23], or oviduct [24]), urinary system (e.g., kidney [25, 26], ureter [25], urethra [27], or bladder [28, 29]), digestive system (e.g., esophagus [30], stomach [9], duodenum [31], jejunum [32], ileum [33], colon [9], pancreas [8, 34], liver [35], or gallbladder [36]), respiratory

Table 1.1 Telocyte distribution in various organs from different species and TC-related diseases

Organs	Species	Diseases
Cardiac system		
Epicardium	Human	Myocardial infarction
Endocardium	Mouse, rat	
Myocardium	Human, mouse, rat	
Heart valves	Human	
Respiratory system		
Lung and trachea	Human, mouse	COPD
Pleura	Human, mouse	
Digestive system		
Esophagus	Human, rat	Crohn's disease
Stomach	Human	PEComas and GISTs
Duodenum	Human, rat	Ulcerative colitis
Ileum	Mouse, chicken	Liver fibrosis
Jejunum	Rat	Gallstone
Colon	Human	
Liver	Mouse	
Pancreas	Human, *Octodon degus*	
Gallbladder	Human	
Reproductive system		
Prostate	Human, gerbils	Preeclampsia
Penis	Guinea pig	Acute salpingitis
Uterus	Human, rat	
Oviduct	Chinese soft-shelled turtle, *Pelodiscus sinensis*	
Placenta	Human, mouse	
Urinary system		
Kidney	Human, rat	Renal tubule IRI
Ureter	Human, rat	
Urethra	Human	
Bladder	Human, rat	

(continued)

Table 1.1 (continued)

Organs	Species	Diseases
Vessels		
Artery	Rat, mouse, guinea pig	Atrial amyloidosis
Vein	Human, rat, rabbit	
Lymphatics	Rat	
Glands		
Salivary glands	Human	Breast cancer
Mammary glands	Human	
Parotid glands	Human, rat	
Others		
Mesentery	Rat	
Skin	Human	Skin carcinomas, systemic sclerosis
Spleen	Rat	Sjögren's syndrome
Eyes	Mouse	Psoriasis
Skeletal muscle	Human, rat	
Meninges and choroid plexus	Human, rat, mouse	
Neuromuscular spindles	Human	
Trigeminal ganglion	Human	
Bone marrow	Mouse	
Fascia lata	Human	
Temporomandibular joint disc	Rat	

system (e.g., lung and trachea [37, 38], pleura [39]), vessels (e.g., mesenteric artery [40], aorta [41], coronary arteries [42], internal thoracic arteries [42], carotid arteries [42], portal vein [43], pulmonary vein [44], or lymphatics [45]), glands (e.g., parotid glands [46], mammary glands [47, 48], minor salivary glands [49]), skin [50, 51], skeletal muscle [52], trigeminal ganglion [53], meninges and choroid plexus [54], neuromuscular spindles [55], eyes [56], temporomandibular joint disc [57], bone marrow [58], spleen [59], fascia lata [60], or mesentery [61]. Meanwhile, the species involved in researches are wide, including human [13, 23, 39], rat [31, 38, 62], mouse [14, 37, 56], gerbils [19], degus [63], pig [28], chicken [33], rabbit [43], sheep [45], cow [64], cat [64], dog [64], or the Chinese soft-shelled turtle [65]. In pathological condition, TCs play a potential role in different diseases, such as atrial amyloidosis [66], myocardial infarction [67], skin carcinomas [68], breast cancer [69], PEComas (perivascular epithelioid cell tumors) and GISTs [70], systemic sclerosis [71], gallstone [72], Crohn's disease [73], preeclampsia [74], ulcerative colitis [75], liver fibrosis [76], Sjögren's syndrome [49], acute salpingitis [77], psoriasis [78], chronic obstructive pulmonary disease (COPD) [79], or ischemia-reperfusion injury (IRI) in renal tubules [80]. All these identified TCs can make up a 3D network and intermingle with surrounding cells (e.g., fibroblasts, muscle cells, immune cells, adipocytes, or nerve cells). Therefore, we make a conclusion that TCs are a distinct cell type in the connective tissue.

1.4 Morphology

TCs have a small-sized body with several extremely long and thin telopodes, which can be remembered easily at the first glance. On TEM image, the average size of cellular body is 9.39±3.26 μm (Fig. 1.2). The shape of the cellular body is associated with the number of telopode, such as piriform with one telopode, spindle with two telopodes, triangular with three telopodes, and stellate or polygonal with over three telopodes. The oval-shaped nucleus has no obvious nucleolus, and the clusters of heterochromatin are attached to the nuclear envelop, which makes up 25 % of the body [1]. The periphery of nucleus is surrounded by a thin rim of the cytoplasm. The perinuclear cytoplasm contains several mitochondria, a small Golgi apparatus, few rough and smooth endoplasmic reticulum, and cytoskeletal elements. The mitochondrium is rich and makes up 2 % of the cell body. The small Golgi apparatus and smooth endoplasmic reticulum are less than 1–2 % of cell volume and mainly are detected in dilations of cell processes [79]. Ultrastructurally, TCs contain many intermediate filaments for cytoskeleton. Several caveolae are located on the cell membrane, but the basal lamina is absent. As it is measured, the caveolae make up 2–3 % of the cell body and present ~0.5 caveolae/μm of cell membrane length [1, 50].

The number of telopode is varying in TCs. Most of them present two to three telopodes on a single section relying on the site and angle of section, because the 3D convolutions prevent them to be observed at their full length in a 2D section [1]. The length of telopode varies from tens to hundreds μm on the TEM image. The average thickness is difficult to define as for its moniliform appearance with alternation of thin segments and dilations. The thin segments are named as podomeres, which are

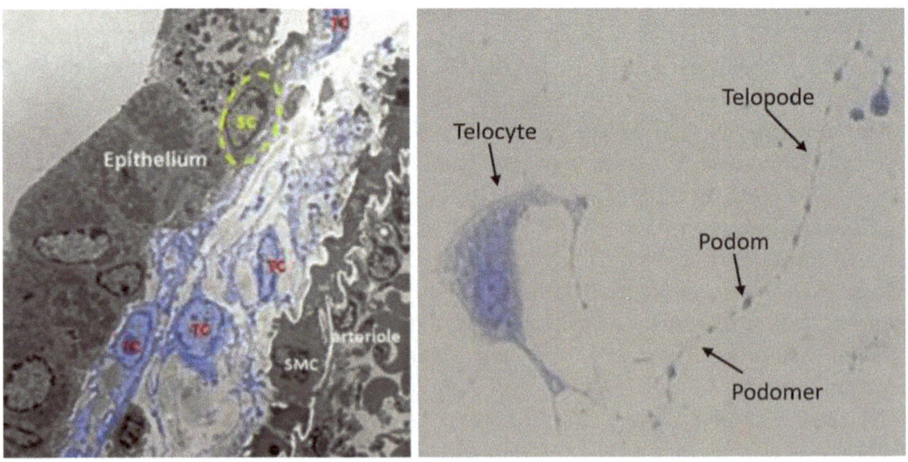

Fig. 1.2 The typical morphology of telocytes in situ and in vitro. The shape of the cellular body is piriform, spindle, triangular, or stellate. The cellular process is called telopodes with alternation of podomeres and podoms

only 0.10±0.05 μm in thickness. Meanwhile, the dilated segments are named as podoms and detected as Ca2+ uptake/release units. There are mitochondria, rough and smooth endoplasmic reticulum, and caveolae in podoms on the ultrastructural image [12, 79]. Telopode is the typical feature to distinguish TCs from neuronal dendrites, ICCs, fibroblasts, mesenchymal stem cells, and other interstitial cells, and it is considered to be the longest prolongation in the tissue despite of some axons of special neuron. TCs can interconnect with homo- or heterogenous cells to form a 3D network through intercellular junctions (such as gap or adherens junctions), budding shed vesicles and exosomes, and dichotomous branching pattern [81–83].

1.5 Functions

TCs make homo- or heterocellular contacts to form the structure of 3D network and thus provide a mechanical support throughout the tissue. Faussone-Pellegrini et al. [61] found that TCs had many processes and were attached to neighboring ones with so-called desmosome junctions to form a continuous 3D network in rat mesentery. Caveolae and attachment plaques lead to connecting cellular processes with the extracellular matrix. Thus, it is hypothesized that the TCs' network is resistant and deformable following intestinal loop repletion/depletion and controls mesenteric blood vessel rheology. Similarly, the TCs' network is considered to be resistant to and deformable following intestine movements in the myenteric plexus of gastrointestinal muscle coat [9]. In adult human neuromuscular spindles, TCs are detected to be located in the inner and outermost layer of the capsule and may provide mechanical support to control muscle tone and motor activity with their special microenvironment [55]. In addition, the 3D network formed by TCs in human urinary bladder provides a structural support to protect from anomalous deformation of the bladder wall with its distension and relaxation [29].

Another important role for TCs in 3D network is to be active and transmit intercellular signaling. Recently, accumulating evidence shows that TCs can establish direct cell-to-cell contact or a juxtacrine and/or paracrine association by budding shed vesicles, exosomes, or other micromolecule in different organs. Popescu et al. [84] showed that TCs formed complex and atypical junctions with homo- or heterogenous cells in the human heart. Meanwhile, small dense structures (10–15 nm nanocontacts) which are able to directly connect TCs with cardiomyocytes are identified in rat heart [81]. They can also take part in a juxtacrine intercellular signaling process or facilitate a paracrine signaling process [82]. The release of shed vesicles and exosomes from telopodes is noted to transport different macromolecular signals to the surrounding cells and regulate physiological and pathological process in various organs [34, 82, 85]. As it is reported, TCs express estrogen (ER) and progesterone (PR) receptors to be hormonal sensors to participate in myometrial contractions and fallopian tube motility via gap junctions or juxtacrine and/or paracrine mechanisms [86, 87]. Also, TCs secrete vascular endothelial growth factor (VEGF), nitric oxide (NO), IL-6, some chemokines, and even microRNAs to affect and control surrounding cells [67, 88]. In the gastrointestinal muscle coat, it is suggested that TCs intertangle with ICC to participate in spreading the slow waves generated by ICC [9]. Besides, Popescu et al. [89] proved that T-type calcium channels were present in TCs in human myometrium and played a role in activating endogenous bioelectric signals to regulate myometrium contractions.

The role of TCs in tissue regeneration or repair, considered as a vital function for TCs, has been identified in various organs in vivo or in vitro. The stem cell-based tissue regeneration therapy is thought to be a potential and promising method to treat some troublesome diseases in different tissues and organs, such as the heart [90], liver [91], lung [92], kidney [93], and nerve [94]. However, a deficiency in the interstitial to support for proliferation and differentiation of exogenous stem cells in tissues leads to the decline in the treatment effect of stem cell therapy. Now, the growing evidences show that there is a close contact between TCs and stem cells and progenitor cells in stem cell niches in various tissues and organs [35, 51, 54, 56, 80, 84, 85, 95, 96]. The 3D interstitial network constructed by TCs not only provided a mechanical support for stem cells and progenitor cells but also promoted proliferation, differentiation, maturation, and migration of stem cells and progenitor cells through atypical junctions and juxtacrine or paracrine [22, 52, 83, 85, 97].

Noteworthy, Popescu et al. [98] fully elaborated that TCs acted as a stromal supporting cells and connected with cardiac stem cells (CSCs) and cardiomyocyte progenitors (CMPs) in the cardiac stem cell niche (CSCN) to sustain a continuous cardiac renewal process. It is also reported that cardiac TCs can secrete several cytokines and chemokines and even microRNAs to CSCs to regulate stem cell proliferation and differentiation [88, 99]. In addition, TCs are supposed to enhance angiogenesis in myocardial infarction models through secreting VEGF and NOS2 to promote regenerating and repairing injured myocardium [67, 100, 101]. Besides, it is speculated that TCs itself may be the subpopulation of epicardium-derived progenitor cells because both of them express the same surface molecules such as c-kit and PDGFR-β [102]. Thus, cardiac TCs are considered as a source of progenitor cells in myocardial regeneration and repair. Similarly, the relationship between TCs and stem cells or progenitor cells is also identified in the lung [85], liver [35], eyes [56], skin [50], skeletal muscle [52], heart valves [17], and meninges and choroid plexus [54], but the functional mechanism needs to be further clarified.

TCs also play a role in immune surveillance and immune homeostasis. It is found that TCs present a contact with different immune cell in tissues [33, 77, 103]. In vitro, TCs can activate and regulate murine macrophages by secreting cytokines (e.g., IL-6, TNFα, IL1-R1 and IL-10) [103]. The global analysis of gene expression of different chromosomes in TCs also provides a support to the role of TCs in immune modulation [104, 105]. TCs are supposed to act as a mechanical sensor to participate in contraction modulation. Popescu et al. [106] showed that the growth rate of TCs from pregnant myometrium presented more active with low-level laser stimulation (LLLS) than that from nonpregnant myometrium. The extension of telopodes is inhibited by mibefradil which is known to inhibit the bioelectrical signal and uterine contractile forces. Thus, it is considered that TCs may involve in mechanical modulations to regulate and control uterine contraction. Besides TCs are located in neuromuscular spindle which is an encapsulated proprioceptor and contacts with striated muscle fibers, nerves, and blood vessels by cell-to-cell signaling. It also should take into account that TCs take part in perceiving stimulation to control muscle tone and activity [55].

1.6 Omics Science

The gene expression profile of TCs is firstly identified by comparing with mesenchymal stem cells and fibroblasts in mice by DNA microarray [107]. Differentially expressed genes in these three cells provide an evidence that TCs are a distinct interstitial cell type and have developmental and functional difference from mesenchymal stem cells and fibroblasts. Several upregulated genes are identified in TCs, e.g., connective tissue growth factor (CTGF), transgelin (Tagln), nidogen 1 (Nid1), tissue inhibitor of metalloproteinase 3 (TIMP3), collagen type IV, alpha (Col4a, Col4a6, Col45), matrix metallopeptidase 10 (Mmp10), matrix metallopeptidase 3 (Mmp3), and retinol-binding protein 1 (RBP1). All of these have been demonstrated to play an important role in tissue remodeling and repair. Subsequently, our team integrates gene expression profiles and focuses on the different chromosomes to mine the significantly expressed genes in TCs by comparing with mesenchymal stem cells, fibroblasts, alveolar type II cells, airway epithelial cells, and lymphocytes [105]. Initially, the genes in chromosome 1 are analyzed and there are 14 upregulated and 39 downregulated differentially expressed genes which are identified on chromosome 1 in TCs compared with the other seven cells (Fig. 1.3). Among

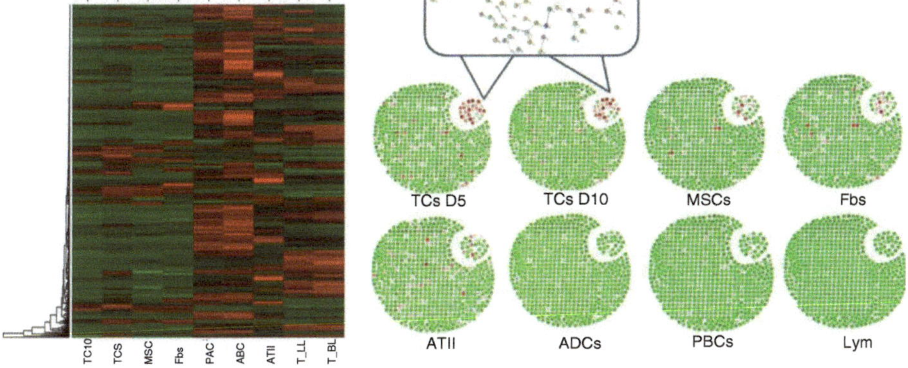

Fig. 1.3 Hierarchical cluster analysis of the differentially expressed genes of telocytes in chromosome1. Telocytes were a peculiar cell type that was distinguished from mesenchymal stem cells (*MSCs*), fibroblasts (*Fbs*), alveolar type II cells (*ATII*), proximal airway cells (*PACs*), airway basal cells (*ABCs*), and lymphocytes from the lungs (*T-LL*) and bronchial lymph nodes (*T-BL*)

these differentially expressed genes, three genes, Capn2, Fhl2, and Qsox1, are over-expressed and contribute to speculating that TCs may have a close association with morphogenesis and local tissue homeostasis. Meanwhile the function for immune modulation in tissue inflammatory diseases is proposed. Subsequently, the variation of differentially expressed genes is identified in chromosomes 2, 3, 4, 17, 18, and X, respectively [104, 108–110]. This provides a new way to explore and identify the biological functions of TCs and reveals the significant relationship between TCs and pulmonary diseases.

MicroRNAs (miRs) are small, single-strand RNA molecules and can inhibit gene translation to control target protein expression [111]. Popescu et al. [112] screened the microRNA signatures in cardiac cells and detected the difference in microRNA expression between TCs and cardiomyocytes. TCs present the mesenchymal origin by significantly up-expressing miR-21, miR-22, miR-29, and miR-199a-5p and down-expressing miR-1, miR-133a, and miR-208a. Besides, the expression of miR-193, which has been considered to regulate c-kit protein expression, is absent in TCs by comparing with cardiomyocytes, which concurs with the fact that TCs are a distinct interstitial cell type with c-kit-positive expression.

The proteomic analysis of TCs, by comparing with microvascular endothelial cells and fibroblasts, is conducted to identify the differentially expressed proteins by using isobaric tag for relative and absolute quantification (iTRAQ) labeling (Fig. 1.4). Xiangdong Wang et al. [113] firstly performed a proteomic analysis of human lung TCs by comparing with fibroblasts and discovered that myosin-4, periplakin, and oxidoreductase activity-associated proteins are upregulated in TCs.

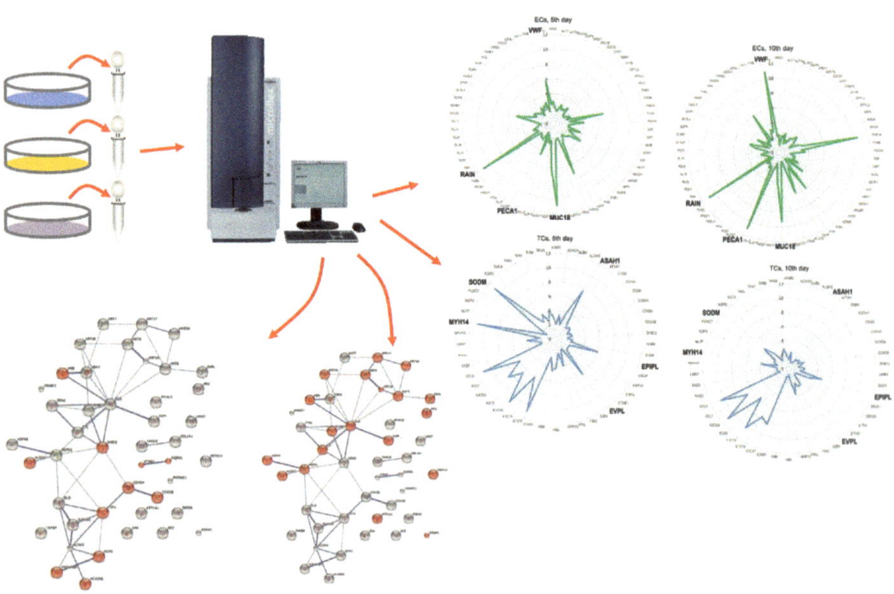

Fig. 1.4 Interaction networks of differentially expressed protein in telocytes were analyzed with STRING. The oxidation-reduction process and extracellular vesicular exosome were identified to reveal the role of telocytes in tissues

These differentially expressed genes provide a support to the assumption that TCs play an important role in tissue repair and remodeling, mechanical sensing, and microenvironment homeostasis. Meanwhile, the up-expressed proteins in TCs, such as mitochondrial thioredoxin-dependent peroxide reductase, protein disulfide-isomerase A3, myosin-14, myosin-10, filamin-B, sodium/potassium-transporting ATPase subunit a-1 and keratin, and type II cytoskeletal 1, show a consistence with those in mammalian extracellular vesicle proteome. Thus, it is assumed that TCs release these proteins via vesicles to transmit intercellular signaling, nurse stem cell niche, and maintain extracellular environment homeostasis. Subsequently, the difference in protein expression is also analyzed in TCs by comparing with endothelial cells using the same method [114]. SODM (SOD2), acid ceramidase, and envoplakin are up-expressed in TCs, while microtubule-associated protein RP/EB family member 1, MUC18, cysteine-rich protein 2, von Willebrand factor, and platelet endothelial cell adhesion molecule are up-expressed in endothelial cells. Thus, the differentially expressed proteins in TCs further confirm the role of TCs in tissue homeostasis, stem cell differentiation, and morphogenetic bioelectrical signaling.

1.7 Interaction with Other Cells

TCs are reported to contact with various cells in the tissue (Fig. 1.5). The synapse-like contacts between TCs and neighboring cells are detected with ultrastructural evidence. The new term "stromal synapse" is firstly introduced to describe this new connective connection which seems like chemical juxtacrine synapse with exosome secretion [115]. The stromal synapses are found to connect TCs with immunoactive

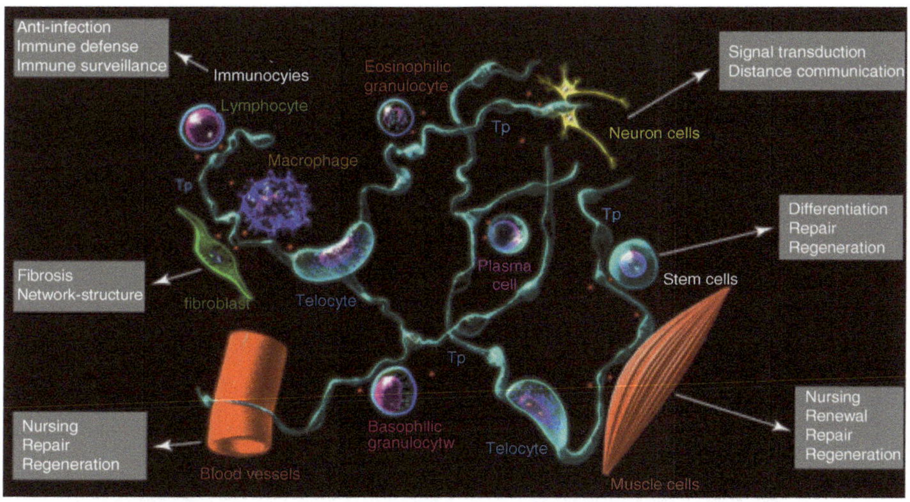

Fig. 1.5 The interaction of telocytes with other neighboring cells and tissues, including stem cells, immune cells, fibroblast, blood vessels, and nerve ends. Telocytes interconnect with homo- or heterogenous cells to form a 3D network through establishing intercellular junction (such as gap or adherens junctions), budding shed vesicles and exosomes, and releasing cytokines

cells in various organs. In rat myometrium, it is reported that TCs establish the connection with eosinophil by releasing extracellular vesicles. Thus, the assumption that vesicle-based antigen signaling exchange works between TCs and eosinophil to regulate immune microenvironment is proposed. The contact between TCs and lymphocytes or plasma cells is also reported, which gives a support to the role of TCs in immunomodulation [115]. Meanwhile, there is a close positional and functional link between TCs and macrophages in human fallopian tube [115, 116]. It is consistent with the observation that TCs modulate the morphology, viability, and cytokine secretion of macrophages in vitro [103]. Besides, TCs are found to contact with mast cells in human mammary gland stroma [115].

In skeletal muscle interstitium, TCs establish a close contact with blood capillaries, nerve fibers, satellite cells, and myocytes and are supposed to participate in muscle regeneration [52]. TCs are also identified to contact with smooth muscle cells, blood vessels, and even nerve fiber via cell processes or telopodes in the myenteric plexus of chicken ileum [33]. Equally, TCs established close spatial relationships with myocardial cells, blood vessels, and nerve endings in the heart [13, 14]. The spatial connection between TCs and nerve endings and blood capillaries is also discovered in the lung and trachea under TEM [37]. It is firmly supported that TCs can protect human pulmonary microvascular endothelial cells from endotoxin by secreting VEGF and EGF in vitro [117]. Pancreatic TCs are also reported to be in close vicinity of blood vessels, nerves, and pancreatic acinar cells and ducts [34]. In addition, skin TCs are closely related to fibroblasts, mast cells, adipocytes, and connective fiber bundles [50].

The stem cell niche provides a microenvironment for stem cells and progenitor cells to regulate and promote stem cell survival, commitment, and differentiation [79]. A considerable number of observation shows that TCs exist in stem cell niche and participate in nursing stem cells and progenitor cells for tissue regeneration and repair. The electron microscopy showed that TCs offer physical and informational support for stem cells and progenitors in cardiac stem cell niche to induce cell differentiation and cardiac renewal process [98, 118]. Equally, TCs extend their telopodes to surround putative stem cells to form the TC-SC niches in the lung according to ultrastructural feature. Further, they contact with each other via nanocontacts and shed vesicles [85]. It is also reported that TCs have a close cooperation with skeletal muscle stem cells under TEM, and differentiative capacity of skeletal muscle stem cells is enhanced when cocultured with TCs in vitro [52, 95]. This provides a positive support to the functional role of TCs with stem cells. Besides, TCs establish atypical heterocellular junctions with stem cells in the skin, liver, and urinary system, to name just a few [51, 96, 102].

1.8 Methods

Now, there are many methods to identify TCs in situ or in vitro, but TEM is still the "gold standard" to distinguish TCs from other interstitial cells in different tissues. TEM not only can display the special feature of morphology but also help to recognize and define the connection or junction between TCs and neighboring cells, such as stem cells, myocardial cells, immune cells, and so on [119]. The most advanced electron microscope technology-focused ion beam scanning electron microscopy (FIB-SEM) tomography is even introduced to study the 3D architecture of TCs' network [120]. The 3D reconstruction of TCs by FIB-SEM tomography supplied a visual level to reveal the typical feature of TCs and the putative connection between TCs and other cells.

Immunohistochemistry is another key method in the study of TCs, but there is no unique and exclusive immunohistochemical label to identify TCs up to now. Meanwhile, TCs present different immunohistochemical markers in different organs and even in different tissues from the same organ. Recently, TCs have been reported to be positive for c-kit/CD117, CD34, vimentin, caveolin-1, platelet-derived growth factor receptor α (PDGFRα), vascular endothelial growth factor (VEGF), inducible nitric oxide synthase (iNOS), and so on [13, 23, 48, 59, 121]. TCs co-express c-kit, CD34, vimentin, nestin, and S-100 in the epicardium, while it is positive for c-kit, CD34, vimentin, and EGFR and negative for nestin, desmin, CD13, and S-100 in the myocardium [13, 122]. Furthermore, three subtypes of TCs are identified in human urinary bladder by TEM and immunohistochemistry. TCs' subtype expresses PDGFRα/calret positive and αSMA/CD34/c-Kit negative or PDGFRα/calret/αSMA positive and CD34/c-Kit negative in the different locations of the sub-urothelium, respectively. The third TC subtype is undoubtedly identified by expressing CD34/calret positive and c-Kit/PDGFRα/αSMA negative in the submucosa and detrusor of urinary bladder [29]. These TCs' subtypes identified in different tissues give us a new sight to elaborate the tissue special role of TCs in various organs.

It shows that several kinds of dyes have been used to observe and identify TCs in tissues. Methylene blue vital staining is widely applied to confirm the existence of TCs in cell cultures [123]. TCs have also been detected to present an affinity to crystal violet and silver impregnation, which are seldom selected to study [15, 124]. Janus green B vital staining and MitoTracker green FM, both of which have a high affinity for mitochondria, have been introduced to assess viability and localize mitochondria in TCs and its telopodes [37]. Flow cytometry is used to identify and screen the phenotype of TCs in vitro and helps to subdivide different subtypes of TCs. As it is reported, there are 64 % of the primary cultured cardiac TCs to express CD117/CD34 positive, which shows a great difference from bone mesenchymal stem cells, cardiac fibroblasts, and cardiomyocytes about telomerase activity [125]. Time-lapse video microscopy is also applied to monitor the behavior of TCs from nonpregnant myometrium in cell culture. It shows a real-time observation to interaction between TCs and myocytes in vitro [126].

1.9 Conclusion

The existence of TCs is continuously identified in various organs from different species. TEM is considered the only method to distinguish TCs from other interstitial cells, such as fibroblasts, ICCs, or stem cells. The typically ultrastructural characteristics of TCs are considered as a tiny body with several long and thin telopodes which extend with alternation of podomeres and podoms. The role of TCs has been speculated to involve in mechanical support, intercellular signaling, tissue renewal or repair, immune surveillance, and mechanical sensor in physical and pathological conditions. More and more advanced technologies are introduced to reveal the essence and function of TCs in situ or in vitro, especially the omics science. However, there are still lots of problems to be solved, e.g., whether TCs are a population of homogeneous cells or TC subpopulation needs to be divided and identified; what are the TC-specific biomarkers to be detected; how the role of TCs in different diseases should be explored although TEM shows the change of TCs in quantity and morphology; and what are direct evidences to demonstrate the functions of TCs in various conditions. Thus, TCs as a new type of interstitial cells will provide us a brand-new sight and foresight treatment to various diseases in the future.

References

1. Popescu LM, Faussone-Pellegrini MS. TELOCYTES - a case of serendipity: the winding way from Interstitial Cells of Cajal (ICC), via Interstitial Cajal-Like Cells (ICLC) to TELOCYTES. J Cell Mol Med. 2010;14(4):729–40.
2. Thuneberg L. One hundred years of interstitial cells of Cajal. Microsc Res Tech. 1999;47(4): 223–38.
3. Sanders KM. A case for interstitial cells of Cajal as pacemakers and mediators of neurotransmission in the gastrointestinal tract. Gastroenterology. 1996;111(2):492–515.
4. Thuneberg L. Interstitial cells of Cajal: intestinal pacemaker cells? Adv Anat Embryol Cell Biol. 1982;71:1–130.
5. Faussone PM, Cortesini C, Romagnoli P. Ultrastructure of the tunica muscularis of the cardial portion of the human esophagus and stomach, with special reference to the so-called Cajal's interstitial cells. Arch Ital Anat Embriol. 1977;82(2):157–77.
6. Min KW, Leabu M. Interstitial cells of Cajal (ICC) and gastrointestinal stromal tumor (GIST): facts, speculations, and myths. J Cell Mol Med. 2006;10(4):995–1013.
7. Popescu LM, Ciontea SM, Cretoiu D. Interstitial Cajal-like cells in human uterus and fallopian tube. Ann N Y Acad Sci. 2007;1101:139–65.
8. Popescu LM, et al. Interstitial cells of Cajal in pancreas. J Cell Mol Med. 2005;9(1): 169–90.
9. Pieri L, Vannucchi MG, Faussone-Pellegrini MS. Histochemical and ultrastructural characteristics of an interstitial cell type different from ICC and resident in the muscle coat of human gut. J Cell Mol Med. 2008;12(5B):1944–55.
10. Kostin S, Popescu LM. A distinct type of cell in myocardium: interstitial Cajal-like cells (ICLCs). J Cell Mol Med. 2009;13(2):295–308.
11. Hutchings G, et al. Myometrial interstitial cells and the coordination of myometrial contractility. J Cell Mol Med. 2009;13(10):4268–82.
12. Faussone PM, Popescu LM. Telocytes. Biomol Concepts. 2011;2(6):481–9.
13. Popescu LM, et al. Telocytes in human epicardium. J Cell Mol Med. 2010;14(8):2085–93.
14. Gherghiceanu M, Manole CG, Popescu LM. Telocytes in endocardium: electron microscope evidence. J Cell Mol Med. 2010;14(9):2330–4.
15. Kostin S. Myocardial telocytes: a specific new cellular entity. J Cell Mol Med. 2010;14(7): 1917–21.

16. Hinescu ME, Popescu LM. Interstitial Cajal-like cells (ICLC) in human atrial myocardium. J Cell Mol Med. 2005;9(4):972–5.
17. Yang Y, et al. Telocytes in human heart valves. J Cell Mol Med. 2014;18(5):759–65.
18. Van der Aa F, et al. Interstitial cells in the human prostate: a new therapeutic target? Prostate. 2003;56(4):250–5.
19. Corradi LS, et al. Structural and ultrastructural evidence for telocytes in prostate stroma. J Cell Mol Med. 2013;17(3):398–406.
20. Hashitani H, Suzuki H. Identification of interstitial cells of Cajal in corporal tissues of the guinea-pig penis. Br J Pharmacol. 2004;141(2):199–204.
21. Duquette RA, et al. Vimentin-positive, c-kit-negative interstitial cells in human and rat uterus: a role in pacemaking? Biol Reprod. 2005;72(2):276–83.
22. Cretoiu SM, Cretoiu D, Popescu LM. Human myometrium – the ultrastructural 3D network of telocytes. J Cell Mol Med. 2012;16(11):2844–9.
23. Suciu L, et al. Telocytes in human term placenta: morphology and phenotype. Cells Tissues Organs. 2010;192(5):325–39.
24. Dixon RE, et al. Chlamydia infection causes loss of pacemaker cells and inhibits oocyte transport in the mouse oviduct. Biol Reprod. 2009;80(4):665–73.
25. Zheng Y, et al. Telocytes in the urinary system. J Transl Med. 2012;10:188.
26. Qi G, et al. Telocytes in the human kidney cortex. J Cell Mol Med. 2012;16(12):3116–22.
27. Sergeant GP, et al. Interstitial cells of Cajal in the urethra. J Cell Mol Med. 2006;10(2):280–91.
28. Davidson RA, McCloskey KD. Morphology and localization of interstitial cells in the guinea pig bladder: structural relationships with smooth muscle and neurons. J Urol. 2005;173(4):1385–90.
29. Vannucchi MG, et al. Telocytes subtypes in human urinary bladder. J Cell Mol Med. 2014;18(10):2000–8.
30. Chen X, et al. Telocytes in human oesophagus. J Cell Mol Med. 2013;17(11):1506–12.
31. Cantarero CI, Luesma BM, Junquera EC. Identification of telocytes in the lamina propria of rat duodenum: transmission electron microscopy. J Cell Mol Med. 2011;15(1):26–30.
32. Cretoiu D, et al. Telocytes, a distinct type of cell among the stromal cells present in the lamina propria of jejunum. Histol Histopathol. 2012;27(8):1067–78.
33. Yang P, et al. Ultrastructural identification of telocytes in the muscularis of chicken ileum. Exp Ther Med. 2015;10(6):2325–30.
34. Nicolescu MI, Popescu LM. Telocytes in the interstitium of human exocrine pancreas: ultrastructural evidence. Pancreas. 2012;41(6):949–56.
35. Xiao J, et al. Telocytes in liver: electron microscopic and immunofluorescent evidence. J Cell Mol Med. 2013;17(12):1537–42.
36. Hinescu ME, et al. Interstitial Cajal-like cells in human gallbladder. J Mol Histol. 2007;38(4):275–84.
37. Zheng Y, et al. Telocytes in trachea and lungs. J Cell Mol Med. 2011;15(10):2262–8.
38. Rusu MC, et al. Tracheal telocytes. J Cell Mol Med. 2012;16(2):401–5.
39. Hinescu ME, et al. Telocytes in pleura: two- and three-dimensional imaging by transmission electron microscopy. Cell Tissue Res. 2011;343(2):389–97.
40. Pucovsky V, Moss RF, Bolton TB. Non-contractile cells with thin processes resembling interstitial cells of Cajal found in the wall of guinea-pig mesenteric arteries. J Physiol. 2003;552(Pt 1): 119–33.
41. Zhang HQ, et al. Morphological evidence of telocytes in mice aorta. Chin Med J (Engl). 2015;128(3):348–52.
42. Li H, et al. Scanning electron microscope evidence of telocytes in vasculature. J Cell Mol Med. 2014;18(7):1486–9.
43. Harhun MI, et al. Function of interstitial cells of Cajal in the rabbit portal vein. Circ Res. 2004;95(6):619–26.
44. Gherghiceanu M, et al. Interstitial Cajal-like cells (ICLC) in myocardial sleeves of human pulmonary veins. J Cell Mol Med. 2008;12(5A):1777–81.
45. McCloskey KD, et al. Kit-like immunopositive cells in sheep mesenteric lymphatic vessels. Cell Tissue Res. 2002;310(1):77–84.

46. Nicolescu MI, et al. Telocytes in parotid glands. Anat Rec (Hoboken). 2012;295(3):378–85.
47. Popescu LM, Andrei F, Hinescu ME. Snapshots of mammary gland interstitial cells: methylene-blue vital staining and c-kit immunopositivity. J Cell Mol Med. 2005;9(2): 476–7.
48. Petre N, et al. Telocytes of the mammary gland stroma. Folia Morphol 2016;75(2):224–31.
49. Alunno A, et al. Telocytes in minor salivary glands of primary Sjogren's syndrome: association with the extent of inflammation and ectopic lymphoid neogenesis. J Cell Mol Med. 2015;19(7):1689–96.
50. Rusu MC, et al. Skin telocytes. Ann Anat. 2012;194(4):359–67.
51. Ceafalan L, et al. Telocytes in human skin – are they involved in skin regeneration? J Cell Mol Med. 2012;16(7):1405–20.
52. Popescu LM, et al. Identification of telocytes in skeletal muscle interstitium: implication for muscle regeneration. J Cell Mol Med. 2011;15(6):1379–92.
53. Rusu MC, et al. The human trigeminal ganglion: c-kit positive neurons and interstitial cells. Ann Anat. 2011;193(5):403–11.
54. Popescu BO, et al. Telocytes in meninges and choroid plexus. Neurosci Lett. 2012;516(2): 265–9.
55. Diaz-Flores L, et al. Telocytes in neuromuscular spindles. J Cell Mol Med. 2013;17(4): 457–65.
56. Luesma MJ, Gherghiceanu M, Popescu LM. Telocytes and stem cells in limbus and uvea of mouse eye. J Cell Mol Med. 2013;17(8):1016–24.
57. Rusu MC, Loreto C, Manoiu VS. Network of telocytes in the temporomandibular joint disc of rats. Acta Histochem. 2014;116(4):663–8.
58. Li H, et al. Telocytes in mice bone marrow: electron microscope evidence. J Cell Mol Med. 2014;18(6):975–8.
59. Chang Y, et al. Telocytes in the spleen. PLoS One. 2015;10(9):e0138851.
60. Dawidowicz J, et al. Electron microscopy of human fascia lata: focus on telocytes. J Cell Mol Med. 2015;19(10):2500–6.
61. Hinescu ME, et al. Interstitial Cajal-like cells in rat mesentery: an ultrastructural and immunohistochemical approach. J Cell Mol Med. 2008;12(1):260–70.
62. Rusu MC, et al. Esophageal telocytes and hybrid morphologies. Cell Biol Int. 2012;36(12): 1079–88.
63. Bosco C, et al. Ganglionar nervous cells and telocytes in the pancreas of Octodon degus: extra and intrapancreatic ganglionar cells and telocytes in the degus. Auton Neurosci. 2013;177(2):224–30.
64. Metzger R, et al. Cajal-like cells in the upper urinary tract: comparative study in various species. Pediatr Surg Int. 2005;21(3):169–74.
65. Ullah S, et al. Identification and characterization of telocytes in the uterus of the oviduct in the Chinese soft-shelled turtle, Pelodiscus sinensis: TEM evidence. J Cell Mol Med. 2014;18(12):2385–92.
66. Mandache E, et al. Telocytes in human isolated atrial amyloidosis: ultrastructural remodelling. J Cell Mol Med. 2010;14(12):2739–47.
67. Manole CG, et al. Experimental acute myocardial infarction: telocytes involvement in neo-angiogenesis. J Cell Mol Med. 2011;15(11):2284–96.
68. Mirancea N, et al. Infrastructure of the telocytes from tumor stroma in the skin basal and squamous cell carcinomas. Rom J Morphol Embryol. 2013;54(4):1025–37.
69. Mou Y, et al. Immunohistochemical characterization and functional identification of mammary gland telocytes in the self-assembly of reconstituted breast cancer tissue in vitro. J Cell Mol Med. 2013;17(1):65–75.
70. Ardeleanu C, Bussolati G. Telocytes are the common cell of origin of both PEComas and GISTs: an evidence-supported hypothesis. J Cell Mol Med. 2011;15(12):2569–74.
71. Manetti M, et al. Evidence for progressive reduction and loss of telocytes in the dermal cellular network of systemic sclerosis. J Cell Mol Med. 2013;17(4):482–96.
72. Matyja A, et al. Telocytes: new insight into the pathogenesis of gallstone disease. J Cell Mol Med. 2013;17(6):734–42.

73. Milia AF, et al. Telocytes in Crohn's disease. J Cell Mol Med. 2013;17(12):1525–36.
74. Bosco C, et al. A putative role for telocytes in placental barrier impairment during preeclampsia. Med Hypotheses. 2015;84(1):72–7.
75. Manetti M, et al. Telocytes are reduced during fibrotic remodelling of the colonic wall in ulcerative colitis. J Cell Mol Med. 2015;19(1):62–73.
76. Fu S, et al. Telocytes in human liver fibrosis. J Cell Mol Med. 2015;19(3):676–83.
77. Yang J, et al. Ultrastructure damage of oviduct telocytes in rat model of acute salpingitis. J Cell Mol Med. 2015;19(7):1720–8.
78. Manole CG, Gherghiceanu M, Simionescu O. Telocyte dynamics in psoriasis. J Cell Mol Med. 2015;19(7):1504–19.
79. Zheng Y, Bai C, Wang X. Telocyte morphologies and potential roles in diseases. J Cell Physiol. 2012;227(6):2311–7.
80. Li L, et al. Renal telocytes contribute to the repair of ischemically injured renal tubules. J Cell Mol Med. 2014;18(6):1144–56.
81. Gherghiceanu M, Popescu LM. Heterocellular communication in the heart: electron tomography of telocyte-myocyte junctions. J Cell Mol Med. 2011;15(4):1005–11.
82. Mandache E, Popescu LM, Gherghiceanu M. Myocardial interstitial Cajal-like cells (ICLC) and their nanostructural relationships with intercalated discs: shed vesicles as intermediates. J Cell Mol Med. 2007;11(5):1175–84.
83. Popescu LM, et al. Cardiac renewing: interstitial Cajal-like cells nurse cardiomyocyte progenitors in epicardial stem cell niches. J Cell Mol Med. 2009;13(5):866–86.
84. Gherghiceanu M, Popescu LM. Cardiac telocytes – their junctions and functional implications. Cell Tissue Res. 2012;348(2):265–79.
85. Popescu LM, et al. Telocytes and putative stem cells in the lungs: electron microscopy, electron tomography and laser scanning microscopy. Cell Tissue Res. 2011;345(3):391–403.
86. Cretoiu D, et al. Interstitial Cajal-like cells (ICLC) as steroid hormone sensors in human myometrium: immunocytochemical approach. J Cell Mol Med. 2006;10(3):789–95.
87. Cretoiu SM, et al. Interstitial Cajal-like cells of human Fallopian tube express estrogen and progesterone receptors. J Mol Histol. 2009;40(5–6):387–94.
88. Cismasiu VB, Popescu LM. Telocytes transfer extracellular vesicles loaded with microRNAs to stem cells. J Cell Mol Med. 2015;19(2):351–8.
89. Cretoiu SM, et al. Isolated human uterine telocytes: immunocytochemistry and electrophysiology of T-type calcium channels. Histochem Cell Biol. 2015;143(1):83–94.
90. Quaife-Ryan GA, et al. Resetting the epigenome for heart regeneration. Semin Cell Dev Biol. 2016.
91. Sadri AR, Jeschke MG, Amini-Nik S. Advances in liver regeneration: revisiting hepatic stem/progenitor cells and their origin. Stem Cells Int. 2016;2016:7920897.
92. Akram KM, et al. Lung regeneration: endogenous and exogenous stem cell mediated therapeutic approaches. Int J Mol Sci. 2016;17(1):128.
93. Suzuki E, et al. Adult stem cells as a tool for kidney regeneration. World J Nephrol. 2016;5(1):43–52.
94. Gordon T, Borschel GH. The use of the rat as a model for studying peripheral nerve regeneration and sprouting after complete and partial nerve injuries. Exp Neurol. 2016.
95. Bojin FM, et al. Telocytes within human skeletal muscle stem cell niche. J Cell Mol Med. 2011;15(10):2269–72.
96. Wang F, et al. Telocytes in liver regeneration: possible roles. J Cell Mol Med. 2014;18(9):1720–6.
97. Bani D, et al. Telocytes as supporting cells for myocardial tissue organization in developing and adult heart. J Cell Mol Med. 2010;14(10):2531–8.
98. Gherghiceanu M, Popescu LM. Cardiomyocyte precursors and telocytes in epicardial stem cell niche: electron microscope images. J Cell Mol Med. 2010;14(4):871–7.
99. Albulescu R, et al. The secretome of myocardial telocytes modulates the activity of cardiac stem cells. J Cell Mol Med. 2015;19(8):1783–94.
100. Zhao B, et al. Intramyocardial transplantation of cardiac telocytes decreases myocardial infarction and improves post-infarcted cardiac function in rats. J Cell Mol Med. 2014;18(5):780–9.
101. Zhao B, et al. Cardiac telocytes were decreased during myocardial infarction and their therapeutic effects for ischaemic heart in rat. J Cell Mol Med. 2013;17(1):123–33.

102. Bei Y, et al. Telocytes in regenerative medicine. J Cell Mol Med. 2015;19(7):1441–54.
103. Chi C, et al. In vitro morphology, viability and cytokine secretion of uterine telocyte-activated mouse peritoneal macrophages. J Cell Mol Med. 2015;19(12):2741–50.
104. Wang J, et al. Global analyses of Chromosome 17 and 18 genes of lung telocytes compared with mesenchymal stem cells, fibroblasts, alveolar type II cells, airway epithelial cells, and lymphocytes. Biol Direct. 2015;10:9.
105. Sun X, et al. Differences in the expression of chromosome 1 genes between lung telocytes and other cells: mesenchymal stem cells, fibroblasts, alveolar type II cells, airway epithelial cells and lymphocytes. J Cell Mol Med. 2014;18(5):801–10.
106. Campeanu RA, et al. Near-infrared low-level laser stimulation of telocytes from human myometrium. Lasers Med Sci. 2014;29(6):1867–74.
107. Zheng Y, et al. Genetic comparison of mouse lung telocytes with mesenchymal stem cells and fibroblasts. J Cell Mol Med. 2013;17(4):567–77.
108. Song D, et al. Comparison of Chromosome 4 gene expression profile between lung telocytes and other local cell types. J Cell Mol Med. 2016;20(1):71–80.
109. Zhu Y, et al. Global comparison of chromosome X genes of pulmonary telocytes with mesenchymal stem cells, fibroblasts, alveolar type II cells, airway epithelial cells, and lymphocytes. J Transl Med. 2015;13(1):318.
110. Zheng M, et al. Variations of chromosomes 2 and 3 gene expression profiles among pulmonary telocytes, pneumocytes, airway cells, mesenchymal stem cells and lymphocytes. J Cell Mol Med. 2014;18(10):2044–60.
111. Tuna M, Machado AS, Calin GA. Genetic and epigenetic alterations of microRNAs and implications for human cancers and other diseases. Genes Chromosomes Cancer. 2016;55(3): 193–214.
112. Cismasiu VB, Radu E, Popescu LM. miR-193 expression differentiates telocytes from other stromal cells. J Cell Mol Med. 2011;15(5):1071–4.
113. Zheng Y, et al. Comparative proteomic analysis of human lung telocytes with fibroblasts. J Cell Mol Med. 2014;18(4):568–89.
114. Zheng Y, et al. Protein profiling of human lung telocytes and microvascular endothelial cells using iTRAQ quantitative proteomics. J Cell Mol Med. 2014;18(6):1035–59.
115. Popescu LM, et al. The connective connection: interstitial cells of Cajal (ICC) and ICC-like cells establish synapses with immunoreactive cells. Electron microscope study in situ. J Cell Mol Med. 2005;9(3):714–30.
116. Popescu LM, et al. Novel type of interstitial cell (Cajal-like) in human fallopian tube. J Cell Mol Med. 2005;9(2):479–523.
117. Zheng Y, et al. Human lung telocytes could promote the proliferation and angiogenesis of human pulmonary microvascular endothelial cells in vitro. Mol Cell Ther. 2014;2:3.
118. Popescu LM, Fertig ET, Gherghiceanu M. Reaching out: junctions between cardiac telocytes and cardiac stem cells in culture. J Cell Mol Med. 2016;20(2):370–80.
119. Cantarero I, et al. Transmission electron microscopy as key technique for the characterization of telocytes. Curr Stem Cell Res Ther. 2016;11(5):410–4.
120. Cretoiu D, et al. FIB-SEM tomography of human skin telocytes and their extracellular vesicles. J Cell Mol Med. 2015;19(4):714–22.
121. Vannucchi MG, et al. Telocytes express PDGFRalpha in the human gastrointestinal tract. J Cell Mol Med. 2013;17(9):1099–108.
122. Hinescu ME, et al. Interstitial Cajal-like cells (ICLC) in atrial myocardium: ultrastructural and immunohistochemical characterization. J Cell Mol Med. 2006;10(1):243–57.
123. Zhou J, et al. Telocytes accompanying cardiomyocyte in primary culture: two- and three-dimensional culture environment. J Cell Mol Med. 2010;14(11):2641–5.
124. Cretoiu SM, Popescu LM. Telocytes revisited. Biomol Concepts. 2014;5(5):353–69.
125. Li YY, et al. Comparative analysis of telomerase activity in CD117(+) CD34(+) cardiac telocytes with bone mesenchymal stem cells, cardiac fibroblasts and cardiomyocytes. Chin Med J (Engl). 2015;128(14):1942–7.
126. Cretoiu SM, et al. Telocytes: ultrastructural, immunohistochemical and electrophysiological characteristics in human myometrium. Reproduction. 2013;145(4):357–70.

Jian Wang is a PhD student on pulmonary medicine in Zhongshan Hospital, Fudan University. His work mainly focuses on application of bioinformatics to identify telocytes' differential gene expression. He has published six articles as the first author, including 3 SCI papers.

Meiling Jin, MD is professor and director of bronchial diseases center in pulmonary department, Zhongshan Hospital, Fudan University. She is also the director of lung function department. She engages in clinical medicine and research for over 20 years and mainly focuses on the diagnosis and treatment for asthma, lung cancer and COPD. She takes charge of several clinical and academic programs. She has published more than 60 papers, including 6 SCI papers.

Zhitu Zhu is Oncology Professor of Jinzhou Medical University, Deputy Director of Jinzhou Medical University Cancer Center. His main research is focused on clinical gastrointestinal cancer, cancer immunology, and molecular and cellular therapies and digestive system cancer, lung cancer, breast cancer, lymphoma and melanoma and other malignant tumor chemotherapy and biological targeted therapy. He is the author of more than 20 scientific publications.

Wenhuan Ma, MD, PhD, Master Tutor, Associate Chief physician, Vice President, Deputy Director of Shanghai Association of Chinese Integrative Medical Respiratory Committee, Beizhan Hospital, Shanghai, China.

Xiangdong Wang is a Distinguished Professor of Medicine at Fudan University, Director of Shanghai Institute of Clinical Bioinformatics, Director of Fudan University Center of Clinical Bioinformatics, Deputy Director of Shanghai Respiratory Research Institute, Adjunct Professor of Clinical Bioinformatics at Lund University, and Visiting Professor of King's College London. He serves as an Executive Dean of Institute of Clinical Science and Director of Biomedical Research Center, Fudan University, Zhongshan Hospital. His main research is focused on clinical bioinformatics, disease-specific biomarkers, lung chronic diseases, cancer immunology, and molecular and cellular therapies. He is the author of more than 200 scientific publications with the impact factor about 600, citation number about 6642, h-index 49, and i10-index 181 and cited journal impact factor about 5000.

Chapter 2
Decoding Telocytes

Junjie Xiao and Yihua Bei

Abstract Telocytes (TCs) are a novel type of interstitial cells, with extremely long and thin cellular prolongations termed telopodes (Tps). TCs were first identified by Popescu et al. and described their finding as "cells with telopodes." The presence of TCs has been reported in the majority of tissues and organs (for details please visit www.telocytes.com). TCs have been ignored or overlooked for a long time due to our inability to observe these cells via a light microscopy. TCs represent a distinct cell population, different from other types of interstitial cells, based on their distinct (ultra)structure, immunophenotype, microRNA profile, gene feature, proteome signature, and secretome features. As TCs have been suggested as new cellular targets for forthcoming therapies, developing specific methods to modulate TC numbers represents an important objective.

2.1 Introduction

Telocytes (TCs) are a type of interstitial cell, identified based on the specific presence of extremely long and thin cellular prolongations named telopodes (Tps) [1–3]. Telocytes were first identified by Popescu et al. who coined the name TCs on the basis of their physical appearance (using the Greek prefix "telos," meaning goal, end, or fulfillment) and defined this cell, in its simplest form, as "cells with telopodes" [1].

While TCs have been reported in the majority of tissues and organs, this cell type is often overlooked due to the fact that Tps are below the resolving power of the light microscope, which prevents us from visualizing this cell using conventional light microscopy [4–15].

There has been considerable attention placed on discrimination TCs from other types of interstitial cells [16–27]. Currently, it is widely accepted that TCs represent a distinct cell population from other types of interstitial cells based on their distinct

J. Xiao (✉) • Y. Bei
Regeneration and Ageing Lab, Experimental Center of Life Sciences, School of Life Science, Shanghai University, 333 Nan Chen Road, Shanghai 200444, China
e-mail: junjiexiao@live.cn

X. Wang, D. Cretoiu (eds.), *Telocytes*, Advances in Experimental Medicine and Biology, 913, DOI 10.1007/978-981-10-1061-3_2

(ultra)structure, immunophenotype, microRNA (miRNA, miR) profile, gene feature, proteome signature, and secretome features [3]. Thus, in this chapter, we attempt to detail and describe the nature of TCs.

2.2 (Ultra)structure of TCs

To identify TCs accurately, transmission electron microscopy (TEM) is generally considered as the golden standard [9, 25, 26, 28–45]. TCs are cells with extremely long cellular extensions (Tps) with an alternation of thin segments (podomers) and dilated, cistern-like portions (podoms) (Fig. 2.1) [3]. Strikingly, TCs have long and thin Tps and an irregular cell body with scarce cytoplasm, accommodating the Golgi apparatus, mitochondria, and few endoplasmic reticulums [46, 47]. The shape of TCs' cell body is determined by the number of Tps and can be piriform, spindle, triangular, or stellate [3]. Alternating podomers and podoms afford the typical "bead-on-a-string" appearance of Tps [48]. The podoms accommodate endoplasmic reticulum, caveolae, and mitochondria [3]. The presence of mitochondria in the cell body and particularly in the podoms of Tps has been demonstrated by Janus green B staining and MitoTracker Green FM [3]. Through their long prolongations, they form a three-dimensional (3D) network to maintain organism homeostasis [49, 50]. By contrast, fibroblasts have a large and pleomorphic cell body, with prominent Golgi complex, well-developed rough endoplasmic reticulum, and few, short, and large caliber cell processes, making them easily recognizable via light microscope [3]. Recently, three-dimensional imaging of human cardiac TCs has been revealed using a focused ion beam scanning electron microscopy (FIB-SEM) [51].

As such, TCs differ from other types of interstitial cells based on their distinctive ultrastructure [3].

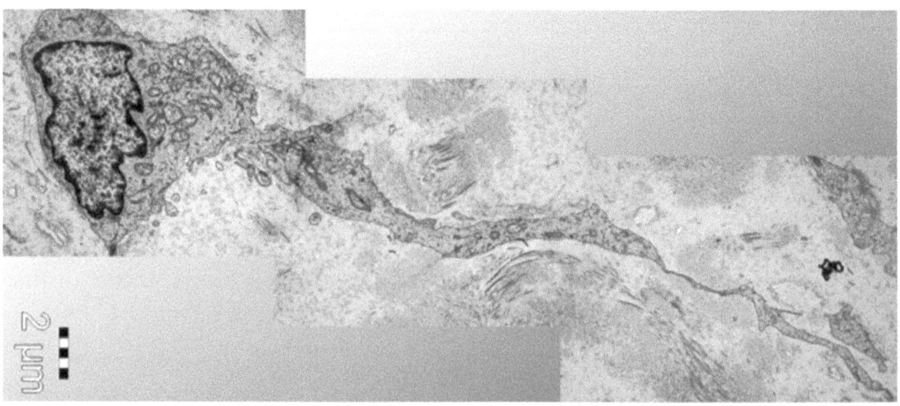

Fig. 2.1 Three electron microscopy images of a telocyte in human heart valve. A telocyte (TC) with typical telopode (Tp) that emerges from the small cell body; scale bars: 2 μm (Courtesy of Dr. LM. Popescu, Department of Ultrastructural Pathology, Victor Babes National Institute of Pathology, Bucharest, Romania. Reproduced with permission from Bei et al. [2])

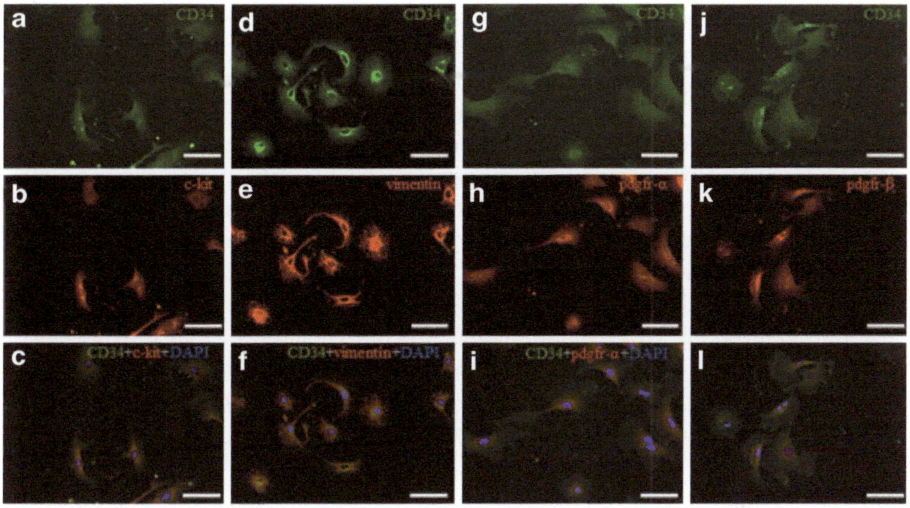

Fig. 2.2 Double immunofluorescence staining for CD34/c-kit, CD34/vimentin, CD34/PDGFR-α, and CD34/PDGFR-β of cardiac telocytes (TCs) primary culture. Cardiac TCs are positive for (**a**) CD34 (*green*) and (**b**) c-kit (*red*). Co-localization of CD34 and c-kit is obvious in TCs (**c**). (**d**) CD34 (*green*) and (**e**) vimentin (*red*). Co-localization of CD34 and vimentin is obvious in TCs (**f**). (**g**) CD34 (*green*) and (**h**) PDGFR-α (*red*). Co-localization of CD34 and PDGFR-α is obvious in TCs (**i**). (**j**) CD34 (*green*) and (**k**) PDGFR-β (*red*). Co-localization of CD34 and PDGFR-β is obvious in TCs (**l**)

2.3 Immunophenotype

Immunostaining can also represent a modality by which TCs can be distinguished from other types of interstitial cells [52, 53]. Double immunolabeling is a useful method for semiquantitative analysis for TCs [54]. Unfortunately, currently no single specific immunophenotype for TCs has been established [3]. Moreover, there may exist organ- or tissue-specific immunophenotypes of TCs [3].

Cardiac TCs are the best-characterized TC subtype [55–58]. Cardiac TCs reportedly express hematopoietic stem cell marker CD34, mesenchymal cell marker vimentin, myocardial stem cell markers sca-1 and c-kit, and embryonic stem cell-associated gene of Nanog [59]. CD34 is mainly expressed in the proximal prolongations and the cell body, while vimentin is only expressed in the prolongations [59]. By contrast, sca-1 is mainly exhibited in the initial part of the Tps and the cell body, while Nanog is only expressed in the cell body [59]. Moreover, c-kit is highly expressed in the cell body and Tps of TCs [59]. However, no direct comparison with cardiac fibroblasts or other types of interstitial cells was performed in that study [59]. Thus, initially, CD34/c-kit and CD34/vimentin have been considered as appropriate immunomarkers for TCs (Fig. 2.2) [3].

Interestingly, TCs are also immunohistochemically negative for CD90/Thy-1 (fibroblasts), procollagen 1, CD31/PECAM-1 (endothelial cells), CD11c (dendritic cells and macrophages), and α-SMA (myofibroblasts, pericytes, and vascular smooth muscle cells) [3, 60]. TCs have also been shown to be negative for CD68 and other markers related to immune functions including CD1a and CD62-P, indicating a clear difference between TCs and macrophages [3, 60]. TCs are sometimes negative by immunohistochemistry for c-kit/CD117 (mast cells) [60]. Thus, CD34-positive and CD31-negative immunostaining is frequently used to identify TCs [3]. However, endothelial progenitor cells (EPCs) also express CD34 and c-kit, which may confound this immunohistochemical approach [60]. Moreover, controversial data from the presence or absence of c-kit expression in TCs has been reported [60]. A "switching" phenotype behavior described in other cell populations or differences in the technical procedures may explain these conflicting results [60]. Importantly, while CD34 immunolabeling does not allow an unequivocal identification of TCs since it does not label all TCs, it is still the best available choice when used in combination with vimentin or c-kit [60]. Importantly, a single positive signal for CD34 or c-kit is insufficient to confirm a TC [60]. Therefore, hitherto, double-positive immunostaining for CD34/vimentin is regarded as a most useful marker for TCs [60].

Platelet-derived growth factor (PDGF), a major mitogen of mesenchymal cells, can induce the migratory and mitogenic response of mesenchymal stem cells [61]. PDGF-B polypeptide is secreted by endothelial cells and promotes the formation of the surrounding muscular wall. PDGF receptor-beta (PDGFR-β), as the receptor of PDGF-B, is important for vascular stability and also plays an important role in the promotion of migration and proliferation of mesenchymal stem cells [61]. Interestingly, the PDGFR-β-positive TCs have been identified in the skeletal muscle interstitium and potentially regulate the process of microvessel cell recruitment during vascular remodeling and angiogenesis [62, 63]. Thus, double-positive immunostaining with CD34/PDGFR-β is also regarded as a useful marker for the identification of TCs (Fig. 2.2) [60].

Besides CD34/PDGFR-β double immunolabeling, CD34/PDGFR-α has also been suggested as a marker for TCs [64]. Initially, CD34/PDGFR-α double-positive immunostaining was suggested as a specific marker for TCs in various segments of the human gut including the esophagus, corpus and antrum, gastric fundus, and large and small intestine [64]. However, hepatic TCs were also found to be CD34/PDGFR-α double positive in mice and human [36, 65–67]. Recently, we isolated cardiac TCs from adult mice and confirmed TCs by double-positive immunostaining for CD34/vimentin and CD34/PDGFR-β [68]. We provided direct evidence that isolated cardiac TCs are double positive for CD34/PDGFR-α [68]. Moreover, quantitatively, CD34/PDGFR-α-positive cells accounted for almost one third of cardiac TCs enriched in rat cardiac interstitial cell populations [68]. Thus, CD34/PDGFR-α double immunolabeling also helps in the identification of TCs (Fig. 2.2). However, endothelial cells also express both CD34 and PDGFR-α [64]. Therefore, the combination of CD34 or PDGFR-α immunolabeling with an endothelial-specific marker (i.e., CD31/PECAM-1) may help identify TCs.

It is worth noting that TCs have been proposed to represent a source of cardiac mesenchymal cells, a hypothesis supported by CD29-positive staining in these cell types [69, 70]. Notably, CD29 is also a well-known marker of mesenchymal stem cells [70]. In addition, as PDGFR-α is expressed in mesenchymal stem cells, the presence of PDGFR-α expression in TCs reinforces the hypothesis that TCs represent the adult stromal mesenchymal cells [60].

Importantly, immunohistochemical profile of TCs may be different between tissues and organs [60]. A recent study using immunostaining indicated that TCs in the spleen are positive for vimentin, CD34, Nanog, and sca-1 but negative for c-kit [71]. In addition to the organ- and tissue-specific immunophenotype, TC subtypes also exist and they might play region-specific roles [72]. Three subtypes of TCs have been identified in human urinary bladder [72]. The first subtype of TCs is immediately beneath the urothelium and is PDGFR-α/calret positive and α-SMA/CD34/c-kit negative [72]. Another subtype of TCs is deeper in the sub-urothelium and is α-SMA positive, which is similar to myofibroblasts [72]. However, myofibroblasts are PDGFR-α/calret negative [72]. A third subtype of TCs exists in both the submucosa and detrusor and they are calret positive but c-kit/PDGFR-α/α-SMA negative [72]. Thus, TC subtypes may adapt their phenotype and morphology based on its organ and tissue localization.

In summary, CD34/vimentin double positive and CD34/PDGFR-β double positive enable the identification of TCs. In addition, CD34 positive/CD31 negative and PDGFRα positive/CD31 negative are also markers for TCs [73].

Interestingly, Podoplanin (D2-40) has also recently been identified as a reliable marker of TCs in the bladder [74]. However, considering that Podoplanin (D2-40) is also expressed in many other normal cells including chondrocytes, osteocytes, and follicular dendritic cells, this marker would need to be combined with other markers to confirm the identity of TCs [74].

2.4 miRNA Imprint

miRNAs are a novel class of endogenous noncoding RNAs, which play critical roles in the regulation of gene expressions at the posttranscriptional level [75–77]. A single miRNA can target hundreds of genes, and a single gene is often regulated by multiple miRNAs [78–80]. It has been suggested that nearly 60 % genes are regulated by miRNAs thereby impacting proliferation, differentiation, apoptosis, metabolism, and tissue remodeling [75–77, 81]. Interestingly, miRNAs also exhibit a tissue- and cell-type-specific distribution and roles [78–80]. Cardiac-specific miRNAs, including miR-208a and miR-133a, have been shown to be expressed in cardiomyocytes but not in cardiac TCs [82]. By contrast, miR-21, miR-29, and miR-199a-5p are highly expressed in fibroblasts [82]. Importantly, TCs do not express miR-193, while fibroblasts express miR-193 [82]. While this finding suggests that miR-193 may be a surrogate marker of TCs, further studies are needed to clarify the role of this small molecule in TCs [82]. More specifically, an miRNA analysis of primary isolated cardiac TCs and cardiac fibroblasts using miRNA arrays could substantially improve our molecular understanding of these cell types.

2.5 Genome Features

Using an Agilent Mouse 4 × 44K Gene Expression Array, the gene expression profile of mouse lung TCs were compared to the profile observed in mesenchymal stem cells and fibroblasts [83]. Over 2,000 and 4,000 different genes have been found to be upregulated and downregulated, respectively, in TCs as compared to either mesenchymal stem cells or fibroblasts [83]. Interestingly, *Cdh2*, *Cyba*, *Rnf128*, *Dpysl3*, *Fstl1*, *Rbp1*, *Gm12892*, *Cdh2*, *Aldh1a1*, and *Gm5864* were found upregulated >100 times in TCs versus fibroblasts, while *Rbp1* and *Glipr1* were found upregulated >100 times in TCs versus mesenchymal stem cells [83]. A gene ontology (GO) analysis revealed that differentially expressed genes in TCs are involved in development and tissue and/or organ morphogenesis [83]. In addition, numerous genes were found to be specifically upregulated or downregulated in TCs when compared to both fibroblasts and mesenchymal stem cells [83]. Commonly upregulated genes include *Ctgf*, *Mmp10*, *Mmp3*, *Col4a4*, *Col4a6*, *Col4a5*, *Unc13b*, *Mapk13*, *lgsf9*, *Glipr1*, *Clic5*, *Myh14*, *Aldh1a1*, *Aldh1a2*, *Rbp1*, *Gprc5c*, *Gsta3*, *Plac9*, *Fgd3*, *Dok2*, and *Scnn1a* [83]. Commonly downregulated genes include *Car6*, *Odz4*, *Oz/ten-m*, *Cdsn*, *Hoxc6*, and *Lfi203* [83].

Following expression profiling, the patterns of mouse lung TC-specific gene profiles on chromosome 1 were compared against other types of cells including mesenchymal stem cells, fibroblasts, alveolar type II cells, airway epithelial cells, and lymphocytes [84]. The gene expression profile data from pulmonary TCs on day 5 and day 10, compared with mesenchymal stem cells and fibroblasts, were obtained from previous studies. The alveolar type II cells, airway basal cells (ABCs), proximal airway cells (PACs), lymphocytes from bronchial lymph nodes (T-BL), and lymphocytes from lung (T-LL) gene expression profile data were collected from the National Center for Biotechnology Information (NCBI) Gene Expression Omnibus database (GSE6846, GSE27379, GSE28651). String Network analysis (www.string-db.org) was used to analyze the physical and functional associations of selected genes of TCs in chromosome 1. TC-specific genes on chromosome 1 are selected as genes enriched in TCs on day 5 and/or day 10 as compared to other cells types previously described. Approximately 10–20 % of genes demonstrated a similar pattern in TCs, mesenchymal stem cells, fibroblasts, and alveolar type II cells, while few similarities were identified between TCs and ABCs, PACs, T-BL, or T-LL, respectively. Interestingly, among all genes analyzed, *Capn2*, *Fhl2*, and *Qsox*1 were found to be significantly overexpressed in TCs when compared to other cell types evaluated. Capn2, a member of non-lysosomal calcium-activated neutral proteases, is involved in cell migration. The enrichment of Capn2 in TCs suggests that TCs may participate in tissue/organ homoeostasis and morphogenesis. Fhl2 is able to interact with plasma membrane integrin. The enrichment of Fhl2 in TCs suggests that TCs might be associated with the regulation of tissue inflammation, immune response, injury, repair, and cell movement. Qsox1 controls cell cycle, oxidative protein folding, and extracellular matrix remodeling. Based on the gene profiles on chromosome 1, these results suggest that TCs may have the capacity to induce cell expansion, proliferation, and movement [84].

Following the identification of TC-specific gene profiles on chromosome 1, the TC-specific gene expression profiles of chromosomes 2 and 3 were evaluated [85]. A total of 26 or 80 TC genes on chromosomes 2 and 13 or 59 TC genes on chromosome 3 were found to be upregulated or downregulated, respectively, as compared with other cell types [85]. On chromosome 2, myosin light polypeptide 9 (Myl9) is highly expressed in TCs, while phospholipid transfer protein (Pltp) is expressed in TCs at extremely low abundance, which indicates that TCs may modulate inflammation in the lung [85]. On chromosome 3, SH3-domain GRB2-like B1 (Sh3glb1), transmembrane 4 superfamily member 1 (Tm4sf1), and colony-stimulating factor 1 (Csf1) are highly expressed in TCs, while phosphodiesterase 5A (PDE5A) is downregulated [85]. Myl9 plays a central role in cell adhesion, migration, and division, while Pltp regulates lipid metabolism, immune response, and lipopolysaccharide binding. Sh3glb1 participates in cell proliferation and survival related to cellular stress. Tm4sf1 is highly expressed in various carcinomas, while Csf1 regulates cancer metastasis and invasion. PDE5A is closely related to acute and chronic interstitial lung diseases. Based on the gene profiles of TCs on chromosome 2 and 3, the possible functional role affected by TCs could be inferred [85].

The TC-specific gene expression profile on chromosome 4 was also summarized [86]. A total of 17 genes were upregulated, while 56 genes were found to be downregulated in chromosome 4 of TCs as compared to other cell types [86]. A-kinase anchor protein 2 (Akap2), G protein-coupled receptor 153 (Gpr153), Syndecan 3 (Sdc3), and TBC1 domain family membrane 2A (Tbc1d2) were found to be overexpressed between one- and fourfold, while sushi, von Willebrand factor type A, EGF, and pentraxin domain containing 1 (Svep1) were upregulated greater than fourfold [86]. Akap 2 regulates G protein signaling and actin filament dynamics. As an orphan receptor, the functional role of Gpr153 remains unclear. Sdc3 affects the actin cytoskeleton and controls the cell shape. Tbc1d2 controls cytoplasmic vesicle trafficking and cell junctions. Svep1 participates in cell adhesion [86].

The specific genes and functional networks of TCs found on chromosome 17 and 18 were also summarized [87]. A total of 16 and 10 of TC-specific genes are upregulated and 68 and 22 are downregulated in chromosome 17 and 18, respectively, as compared to other cell types [87]. Mapk14, Trem2, and MCFD2 are upregulated in TCs, while E4F1 and programmed cell death 2 (PDCD2) are downregulated [87]. Mapk14, encoded by the p38 gene, can promote the production of pro-inflammatory cytokines and also control the response to stress and metabolic pathways. Trem2 negatively regulates cytokine synthesis and plays as an inhibitory regulator in inflammatory response. MCFD2 controls the transport of FV and FVIII from the endoplasmic reticulum to the Golgi apparatus. E4F1, a key posttranslational regulator of p53, is able to regulate mammalian embryonic and somatic cell proliferation and survival. PDCD2 overexpression has been shown to induce apoptosis in human cell lines [87].

Following the identification of TC-specific gene profiles on chromosomes 17 and 18, the specific genes and functional networks of TCs in chromosome X were summarized [88]. *Flna*, *Msn*, *Cfp*, *Col4a5*, *Mum1l1*, *Rnf128*, *Syn1*, and *Srpx2* are upregulated, while *Abcb7*, *Atf1*, *Ddx26b*, *Drp2*, *Fam122b*, *Gyk*, *Lrak1*, *Lamp2*,

Mecp2, *Ndufb11*, *Ogt*, *Pdha1*, *Pola1*, *Rab9*, *Rbmx2*, *Rhox9*, *Thoc2*, *Vbp1*, *Dkc1*, *Nkrf*, *Piga*, *Tmlhe*, and *Tsr2* are downregulated in TCs as compared with other cells [88]. Based on these genes, pulmonary TCs are believed to modulate cellular growth and migration, protect cells from stress, regulate immune responses, and control tissue remodeling, repair, and vessel formation [88].

In conclusion, more studies should be conducted to determine the functional roles of TC-specific genes and to further explore the function role of TCs. As tissue- and cellular-specific gene expression patterns exist, investigators should consider evaluating the similarities and differences in the expression between TCs from different tissues/organs using array-based or RNA seq experiments.

2.6 Proteome

Defining the TC proteome is a priority to help establish that TCs are a distinct cell type [89]. A total of 39 proteins, most notably myosin-14, were found to be upregulated in lung TCs following 5 days of culture. There were also 25 proteins, including collagen alfa 3 (VI) chain, secernin-1, and fascin, shown to be downregulated, as compared to fibroblasts [89]. Based on the 39 proteins identified in TCs, the most affected biological processes were thought to be metabolic processes, cellular processes, development processes, cellular component organization, precursor metabolites and energy generation, immune system processes, cell communication, transport, cell adhesion, cell cycle, system processes, homeostatic processes, biological processes regulation, and responses to stimulus [89]. Meanwhile, a total of 24 proteins, including superoxide dismutase and prostacyclin synthase 6A, were found to be upregulated in lung TCs cultured for 10 days, while 40 proteins, including microtubule-associated protein RP/EB family member 1 and collagen α-3 (VI) chain, were downregulated, relative to fibroblasts [89]. Interestingly, myosin-14 remained overexpressed in TCs over the course of 10 days in cell culture [89]. Based on the changes in protein expression, the most affected biological processes were suggested to be structural molecule activity and catalytic activity and binding [89]. As myosin-14 is reported to be associated with sensory perception, TCs are candidates for a mechanical sensing and mechanochemical conversion task [89]. Interestingly, in TCs, the differentially identified proteins participate in cell signaling and energy and metabolic pathways [89].

iTRAQ quantitative proteomics has also been performed to profile human lung TCs and microvascular endothelial cells [90]. A total of 38 proteins including myosin-14, superoxide dismutase (Sod2), acid ceramidase (AC), envoplakin, and epiplakin were upregulated in TCs after 5 days culture, while 60 proteins including MUC18, Ras-interacting protein 1, BTB/POZ domain-containing protein, peptidyl prolyl cis/trans isomerase, nestin, and von Willebrand factor were downregulated relative to microvascular endothelial cells [90]. Based on the changed proteins expression, the most affected biological processes were thought to be catalytic and structural molecular activity [90]. Meanwhile, a total of 26 proteins, including prostacyclin synthase, epiplakin, and superoxide dismutase, were upregulated in TCs following 10 days in culture. Conversely, 56 proteins, including microtubule-associated RP/EB family member 1, cysteine-rich protein 2, von Willebrand factor, platelet endothelial cell adhesion molecule peptidyl prolyl cis/trans isomerase, and cell surface glycoprotein MUC18, were downregulated as compared to microvascular endothelial cells [90]. Based on these protein changes, the most affected biological processes were identified as processes affecting catalytic activity, receptor activity, transporter activity, and structural molecule activity [90].

2.7 Secretome Features

RayBio Human Cytokine Antibody Array V analyses were performed using the supernatant from cultured human dermal TCs and fibroblasts [91]. A total of 79 cytokines were identified and epithelial-derived neutrophil-activating peptide 78 (ENA-78) and granulocyte chemotactic protein 2 (GCP-2) were found to be significantly higher in the supernatant from TCs as compared to fibroblasts. Moreover, 37 cytokines were significantly lower in the supernatant derived from TCs as compared to fibroblasts, including interleukin 5, monocyte chemotactic protein-3 (MCP-3), MCP-4, angiogenin, macrophage inflammatory protein-3, and thrombopoietin. All results were confirmed by ELISA analysis [91]. ENA-78 participates in angiogenesis, which is consistent with previously described role of TCs in neoangiogenesis within the heart. Interestingly, GCP-2 has also been correlated with enhanced angiogenesis and reduction of infarct size, which, similarly, is in keeping with the reported functional role for cardiac TCs [91].

The protein secretory profile was also analyzed using high-sensitivity on-chip electrophoresis, surface-enhanced laser desorption/ionization time-of-flight mass spectrometry, and multiple analysis by Luminex xMAP by using supernatants from mouse-cultured cardiac TCs and 3T3 fibroblasts [92]. Interleukin (IL)-6, VEGF, macrophage inflammatory protein 1a (MIP-1a), MIP-2, and MCP-1 were reportedly enriched in the supernatants from mouse-cultured cardiac TCs, as compared to those from 3T3 fibroblasts [92]. Interestingly, MIP-1a, MIP-2, and MCP-1 are all induced by IL-6 signaling, while the latter is associated with cell proliferation [92]. Moreover, the presence of IL-6 and VEGF in TC supernatants suggests a potential regulatory role of TCs in the control of cell growth/myocyte differentiation and angiogenesis [92]. In addition, cardiac TCs have been suggested to deliver signals to cardiac progenitor cells, which enable them to differentiate into cardiomyocytes and participate in cardiac regeneration [92].

Interestingly, Tps have been reported to release three types of extracellular vesicles including exosomes, ectosomes, and multivesicular cargos [93–96]. However, the nature of these extracellular vesicles remains unclear [93, 94]. As such, further studies are critically required to compare the extracellular vesicles secreted by TCs to those secreted by other types of cells.

2.8 Electrophysiology

Distinct electrophysiological properties are also important for the identification and confirmation of TCs as a distinct type of interstitial cells [22, 97]. Large conductance Ca^{2+}-activated K^+ current (BK_{Ca}), transient outward K^+ current (I_{to}), inwardly rectifying K^+ current (IK_{ir}), and ATP-sensitive potassium current (K_{ATP}) have been identified in human atrial fibroblasts [98]. Following a comparison of the electrophysiological properties between human atrial TCs and human atrial fibroblasts, BK_{Ca} and IK_{ir}, but not I_{to} and K_{ATP}, were found to be expressed in human atrial TCs [97]. Interestingly, Kv4.3 (responsible for I_{to}) expression in TCs is barely above background levels, which supports the fact that I_{to} current is undetectable in human atrial TCs [97]. Moreover, human ventricular TCs also express BK_{Ca} and IK_{ir}, but not I_{to} and K_{ATP} [97]. Thus, further studies are needed to compare the currents between TCs and fibroblasts. Moreover, the difference of electrophysiological properties between human atrial TCs and human ventricular TCs should also be evaluated.

2.9 Conclusion

TCs have been recognized as a distinct novel type of interstitial cells [99, 100]. Functionally, TCs are mainly involved in controlling intercellular signaling, either by direct contact (junctions) with surrounding elements or at long distance by release of extracellular vesicles including exosomes, ectosomes, and multivesicular cargos [101]. Fibroblasts, by contrast, are oriented to collagen and other matrix component synthesis [60]. The functional difference between TCs and fibroblasts suggests that TCs are distinct cellular entities from fibroblasts. However, in spite of the growing literature on TCs, it is still difficult to clearly identify their roles. Numerous studies have suggested that TCs mediate (1) a mechanical support, (2) immune surveillance, (3) paracrine or juxtacrine intercellular signaling, (4) cell precursors for a mixture of many cell types with common mesenchymal origin, (5) guiding and nursing cells, and (6) neoangiogenesis [29, 60, 102–128].

There are, however, numerous issues relating to our understanding of TCs that need to be resolved. Firstly, the lncRNA profile of TCs as compared with other types of interstitial cells needs to be determined. RNA seqs or lncRNA arrays will provide more compelling evidence to discriminate between TCs and other types of interstitial cells. Secondly, metabonomic studies would help reveal differences in the metabonomic between TCs and other types of interstitial cells. Thirdly, a single specific immunophenotype for TCs or at least specific for the TCs of a given organ must be established with the help of modern proteome studies. Fourthly, as several TC subtypes share the same ultrastructural features but display totally different immunophenotypes, comparisons of the region-specific roles of TCs should be performed. Finally, as TCs have been suggested to represent novel cellular targets for forthcoming therapies, developing specific methods to modulate TCs may be very helpful [129].

Acknowledgments This work was supported by the grants from National Natural Science Foundation of China (81570362 and 81200169 to JJ Xiao, 81400647 to Y Bei), Innovation Program of Shanghai Municipal Education Commission (13YZ014 to JJ Xiao), Innovation fund from Shanghai University (sdcx2012038 to JJ Xiao), and Program for the integration of production, teaching, and research for University Teachers supported by Shanghai Municipal Education Commission (year 2014, to JJ Xiao).

Conflicts of Interest The authors declare there are no conflicts of interest.

References

1. Popescu LM, Faussone-Pellegrini MS. TELOCYTES – a case of serendipity: the winding way from Interstitial Cells of Cajal (ICC), via Interstitial Cajal-Like Cells (ICLC) to TELOCYTES. J Cell Mol Med. 2010;14(4):729–40.
2. Bei Y, Wang F, Yang C, Xiao J. Telocytes in regenerative medicine. J Cell Mol Med. 2015;19(7):1441–54.
3. Cretoiu SM, Popescu LM. Telocytes revisited. Biomol Concepts. 2014;5(5):353–69.
4. Li YY, Lu SS, Xu T, Zhang HQ, Li H. Comparative analysis of telomerase activity in CD117(+) CD34(+) cardiac telocytes with bone mesenchymal stem cells, cardiac fibroblasts and cardiomyocytes. Chin Med J. 2015;128(14):1942–7.
5. Hatta K, Huang ML, Weisel RD, Li RK. Culture of rat endometrial telocytes. J Cell Mol Med. 2012;16(7):1392–6.
6. Dawidowicz J, Szotek S, Matysiak N, Mielanczyk L, Maksymowicz K. Electron microscopy of human fascia lata: focus on telocytes. J Cell Mol Med. 2015;19(10):2500–6.
7. Ullah S, Yang P, Zhang L, Zhang Q, Liu Y, Chen W, Waqas Y, Le Y, Chen B, Chen Q. Identification and characterization of telocytes in the uterus of the oviduct in the Chinese soft-shelled turtle, Pelodiscus sinensis: TEM evidence. J Cell Mol Med. 2014;18(12): 2385–92.
8. Cantarero Carmona I, Luesma Bartolome MJ, Junquera Escribano C. Identification of telocytes in the lamina propria of rat duodenum: transmission electron microscopy. J Cell Mol Med. 2011;15(1):26–30.
9. Milia AF, Ruffo M, Manetti M, Rosa I, Conte D, Fazi M, Messerini L, Ibba-Manneschi L. Telocytes in Crohn's disease. J Cell Mol Med. 2013;17(12):1525–36.
10. Popescu LM, Manole CG, Gherghiceanu M, Ardelean A, Nicolescu MI, Hinescu ME, Kostin S. Telocytes in human epicardium. J Cell Mol Med. 2010;14(8):2085–93.
11. Yang Y, Sun W, Wu SM, Xiao J, Kong X. Telocytes in human heart valves. J Cell Mol Med. 2014;18(5):759–65.
12. Mandache E, Gherghiceanu M, Macarie C, Kostin S, Popescu LM. Telocytes in human isolated atrial amyloidosis: ultrastructural remodelling. J Cell Mol Med. 2010;14(12):2739–47.
13. Chen X, Zheng Y, Manole CG, Wang X, Wang Q. Telocytes in human oesophagus. J Cell Mol Med. 2013;17(11):1506–12.
14. Suciu L, Popescu LM, Gherghiceanu M, Regalia T, Nicolescu MI, Hinescu ME, Faussone-Pellegrini MS. Telocytes in human term placenta: morphology and phenotype. Cells Tissues Organs. 2010;192(5):325–39.
15. Zhang H, Zhong S, Ge T, Peng S, Yu P, Zhou Z, Guo X. Telocytes in ileum of the Chinese giant salamander: ultrastructural evidence. J Cell Mol Med. 2016;20:568.
16. Rusu MC, Nicolescu MI, Jianu AM, Lighezan R, Manoiu VS, Paduraru D. Esophageal telocytes and hybrid morphologies. Cell Biol Int. 2012;36(12):1079–88.
17. Manetti M, Guiducci S, Ruffo M, Rosa I, Faussone-Pellegrini MS, Matucci-Cerinic M, Ibba-Manneschi L. Evidence for progressive reduction and loss of telocytes in the dermal cellular network of systemic sclerosis. J Cell Mol Med. 2013;17(4):482–96.

18. Bojin FM, Gavriliuc OI, Cristea MI, Tanasie G, Tatu CS, Panaitescu C, Paunescu V. Telocytes within human skeletal muscle stem cell niche. J Cell Mol Med. 2011;15(10):2269–72.
19. Cretoiu D, Cretoiu SM, Simionescu AA, Popescu LM. Telocytes, a distinct type of cell among the stromal cells present in the lamina propria of jejunum. Histol Histopathol. 2012;27(8):1067–78.
20. Matyja A, Gil K, Pasternak A, Sztefko K, Gajda M, Tomaszewski KA, Matyja M, Walocha JA, Kulig J, Thor P. Telocytes: new insight into the pathogenesis of gallstone disease. J Cell Mol Med. 2013;17(6):734–42.
21. Yang P, Ahmad N, Hunag Y, Ullah S, Zhang Q, Waqas Y, Liu Y, Li Q, Hu L, Chen Q. Telocytes: novel interstitial cells present in the testis parenchyma of the Chinese soft-shelled turtle Pelodiscus sinensis. J Cell Mol Med. 2015;19(12):2888–99.
22. Cretoiu SM, Cretoiu D, Marin A, Radu BM, Popescu LM. Telocytes: ultrastructural, immunohistochemical and electrophysiological characteristics in human myometrium. Reproduction. 2013;145(4):357–70.
23. Enciu AM, Popescu LM. Telopodes of telocytes are influenced in vitro by redox conditions and ageing. Mol Cell Biochem. 2015;410(1–2):165–74.
24. Rusu MC, Jianu AM, Mirancea N, Didilescu AC, Manoiu VS, Paduraru D. Tracheal telocytes. J Cell Mol Med. 2012;16(2):401–5.
25. Yang P, Liu Y, Ahmed N, Ullah S, Liu YI, Chen Q. Ultrastructural identification of telocytes in the muscularis of chicken ileum. Exp Ther Med. 2015;10(6):2325–30.
26. Yang J, Chi C, Liu Z, Yang G, Shen ZJ, Yang XJ. Ultrastructure damage of oviduct telocytes in rat model of acute salpingitis. J Cell Mol Med. 2015;19(7):1720–8.
27. Diaz-Flores L, Gutierrez R, Garcia MP, Saez FJ, Aparicio F, Diaz-Flores Jr L, Madrid JF. Uptake and intracytoplasmic storage of pigmented particles by human CD34+ stromal cells/telocytes: endocytic property of telocytes. J Cell Mol Med. 2014;18(12):2478–87.
28. Gevaert T, De Vos R, Van Der Aa F, Joniau S, van den Oord J, Roskams T, De Ridder D. Identification of telocytes in the upper lamina propria of the human urinary tract. J Cell Mol Med. 2012;16(9):2085–93.
29. Mirancea N, Morosanu AM, Mirancea GV, Juravle FD, Manoiu VS. Infrastructure of the telocytes from tumor stroma in the skin basal and squamous cell carcinomas. Rom J Morphol Embryoly = Rev Roum Morphol Embryol. 2013;54(4):1025–37.
30. Li H, Lu S, Liu H, Ge J, Zhang H. Scanning electron microscope evidence of telocytes in vasculature. J Cell Mol Med. 2014;18(7):1486–9.
31. Bani D, Formigli L, Gherghiceanu M, Faussone-Pellegrini MS. Telocytes as supporting cells for myocardial tissue organization in developing and adult heart. J Cell Mol Med. 2010;14(10):2531–8.
32. Vannucchi MG, Bani D, Faussone-Pellegrini MS: Telocytes contribute as cell progenitors and differentiation inductors in tissue regeneration. Curr Stem Cell Res Ther. 2016;11(5):383–9.
33. Yang XJ, Yang J, Liu Z, Yang G, Shen ZJ. Telocytes damage in endometriosis-affected rat oviduct and potential impact on fertility. J Cell Mol Med. 2015;19(2):452–62.
34. Rusu MC, Pop F, Hostiuc S, Curca GC, Jianu AM, Paduraru D. Telocytes form networks in normal cardiac tissues. Histol Histopathol. 2012;27(6):807–16.
35. Gherghiceanu M, Manole CG, Popescu LM. Telocytes in endocardium: electron microscope evidence. J Cell Mol Med. 2010;14(9):2330–4.
36. Xiao J, Wang F, Liu Z, Yang C. Telocytes in liver: electron microscopic and immunofluorescent evidence. J Cell Mol Med. 2013;17(12):1537–42.
37. Popescu BO, Gherghiceanu M, Kostin S, Ceafalan L, Popescu LM. Telocytes in meninges and choroid plexus. Neurosci Lett. 2012;516(2):265–9.
38. Alunno A, Ibba-Manneschi L, Bistoni O, Rosa I, Caterbi S, Gerli R, Manetti M. Telocytes in minor salivary glands of primary Sjogren's syndrome: association with the extent of inflammation and ectopic lymphoid neogenesis. J Cell Mol Med. 2015;19(7):1689–96.
39. Diaz-Flores L, Gutierrez R, Saez FJ, Diaz-Flores Jr L, Madrid JF. Telocytes in neuromuscular spindles. J Cell Mol Med. 2013;17(4):457–65.

40. Nicolescu MI, Bucur A, Dinca O, Rusu MC, Popescu LM. Telocytes in parotid glands. Anat Rec. 2012;295(3):378–85.
41. Hinescu ME, Gherghiceanu M, Suciu L, Popescu LM. Telocytes in pleura: two- and three-dimensional imaging by transmission electron microscopy. Cell Tissue Res. 2011;343(2): 389–97.
42. Qi G, Lin M, Xu M, Manole CG, Wang X, Zhu T. Telocytes in the human kidney cortex. J Cell Mol Med. 2012;16(12):3116–22.
43. Nicolescu MI, Popescu LM. Telocytes in the interstitium of human exocrine pancreas: ultrastructural evidence. Pancreas. 2012;41(6):949–56.
44. Zheng Y, Zhu T, Lin M, Wu D, Wang X. Telocytes in the urinary system. J Transl Med. 2012;10:188.
45. Zheng Y, Li H, Manole CG, Sun A, Ge J, Wang X. Telocytes in trachea and lungs. J Cell Mol Med. 2011;15(10):2262–8.
46. Zhang HQ, Lu SS, Xu T, Feng YL, Li H, Ge JB. Morphological evidence of telocytes in mice aorta. Chin Med J. 2015;128(3):348–52.
47. Li H, Zhang H, Yang L, Lu S, Ge J. Telocytes in mice bone marrow: electron microscope evidence. J Cell Mol Med. 2014;18(6):975–8.
48. Cantarero I, Luesma MJ, Junquera C. The primary cilium of telocytes in the vasculature: electron microscope imaging. J Cell Mol Med. 2011;15(12):2594–600.
49. Rusu MC, Mirancea N, Manoiu VS, Valcu M, Nicolescu MI, Paduraru D. Skin telocytes. Ann Anat = Anat Anz: Off Organ Anat Ges. 2012;194(4):359–67.
50. Corradi LS, Jesus MM, Fochi RA, Vilamaior PS, Justulin Jr LA, Goes RM, Felisbino SL, Taboga SR. Structural and ultrastructural evidence for telocytes in prostate stroma. J Cell Mol Med. 2013;17(3):398–406.
51. Cretoiu D, Hummel E, Zimmermann H, Gherghiceanu M, Popescu LM. Human cardiac telocytes: 3D imaging by FIB-SEM tomography. J Cell Mol Med. 2014;18(11):2157–64.
52. Diaz-Flores L, Gutierrez R, Garcia MP, Gonzalez M, Saez FJ, Aparicio F, Diaz-Flores Jr L, Madrid JF. Human resident CD34+ stromal cells/telocytes have progenitor capacity and are a source of alphaSMA+ cells during repair. Histol Histopathol. 2015;30(5):615–27.
53. Mou Y, Wang Y, Li J, Lu S, Duan C, Du Z, Yang G, Chen W, Zhao S, Zhou J, et al. Immunohistochemical characterization and functional identification of mammary gland telocytes in the self-assembly of reconstituted breast cancer tissue in vitro. J Cell Mol Med. 2013;17(1):65–75.
54. Cretoiu SM, Radu BM, Banciu A, Banciu DD, Cretoiu D, Ceafalan LC, Popescu LM. Isolated human uterine telocytes: immunocytochemistry and electrophysiology of T-type calcium channels. Histochem Cell Biol. 2015;143(1):83–94.
55. Tao L, Wang H, Wang X, Kong X, Li X. Cardiac Telocytes. Curr Stem Cell Res Ther. 2016;11(5):404–9.
56. Gherghiceanu M, Popescu LM. Cardiac telocytes – their junctions and functional implications. Cell Tissue Res. 2012;348(2):265–79.
57. Kostin S. Myocardial telocytes: a specific new cellular entity. J Cell Mol Med. 2010;14(7): 1917–21.
58. Bei Y, Zhou Q, Sun Q, Xiao J. Telocytes in cardiac regeneration and repair. Sem Cell Dev Biol. 2016;55(1):14–21.
59. Chang Y, Li C, Lu Z, Li H, Guo Z. Multiple immunophenotypes of cardiac telocytes. Exp Cell Res. 2015;338(2):239–44.
60. Roatesi I, Radu BM, Cretoiu D, Cretoiu SM. Uterine telocytes: a review of current knowledge. Biol Reprod. 2015;93(1):10.
61. Tokunaga A, Oya T, Ishii Y, Motomura H, Nakamura C, Ishizawa S, Fujimori T, Nabeshima Y, Umezawa A, Kanamori M, et al. PDGF receptor beta is a potent regulator of mesenchymal stromal cell function. J Bone Miner Res: Off J Am Soc Bone Miner Res. 2008;23(9):1519–28.
62. Popescu LM, Manole E, Serboiu CS, Manole CG, Suciu LC, Gherghiceanu M, Popescu BO. Identification of telocytes in skeletal muscle interstitium: implication for muscle regeneration. J Cell Mol Med. 2011;15(6):1379–92.

63. Suciu LC, Popescu BO, Kostin S, Popescu LM. Platelet-derived growth factor receptor-beta-positive telocytes in skeletal muscle interstitium. J Cell Mol Med. 2012;16(4):701–7.
64. Vannucchi MG, Traini C, Manetti M, Ibba-Manneschi L, Faussone-Pellegrini MS. Telocytes express PDGFR alpha in the human gastrointestinal tract. J Cell Mol Med. 2013;17(9):1099–108.
65. Wang F, Bei Y, Zhao Y, Song Y, Xiao J, Yang C. Telocytes in pregnancy-induced physiological liver growth. Cell Physiol Biochem: Int J Exp Cell Physiol Biochem Pharmacol. 2015;36(1):250–8.
66. Wang F, Song Y, Bei Y, Zhao Y, Xiao J, Yang C. Telocytes in liver regeneration: possible roles. J Cell Mol Med. 2014;18(9):1720–6.
67. Liu J, Cao Y, Song Y, Huang Q, Wang F, Yang W, Yang C. Telocytes in liver. Curr Stem Cell Res Ther. 2016;11(5):415–9.
68. Zhou Q, Wei L, Zhong C, Fu S, Bei Y, Huica RI, Wang F, Xiao J. Cardiac telocytes are double positive for CD34/PDGFR-alpha. J Cell Mol Med. 2015;19(8):2036–42.
69. Bei Y, Zhou Q, Fu S, Lv D, Chen P, Chen Y, Wang F, Xiao J. Cardiac telocytes and fibroblasts in primary culture: different morphologies and immunophenotypes. PLoS One. 2015;10(2):e0115991.
70. Petre N, Rusu MC, Pop F, Jianu AM. Telocytes of the mammary gland stroma. Folia Morphol. 2015. doi: 10.5603/FM.a2015.0123.
71. Chang Y, Li C, Gan L, Li H, Guo Z. Telocytes in the spleen. PLoS One. 2015;10(9):e0138851.
72. Vannucchi MG, Traini C, Guasti D, Del Popolo G, Faussone-Pellegrini MS. Telocytes subtypes in human urinary bladder. J Cell Mol Med. 2014;18(10):2000–8.
73. Diaz-Flores L, Gutierrez R, Garcia MP, Saez FJ, Diaz-Flores Jr L, Valladares F, Madrid JF. CD34+ stromal cells/fibroblasts/fibrocytes/telocytes as a tissue reserve and a principal source of mesenchymal cells. Location, morphology, function and role in pathology. Histol Histopathol. 2014;29(7):831–70.
74. Povysil C, Kana M, Zamecnik L, Valova Z, Hanus T. Podoplanin (D2-40) is a reliable marker of urinary bladder myofibroblasts (telocytes). Folia Biol. 2014;60(6):286–9.
75. Melman YF, Shah R, Danielson K, Xiao J, Simonson B, Barth A, Chakir K, Lewis GD, Lavender Z, Truong QA, et al. Circulating MicroRNA-30d is associated with response to cardiac resynchronization therapy in heart failure and regulates cardiomyocyte apoptosis: a translational pilot study. Circulation. 2015;131(25):2202–16.
76. Li J, Xu J, Cheng Y, Wang F, Song Y, Xiao J. Circulating microRNAs as mirrors of acute coronary syndromes: MiRacle or quagMire? J Cell Mol Med. 2013;17(11):1363–70.
77. Xu J, Zhao J, Evan G, Xiao C, Cheng Y, Xiao J. Circulating microRNAs: novel biomarkers for cardiovascular diseases. J Mol Med (Berlin, Germany). 2012;90(8):865–75.
78. Xiao J, Bei Y, Liu J, Dimitrova-Shumkovska J, Kuang D, Zhou Q, Li J, Yang Y, Xiang Y, Wang F, et al. miR-212 downregulation contributes to the protective effect of exercise against non-alcoholic fatty liver via targeting FGF-21. J Cell Mol Med. 2016;20(2):204–16.
79. Liu X, Xiao J, Zhu H, Wei X, Platt C, Damilano F, Xiao C, Bezzerides V, Bostrom P, Che L, et al. miR-222 is necessary for exercise-induced cardiac growth and protects against pathological cardiac remodeling. Cell Metab. 2015;21(4):584–95.
80. Tao L, Bei Y, Zhou Y, Xiao J, Li X. Non-coding RNAs in cardiac regeneration. Oncotarget. 2015;6(40):42613–22.
81. Lv D, Liu J, Zhao C, Sun Q, Zhou Q, Xu J, Xiao J. Targeting microRNAs in pathological hypertrophy and cardiac failure. Mini Rev Med Chem. 2015;15(6):475–8.
82. Cismasiu VB, Radu E, Popescu LM. miR-193 expression differentiates telocytes from other stromal cells. J Cell Mol Med. 2011;15(5):1071–4.
83. Zheng Y, Zhang M, Qian M, Wang L, Cismasiu VB, Bai C, Popescu LM, Wang X. Genetic comparison of mouse lung telocytes with mesenchymal stem cells and fibroblasts. J Cell Mol Med. 2013;17(4):567–77.
84. Sun X, Zheng M, Zhang M, Qian M, Zheng Y, Li M, Cretoiu D, Chen C, Chen L, Popescu LM, et al. Differences in the expression of chromosome 1 genes between lung telocytes and

other cells: mesenchymal stem cells, fibroblasts, alveolar type II cells, airway epithelial cells and lymphocytes. J Cell Mol Med. 2014;18(5):801–10.
85. Zheng M, Sun X, Zhang M, Qian M, Zheng Y, Li M, Cretoiu SM, Chen C, Chen L, Cretoiu D, et al. Variations of chromosomes 2 and 3 gene expression profiles among pulmonary telocytes, pneumocytes, airway cells, mesenchymal stem cells and lymphocytes. J Cell Mol Med. 2014;18(10):2044–60.
86. Song D, Cretoiu D, Zheng M, Qian M, Zhang M, Cretoiu SM, Chen L, Fang H, Popescu LM, Wang X. Comparison of Chromosome 4 gene expression profile between lung telocytes and other local cell types. J Cell Mol Med. 2016;20(1):71–80.
87. Wang J, Ye L, Jin M, Wang X. Global analyses of Chromosome 17 and 18 genes of lung telocytes compared with mesenchymal stem cells, fibroblasts, alveolar type II cells, airway epithelial cells, and lymphocytes. Biol Direct. 2015;10:9.
88. Zhu Y, Zheng M, Song D, Ye L, Wang X. Global comparison of chromosome X genes of pulmonary telocytes with mesenchymal stem cells, fibroblasts, alveolar type II cells, airway epithelial cells, and lymphocytes. J Transl Med. 2015;13(1):318.
89. Zheng Y, Cretoiu D, Yan G, Cretoiu SM, Popescu LM, Wang X. Comparative proteomic analysis of human lung telocytes with fibroblasts. J Cell Mol Med. 2014;18(4):568–89.
90. Zheng Y, Cretoiu D, Yan G, Cretoiu SM, Popescu LM, Fang H, Wang X. Protein profiling of human lung telocytes and microvascular endothelial cells using iTRAQ quantitative proteomics. J Cell Mol Med. 2014;18(6):1035–59.
91. Kang Y, Zhu Z, Zheng Y, Wan W, Manole CG, Zhang Q. Skin telocytes versus fibroblasts: two distinct dermal cell populations. J Cell Mol Med. 2015;19(11):2530–9.
92. Albulescu R, Tanase C, Codrici E, Popescu DI, Cretoiu SM, Popescu LM. The secretome of myocardial telocytes modulates the activity of cardiac stem cells. J Cell Mol Med. 2015;19(8):1783–94.
93. Fertig ET, Gherghiceanu M, Popescu LM. Extracellular vesicles release by cardiac telocytes: electron microscopy and electron tomography. J Cell Mol Med. 2014;18(10):1938–43.
94. Cretoiu D, Gherghiceanu M, Hummel E, Zimmermann H, Simionescu O, Popescu LM. FIB-SEM tomography of human skin telocytes and their extracellular vesicles. J Cell Mol Med. 2015;19(4):714–22.
95. Cismasiu VB, Popescu LM. Telocytes transfer extracellular vesicles loaded with microRNAs to stem cells. J Cell Mol Med. 2015;19(2):351–8.
96. Smythies J, Edelstein L. Telocytes, exosomes, gap junctions and the cytoskeleton: the makings of a primitive nervous system? Front Cell Neurosci. 2014;7:278.
97. Sheng J, Shim W, Lu J, Lim SY, Ong BH, Lim TS, Liew R, Chua YL, Wong P. Electrophysiology of human cardiac atrial and ventricular telocytes. J Cell Mol Med. 2014;18(2):355–62.
98. Sheng J, Shim W, Wei H, Lim SY, Liew R, Lim TS, Ong BH, Chua YL, Wong P. Hydrogen sulphide suppresses human atrial fibroblast proliferation and transformation to myofibroblasts. J Cell Mol Med. 2013;17(10):1345–54.
99. Rusu MC, Loreto C, Manoiu VS. Network of telocytes in the temporomandibular joint disc of rats. Acta Histochem. 2014;116(4):663–8.
100. Padhi S, Sarangi R, Mallick S. Pancreatic extragastrointestinal stromal tumors, interstitial Cajal like cells, and telocytes. JOP: J Pancreas. 2013;14(1):1–14.
101. Diaz-Flores L, Gutierrez R, Diaz-Flores L, Jr., Gomez MG, Saez FJ, Madrid JF. Behaviour of telocytes during physiopathological activation. Sem Cell Dev Biol. 2016;55(1):50–61.
102. Zhao B, Chen S, Liu J, Yuan Z, Qi X, Qin J, Zheng X, Shen X, Yu Y, Qnin TJ, et al. Cardiac telocytes were decreased during myocardial infarction and their therapeutic effects for ischaemic heart in rat. J Cell Mol Med. 2013;17(1):123–33.
103. Suciu L, Nicolescu MI, Popescu LM. Cardiac telocytes: serial dynamic images in cell culture. J Cell Mol Med. 2010;14(11):2687–92.
104. Gherghiceanu M, Popescu LM. Cardiomyocyte precursors and telocytes in epicardial stem cell niche: electron microscope images. J Cell Mol Med. 2010;14(4):871–7.
105. Zhou J, Wang Y, Zhu P, Sun H, Mou Y, Duan C, Yao A, Lv S, Wang C. Distribution and characteristics of telocytes as nurse cells in the architectural organization of engineered heart tissues. Sci China Life Sci. 2014;57(2):241–7.

106. Edelstein L, Smythies J. Epigenetic aspects of telocytes/cordocytes: jacks of all trades, masters of most. Front Cell Neurosci. 2014;8:32.
107. Manole CG, Cismasiu V, Gherghiceanu M, Popescu LM. Experimental acute myocardial infarction: telocytes involvement in neo-angiogenesis. J Cell Mol Med. 2011;15(11): 2284–96.
108. Richter M, Kostin S. The failing human heart is characterized by decreased numbers of telocytes as result of apoptosis and altered extracellular matrix composition. J Cell Mol Med. 2015;19(11):2597–606.
109. Bosco C, Diaz E, Gutierrez R, Gonzalez J, Perez J. Ganglionar nervous cells and telocytes in the pancreas of Octodon degus: extra and intrapancreatic ganglionar cells and telocytes in the degus. Auton Neurosci: Basic Clin. 2013;177(2):224–30.
110. Zheng Y, Chen X, Qian M, Zhang M, Zhang D, Bai C, Wang Q, Wang X. Human lung telocytes could promote the proliferation and angiogenesis of human pulmonary microvascular endothelial cells in vitro. Mol Cell Ther. 2014;2:3.
111. Cretoiu SM, Cretoiu D, Popescu LM. Human myometrium – the ultrastructural 3D network of telocytes. J Cell Mol Med. 2012;16(11):2844–9.
112. Smythies J. Intercellular signaling in cancer-the SMT and TOFT hypotheses, exosomes, telocytes and metastases: is the messenger in the message? J Cancer. 2015;6(7):604–9.
113. Zhao B, Liao Z, Chen S, Yuan Z, Yilin C, Lee KK, Qi X, Shen X, Zheng X, Quinn T, et al. Intramyocardial transplantation of cardiac telocytes decreases myocardial infarction and improves post-infarcted cardiac function in rats. J Cell Mol Med. 2014;18(5):780–9.
114. Manetti M, Rosa I, Messerini L, Guiducci S, Matucci-Cerinic M, Ibba-Manneschi L. A loss of telocytes accompanies fibrosis of multiple organs in systemic sclerosis. J Cell Mol Med. 2014;18(2):253–62.
115. Boos AM, Weigand A, Brodbeck R, Beier JP, Arkudas A, Horch RE. The potential role of telocytes in tissue engineering and regenerative medicine. Sem Cell Dev Biol. 2016;55(1):70–8.
116. Zheng Y, Bai C, Wang X. Potential significance of telocytes in the pathogenesis of lung diseases. Expert Rev Respir Med. 2012;6(1):45–9.
117. Bosco C, Diaz E, Gutierrez R, Gonzalez J, Parra-Cordero M, Rodrigo R, Barja P. A putative role for telocytes in placental barrier impairment during preeclampsia. Med Hypotheses. 2015;84(1):72–7.
118. Popescu LM, Fertig ET, Gherghiceanu M. Reaching out: junctions between cardiac telocytes and cardiac stem cells in culture. J Cell Mol Med. 2016;20(2):370–80.
119. Faussone-Pellegrini MS, Bani D. Relationships between telocytes and cardiomyocytes during pre- and post-natal life. J Cell Mol Med. 2010;14(5):1061–3.
120. Li L, Lin M, Li L, Wang R, Zhang C, Qi G, Xu M, Rong R, Zhu T. Renal telocytes contribute to the repair of ischemically injured renal tubules. J Cell Mol Med. 2014;18(6):1144–56.
121. Edelstein L, Smythies J. The role of telocytes in morphogenetic bioelectrical signaling: once more unto the breach. Front Mol Neurosci. 2014;7:41.
122. Zhou J, Zhang Y, Wen X, Cao J, Li D, Lin Q, Wang H, Liu Z, Duan C, Wu K, et al. Telocytes accompanying cardiomyocyte in primary culture: two- and three-dimensional culture environment. J Cell Mol Med. 2010;14(11):2641–5.
123. Popescu LM, Curici A, Wang E, Zhang H, Hu S, Gherghiceanu M. Telocytes and putative stem cells in ageing human heart. J Cell Mol Med. 2015;19(1):31–45.
124. Popescu LM, Gherghiceanu M, Suciu LC, Manole CG, Hinescu ME. Telocytes and putative stem cells in the lungs: electron microscopy, electron tomography and laser scanning microscopy. Cell Tissue Res. 2011;345(3):391–403.
125. Luesma MJ, Gherghiceanu M, Popescu LM. Telocytes and stem cells in limbus and uvea of mouse eye. J Cell Mol Med. 2013;17(8):1016–24.
126. Manetti M, Rosa I, Messerini L, Ibba-Manneschi L. Telocytes are reduced during fibrotic remodelling of the colonic wall in ulcerative colitis. J Cell Mol Med. 2015;19(1):62–73.
127. Ardeleanu C, Bussolati G. Telocytes are the common cell of origin of both PEComas and GISTs: an evidence-supported hypothesis. J Cell Mol Med. 2011;15(12):2569–74.

128. Diaz-Flores L, Gutierrez R, Garcia MP, Gonzalez M, Diaz-Flores L, Jr., Madrid JF. Telocytes as a source of progenitor cells in regeneration and repair through granulation tissue. Curr Stem Cell Res Ther. 2016;11(5):395–403.
129. Campeanu RA, Radu BM, Cretoiu SM, Banciu DD, Banciu A, Cretoiu D, Popescu LM. Near-infrared low-level laser stimulation of telocytes from human myometrium. Lasers Med Sci. 2014;29(6):1867–74.

Chapter 3
Extracellular Microvesicles (ExMVs) in Cell to Cell Communication: A Role of Telocytes

Mariusz Z. Ratajczak, Daniel Ratajczak, and Daniel Pedziwiatr

Abstract There are several mechanisms by which cells communicate with each other. Evidence accumulates that the evolutionary oldest mechanisms of cell-cell communication involves extracellular microvesicles (ExMVs). Generally, these circular membrane fragments enriched for mRNA, miRNA, proteins, and bioactive lipids are released by exocytosis from endosomal compartment or are directly formed by budding from cell surface membranes. ExMVs from endosomal compartment called exosomes are smaller in size ~100 nM as compared to larger ones released from cell membranes that are in size up to 1 μM. In this chapter we will present an emerging link between ExMVs and recently identified novel cell-cell communication network involving a new type of cell known as telocyte. Mounting evidence accumulates that telocytes mediate several of their biological effects in several organs by releasing ExMVs enriched in mRNA, miRNA, proteins, and several biological mediators to the target cells.

3.1 Introduction

Cells communicate with each other and exchange biological information by employing different mechanisms [4, 14, 15, 17, 20–22]. The most important cell-cell communication systems are based on (i) secreted growth factors, cytokines, chemokines, and small molecular mediators (e.g., extracellular nucleotides, bioactive lipids, ROS, and nitric oxide ions), (ii) cell to cell adhesion contacts mediated by sets of specialized adhesion molecule-ligand interactions, (iii) exchanging information by means of tunneling nanotubules, and (iv) what is a subject of this chapter by circular membrane fragments called extracellular microvesicles (ExMVs), a mechanism that for many years has been largely overlooked [4, 14, 20, 21, 23].

ExMVs are small circular membrane fragments secreted from the endosomal compartment known as exosomes or shed from the cell surface by blebbing of cell

M.Z. Ratajczak (✉) • D. Ratajczak • D. Pedziwiatr
Stem Cell Institute at James Graham Brown Cancer Center, University of Louisville, 500 S. Floyd Street, Rm. 107, Louisville, KY 40202, USA
e-mail: mzrata01@louisville.edu

X. Wang, D. Cretoiu (eds.), *Telocytes*, Advances in Experimental Medicine and Biology, 913, DOI 10.1007/978-981-10-1061-3_3

surface membrane and play an important role in cell-cell communication [10, 20]. This intriguing ExMV-mediated communication system emerged very early during evolution and most likely served as a template for the further development of cell-cell interaction mechanisms involving soluble bioactive mediators and fine-tuned ligand-receptor interactions. ExMVs as mediators of physiological cell-cell communication are different from apoptotic bodies and other cell fragments that emerge in conditions related to irreversible cell damage [4, 20]. Nevertheless, their overall small size and similarity to cellular debris/fragments or apoptotic bodies is one of the reasons that biological significance of ExMVs for many years has been somehow underappreciated.

However, recent augmenting evidence accumulates to show that these tiny membrane fragments orchestrate several biological responses [4, 14, 15, 17, 20–23]. ExMVs contain numerous cell surface proteins and lipids similar to those present in the membranes of the cells from which they originate [12, 19]. As demonstrated, ExMVs may stimulate target cells directly by surface-expressed ligands acting as a kind of "signaling device" [12, 19]. They may also transfer cell surface receptors between various cells [16, 18]. These receptors after transfer may remain functional and change a surface phenotype of the target cells [19]. Furthermore, since they engulf some cytoplasm during membrane blebbing, they may also contain intracellular proteins, mRNA, and miRNA, and as we have demonstrated, they are involved in horizontal transfer of functional mRNA species between cells [5, 16, 18]. ExMVs are also enriched in bioactive lipids (e.g., sphingosine-1-phosphate, ceramide-1-phosphate) and extracellular signaling nucleotides (e.g., ATP, ADP, AMP, adenosine) that all may induce biological responses of the target cells after exposure to ExMVs [20, 21].

More importantly what is highly relevant for this chapter is that ExMVs have been recently implied to be involved in biological effects mediated by novel interstitial (stromal) cell type known as telocytes [5–9]. These intriguing cells are present in different organs including, e.g., the heart, kidney, lung, esophagus, intestine, reproductive system, and skin [7, 10]. Telocytes are $CD34^+/PDGF\alpha^+$ cells and are characterized by small cell bodies (9–15 μM) that give rise to extremely long one to five thin tubular processes as compared to cell body, called telopodes [7]. Electron microscopy studies revealed that telopodes are up to 100 μM long yet 80–300 nm in diameter and are not homogenous but consist of short dilatations known as podomes (250–300 nm) and long thin tubes that connect podomes known as podomers (~80 nm) [7, 11]. The overall size of telocyte can reach up to 1,000 μM [7, 11]. Enlarged fragments of telopodes – podomes – are abundant in mitochondria, endoplasmatic reticulum, and planar and invaginated surface membrane lipid rafts [5–9]. As it will be discussed in this chapter, telocyte-derived ExMVs play an important role in several biological effects of telocytes related to physiological adult body homeostasis and tissue/organ regeneration and may even be involved in some pathological processes [10].

Overall, telocytes represent an evolutionary conserved type of interstitial cells and have been described in multiple species including fish, reptiles, birds, and mammals [7]. Telocytes have been also reported to express some stem cell markers including c-kit and Sca-1 (mouse) and even express the intranuclear stem cell pluripotency marker Oct-4 [7, 22]. The telocyte markers, however, may vary between tissues and anatomical location of these cells in a given tissue [7]. This demonstrates somehow some degree of existing heterogeneity among these cells that may be dictated and epigenetically enforced by the microenvironment in which these cells reside.

3.2 The Mechanisms Involving ExMV Release from the Cells

It is well documented that telocytes identified in many tissues secrete ExMVs [2, 5–9]. However, an open question remains if telocyte-derived ExMVs are released into extracellular space as it occurs in all types of cells secreting these circular fragments or in some cases telocytes as postulated may additionally form "bridging nanostructures" to the target cells (e.g., stem cells residing in stem cell niches) as a route for the transfer of exosomes [7, 22, 23]. This latter intriguing possibility has been described, for example, during exosome delivery from telocytes to lung stem cells [7, 22].

Nevertheless, the most common mechanism of ExMV signaling and biological cargo delivery to the cells is mediated by their release from cells by (i) exocytosis from multivesicular bodies or (ii) by shedding from the cell surface membrane [2, 7]. The process of ExMV formation is energy consumption dependent, and it is still not very well known from a mechanistic point of view, although some of the crucial steps have been already identified [1, 3, 12, 13, 23].

The above mentioned endosomal cell membrane compartment-derived smaller exosomes are released from cells during exocytosis often together with proteins secreted from the Golgi apparatus. The first step in the creation of multivesicular bodies enriched for intraluminal vesicles, which are precursors of exosomes, requires involvement of the so-called endosomal sorting complex required for transport (ESCRT) machinery [7]. After intraluminal vesicles are formed, in a next step, multivesicular bodies may fuse with lysosomes, and their content becomes degraded or they may fuse with plasma membrane to release intraluminal vesicles (exosomes) from the cells into extracellular space [12, 13]. This process requires involvement of Rab GTPases (e.g., Rab 27a, Rab 27b, and Rab 11) [13]. Exosomes were reported to express some characteristic surface proteins including Alix, CD63, CD9, CD81, HSP70, and TSG101 (ESCRT machinery) [1]. In contrast cell surface-derived larger ExMVs are shed from the cell surface membrane by blebbing in response to cell stimulation that leads to a cytosolic Ca^{2+} increase that promotes changes in the structure of cell membrane [10, 19]. Blebbing of cell membrane and formation of ExMVs occur mostly in cholesterol-enriched fragments of cell membrane known as lipid rafts [19].

The cell cytoplasmic membrane consists of a phospholipid bilayer with embedded proteins that is held together via non-covalent interactions between the hydrophobic tails. Moreover, the cytoplasmic membrane has an asymmetric distribution of phospholipids including aminophospholipids, phosphatidylserine, and phosphatidylethanolamine that as demonstrated are specifically sequestered in the inner membrane leaflet [1, 3, 12, 13]. This transmembrane lipid distribution is under the control of three phospholipidic pumps: (i) flippase, (ii) floppase, and (iii) lipid scramblase. The latter phospholipidic pump is responsible for nonspecific redistribution of lipids across the cytoplasmic membrane [10].

It is known that the phospholipid molecules in the cell membrane are in a liquid crystalline state and contain distinguished combinations of glycosphingolipids and protein receptors organized into glycoprotein microdomains that are known in literature as membrane lipid rafts [19]. There are two described types of lipid rafts in cell membranes: (i) planar lipid rafts and (ii) invaginated lipid rafts, called caveolae [19]. Lipid rafts are cholesterol-enriched microdomains in the cell membrane, and cholesterol can be envisioned as a kind of molecular glue that holds the components of lipid rafts together and is important for their integrity [19]. It has been proposed that the loss of phospholipid asymmetry of the cytoplasmic membrane, which leads to phosphatidylserine exposure on cell surface, and a transient phospholipidic imbalance between the externals, at the expense of the inner leaflet caused by lipid scramblase, results in blebbing of the plasma membrane and ExMVs releasing from the areas enriched in lipid rafts [1, 10, 19].

As described for several types of cells during the blebbing process of the cytoplasm membrane, a fragment of cytoplasm that contains mRNA, miRNA, proteins, and even organelles (e.g., mitochondria) is encapsulated into ExMVs [11]. Evidence accumulates that this process of enrichment for mRNA or miRNA species is not random but somehow regulated by proteins involved in mRNA and miRNA storage, transport, and processing [1, 10, 19].

Telocytes as a cell also producing ExMVs are somehow unique. It has been described that telocytes may secrete three types of ExMVs including not only classical (i) endosomal cell membrane compartment-derived exosomes (45 ± 8 nm) and (ii) larger ExMVs (ectosomes) corresponding to small ExMVs shed from the cell membranes (128 ± 28 nm) but also a novel type of ExMVs (iii) described as multivesicular cargo (1 ± 0.4 μM) that are large ExMVs containing tightly packed endomembrane-bound smaller vesicles (145 ± 35 nm) [7]. This interesting new type of ExMVs has been described initially to be secreted by telocytes in myocardium and as postulated involved in paracrine effects of these cells residing in a normal heart where they form tridimensional structure connected with all the types of cells present in this organ including cardiac stem cells and cardiomyocyte progenitors [2, 7].

It is known that ExMVs that originate from the blebbing of cell membrane express several receptors present on the cell surface, and future studies are needed to characterize these receptors on telocyte-derived ExMVs. This will facilitate their detection in the tissues as well as allow to assess if they can contribute to the pool of circulating ExMVs in peripheral blood and lymph [14, 17].

3.3 The Physiological Effects of Telocyte-Derived ExMVs

Mounting evidence accumulates that ExMVs are mediators for several long-distance paracrine functions of telocytes residing in adult organs. Telocytes may affect biology of several cell types including differentiated somatic cells as well as tissue-residing stem cells [7, 16–18]. As reported they are in contact with stem cell niches, blood capillaries, and nerve bundles as well as collagen and elastic fibers [7]. Thus, telocytes most likely regulate blood rheology, muscle tonus, as well as their motoric activity [5–11]. Further work is needed to elucidate how these interactions are regulated via paracrine signals from telocytes including the release of ExMVs.

ExMV-mediated paracrine effects are based, as described for other types of cells [16, 18], on direct stimulation of target cells by ExMV-expressed signaling molecules or horizontal transfer between cells of mRNA and proteins [5, 16]. However, in contrast to telocytes themselves [7], telocyte-derived ExMVs have not been characterized yet for their content of mRNA and miRNA species or proteins. Nevertheless, taking into consideration that telocytes express several miRNAs that possess pro-angiopoietic potential (e.g., miR-126, miR-130, let-7e, miR-100), the horizontal transfer of these miRNAs via ExMVs to the target cells may promote angiogenesis in the damaged tissues. A similar role has been also postulated for ExMVs secreted by other types of cells [3, 16]. Moreover, a fact that telocytes play a role in several organs in ameliorating oxidative stress and aging and stimulate proliferation and inhibit apoptosis lends support for further studies to identify secretome of telocytes along with molecular composition of ExMVs that are involved in all of these processes. Such studies could identify important factors involved in keeping the homeostasis of adult tissues. We could envision that similarly for other ExMV-producing cells [3, 21, 23], telocytes may augment ExMV secretion in response to hypoxia, inflammation, and tissue/organ damage after stimulation by some inflammatory cytokines as well as after exposure to activated components of complement and coagulation cascades. Elucidation of these possibilities may better explain involvement of telocytes in tissue/organ regeneration [7].

Similarly, since telocytes have been reported to play a role in immunosurveillance and interact with cells being involved in innate and acquired immunity, further studies are needed to show how much this interaction involves telocyte-ExMVs and vice versa if telocytes may respond to ExMVs secreted by cells involved in immune responses. We have to remember that ExMV-directed crosstalk between cells is a two-way street [24], and we have to consider that telocytes most likely also respond to ExMVs secreted by surrounding cells.

In frame of the last possibility, it is tempting to hypothesize that telocytes could be involved in the distribution and trafficking of ExMVs secreted by other types of cells and could deliver such "third-party" ExMVs to the target cells via telopodes and the abovementioned "bridging nanostructures" [22]. Moreover, there is no doubt that the development of pharmacological strategies to modulate the secretion of ExMVs from telocytes may turn out to be an important means to enhance at the paracrine level important influences of these cells in maintaining tissue homeostasis.

3.4 The Role of Telocyte-Derived ExMVs in Pathology

Telocytes may ameliorate several pathological processes. For example, since telocyte numbers decrease in infarcted myocardium [7], the therapeutic injection of exogenous telocytes to a damaged heart reduced infarct size and leads to improved heart function [7]. Based on this it is a possibility that telocyte-derived ExMVs could exert a very similar effect and replace treatment by intact telocytes. To support this latter notion, mesenchymal stem cell-derived ExMVs have been demonstrated to have a similar biological therapeutic efficacy as intact cells [3]. On the other hand, one can envision that telocyte-derived induced pluripotent stem cells (iPSCs), as cells endowed with telocyte-characteristic epigenetic memory, could be potentially employed as a source of therapeutic ExMVs [20, 21]. Such ExMVs could also be employed in several other clinical situations where telocytes have been demonstrated to play a positive therapeutic role, e.g., in lung pathology, liver regeneration, ameliorating scleroderma, and improving the function of the digestive and reproductive systems [7].

However, telocyte function is mostly related to their supportive role in organ and tissue homeostasis; it has been recently reported that telocytes may be responsible for the origin of some malignancies. Accordingly, telocytes may promote proliferation of breast cancer cells and inhibit their apoptosis [22]. Obviously, in this particular case the pro-proliferative effect of telocytes has been activated at the wrong time and the wrong place. How much this effect depends on ExMVs and how valid this effect is in other types of malignancies require further studies.

3.5 Conclusions and Future Directions

Evidence has accumulated that paracrine effects of telocytes in adult tissues largely depend on secretion of ExMVs. In this respect telocytes are somehow unique cells since, in addition to classical exosomes and cytoplasmic cell membrane-derived ExMVs, they also secrete large ExMV characteristic for telocytes known as multicellular cargo vesicles [7].

Further studies are necessary to identify employing mRNA and miRNA arrays, proteomics, and lipidomic analysis molecular "cargo" present in telocyte-derived ExMVs as well as to elucidate and understand mechanisms that promote their secretion. Taking into consideration an important role telocytes play in organ and tissue homeostasis development of pharmacological strategies to modulate/enhance secretion of ExMVs, these cells may lead to better treatment strategies in all these situations where telocytes have been demonstrated to be of benefit. Finally, it is tempting to postulate the established telocyte-derived induced pluripotent stem cells (iPSC) that could be employed as ExMV-producing cell lines [20, 21]. Such iPSC immortalized cell lines would be endowed with telocyte-epigenetic memory, and ExMVs harvested from these cells could be employed for therapeutic purposes. All these intriguing possibilities are depicted in Fig. 3.1. We may expect that the next few years will provide us with more information about the paracrine effects of telocytes in regulating body homeostasis and will also lead to the development of therapeutic strategies to employ these cells or telocyte-based ExMVs in the clinic.

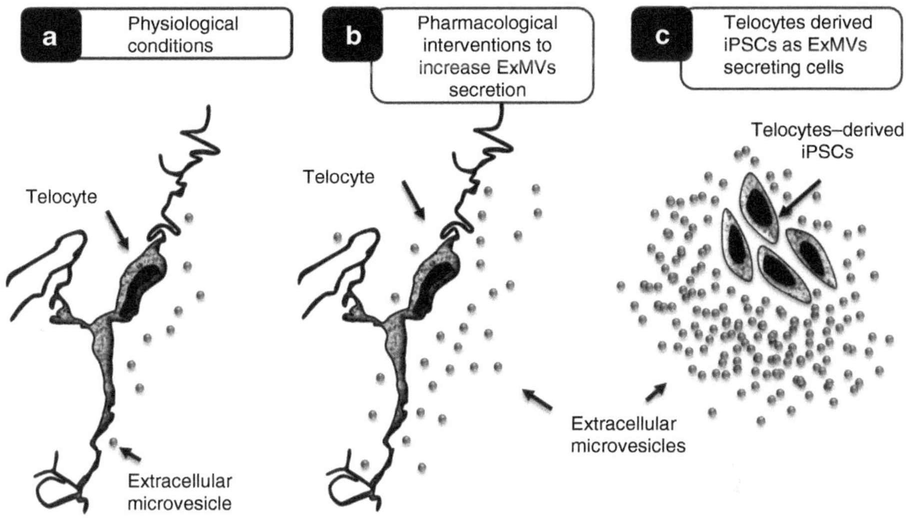

Fig. 3.1 Future potential strategies of telocyte-derived ExMVs in modulating bioactive function of telocytes. Panel (**a**): Telocytes secrete among paracrine factors circular membrane fragments known as ExMVs. Panel (**b**): Development of pharmacological strategies to augment secretion of ExMVs from telocytes may enhance involvement of these cells in maintaining tissue/organ homeostasis. Panel (**c**): Telocyte-derived induced pluripotent stem cells (iPSCs) endowed with telocyte-epigenetic memory could be a source of therapeutic ExMVs employed in situations where beneficial effects of telocytes have been demonstrated (e.g., impaired function of myocardium after heart infarct)

Acknowledgments This work was supported by NIH grants 2R01 DK074720 and R01HL112788, the Stella and Henry Endowment, and Maestro grant 2011/02/A/NZ4/00035 to MZR.

References

1. Adell MA, Vogel GF, Pakdel M, Müller M, Lindner H, Hess MW, Teis D. Coordinated binding of Vps4 to ESCRT-III drives membrane neck constriction during MVB vesicle formation. J Cell Biol. 2014;205(1):33–49.
2. Albulescu R, Tanase C, Codrici E, Popescu DI, Cretoiu SM, Popescu LM. The secretome of myocardial telocytes modulates the activity of cardiac stem cells. J Cell Mol Med. 2015;19(8):1783–94.
3. Bruno S, Grange C, Collino F, Deregibus MC, Cantaluppi V, Biancone L, Tetta C, Camussi G. Microvesicles derived from mesenchymal stem cells enhance survival in a lethal model of acute kidney injury. PLoS One. 2012;7(3):e33115.
4. Camussi G, Quesenberry PJ. Perspectives on the potential therapeutic uses of vesicles. Exosomes Microvesicles. 2013;1(6). doi:10.5772/57393.
5. Cismaşiu VB, Popescu LM. Telocytes transfer extracellular vesicles loaded with microRNAs to stem cells. J Cell Mol Med. 2015;19(2):351–8.
6. Cretoiu D, Gherghiceanu M, Hummel E, Zimmermann H, Simionescu O, Popescu LM. FIB-SEM tomography of human skin telocytes and their extracellular vesicles. J Cell Mol Med. 2015;19(4):714–22.

7. Cretoiu SM, Popescu LM. Telocytes revisited. Biomol Concepts. 2014;5(5):353–69.
8. Creţoiu SM, Creţoiu D, Popescu LM. Human myometrium – the ultrastructural 3D network of telocytes. J Cell Mol Med. 2012;16(11):2844–9.
9. Cretoiu SM, Cretoiu D, Marin A, Radu BM, Popescu LM. Telocytes: ultrastructural, immunohistochemical and electrophysiological characteristics in human myometrium. Reproduction. 2013;145(4):357–70.
10. Enjeti AK, Lincz LF, Seldon M. Microparticles in health and disease. Semin Thromb Hemost. 2008;34(7):683–91.
11. Fertig ET, Gherghiceanu M, Popescu LM. Extracellular vesicles release by cardiac telocytes: electron microscopy and electron tomography. J Cell Mol Med. 2014;18(10):1938–43.
12. Hugel B, Martínez MC, Kunzelmann C, Freyssinet JM. Membrane microparticles: two sides of the coin. Physiology (Bethesda). 2005;20:22–7.
13. Ostrowski M, Carmo NB, Krumeich S, Fanget I, Raposo G, Savina A, Moita CF, Schauer K, Hume AN, Freitas RP, Goud B, Benaroch P, Hacohen N, Fukuda M, Desnos C, Seabra MC, Darchen F, Amigorena S, Moita LF, Thery C. Rab27a and Rab27b control different steps of the exosome secretion pathway. Nat Cell Biol. 2010;12(1):19–30. sup pp 1–13.
14. Quesenberry PJ, Aliotta JM. The paradoxical dynamism of marrow stem cells: considerations of stem cells, niches, and microvesicles. Stem Cell Rev. 2008;4(3):137–47.
15. Quesenberry PJ, Aliotta J, Deregibus MC, Camussi G. Role of extracellular RNA-carrying vesicles in cell differentiation and reprogramming. Stem Cell Res Ther. 2015;6:153.
16. Ratajczak J, Miekus K, Kucia M, Zhang J, Reca R, Dvorak P, Ratajczak MZ. Embryonic stem cell-derived microvesicles reprogram hematopoietic progenitors: evidence for horizontal transfer of mRNA and protein delivery. Leukemia. 2006;20(5):847–56.
17. Ratajczak J, Wysoczynski M, Hayek F, Janowska-Wieczorek A, Ratajczak MZ. Membrane-derived microvesicles: important and underappreciated mediators of cell-to-cell communication. Leukemia. 2006;20(9):1487–95.
18. Ratajczak J, Kucia M, Mierzejewska K, Marlicz W, Pietrzkowski Z, Wojakowski W, Greco NJ, Tendera M, Ratajczak MZ. Paracrine proangiopoietic effects of human umbilical cord blood-derived purified CD133+ cells – implications for stem cell therapies in regenerative medicine. Stem Cells Dev. 2013;22(3):422–30.
19. Ratajczak MZ, Adamiak M. Membrane lipid rafts, master regulators of hematopoietic stem cell retention in bone marrow and their trafficking. Leukemia. 2015;29(7):1452–7.
20. Ratajczak MZ, Kucia M, Jadczyk T, Greco NJ, Wojakowski W, Tendera M, Ratajczak J. Pivotal role of paracrine effects in stem cell therapies in regenerative medicine: can we translate stem cell-secreted paracrine factors and microvesicles into better therapeutic strategies? Leukemia. 2012;26(6):1166–73.
21. Ratajczak MZ, Jadczyk T, Pędziwiatr D, Wojakowski W. New advances in stem cell research: practical implications for regenerative medicine. Pol Arch Med Wewn. 2014;124(7–8):417–26.
22. Smythies J. Intercellular signaling in cancer-the SMT and TOFT hypotheses, exosomes, telocytes and metastases: is the messenger in the message? J Cancer. 2015;6(7):604–9.
23. Smythies J, Edelstein L. Telocytes, exosomes, gap junctions and the cytoskeleton: the makings of a primitive nervous system? Front Cell Neurosci. 2014;7:278.
24. Wysoczynski M, Ratajczak MZ. Lung cancer secreted microvesicles: underappreciated modulators of microenvironment in expanding tumors. Int J Cancer. 2009;125(7):1595–603.

Mariusz Z. Ratajczak is a Professor of Medicine, the Henry M. and Stella M. Hoenig Endowed Chair in Cancer Biology, and the Director of the Developmental Biology Research Program at the University of Louisville's James Graham Brown Cancer Center. Dr. Ratajczak earned his M.D. at the Pomeranian

Medical University and his Ph.D. and D.Sci. and habilitation at the Center for Clinical and Experimental Medicine of Polish Academy of Sciences in Warsaw, Poland, and completed his training at the University of Pennsylvania, Philadelphia, USA. He is an internationally known specialist in the field of adult stem cell biology.

Chapter 4
Telocytes in Chronic Inflammatory and Fibrotic Diseases

Lidia Ibba-Manneschi, Irene Rosa, and Mirko Manetti

Abstract Telocytes are a peculiar stromal (interstitial) cell type implicated in tissue homeostasis and development, as well as in the pathophysiology of several disorders. Severe damage and reduction of telocytes have been reported during fibrotic remodeling of multiple organs in various diseases, including scleroderma, Crohn's disease, ulcerative colitis, and liver fibrosis, as well as in chronic inflammatory lesions like those of primary Sjögren's syndrome and psoriasis. Owing to their close relationship with stem cells, telocytes are also supposed to contribute to tissue repair/regeneration. Indeed, telocytes are universally considered as "connecting cells" mostly oriented to intercellular signaling. On the basis of recent promising experimental findings, in the near future, telocyte transplantation might represent a novel therapeutic opportunity to control the evolution of chronic inflammatory and fibrotic diseases. Notably, there is evidence to support that telocytes could help in preventing abnormal activation of immune cells and fibroblasts, as well as in attenuating the altered matrix organization during the fibrotic process. By targeting telocytes alone or in tandem with stem cells, we might be able to promote regeneration and prevent the evolution to irreversible tissue injury. Besides exogenous transplantation, exploring pharmacological or non-pharmacological methods to enhance the growth and/or survival of telocytes could be an additional therapeutic strategy for many disorders.

4.1 Introduction

During both development and reparative processes of tissues and organs, the stromal compartment takes center stage not only by providing mechanical support and protection to parenchymal cells but also as a pivotal regulator of different cell activities, including proliferation, survival, differentiation, and metabolism [1, 2]. Accordingly,

L. Ibba-Manneschi • I. Rosa • M. Manetti, PhD (✉)
Section of Anatomy and Histology, Department of Experimental and Clinical Medicine, University of Florence, Largo Brambilla 3, Florence 50134, Italy
e-mail: mirkomanetti@yahoo.it; mirko.manetti@unifi.it

X. Wang, D. Cretoiu (eds.), *Telocytes*, Advances in Experimental Medicine and Biology, 913, DOI 10.1007/978-981-10-1061-3_4

abnormalities in the stromal compartment may deeply impair tissue homeostasis, thus representing a key step in the development and progression of multiple pathologic conditions, such as chronic inflammatory, fibrotic, and neoplastic diseases.

It is now well established that the development and perpetuation of chronic inflammation are the consequence of a complex interplay between immune cells and nonimmune, tissue-resident, stromal (interstitial) cells. Actually, stromal cells are no longer believed to be "innocent bystanders" as a growing evidence suggests that they are active players in the induction and maintenance of the inflammatory process [3, 4]. For instance, stromal cells isolated from an inflammatory microenvironment are capable to drive the migration of immune cells in vitro [5]. In addition, under injury conditions tissue-resident stromal cells may secrete a variety of soluble mediators, including cytokines and chemokines that allow the transformation of diffuse immune infiltrates into highly organized tertiary lymphoid structures, namely, germinal center-like structures [6–8]. Nevertheless, chronic inflammation may often evolve into fibrosis, a condition also characterized by profound changes in the stromal compartment leading to progressive destruction of the normal tissue architecture and consequent organ dysfunction [9]. As far as the fibrotic process is concerned, fibroblasts are considered the principal effector stromal cells, as their chronic dysregulation may result in resistance to proapoptotic stimuli, increased transition to myofibroblasts, and excessive production and deposition of collagens and other extracellular matrix (ECM) components [10, 11].

Indeed, "classical" stromal cells include fibroblasts, myofibroblasts, dendritic cells, macrophages, vascular endothelial cells, and pericytes among others. In this context, stromal cells bearing very long cellular extensions have been long neglected and simplistically labeled as fibroblasts. However, in recent years, this view has rapidly changed because of the identification of an additional type of stromal cells with peculiar phenotypic features in a variety of human and animal tissues and organs [12–14]. These cells, named telocytes (telos, i.e., provided with long-distance cell projections), display a small cell body and extremely long and thin prolongations, termed "telopodes," and appear definitely distinct from the "classical" fibroblasts [12–16]. "Cells with telopodes" is the shortest definition of telocytes [14]. Telopodes typically exhibit a moniliform aspect characterized by the alternation of thin segments (podomers) and small dilated regions (podoms) accommodating the mitochondria, endoplasmic reticulum, and caveolae [12–14]. This peculiar ultrastructural phenotype observed under transmission electron microscopy (TEM) is currently considered as the most reliable hallmark for the identification of telocytes, which do not possess unique immunophenotypic characteristics [13, 14]. Although different markers have been proposed, at present a combination of CD34 and platelet-derived growth factor receptor α (PDGFRα) seems the best available choice for the immunohistochemical identification of telocytes under light and fluorescence microscopy (Fig. 4.1) [13, 14, 17–19]. Indeed, by immunoelectron microscopy, it could be demonstrated that the CD34-positive interstitial cells are ultrastructurally identifiable as telocytes (Fig. 4.1) [20]. A strong expression of CD34 and PDGFRα antigens has been firmly reported in telocytes from different organs [13, 14, 17–29]. Conversely, other markers resulted in

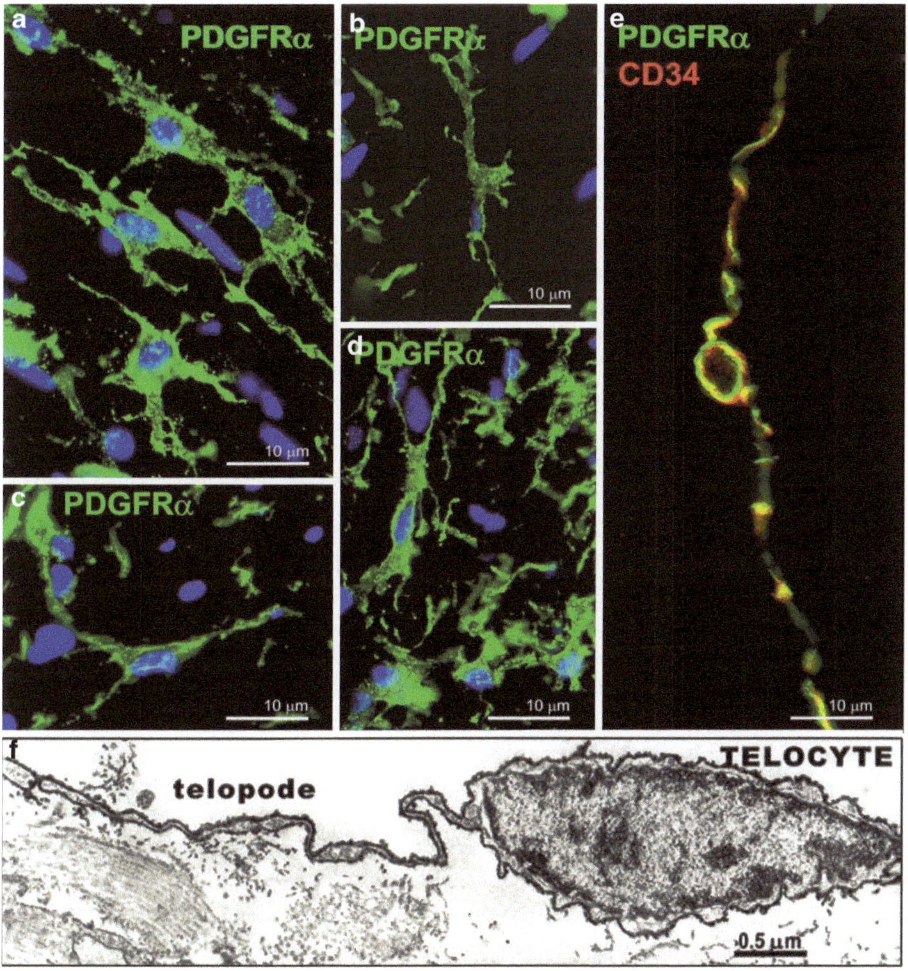

Fig. 4.1 Morphology and immunophenotype of telocytes in the human gastrointestinal tract. (**a**–**d**) PDGFRα-immunoreactivity (nuclei are blue stained with DAPI); (**e**) PDGFRα/CD34 double labeling; (**f**) CD34-immunoelectro-labeling. (**a, b**) Muscle layers (small intestine). Intramuscular PDGFRα-positive telocytes display two long telopodes and several short processes starting from the nucleated portion. (**c**) Submucosa (small intestine). PDGFRα-positive telocytes show a triangular body and three long and varicose telopodes. (**d**) Myenteric plexus region (small intestine). PDGFRα-positive telocytes display an oval body and several telopodes running in every direction. (**e**) A PDGFRα/CD34-positive telocyte at the border of a circular muscle bundle (small intestine) shows a small nucleated body and two long and thin telopodes starting from the opposite poles of the cell and with podomers and podoms clearly identifiable. (**f**) CD34-immunoelectro-labeling is present on the surface of a telocyte in the small intestine. The labeling appears as an electron-dense material distributed all along the plasma membrane, from which spherules protrude outside (Adapted with permission from [18, 20])

a weakly and inconstantly positive immunostaining of telocytes [13, 24]. Besides their ultrastructural and immunophenotypic features, a growing number of studies suggest that telocytes possess gene expression and proteomic profiles, as well as microRNA signature, that are definitely distinct from those of "classical" fibroblasts [14, 30–38].

According to their location in different organs and tissues, telocytes have been proposed to exert different functions and participate in a wide range of physiological processes. By their extremely long, tortuous, and overlapping telopodes, these cells interconnect to form a three-dimensional network that may function as a scaffold to define the correct tissue organization during prenatal life or repair/renewal in postnatal life, thus making a substantial contribution to the maintenance of local tissue homeostasis [13, 14, 18, 39, 40]. For instance, their importance in organ morphogenesis is supported by the evidence that during mouse heart development, telocytes act as mediators for heart compaction from embryonic myocardial trabeculae [41]. In different organs, telocytes occupy a strategic position in relation to stem cell niches, blood capillaries, and nerve bundles [14, 18, 22, 23, 26, 27, 42–46]. Telopodes also establish heterocellular contacts with other cell types including mast cells, basophils, lymphocytes, plasma cells, macrophages, or fibroblasts and noncellular elements, such as collagen and elastic fibers [14, 47, 48]. Furthermore, it appears that telocytes may participate in intercellular signaling not only by cell-to-cell contacts but also in a paracrine manner via the release of at least three different types of extracellular vesicles, namely, exosomes, ectosomes, and multivesicular cargos [14, 26, 49–54]. These vesicles might function as intercellular shuttles for the transfer of biological signals, including microRNAs, to neighboring cells [14]. Thus, differently from fibroblasts which, functionally, are mainly involved in the synthesis of collagen and other ECM components, it is believed that telocytes act as "connecting cells" being mostly oriented to intercellular signaling. As recently proposed, telocytes might even be considered as active players in immunomodulation and immune surveillance, acting like "local data suppliers" for the immune response [14, 28, 46, 47]. Increasing evidence also suggests that telocytes might act as "nurse" cells for adjacent tissue-resident stem cells and cooperate with them to promote tissue regeneration and/or repair [14, 21, 36, 44, 49, 55]. Moreover, telocytes have been suggested to participate in a variety of processes, such as stimulation of angiogenesis, inhibition of oxidative stress, and prevention of cellular aging [14, 34, 35]. An electrophysiologic activity of telocytes has also been demonstrated in organs like the myometrium and the heart [14, 56, 57]. Finally, in the gastrointestinal tract, telocytes have been proposed to participate in the regulation of neurotransmission and gut motility, presumably by spreading the slow waves generated by the pacemaker interstitial cells of Cajal (ICC) [13, 14, 18, 20].

Owing to the aforementioned intriguing roles proposed for telocytes, their possible involvement in different pathologic processes is being increasingly investigated [14, 22, 23, 27–29, 58–66]. In this regard, the present chapter will focus on the most recent findings concerning the implication of telocytes in a variety of chronic inflammatory, autoimmune, and fibrotic diseases.

4.2 Telocytes in Systemic Sclerosis: A Prototypic Multisystem Fibrotic Disorder

Systemic sclerosis (SSc), or scleroderma, is a chronic connective tissue disease characterized by extensive microvascular injury, immune system dysregulation, and progressive fibrosis affecting the skin and a variety of internal organs, especially the lungs, heart, and gastrointestinal tract [67, 68]. Endothelial cell damage/activation is supposed to be the initial event which together with inflammatory and autoimmune reactions leads to the chronic activation and transdifferentiation of fibroblasts into myofibroblasts, finally resulting in a deregulated wound healing process and severe tissue fibrosis with consequent multiple organ failure [67–69]. Indeed, visceral organ fibrosis is responsible for significant morbidity and is also a major cause of death in patients with SSc. Moreover, the concomitant fibroproliferative vasculopathy, characterized by subendothelial deposition of ECM in small and medium-sized arteries and arterioles, as well as the progressive loss of peripheral microvessels may lead to chronic tissue ischemia, clinically manifesting as digital ulceration and gangrene [70].

Two different clinical subsets of SSc are commonly recognized: limited cutaneous SSc (lcSSc) and diffuse cutaneous SSc (dcSSc), which differ in the extent of skin fibrosis, internal organ involvement, autoantibodies, prognosis, and survival [68]. In the early stages of both SSc subsets, the main cutaneous histopathological features are represented by perivascular inflammatory infiltrates, dermal edema, and a variable extent of ECM accumulation in the papillary and reticular dermis [67, 71]. Conversely, in advanced disease, severe dermal fibrotic changes with tightly packed and irregularly distributed collagen bundles, loss of capillaries, occlusion of arterioles, damage of nerve fibers, and atrophy of skin appendages are commonly observed [67, 70–72]. As far as internal organ involvement is concerned, the hallmark pulmonary histopathological lesion is nonspecific interstitial pneumonia characterized by cellular inflammation and fairly uniform interstitial fibrosis, which manifests as progressive thickening of the alveolar septa, ultimately resulting in alveolar airspace obliteration with consequent restrictive lung disease [73]. Cardiac manifestations include myocardial fibrosis, hypertrophy, and disorders of the coronary and conduction systems that can lead to congestive heart failure, arrhythmias, and sudden cardiac death [74]. Gastrointestinal involvement is commonly found in up to 90 % of SSc patients [75] and is characterized by fibrotic lesions in the muscularis mucosae, submucosa, and muscle layers together with smooth muscle cell atrophy [76, 77]. Clinically, gastrointestinal tract dysmotility is a major visceral manifestation, ranging from an asymptomatic form to severe paresis [75].

In SSc, fibroblasts are considered the principal effector cells [69]. In fact, it is well known that these stromal cells are chronically activated by a number of profibrotic cytokines, growth factors, and stimulatory autoantibodies and that they transform into apoptosis-resistant myofibroblasts which produce an excessive amount of collagens and other ECM components [67–69]. Abnormal SSc fibroblasts are believed to develop from a subset of cells that have escaped from normal control mechanisms. Indeed, fibroblasts from clinically affected SSc skin still continue to produce excessive amounts of ECM proteins in vitro, suggesting that once activated, these cells establish a constitutive self-activation system [67, 78]. Moreover, there is substantial evidence that vascular endothelial cells, pericytes, and cells of both the innate and adaptive immune systems may contribute to abnormal fibroblast activation and fibrosis in SSc [69, 78, 79].

The recent identification of telocytes as a distinct stromal cell population of human dermis, where they have been proposed to participate to skin homeostasis, remodeling, and regeneration [21, 48], prompted us to investigate the possible involvement of this new interstitial cell type in the pathophysiology of SSc. By an integrated immunohistochemical and ultrastructural approach, we have recently shown for the first time that telocytes display ultrastructural damages, are significantly reduced, and progressively disappear from SSc skin lesions [22]. In normal skin, telocytes are organized to form three-dimensional networks with their telopodes distributed among collagen bundles and elastic fibers throughout the whole dermis (Fig. 4.2). Moreover, telocytes appear concentrated to surround microvessels (Fig. 4.2), nerves, hair follicles, and sebaceous and eccrine sweat glands. As far as SSc skin is concerned, the reduction in telocytes evolves differently according to disease subsets and stages (Fig. 4.3). In particular, in early lcSSc, telocytes are absent from the papillary dermis and reduced in some areas of the reticular dermis. In advanced lcSSc, the loss of telocytes is severe also in the reticular dermis and the connective tissue surrounding skin adnexa. On the contrary, in the early stage of dcSSc, which is characterized by a more rapid disease progression [67], telocytes are very few or absent in both the papillary and reticular dermis, and they almost completely disappear in advanced dcSSc skin lesions (Fig. 4.3). Thus, the progression in telocyte reduction occurs earlier and is more severe in dcSSc than in lcSSc. Interestingly, such a reduction occurs in parallel with the severity of telocyte ultrastructural abnormalities, including swollen mitochondria, cytoplasmic vacuolization, and presence of lipofuscinic bodies, which suggest a cellular degenerative process already present in the early stage of dcSSc and more marked in the advanced stage of both disease subsets (Figs. 4.4 and 4.5). In addition, telocytes often establish intercellular contacts with inflammatory and immune cells in early SSc skin lesions.

Different mechanisms have been proposed to be responsible for the damage and loss of telocytes in clinically affected SSc skin [22]. For instance, it has been suggested that the chronic ischemic microenvironment of fibrotic skin, characterized by low oxygen levels, generation of reactive oxygen species, and scarcity of nutrients, may compromise the telocyte metabolism, thus provoking profound cell sufferance. In fact, in SSc dermis the reduction in telocytes seems to be paralleled by the reduction in microvessels. The hypothesis of an ischemic injury is mainly supported by the presence of telocytes with numerous swollen mitochondria and extensive cytoplasmic vacuolization. Of note, the most severely affected telocytes appear to be those embedded in the fibrotic ECM and those surrounding occluded microvessels, while the telocytes found around patent microvessels still display a normal morphology, even in the advanced disease stages [22]. With disease progression, the more severe and extended damage of telocytes observed under TEM might be

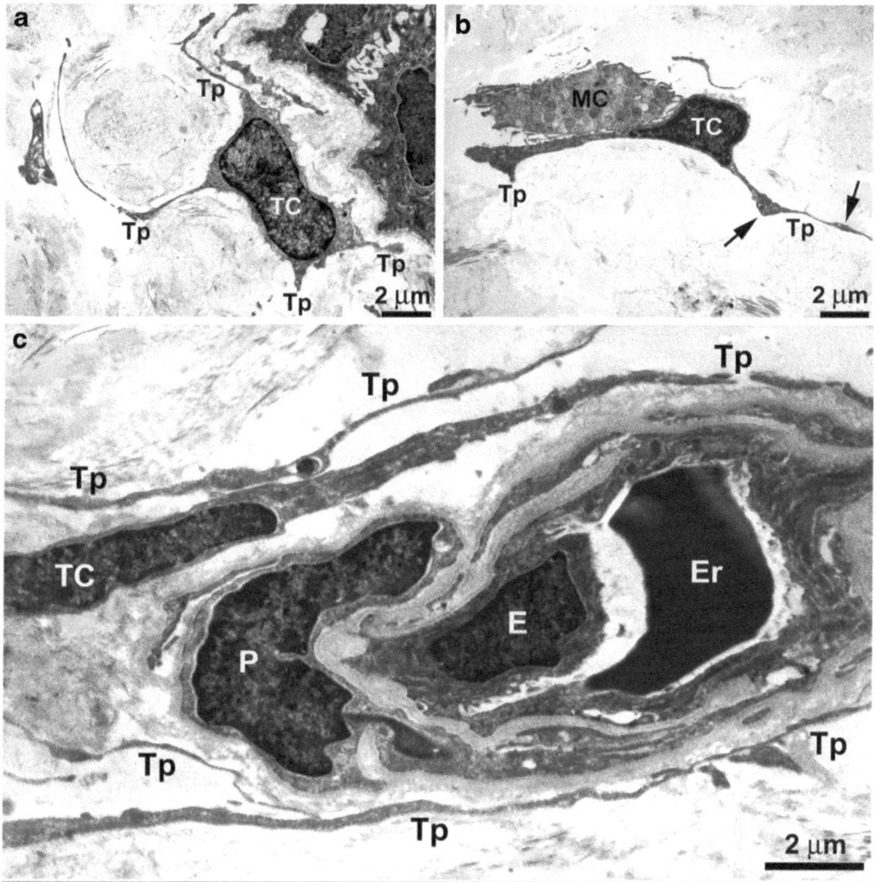

Fig. 4.2 Telocytes in normal skin (transmission electron microscopy). (**a**–**c**) In both the papillary dermis (**a**) and reticular dermis (**b**, **c**), telocytes have a small cell body and very long and thin processes (telopodes) that are collagen embedded or lining elastic fibers. Telocytes lack a basal lamina and have a scarce cytoplasm surrounding the nucleus with few mitochondria and cisternae of the endoplasmic reticulum and a small Golgi apparatus. Telopodes display a moniliform aspect due to the alternation of thin segments (podomers) and dilated segments (podoms) oval or triangular in shape (**b**, *arrows*). (**b**) A telocyte is in contact with a mast cell. (**c**) Telocytes are closely associated with each other. The telopodes of perivascular telocytes encircle the basal lamina of a blood microvessel; pericytes are embedded in the vessel basal lamina (**c**). *TC* telocyte, *Tp* telopode, *MC* mast cell, *E* endothelial cell, *Er* erythrocyte, *P* pericyte (Adapted with permission from Manetti et al. [22])

caused by their entrapment in a poorly permeable ECM due to the overproduction and accumulation of abnormal collagen and elastic fibers by activated fibroblasts/myofibroblasts. Finally, when considering the autoimmune background of SSc [67, 68], the possibility that telocytes might be important/specific target cells of autoantibodies was also taken into account. However, this latter hypothesis seems little supported by the fact that telocytes display a normal morphology in clinically non-involved skin biopsies from SSc patients.

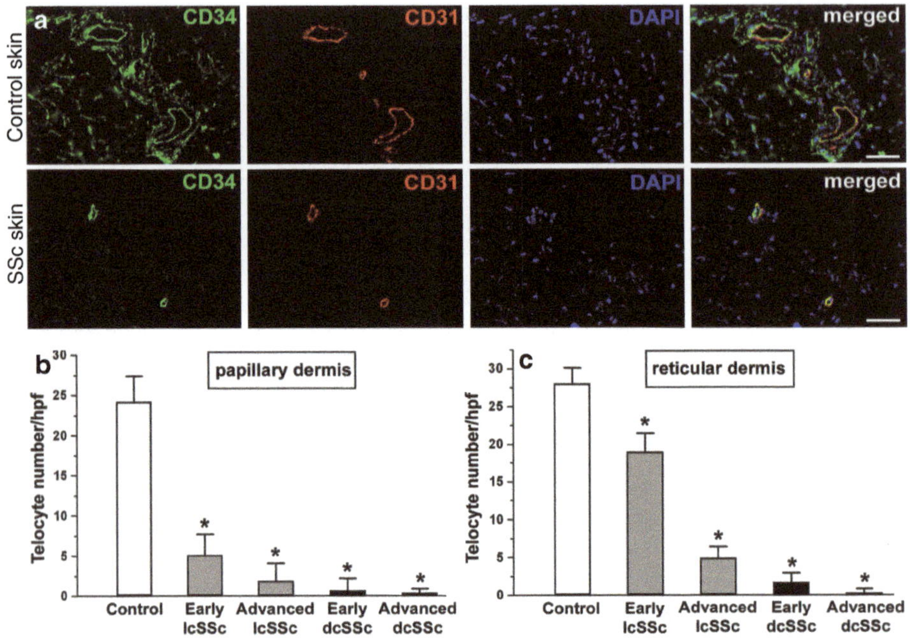

Fig. 4.3 Quantitative analysis of telocytes in skin sections from controls and patients with systemic sclerosis (*SSc*) double immunolabeled for CD34 (*green*) and CD31 (*red*) and counterstained with DAPI (*blue*) for nuclei. (**a**) Representative photomicrographs from control and clinically affected SSc skin samples are shown. Telocytes are identified as CD34-positive/CD31-negative spindle-shaped cells, while microvessels are CD34/CD31 double positive. (**b**, **c**) Telocytes are reduced in both the papillary and reticular dermis of SSc patients throughout different disease stages. Data are represented as mean ± SD. *$P<0.05$ vs control. *lcSSc* limited cutaneous SSc, *dcSSc* diffuse cutaneous SSc (Adapted with permission from Manetti et al. [22])

In a subsequent study, these findings were further extended by the evidence that in SSc, the loss of telocytes is not restricted to the skin, but it is a widespread process affecting multiple visceral organs targeted by the fibrotic process, such as the gastric wall, the myocardium, and the lung (Figs. 4.6 and 4.7) [29].

According to the numerous functions proposed for telocytes [14], several hypotheses have been formulated on the possible pathophysiologic implications that their damage and systemic loss might have in SSc [22, 29]. It has been suggested that by their long telopodes, telocytes might act as supporting cells and form a scaffold to guide the migration of other cells and the correct ECM assembly, thus contributing to define the correct spatial organization of tissues and organs [13, 40, 41]. Indeed, in the stromal compartment of many organs, telopodes are usually collagen embedded or lining elastic fibers. Interestingly, in SSc skin some telocytes were found to surround with their telopodes very large and abnormal aggregates of elastin and collagen fibers, likely in the attempt to limit their spreading into the interstitium. Therefore, it is conceivable that the loss of telocytes could mechanically contribute to the altered three-dimensional organization of the ECM within the stromal

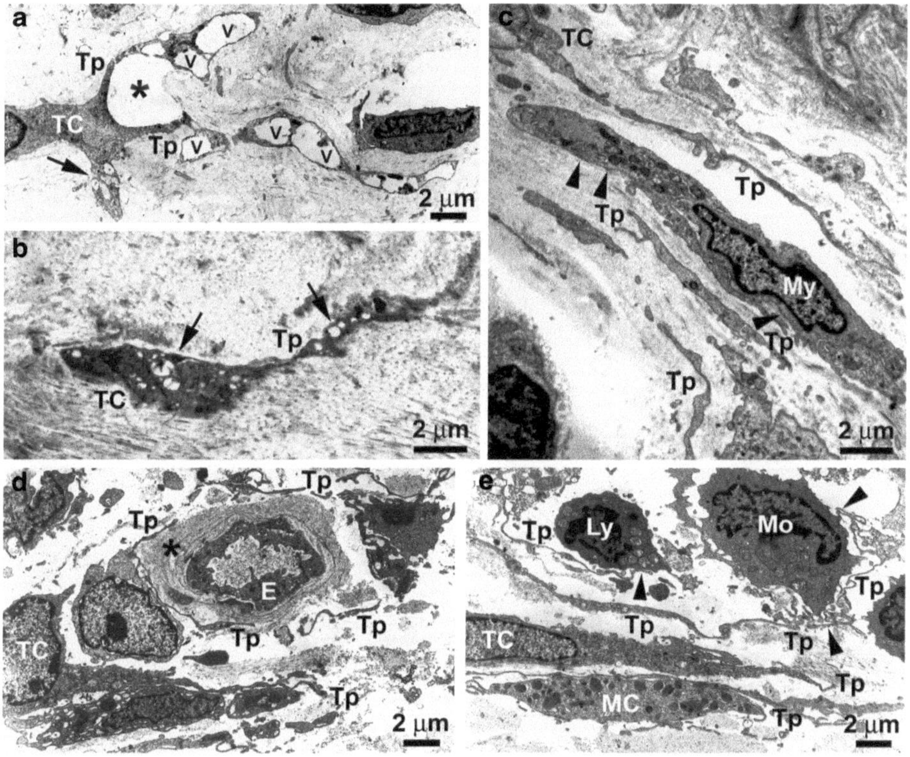

Fig. 4.4 Telocytes in limited cutaneous systemic sclerosis (*lcSSc*) skin (transmission electron microscopy). (**a**) Early lcSSc. A telocyte displaying an enlarged shape, due to the presence of large vacuoles (*v*) in its telopodes, surrounds an area of dermal edema (*asterisk*). Both normal mitochondria and swollen mitochondria with a clear matrix and few cristae (*arrow*) are identifiable in the cytoplasm. (**b**) Advanced lcSSc. A degenerating telocyte entrapped in the fibrotic extracellular matrix shows numerous swollen mitochondria (*arrows*). The cytoplasm is dark and contains vacuoles and lipofuscinic bodies. (**c**) Early lcSSc. Telocytes and telopodes displaying a normal morphology are present in the close vicinity of or even in contact with a myofibroblast which shows a large body rich in rough endoplasmic reticulum, mitochondria, and myofilaments. Subplasmalemmal focal densities are evident (*arrowheads*). (**d**) Early lcSSc. Some telocytes and telopodes with a normal morphology are present around a blood vessel displaying a patent lumen. The vessel basal lamina is markedly thickened (*asterisk*). (**e**) Early lcSSc. Normal telocytes with very long and convoluted telopodes surround a perivascular inflammatory infiltrate composed of monocytes and lymphocytes. Telopodes establish cell-to-cell contacts with inflammatory cells (*arrowheads*). A mast cell is also in contact with telopodes. *TC* telocyte, *Tp* telopode, *My* myofibroblast, *E* endothelial cell, *Ly* lymphocyte, *Mo* monocyte, *MC* mast cell (Reproduced with permission from Manetti et al. [22])

compartment of any organ undergoing fibrotic remodeling, such as the skin, the gastric wall, the myocardium, and the lung [22, 29]. During the fibrotic process, the loss of telocytes might even favor the uncontrolled activation of fibroblasts and their transition to profibrotic myofibroblasts. In fact, telocytes may convert the interstitium into an integrated system that contributes to the maintenance of organ homeostasis.

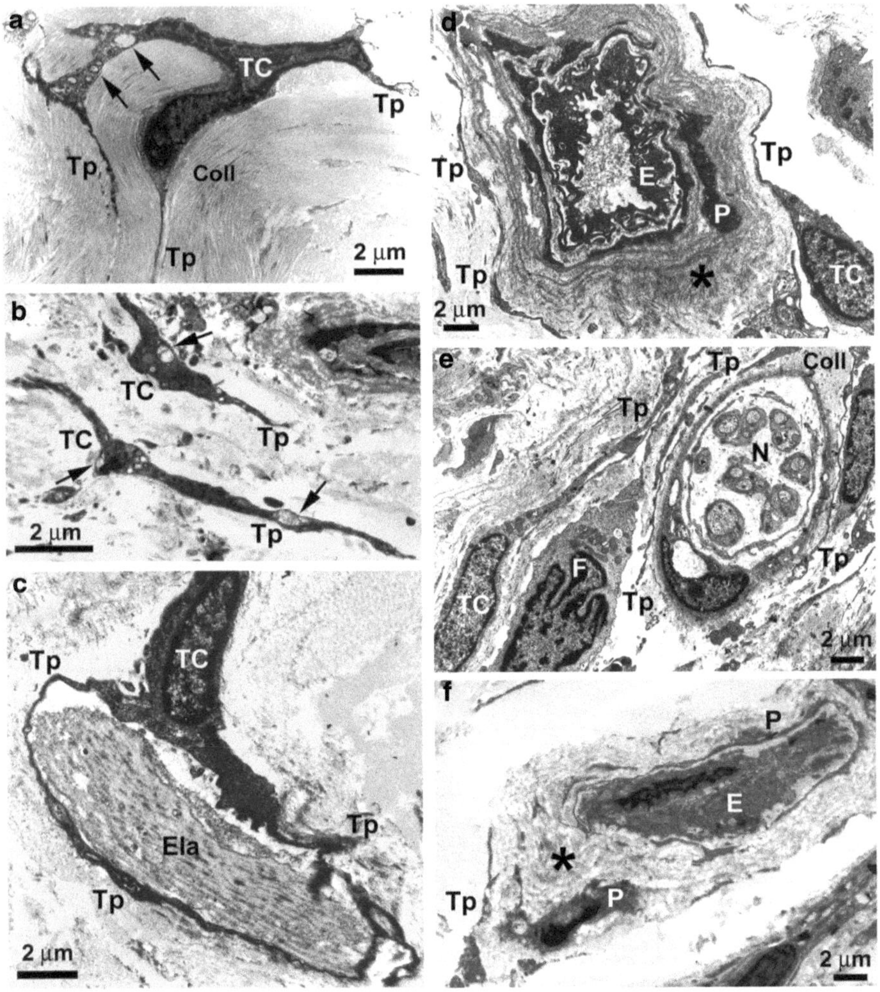

Fig. 4.5 Telocytes in diffuse cutaneous systemic sclerosis (*dcSSc*) skin (transmission electron microscopy). (**a**) Early dcSSc. A telocyte with a small perinuclear cytoplasm and slender telopodes is embedded in a matrix composed of closely packed collagen bundles. Swollen mitochondria and vacuoles (*arrows*) are present in the cytoplasm. (**b**) Advanced dcSSc. Telocytes and telopodes embedded in the fibrotic extracellular matrix show features of degenerating cells. The cytoplasm is dark and contains swollen mitochondria (*arrows*), vacuoles, and lipofuscinic bodies. Many cell debris are evident. (**c**) Early dcSSc. A telocyte displaying a normal morphology embraces with telopodes a large and abnormal elastin fiber. (**d**) Early dcSSc. Normal telocytes surround the thickened basal lamina (*asterisk*) of a blood vessel displaying a patent lumen. (**e**) Early dcSSc. Telocytes with a normal morphology are evident around nerve bundles. Abundant collagen fibers separate telopodes from the nerve bundle. A fibroblast is in the close vicinity of a telocyte and is surrounded by telopodes. (**f**) Advanced dcSSc. Telocytes are not identifiable around an occluded microvessel. Only a few cell debris are observed. The vessel basal lamina is markedly thickened (*asterisk*). *TC* telocyte, *Tp* telopode, *Coll* collagen, *Ela* elastin, *E* endothelial cell, *P* pericyte, *N* nerve, *F* fibroblast (Reproduced with permission from Manetti et al. [22])

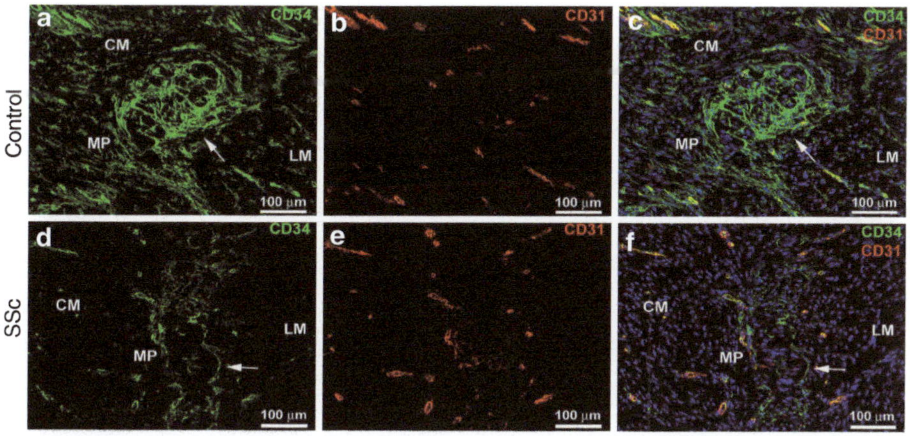

Fig. 4.6 Gastric wall specimens from controls (**a–c**) and patients with systemic sclerosis (*SSc*) (**d–f**). (**a–f**) Double immunofluorescence labeling for CD34 (*green*) and CD31 (*red*) with DAPI (*blue*) counterstain for nuclei. Telocytes are CD34 positive and CD31 negative, while vascular endothelial cells are CD34/CD31 double positive. (**a–c**) In control gastric wall, telocytes form a network around smooth muscle bundles and cells in the circular and longitudinal muscle layers. At the myenteric plexus, telocytes form a complex network enveloping the ganglia (*arrow*) and the nerve strands in the interganglionic region. Telopodes appear intermingled with ganglion cells. (**d–f**) In SSc gastric wall, telocytes are not present in the fibrotic areas of muscle layers. The network of telocytes is discontinuous or even almost completely absent around myenteric plexus ganglia (*arrow*) and nerve strands. *CM* circular muscle layer, *LM* longitudinal muscle layer, *MP* myenteric plexus (Reproduced with permission from Manetti et al. [29])

In particular, these cells seem to be involved in intercellular signaling, either directly by intercellular contacts or indirectly by shedding microvesicles and exosomes or secreting paracrine signaling molecules, including microRNAs [14, 26, 49–54]. Interestingly, intercellular contacts between telocytes and fibroblasts or myofibroblasts have been described in different organs [21, 22, 46, 48], and thus it is tempting to speculate that telocytes could be involved in the maintenance of local tissue homoeostasis by controlling fibroblast/myofibroblast activity. In SSc skin and visceral organs, this control is likely impaired because of the progressive reduction and loss of telocytes [22, 29].

Another attractive hypothesis is that the disappearance of telocytes could impair stem cell-mediated tissue regeneration. Indeed, there is substantial evidence that in several organs, such as the skin, the heart, and the lung, telocytes might cooperate with tissue-resident stem niches to promote regeneration and/or repair [14, 21, 26, 44, 80]. In support to this hypothesis, we and others observed telocytes surrounding stem cell niches in the normal skin [21, 22], but they were rarely seen in affected SSc skin. Furthermore, vascular wall-resident stem cell niches could not be detected in most severely affected skin biopsies, suggesting that telocyte loss might contribute to the depletion of functional stem cell niches with consequent impairment of skin regeneration and/or repair in SSc patients [22].

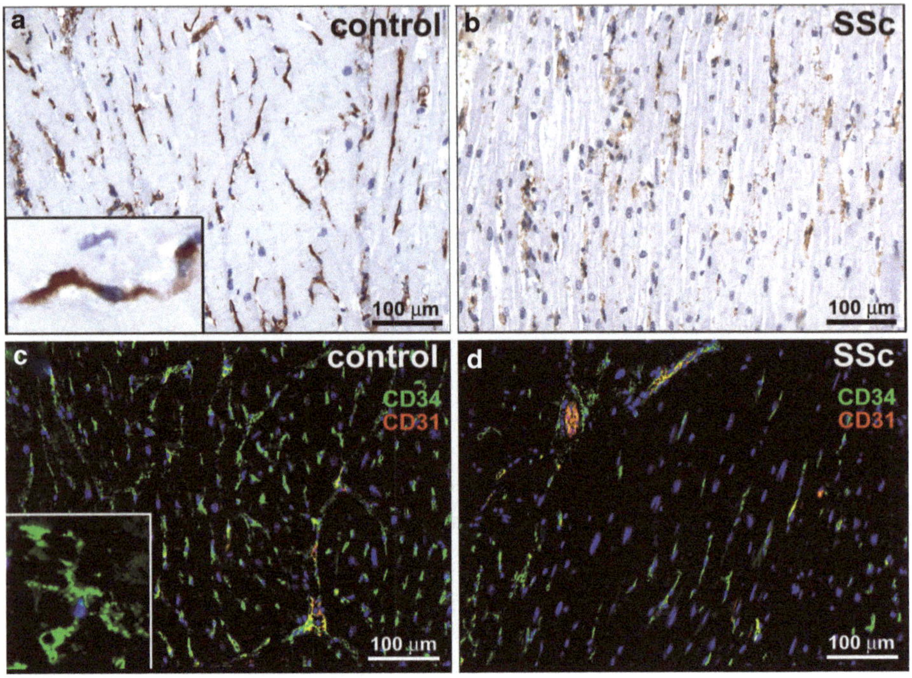

Fig. 4.7 Left ventricular myocardium specimens from controls (**a**, **c**) and patients with systemic sclerosis (*SSc*) (**b**, **d**). (**a**, **b**) CD34 immunoperoxidase labeling with hematoxylin counterstain. (**c**, **d**) Double immunofluorescence labeling for CD34 (*green*) and CD31 (*red*) with DAPI (*blue*) counterstain for nuclei. Myocardial telocytes are CD34 positive and CD31 negative, while vascular endothelial cells are CD34/CD31 double positive. (**a**, **c**) In control myocardium, numerous telocytes are located in the interstitium surrounding the cardiomyocytes. *Insets*: At higher magnification view, myocardial telocytes display a small fusiform cell body with long processes placed between cardiomyocytes. (**b**, **d**) In the fibrotic areas of SSc myocardium, telocytes are almost completely undetectable (Adapted with permission from Manetti et al. [29])

In the myocardium, it has also been shown that telocytes and cardiomyocytes are directly connected and might represent a "functional unit," possibly mediating the electrical coupling of cardiomyocytes [24, 81]. Therefore, it is possible that in SSc heart, the loss of myocardial telocytes (Fig. 4.7) might even be implicated in the arrhythmogenesis and disturbances of the cardiac conduction system [24, 29, 74]. Finally, the loss of the telocyte network observed in the muscularis propria and myenteric plexus of SSc gastric wall (Fig. 4.6) might contribute to gastric dysmotility, clinically manifesting as delayed gastric emptying or gastroparesis [29, 75]. Indeed, in the gastrointestinal tract, telocytes have been proposed to play a role in the regulation of neurotransmission, possibly by spreading the slow waves generated by the ICC [13, 18, 20]. Interestingly, there is also evidence that in SSc, gastrointestinal tract dysmotility may be related to severe damage of the myenteric neural structures and a reduction in the ICC population [29, 77, 82].

4.3 Telocytes in Inflammatory Bowel Diseases

Inflammatory bowel diseases, including Crohn's disease (CD) and ulcerative colitis (UC), are complex disorders in which the interaction of genetic, environmental, and microbial factors drives chronic relapsing and remitting intestinal inflammation that finally leads to extensive tissue fibrosis [83–85]. This is particularly relevant for CD, which may affect the entire gastrointestinal tract with a prevalence of terminal ileum. Indeed, in UC the deposition of the ECM is mainly restricted to the mucosal and submucosal layers of the large bowel, while in CD, fibrosis commonly involves the entire bowel wall, including the mucosa, submucosa, muscularis propria, and subserosa layers, and can result in critical narrowing of the lumen and strictures or stenosis, leading to intestinal obstruction that requires surgery [86]. Accordingly, the frequency of benign stenosis in UC is much lower than in CD, reported as being 3.2–11.2 %, with fibrosis in the submucosa or deeper pointed out as one of the causes [87]. However, increasing evidence indicates that the development of intestinal fibrosis in UC is a neglected problem which has remained largely unexplored [88].

In inflammatory bowel diseases, fibrosis closely follows the distribution and location of inflammation [83, 86, 89]. Of note, there is evidence that chronic exposure of intestinal fibroblasts to inflammatory mediators may drive their transition to activated α-smooth muscle actin (α-SMA)-expressing myofibroblasts, with consequent abnormal collagen production and tissue remodeling [89]. However, it also appears that inflammation subsequently plays a minor role in fibrosis progression, and, accordingly, anti-inflammatory treatment may not be able to limit intestinal fibrosis once excessive ECM deposition has started [86]. Progressive intestinal wall fibrosis ultimately results in a stiff intestine unable to carry out peristalsis, contributing to the abdominal pain and diarrhea commonly experienced by patients with active disease or even in remission. In fact, bowel dysmotility has been well established in both CD and UC [83–85].

Motility of the gastrointestinal tract involves complex processes that require the structural integrity and functionality of different cellular elements. Enteric neurons and glial cells, together with ICC, represent the main regulators of motor functions in the gut wall, ensuring coordinated patterns of smooth muscle cell activity [90, 91]. In particular, the ICC are considered the pacemaker cells and the principal mediators of gut neurotransmission [92]. This complex neural/myogenic network appears to be markedly altered in CD and UC patients, as demonstrated by severe damages of the enteric neural and glial structures as well as a marked reduction in the number of ICC [93–96].

However, in recent years it became evident that also telocytes might be part of this neural/myogenic network. Indeed, it could be demonstrated that both in muscle layers and at the myenteric plexus, telocyte processes form networks intermingling with those of ICC, suggesting that these two cell types might establish cell-to-cell contacts [18]. In particular, since a subset of intramuscular telocytes and ICC seem to be part of a unique network, in which the latter are preferentially in close contact with nerve endings, it has been proposed that telocytes might participate in the regulation of gastrointestinal motility, presumably contributing to the spreading of the slow waves generated by the ICC [13, 18]. Within the gastrointestinal neuromuscular compartment, ICC and telocytes can be easily distinguished on the basis of their different immunophenotypes [18]. In fact, the ICC are positive for c-kit (CD117) and negative for CD34 and PDGFRα [18], while telocytes are positive for CD34 and PDGFRα and negative for c-kit [18]. It has also been ascertained that in the gastrointestinal tract, telocytes correspond to the cells formerly identified as CD34-positive interstitial cells or PDGFRα-positive "fibroblast-like" cells and implicated in the enteric neurotransmission [18, 20, 97, 98].

Considering that gastrointestinal dysmotility with ICC defects is a peculiar pathological feature of both CD and UC [93–96, 99] and that ICC and telocytes are in close relationship within the gut neuromuscular compartment [18], we recently carried out two different studies in which we investigated the presence and distribution of telocytes in surgical specimens obtained from the terminal ileum of CD patients and the colon of UC patients [23, 27].

In CD, the most peculiar histopathological features of the affected intestinal wall segments are represented by discontinuous signs of inflammation and fibrosis, also referred to as "skip lesions." Interestingly, in disease-unaffected specimens from CD patients, telocytes display a distribution similar to control specimens, showing a slender-nucleated body with two or more telopodes and running parallel to each other and/or forming networks throughout the different ileal wall layers, from the mucosa to the subserosa [23]. Conversely, in sections from disease-affected specimens, telocytes disappear, particularly in areas displaying severe fibrosis and architectural derangement of the intestinal wall [23]. In the thickened muscularis mucosae of most severe cases, the few remaining telocytes are mainly located among smooth muscle bundles and cells. Some reactive lymphoid aggregates (e.g., granulomas), especially those surrounded by a prominent and diffuse inflammatory infiltrate, appear completely encircled by telocytes, likely in the attempt to limit their spreading in the connective tissue. Instead, telocytes almost completely disappear around lymphoid aggregates entrapped within the fibrotic tissue containing many α-SMA-positive myofibroblasts [23]. In the muscularis propria of disease-affected CD samples, characterized by a severe derangement of both the circular and longitudinal muscle layers, the telocyte network is preserved among smooth muscle bundles in some areas close to others where telocytes are completely absent. This severe architectural derangement involves also the myenteric plexus, where a discontinuous network of TC is present around ganglia and in the interganglionic region (Fig. 4.8) [23].

As far as UC is concerned, the presence and distribution of telocytes were investigated in full-thickness biopsies of the left colon obtained from UC patients categorized in an early phase or an advanced phase of fibrotic remodeling of the colonic wall [27]. In early fibrotic UC cases, fibrosis affects the muscularis mucosae and submucosa, while the muscularis propria is spared. In particular, the submucosa is characterized by the presence of areas displaying edema and a pattern of incoming fibrosis abruptly mixed with areas displaying established fibrosis with abundant and closely packed collagen bundles. In advanced fibrotic UC cases, an increased ECM deposition is found in the muscularis mucosae, which appears markedly thickened

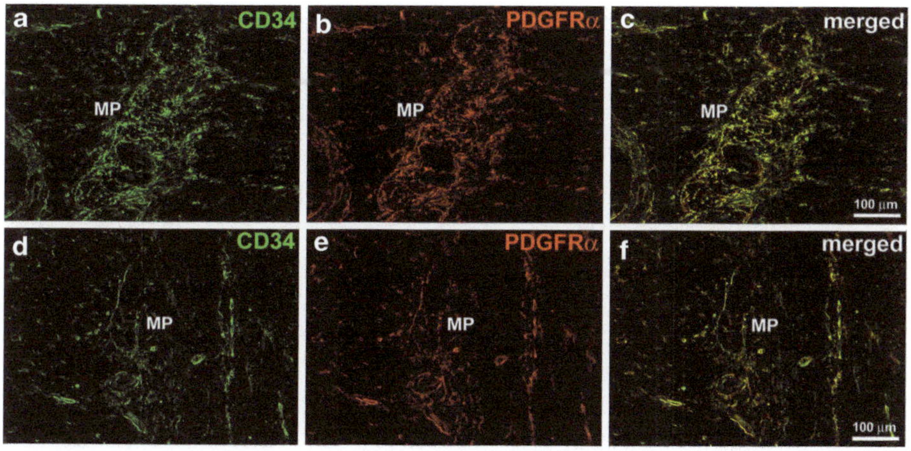

Fig. 4.8 (**a–c**) Control ileal specimens. (**d–f**) Affected ileal specimens from Crohn's disease (*CD*) patients. (**a–f**) Double immunofluorescence labeling for CD34 (*green*) and platelet-derived growth factor receptor α (PDGFRα) (*red*). In the ileal wall, all CD34-positive telocytes are also PDGFRα positive. (**a–c**) In control specimens, telocytes form a broad network surrounding the myenteric plexus. (**d–f**) In severely damaged areas of affected CD specimens, the telocyte network around the myenteric plexus is discontinuous or even completely absent. *MP* myenteric plexus (Reproduced with permission from Milia et al. [23])

and widespread in the submucosa. Moreover, fibrosis extends to involve also wide areas of the circular and longitudinal muscle layers and the myenteric plexus [27]. Interestingly, the fibrotic changes of the colonic wall seem to be paralleled by a severe reduction in telocytes. In fact, a significant reduction in telocytes is found in the muscularis mucosae and submucosa of both early and advanced fibrotic UC colonic wall. Conversely, while a normal distribution of telocytes is observed in the muscularis propria of early fibrotic UC, the network of telocytes is reduced or even completely absent in fibrotic areas of muscle layers and around myenteric ganglia of advanced fibrotic UC cases (Fig. 4.9) [27]. Of note, these data are closely consistent with those reported in "skip lesions" of disease-affected CD specimens, in which telocytes specifically disappear in areas displaying severe fibrosis and architectural derangement of the intestinal wall [23]. Finally, as revealed by CD34/c-kit double immunostaining, in the muscularis propria of both CD and UC, the disappearance of telocytes seems to be paralleled by the loss of the ICC network (Fig. 4.9) [23, 27].

As above discussed for SSc, the loss of telocytes might have different causes and pathophysiologic implications in CD and UC. We suppose that the fibrotic process may entrap telocytes in a poorly permeable ECM, with profound cell sufferance. Moreover, the excessive deposition of ECM and the progressive reduction in telocytes may alter the spatial relationships of telopodes with neighboring immune cells, fibroblasts, smooth muscle cells, ICC, and nervous structures, possibly impairing intercellular signaling and functions. However, whether the loss of telocytes might even precede the onset of fibrosis rather than being merely a consequence of the fibrotic process is difficult to be elucidated. In this context, the

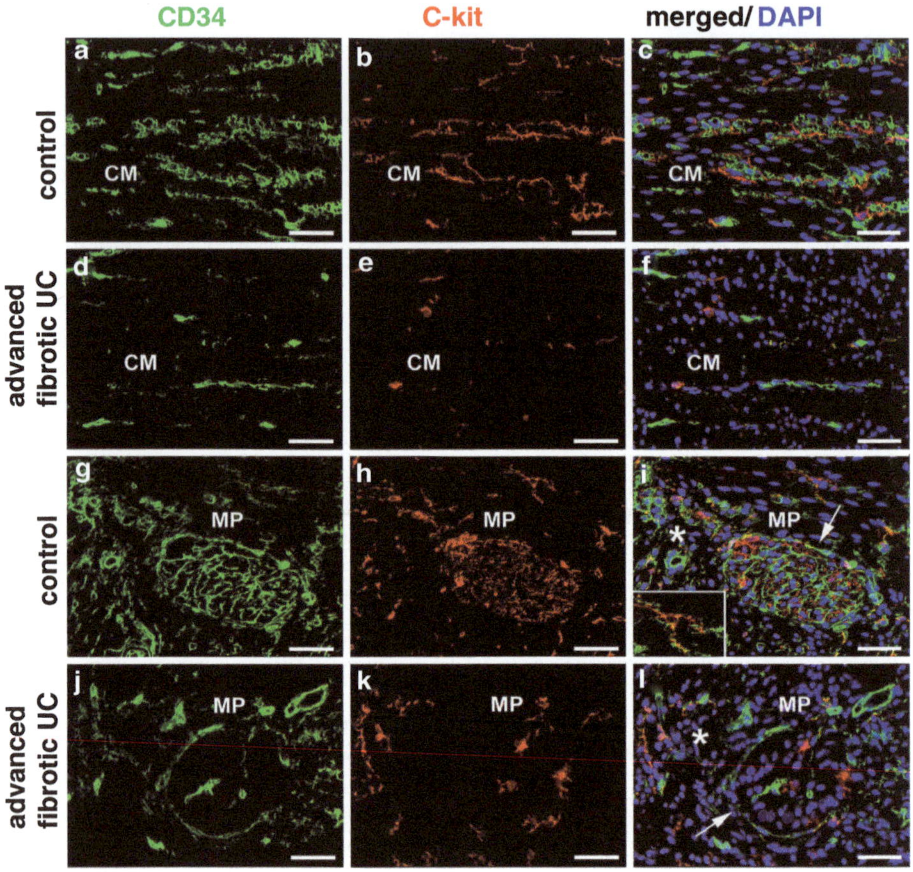

Fig. 4.9 Muscularis propria of colonic wall specimens from controls (**a–c** and **g–i**) and advanced fibrotic ulcerative colitis (*UC*) cases (**d–f** and **j–l**). (**a–l**) Double immunofluorescence labeling for CD34 (*green*) and c-kit/CD117 (*red*) with DAPI (*blue*) counterstain for nuclei. Telocytes are CD34 positive/c-kit negative, whereas interstitial cells of Cajal (*ICC*) are c-kit positive/CD34 negative. (**a–c**) In muscle layers of control colonic sections, telocytes and ICC form interconnected networks among smooth muscle bundles. (**d–f**) In muscle layers of advanced fibrotic UC cases, very few telocytes and ICC can be observed. Representative microphotographs of the circular muscle layer are shown. (**g–i**) Note the abundant networks of telocytes and ICC around ganglia (*arrow* in **i**) and in the interganglionic region (*asterisk* in **i**) of the myenteric plexus of control colonic wall. (*Inset* in **i**): Higher magnification view of an ICC surrounded by telopodes. (**j–l**) In advanced fibrotic UC cases, both telocytes and ICC are scarce around myenteric ganglia (*arrow* in **l**) and in the interganglionic region (*asterisk* in **l**). *CM* circular muscle layer, *MP* myenteric plexus (Reproduced with permission from Manetti et al. [27])

findings that in early fibrotic UC cases telocytes are already reduced in the edematous, less fibrotic areas of submucosa suggest that telocyte loss may be a very precocious event during fibrotic remodeling of the intestinal wall [27]. The progressive loss of telocytes in the intestinal wall could also contribute to the altered three-dimensional organization of the ECM and to the progression of fibrosis, eventually favoring fibroblast-to-myofibroblast transition as proposed in the skin of SSc patients [22]. In support to this last hypothesis, in the colonic submucosa of UC patients, the disappearance of telocytes is paralleled by the increase in the number of α-SMA-positive myofibroblasts [27]. However, as proposed by other authors [17], we cannot completely rule out the possibility that during pathological processes, some telocytes might even change their immunophenotype (e.g., loss of CD34 expression and gain of other markers, such as α-SMA), thus contributing to the increase in the myofibroblast population. Nevertheless, we should consider that CD34/α-SMA double immunostaining did not reveal the presence of double-positive transitioning stromal cells in colonic sections from UC patients [27]. Furthermore, there is clear ultrastructural evidence of telocyte degeneration, rather than activation/transdifferentiation into myofibroblasts, in the setting of tissue fibrosis [22].

In the gut, the three-dimensional network of telocytes has also been proposed to play a specific mechanical and supporting role throughout the different bowel wall layers, being resistant to and deformable following intestine movements [13, 18, 20]. Even more importantly, in the muscularis propria telocytes and ICC may form interconnected networks surrounding smooth muscle bundles and myenteric plexus ganglia [13, 18, 23, 27]. Thus, it has been hypothesized that the reduction in both telocyte and ICC networks within the neuromuscular compartment of the intestinal wall might substantially contribute to gastrointestinal dysmotility in both CD and UC patients [23, 27].

4.4 Telocytes in Liver Fibrosis

Liver fibrosis, primarily as a result of chronic viral hepatitis and fatty liver diseases associated with obesity, is a worldwide burden [100]. Liver fibrosis can further progress to cirrhosis, representing a major cause of morbidity and mortality worldwide, and is responsible for several end-stage liver disease complications, including portal hypertension, impaired metabolic capacity, synthetic dysfunction, and ascites [100, 101]. Hepatic stellate cells are considered as the primary source of the fibrogenic population in the liver [102]. In addition, portal fibroblasts, bone marrow-derived cells, circulating fibrocytes, and fibroblasts deriving from epithelial-mesenchymal transition may also be implicated in the hepatic fibrogenic process [102, 103].

The existence of telocytes in the liver, particularly in the Disse space with a similar density in the four hepatic lobes, has been established [104]. Furthermore, the potential role of telocytes in liver regeneration has been emphasized in a mouse model of partial hepatectomy [105]. In a recent study, the possible involvement of telocytes in liver fibrosis has also been investigated [59]. In particular, it has been reported that telocytes are reduced in the human fibrotic liver, further supporting that within the stromal hepatic compartment telocytes are different from hepatic stellate cells which, instead, are well known to be increased during liver fibrosis [59]. Taking into account the "connecting cell" function of telocytes, the authors suggested that they might be able to control the activity of hepatic stellate cells.

Therefore, telocyte loss might contribute to hepatic stellate cell dysregulation in the fibrotic liver [59]. In addition, the disappearance of telocytes might impair hepatocytes and stem cell-mediated liver regeneration. Indeed, it has been shown that telocytes have a close spatial relationship with hepatic putative stem cells and that they may influence proliferation of hepatocytes and/or the activation of hepatic stem cells [105]. Thus, in human liver fibrosis, the reduction in telocytes might contribute to the depletion of stem cell niches or hepatocyte dysfunction and impair liver regeneration/repair.

4.5 Telocytes in Primary Sjögren's Syndrome

Primary Sjögren's syndrome (pSS) is a systemic autoimmune disorder mainly affecting women during the fourth and fifth decades of life and characterized by chronic inflammation of exocrine glands leading to progressive functional impairment [106, 107]. The histological hallmark of pSS is a focal lymphocytic sialadenitis, and the presence of at least one focus (i.e., an aggregate of at least 50 lymphocytes and plasma cells) in 4 mm^2 of minor salivary gland (MSG) tissue allows the diagnosis of pSS [108]. In the last decades, the pathogenic role of stromal cells in systemic inflammatory/autoimmune disorders including pSS has been extensively investigated [3, 4]. In this context, the unique phenotype, ultrastructural characteristics, tissue distribution, and multiple intercellular connections of telocytes, as well their putative role in local immune surveillance and homeostasis [14, 47], raised the possibility of their involvement in the pathogenesis of inflammatory/autoimmune disorders as pSS.

With regard to salivary glands, to date telocytes have been described in parotid glands and labial MSGs, where these peculiar stromal cells surround secretory and excretory structures, namely, acini and ducts, and are also in close contact with blood vessels [28, 109]. Interestingly, it has been recently highlighted a possible association between telocyte patterns and the extent of glandular inflammation and lymphoid organization in MSGs from patients with pSS [28]. Notably, telocytes are markedly reduced in MSGs of pSS patients with respect to normal and nonspecific chronic sialadenitis (NSCS; i.e., presence of scattered lymphocyte aggregates that do not reach the number of 50 and therefore cannot be classified as foci) MSGs, and such a decrease parallels the worsening of glandular inflammation and the progression of ectopic lymphoid neogenesis. Indeed, periductal telocytes are reduced in the presence of smaller inflammatory foci and completely absent in the presence of germinal center-like structures, thus closely reflecting disease severity [28]. In addition, while in other inflammatory pathological conditions like CD lymphoid aggregates/granulomas are entirely surrounded by telocytes, suggesting a certain attempt to control their spreading [23], the complete absence of telocytes around MSG inflammatory foci underscores that this potential protective mechanism may be impaired in pSS.

In keeping with the proposed function of telocytes as key tissue homeostasis regulators, their local loss might contribute to breaking out of the immune homeostasis contributing to the pathogenesis of focal lymphocytic sialadenitis. This hypothesis is supported by the evidence that telocytes are markedly and specifically reduced in focal lymphocytic sialadenitis but not in NSCS. In this setting, the formation and maintenance of focal lymphocytic sialadenitis as well as the development of ectopic lymphoid structures during chronic inflammation are dependent on the expression of lymphotoxins, cytokines, and chemokines by several cell types [28]. Previous studies reported that in NSCS MSGs, the expression levels of several of these mediators are similar to that of normal MSGs, while they are significantly upregulated in pSS MSGs compared to both normal and NSCS MSGs [28]. These observations suggest that the peculiar inflammatory microenvironment of pSS MSGs might even be one of the causes of local telocyte damage and loss. Moreover, the fact that in pSS MSGs telocytes are preserved around the acini not affected by the inflammatory process allows to speculate that the cross talk between epithelial or infiltrating immune cells and telocytes occurs in a paracrine manner limited to each secretory unit [28]. However, whether the loss of telocytes in pSS may represent either the cause or the consequence of local inflammation remains still unknown. To further gain insights on the possible contribution of telocytes to pSS pathophysiology and the exact mechanisms underlying telocyte cross talk with other MSG cell types and their disappearance during focal lymphocytic sialadenitis, future ultrastructural and functional studies will be required.

4.6 Telocytes in Psoriasis

Psoriasis is a common inflammatory skin condition, mainly considered a keratinization disorder on a genetic background. Indeed, the dermis contribution to the pathogenesis of psoriasis is frequently eclipsed by remarkable epidermal phenomena.

Among stromal cells, several studies have focused on dendritic cells as major participants in the chronic skin inflammatory process that characterizes psoriasis [110, 111]. Besides dendritic cells, the ability of telocytes to establish intercellular communications (either physical or paracrine) with immune cells has been well documented in several organs, including the skin [14, 21, 22, 46, 47]. Thus, in the context of the vast immunology of psoriasis, it has been hypothesized that telocytes could be involved in disease initiation and/or progression [58]. Furthermore, substantial evidence indicates that angiogenesis may also contribute to the pathogenesis and clinical signs of psoriasis [112]. Noteworthy, previous studies have shown that telocytes may be involved in angiogenesis, providing their support in the reparatory process after acute myocardial infarction [50]. Therefore, the possible involvement of telocytes in psoriasis-related angiogenesis has also been investigated [58].

In their recent study, Manole et al. demonstrated a reduction in the number of telocytes in papillary dermis of psoriasis vulgaris, as well as a recovery of these cells after local corticoid therapy [58]. Of note, the density of telocytes with a normal

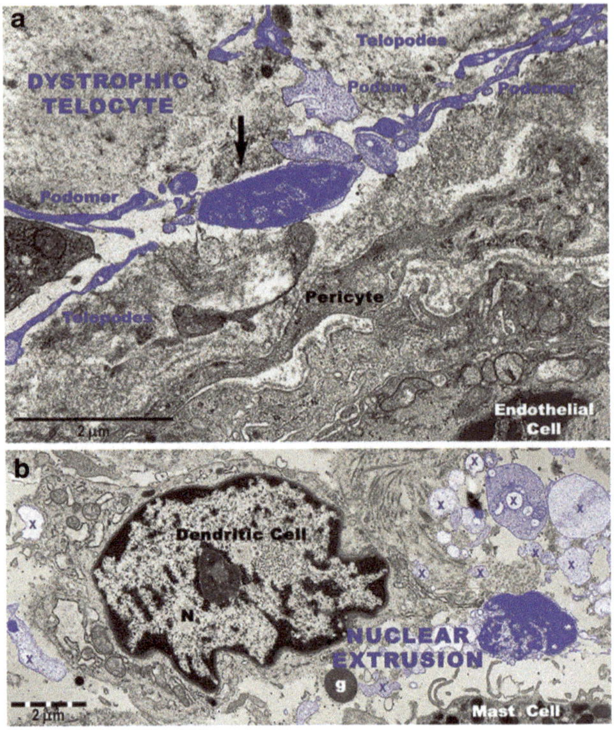

Fig. 4.10 Transmission electron microscopy images show degenerative changes in telocytes (digitally colored in *blue*) from psoriasis skin. (**a**) A telocyte with shriveled nucleus and detached telopodes. The *arrow* indicates dissolution of the cellular membrane and the cytoplasmic content surrounding the nucleus. (**b**) An extruded nucleus and cytoplasmic fragments (*X*) of a telocyte are visible in the vicinity of a dendritic cell. *g* granule (of a mast cell) (Reproduced with permission from Manole et al. [58])

morphology appears comparable in the dermis of uninvolved and treated skin but deeply decreases in the lesional psoriatic papillary dermis. Moreover, the few remaining telocytes exhibit degenerative ultrastructural features. Indeed, in psoriatic lesions, TEM revealed the presence of telocytes with apoptotic nuclei, dystrophic telocytes with fragmented telopodes, and even telocytes with nuclear extrusions, membrane disintegration, and cytoplasmic fragmentation (Fig. 4.10) [58]. Extruded nuclei or apoptotic telocytes were often observed to have close contacts with dendritic cells in the dermis of psoriatic skin. Profound changes in the phenotype of vascular smooth muscle cells (i.e., cells exhibiting a synthetic phenotype with decreased actin filaments and increased rough endoplasmic reticulum) in small blood vessels that lost the protective envelope formed by telocytes were also found. Therefore, it has been suggested that the loss of perivascular telocytes might have important implications in the characteristic vascular pathology of psoriasis [58]. Collectively, the authors proposed that the reduction in telocytes and their interstitial network may significantly influence psoriatic lesion initiation and/or progression, impairing long-distance heterocellular communication. Accordingly, in psoriasis telocytes could be considered as new cellular targets for forthcoming therapies. Besides psoriasis vulgaris, it will be of major importance to clarify the possible involvement of telocytes in other forms of psoriasis, such as pustular and erythrodermic psoriasis.

4.7 Concluding Remarks

Increasing evidence indicates that telocytes are a peculiar interstitial cell type implicated in tissue homeostasis and development [13, 14], as well as in the pathophysiology of several disorders [14, 22, 23, 27–29, 58–66]. In particular, severe damage and a broad reduction of telocytes have been reported during fibrotic remodeling of multiple organs in various diseases, including SSc, CD, UC, and liver fibrosis, as well as in chronic inflammatory lesions like those of pSS and psoriatic skin [14, 22, 23, 27–29, 58, 59]. Although several hypotheses have been proposed, the pathogenetic mechanisms underlying the loss of telocytes and their functional consequences in those disorders need to be further investigated.

Owing to their close relationship with stem cells and/or their capacity to guide or nurse putative progenitor cells in tissue-resident stem cell niches, telocytes are also supposed to contribute to tissue repair/regeneration [14, 55]. Indeed, telocytes are universally considered as "connecting cells" mostly oriented to intercellular signaling within the stromal compartment of almost every organ. Thus, a deeper understanding of how telocytes communicate with neighboring cells and take effect in signaling pathway during tissue repair/regeneration appears crucial to identify novel therapeutic strategies for the aforementioned and, possibly, other disorders. Interestingly, telocytes were found to be reduced during experimental myocardial infarction, particularly in fibrotic areas, and transplantation of cardiac telocytes could decrease the infarction size and improve postinfarcted cardiac function through the reconstruction of the telocyte network and the reduction of cardiac fibrosis [65, 66]. On this basis, in the near future, telocyte transplantation might represent a promising therapeutic opportunity to control the evolution of chronic inflammatory and fibrotic diseases. Notably, there is evidence to support that telocytes could help in preventing abnormal activation of immune cells and fibroblasts, as well as in attenuating the altered ECM organization during the fibrotic process [55]. By targeting telocytes alone or in tandem with stem cells, we might be able to promote regeneration and prevent the evolution to irreversible tissue damage. Finally, besides exogenous transplantation, exploring pharmacological or non-pharmacological methods to enhance the growth and/or survival of telocytes could be an additional therapeutic strategy for many disorders.

References

1. Hay ED. Cell and extracellular matrix: their organization and mutual dependence. Mod Cell Biol. 1983;2:509–48.
2. Doljanski F. The sculpturing role of fibroblast-like cells in morphogenesis. Perspect Biol Med. 2004;47:339–56.
3. Barone F, Nayar S, Buckley CD. The role of non-hematopoietic stromal cells in the persistence of inflammation. Front Immunol. 2013;3:416.
4. Naylor AJ, Filer A, Buckley CD. The role of stromal cells in the persistence of chronic inflammation. Clin Exp Immunol. 2013;171:30–5.

5. McGettrick HM, Smith E, Filer A, et al. Fibroblasts from different sites may promote or inhibit recruitment of flowing lymphocytes by endothelial cells. Eur J Immunol. 2009;39:113–25.
6. Corsiero E, Bombardieri M, Manzo A, et al. Role of lymphoid chemokines in the development of functional ectopic lymphoid structures in rheumatic autoimmune diseases. Immunol Lett. 2012;145:62–7.
7. Pitzalis C, Jones GW, Bombardieri M, et al. Ectopic lymphoid-like structures in infection, cancer and autoimmunity. Nat Rev Immunol. 2014;14:47–62.
8. van de Pavert SA, Mebius RE. New insights into the development of lymphoid tissues. Nat Rev Immunol. 2010;10:664–74.
9. Wynn TA, Ramalingam TR. Mechanisms of fibrosis: therapeutic translation for fibrotic disease. Nat Med. 2012;18:1028–40.
10. Wynn TA. Cellular and molecular mechanisms of fibrosis. J Pathol. 2008;214:199–210.
11. Hinz B, Phan SH, Thannickal VJ, et al. Recent developments in myofibroblast biology: paradigms for connective tissue remodeling. Am J Pathol. 2012;180:1340–55.
12. Popescu LM, Faussone-Pellegrini MS. Telocytes – a case of serendipity: the winding way from interstitial cells of Cajal (ICC), via interstitial Cajal-like cells (ICLC) to telocytes. J Cell Mol Med. 2010;14:729–40.
13. Faussone-Pellegrini MS, Popescu LM. Telocytes. Biomol Concepts. 2011;2:481–9.
14. Cretoiu SM, Popescu LM. Telocytes revisited. Biomol Concepts. 2014;5:353–69.
15. Kang Y, Zhu Z, Zheng Y, et al. Skin telocytes versus fibroblasts: two distinct dermal cell populations. J Cell Mol Med. 2015;19:2530–9.
16. Bei Y, Zhou Q, Fu S, et al. Cardiac telocytes and fibroblasts in primary culture: different morphologies and immunophenotypes. PLoS One. 2015;18(10):e0115991.
17. Díaz-Flores L, Gutiérrez R, García MP, et al. CD34+ stromal cells/fibroblasts/fibrocytes/telocytes as a tissue reserve and a principal source of mesenchymal cells. Location, morphology, function and role in pathology. Histol Histopathol. 2014;29:831–70.
18. Vannucchi MG, Traini C, Manetti M, et al. Telocytes express PDGFRα in the human gastrointestinal tract. J Cell Mol Med. 2013;17:1099–108.
19. Zhou Q, Wei L, Zhong C, et al. Cardiac telocytes are double positive for CD34/PDGFR-α. J Cell Mol Med. 2015;19:2036–42.
20. Pieri L, Vannucchi MG, Faussone-Pellegrini MS. Histochemical and ultrastructural characteristics of an interstitial cell type different from ICC and resident in the muscle coat of human gut. J Cell Mol Med. 2008;12:1944–55.
21. Ceafalan L, Gherghiceanu M, Popescu LM, et al. Telocytes in human skin–are they involved in skin regeneration? J Cell Mol Med. 2012;16:1405–20.
22. Manetti M, Guiducci S, Ruffo M, et al. Evidence for progressive reduction and loss of telocytes in the dermal cellular network of systemic sclerosis. J Cell Mol Med. 2013;17:482–96.
23. Milia AF, Ruffo M, Manetti M, et al. Telocytes in Crohn's disease. J Cell Mol Med. 2013; 17:1525–36.
24. Kostin S, Popescu LM. A distinct type of cell in myocardium: interstitial Cajal-like cells (ICLCs). J Cell Mol Med. 2009;13:295–308.
25. Zheng Y, Li H, Manole CG, et al. Telocytes in trachea and lungs. J Cell Mol Med. 2011;15:2262–8.
26. Popescu LM, Gherghiceanu M, Suciu LC, et al. Telocytes and putative stem cells in the lungs: electron microscopy, electron tomography and laser scanning microscopy. Cell Tissue Res. 2011;345:391–403.
27. Manetti M, Rosa I, Messerini L, et al. Telocytes are reduced during fibrotic remodelling of the colonic wall in ulcerative colitis. J Cell Mol Med. 2015;19:62–73.
28. Alunno A, Ibba-Manneschi L, Bistoni O, et al. Telocytes in minor salivary glands of primary Sjögren's syndrome: association with the extent of inflammation and ectopic lymphoid neogenesis. J Cell Mol Med. 2015;19:1689–96.
29. Manetti M, Rosa I, Messerini L, et al. A loss of telocytes accompanies fibrosis of multiple organs in systemic sclerosis. J Cell Mol Med. 2014;18:253–62.
30. Sun X, Zheng M, Zhang M, et al. Differences in the expression of chromosome 1 genes between lung telocytes and other cells: mesenchymal stem cells, fibroblasts, alveolar type II cells, airway epithelial cells and lymphocytes. J Cell Mol Med. 2014;18:801–10.

31. Zheng M, Sun X, Zhang M, et al. Variations of chromosomes 2 and 3 gene expression profiles among pulmonary telocytes, pneumocytes, airway cells, mesenchymal stem cells and lymphocytes. J Cell Mol Med. 2014;18:2044–60.
32. Wang J, Ye L, Jin M, et al. Global analyses of chromosome 17 and 18 genes of lung telocytes compared with mesenchymal stem cells, fibroblasts, alveolar type II cells, airway epithelial cells, and lymphocytes. Biol Direct. 2015;10:9.
33. Zhu Y, Zheng M, Song D, et al. Global comparison of chromosome X genes of pulmonary telocytes with mesenchymal stem cells, fibroblasts, alveolar type II cells, airway epithelial cells, and lymphocytes. J Transl Med. 2015;13:318.
34. Zheng Y, Cretoiu D, Yan G, et al. Protein profiling of human lung telocytes and microvascular endothelial cells using iTRAQ quantitative proteomics. J Cell Mol Med. 2014;18:1035–59.
35. Zheng Y, Cretoiu D, Yan G, et al. Comparative proteomic analysis of human lung telocytes with fibroblasts. J Cell Mol Med. 2014;18:568–89.
36. Albulescu R, Tanase C, Codrici E, et al. The secretome of myocardial telocytes modulates the activity of cardiac stem cells. J Cell Mol Med. 2015;19:1783–94.
37. Cismasiu VB, Radu E, Popescu LM. miR-193 expression differentiates telocytes from other stromal cells. J Cell Mol Med. 2011;15:1071–4.
38. Zheng Y, Zhang M, Qian M, et al. Genetic comparison of mouse lung telocytes with mesenchymal stem cells and fibroblasts. J Cell Mol Med. 2013;17:567–77.
39. Cretoiu D, Hummel E, Zimmermann H, et al. Human cardiac telocytes: 3D imaging by FIB-SEM tomography. J Cell Mol Med. 2014;18:2157–64.
40. Faussone-Pellegrini MS, Bani D. Relationships between telocytes and cardiomyocytes during pre- and post-natal life. J Cell Mol Med. 2010;14:1061–3.
41. Bani D, Formigli L, Gherghiceanu M, et al. Telocytes as supporting cells for myocardial tissue organization in developing and adult heart. J Cell Mol Med. 2010;14:2531–8.
42. Gherghiceanu M, Popescu LM. Cardiomyocyte precursors and telocytes in epicardial stem cell niche: electron microscope images. J Cell Mol Med. 2010;14:871–7.
43. Luesma MJ, Gherghiceanu M, Popescu LM. Telocytes and stem cells in limbus and uvea of mouse eye. J Cell Mol Med. 2013;17:1016–24.
44. Popescu LM, Gherghiceanu M, Manole CG, et al. Cardiac renewing: interstitial Cajal-like cells nurse cardiomyocyte progenitors in epicardial stem cell niches. J Cell Mol Med. 2009;13:866–86.
45. Gherghiceanu M, Manole CG, Popescu LM. Telocytes in endocardium: electron microscope evidence. J Cell Mol Med. 2010;14:2330–4.
46. Cretoiu D, Cretoiu SM, Simionescu AA, et al. Telocytes, a distinct type of cell among the stromal cells present in the lamina propria of jejunum. Histol Histopathol. 2012;27:1067–78.
47. Popescu LM, Gherghiceanu M, Cretoiu D, et al. The connective connection: interstitial cells of Cajal (ICC) and ICC-like cells establish synapses with immunoreactive cells. Electron microscope study in situ. J Cell Mol Med. 2005;9:714–30.
48. Rusu MC, Mirancea N, Mănoiu VS, et al. Skin telocytes. Ann Anat. 2012;194:359–67.
49. Cismaşiu VB, Popescu LM. Telocytes transfer extracellular vesicles loaded with microRNAs to stem cells. J Cell Mol Med. 2015;19:351–8.
50. Manole CG, Cismasiu V, Gherghiceanu M, et al. Experimental acute myocardial infarction: telocytes involvement in neo-angiogenesis. J Cell Mol Med. 2011;15:2284–96.
51. Fertig ET, Gherghiceanu M, Popescu LM. Extracellular vesicles release by cardiac telocytes: electron microscopy and electron tomography. J Cell Mol Med. 2014;18:1938–43.
52. Smythies J, Edelstein L. Telocytes, exosomes, gap junctions and the cytoskeleton: the makings of a primitive nervous system? Front Cell Neurosci. 2014;7:278.
53. Cretoiu D, Gherghiceanu M, Hummel E, et al. FIB-SEM tomography of human skin telocytes and their extracellular vesicles. J Cell Mol Med. 2015;19:714–22.
54. Cretoiu SM, Cretoiu D, Marin A, et al. Telocytes: ultrastructural, immunohistochemical and electrophysiological characteristics in human myometrium. Reproduction. 2013;145:357–70.
55. Bei Y, Wang F, Yang C, et al. Telocytes in regenerative medicine. J Cell Mol Med. 2015;19:1441–54.

56. Cretoiu SM, Radu BM, Banciu A, et al. Isolated human uterine telocytes: immunocytochemistry and electrophysiology of T-type calcium channels. Histochem Cell Biol. 2015;143: 83–94.
57. Sheng J, Shim W, Lu J, et al. Electrophysiology of human cardiac atrial and ventricular telocytes. J Cell Mol Med. 2014;18:355–62.
58. Manole CG, Gherghiceanu M, Simionescu O. Telocyte dynamics in psoriasis. J Cell Mol Med. 2015;19:1504–19.
59. Fu S, Wang F, Cao Y, et al. Telocytes in human liver fibrosis. J Cell Mol Med. 2015;19: 676–83.
60. Zheng Y, Bai C, Wang X. Potential significance of telocytes in the pathogenesis of lung diseases. Expert Rev Respir Med. 2012;6:45–9.
61. Zheng Y, Bai C, Wang X. Telocyte morphologies and potential roles in diseases. J Cell Physiol. 2012;227:2311–7.
62. Matyja A, Gil K, Pasternak A, et al. Telocytes: new insight into the pathogenesis of gallstone disease. J Cell Mol Med. 2013;17:734–42.
63. Yang J, Chi C, Liu Z, et al. Ultrastructure damage of oviduct telocytes in rat model of acute salpingitis. J Cell Mol Med. 2015;19:1720–8.
64. Richter M, Kostin S. The failing human heart is characterized by decreased numbers of telocytes as result of apoptosis and altered extracellular matrix composition. J Cell Mol Med. 2015;19:2597–606.
65. Zhao B, Chen S, Liu J, et al. Cardiac telocytes were decreased during myocardial infarction and their therapeutic effects for ischaemic heart in rat. J Cell Mol Med. 2013;17:123–33.
66. Zhao B, Liao Z, Chen S, et al. Intramyocardial transplantation of cardiac telocytes decreases myocardial infarction and improves post-infarcted cardiac function in rats. J Cell Mol Med. 2014;18:780–9.
67. Varga J, Abraham D. Systemic sclerosis: a prototypic multisystem fibrotic disorder. J Clin Invest. 2007;117:557–67.
68. Gabrielli A, Avvedimento EV, Krieg T. Scleroderma. N Engl J Med. 2009;360:1989–2003.
69. Bhattacharyya S, Wei J, Varga J. Understanding fibrosis in systemic sclerosis: shifting paradigms, emerging opportunities. Nat Rev Rheumatol. 2011;8:42–54.
70. Manetti M, Guiducci S, Ibba-Manneschi L, et al. Mechanisms in the loss of capillaries in systemic sclerosis: angiogenesis versus vasculogenesis. J Cell Mol Med. 2010;14:1241–54.
71. Manetti M, Guiducci S, Romano E, et al. Differential expression of junctional adhesion molecules in different stages of systemic sclerosis. Arthritis Rheum. 2013;65:247–57.
72. Manneschi LI, Del Rosso A, Milia AF, et al. Damage of cutaneous peripheral nervous system evolves differently according to the disease phase and subset of systemic sclerosis. Rheumatology. 2005;44:607–13.
73. Jain S, Shahane A, Derk CT. Interstitial lung disease in systemic sclerosis: pathophysiology, current and new advances in therapy. Inflamm Allergy Drug Targets. 2012;11:266–77.
74. Meune C, Vignaux O, Kahan A, et al. Heart involvement in systemic sclerosis: evolving concept and diagnostic methodologies. Arch Cardiovasc Dis. 2010;103:46–52.
75. Sallam H, McNearney TA, Chen JD. Systematic review: pathophysiology and management of gastrointestinal dysmotility in systemic sclerosis (scleroderma). Aliment Pharmacol Ther. 2006;23:691–712.
76. Manetti M, Neumann E, Milia AF, et al. Severe fibrosis and increased expression of fibrogenic cytokines in the gastric wall of systemic sclerosis patients. Arthritis Rheum. 2007;56:3442–7.
77. Roberts CG, Hummers LK, Ravich WJ, et al. A case-controlled study of the pathology of esophageal disease in systemic sclerosis (scleroderma). Gut. 2006;55:1697–703.
78. Wei J, Bhattacharyya S, Tourtellotte WG, et al. Fibrosis in systemic sclerosis: emerging concepts and implications for targeted therapy. Autoimmun Rev. 2011;10:267–75.
79. Ebmeier S, Horsley V. Origin of fibrosing cells in systemic sclerosis. Curr Opin Rheumatol. 2015;27:555–62.
80. Popescu LM. The Tandem: telocytes – stem cells. Int J Biol Biomed Eng. 2011;5:83–92.
81. Gherghiceanu M, Popescu LM. Heterocellular communication in the heart: electron tomography of telocyte-myocyte junctions. J Cell Mol Med. 2011;15:1005–11.

82. Manetti M, Milia AF, Benelli G, et al. The gastric wall in systemic sclerosis patients: a morphological study. Ital J Anat Embryol. 2010;115:115–21.
83. Rieder F, Fiocchi C. Intestinal fibrosis in inflammatory bowel disease – current knowledge and future perspectives. J Crohns Colitis. 2008;2:279–90.
84. Baumgard DC, Sandborn WJ. Crohn's disease. Lancet. 2012;380:1590–605.
85. Maul J, Zeitz M. Ulcerative colitis: immune function, tissue fibrosis and current therapeutic considerations. Langenbecks Arch Surg. 2012;397:1–10.
86. Latella G, Sferra R, Speca S, et al. Can we prevent, reduce or reverse intestinal fibrosis in IBD? Eur Rev Med Pharmacol Sci. 2013;17:1283–304.
87. Yamagata M, Mikami T, Tsuruta T, et al. Submucosal fibrosis and basic-fibroblast growth factor-positive neutrophils correlate with colonic stenosis in cases of ulcerative colitis. Digestion. 2011;84:12–21.
88. Gordon IO, Agrawal N, Goldblum JR, et al. Fibrosis in ulcerative colitis: mechanisms, features, and consequences of a neglected problem. Inflamm Bowel Dis. 2014;20:2198–206.
89. Rieder F, Fiocchi C. Intestinal fibrosis in inflammatory bowel disease: progress in basic and clinical science. Curr Opin Gastroenterol. 2008;24:462–8.
90. Quigley EM. What we have learned about colonic motility: normal and disturbed. Curr Opin Gastroenterol. 2010;26:53–60.
91. Wood JD. Enteric nervous system: reflex, pattern generators and motility. Curr Opin Gastroenterol. 2008;24:149–58.
92. Sanders KM. A case for interstitial cells of Cajal as pacemakers and mediators of neurotransmission in the gastrointestinal tract. Gastroenterology. 1996;111:492–515.
93. Wang XY, Zarate N, Soderholm JD, et al. Ultrastructural injury to interstitial cells of Cajal and communication with mast cells in Crohn's disease. Neurogastroenterol Motil. 2007;19: 349–64.
94. Bernardini N, Segnani C, Ippolito C, et al. Immunohistochemical analysis of myenteric ganglia and interstitial cells of Cajal in ulcerative colitis. J Cell Mol Med. 2012;16:318–27.
95. Bassotti G, Antonelli E, Villanacci V, et al. Gastrointestinal motility disorders in inflammatory bowel diseases. World J Gastroenterol. 2014;20:37–44.
96. Ohlsson B, Veress B, Lindgren S, et al. Enteric ganglioneuritis and abnormal interstitial cells of Cajal: features of inflammatory bowel disease. Inflamm Bowel Dis. 2007;13:721–6.
97. Kurahashi M, Nakano Y, Hennig GW, et al. Platelet derived growth factor receptor α-positive cells in the tunica muscularis of human colon. J Cell Mol Med. 2012;16:1397–404.
98. Vanderwinden JM, Rumessen JJ, De Laet MH, et al. CD34 + cells in human intestine are fibroblasts adjacent to, but distinct from, interstitial cells of Cajal. Lab Invest. 1999;79:59–65.
99. Porcher C, Baldo M, Henry M, et al. Deficiency of interstitial cells of Cajal in the small intestine of patients with Crohn's disease. Am J Gastroenterol. 2002;97:118–25.
100. Hernandez-Gea V, Friedman SL. Pathogenesis of liver fibrosis. Annu Rev Pathol. 2011;6: 425–56.
101. Iwaisako K, Taura K, Koyama Y, et al. Strategies to detect hepatic myofibroblasts in liver cirrhosis of different etiologies. Curr Pathobiol Rep. 2014;2:209–15.
102. Friedman SL. Hepatic stellate cells: protean, multifunctional, and enigmatic cells of the liver. Physiol Rev. 2008;88:125–72.
103. Iwaisako K, Jiang C, Zhang M, et al. Origin of myofibroblasts in the fibrotic liver in mice. Proc Natl Acad Sci U S A. 2014;111:E3297–305.
104. Xiao J, Wang F, Liu Z, et al. Telocytes in liver: electron microscopic and immunofluorescent evidence. J Cell Mol Med. 2013;17:1537–42.
105. Wang F, Song Y, Bei Y, et al. Telocytes in liver regeneration: possible roles. J Cell Mol Med. 2014;18:1720–6.
106. Cornec D, Jamin C, Pers JO. Sjögren's syndrome: where do we stand, and where shall we go? J Autoimmun. 2014;51:109–14.
107. Nocturne G, Mariette X. Advances in understanding the pathogenesis of primary Sjögren's syndrome. Nat Rev Rheumatol. 2013;9:544–56.
108. Greenspan JS, Daniels TE, Talal N, et al. The histopathology of Sjögren's syndrome in labial salivary gland biopsies. Oral Surg Oral Med Oral Pathol. 1974;37:217–29.

109. Nicolescu MI, Bucur A, Dinca O, et al. Telocytes in parotid glands. Anat Rec (Hoboken). 2012;295:378–85.
110. Kim J, Krueger JG. The immunopathogenesis of psoriasis. Dermatol Clin. 2015;33:13–23.
111. Chu CC, Di Meglio P, Nestle FO. Harnessing dendritic cells in inflammatory skin diseases. Semin Immunol. 2011;23:28–41.
112. Chua RA, Arbiser JL. The role of angiogenesis in the pathogenesis of psoriasis. Autoimmunity. 2009;42:574–9.

Mirko Manetti Section of Anatomy and Histology, Department of Experimental and Clinical Medicine, University of Florence, Largo Brambilla 3, 50134 Florence, Italy; E-mails: mirkomanetti@yahoo.it or mirko.manetti@unifi.it

Born in Florence, Italy, on April 24, 1978

2003: Bachelor of Biological Sciences with full marks and honors at the University of Florence, Florence, Italy

2004: Abilitation to the profession of Biologist at the University of Florence, Florence, Italy

2006: Visiting Scientist at the Department of Internal Medicine and Rheumatology, Justus-Liebig-University of Giessen, Division of Rheumatology and Clinical Immunology, Kerckhoff-Klinik, Bad Nauheim, Germany.

2007: International fellowship at the Department of Internal Medicine and Rheumatology, Justus-Liebig-University of Giessen, Division of Rheumatology and Clinical Immunology, Kerckhoff-Klinik, Bad Nauheim, Germany

2007: Ph.D. degree in Human Morphology and Morphogenesis, School of Medicine, University of Florence, Florence, Italy

2007–2012: Postdoctoral Research Fellow at the Department of Anatomy, Histology and Forensic Medicine, University of Florence, Florence, Italy

2013–present: Postdoctoral Research Fellow at the Department of Experimental and Clinical Medicine, Section of Anatomy and Histology, University of Florence, Florence, Italy

2014: National Scientific Habilitation for Associate Professor of Human Anatomy

Dr. Manetti's current research interests include autoimmune, chronic inflammatory and connective tissue diseases with a main focus on the pathogenetic mechanisms, histopathological and molecular bases, and animal models of these disorders. Dr. Manetti is author of numerous scientific publications in peer-reviewed international journals and proceedings of national and international congresses and currently serves as referee for numerous peer-reviewed scientific journals.

Chapter 5
Telocytes: New Players in Gallstone Disease

Artur Pasternak, Krzysztof Gil, and Andrzej Matyja

Abstract Cholesterol gallstone disease is highly prevalent in Western countries, particularly in women and some specific ethnic groups. The mechanisms behind the formation of gallstones are not clearly understood, but gallbladder dysmotility seems to be a key factor that triggers the precipitation of cholesterol microcrystals from supersaturated lithogenic bile.

Given that newly described interstitial cells, telocytes, are present in the gallbladder and they are located in close vicinity of smooth muscle cell and neural fibers possibly interfering with gallbladder motility or contractility, authors are trying to summarize the current knowledge on the role of telocytes with respect to disturbed gallbladder function in gallstone disease.

5.1 Introduction

Gallstone disease is often thought to be a major affliction in modern society [11]. However, gallstones must have been known to humans for many years, since they have been found in the gallbladders of Egyptian mummies dating back to 1000 BC [8, 41]. Another piece of evidence is an inscription made by an ancient physician called Soranus of Ephesus from the second century AD, describing the signs of

A. Pasternak, MD, PhD (✉)
First Department of General, Oncological and Gastrointestinal Surgery, Jagiellonian University Medical College, 40 Kopernika St., 31-501 Krakow, Poland
e-mail: artur.pasternak@uj.edu.pl

Department of Anatomy, Jagiellonian University Medical College, 12 Kopernika St., 31-034 Krakow, Poland

K. Gil
Department of Pathophysiology, Jagiellonian University Medical College, 18 Czysta St., 31-121 Krakow, Poland

A. Matyja
First Department of General, Oncological and Gastrointestinal Surgery, Jagiellonian University Medical College, 40 Kopernika St., 31-501 Krakow, Poland

X. Wang, D. Cretoiu (eds.), *Telocytes*, Advances in Experimental Medicine and Biology, 913, DOI 10.1007/978-981-10-1061-3_5

mechanical jaundice. The first precise account of gallstones was given in 1420 by a Florentine pathologist Antonio Benivieni and concerned a woman who died with strong abdominal pain [93].

Strategies and tactics in the treatment of gallstone disease are based on surgical procedures since the second half of nineteenth century when the prominent Scottish surgeon Lawson Tait popularized cholecystostomy operation [94]. However, in 1882 Carl von Langenbuch performed the first cholecystectomy in Berlin and advocated this procedure in favor of cholecystostomy for the surgical treatment of gallstone disease since he realized that the gallbladder should be removed not because it contained stones but because it originated the stones [61, 104]. Langenbuch used an open approach that then remained unchanged for more than 100 years. The first laparoscopic cholecystectomy with use of videoscopy was conducted by Phillipe Mouret in Lyon in 1987 [57]. At the beginning, the new technique of laparoscopic cholecystectomy was perceived by surgeons very skeptically. It gained full acknowledgement as a new surgical technique in 1989–1990 and is now considered worldwide as the "gold standard." Attempts at conservative treatment changed throughout the years, but as it is widely known there is no reliable, secure, and effective method for inoperative treatment of cholelithiasis. There is an increasing international interest to strengthen and progress research areas that can aid in solving disease understanding and perhaps to improve novel methods of treatment and prevention.

5.2 Epidemiology and Etiology of Gallstone Disease

Gallstone disease occurs in 10–15 % in adults in Europe and the United States of America and is one of the most common and most expensive of the digestive disorders that require hospital admission [32, 88]. Its prevalence is especially high in the Scandinavian countries and Chile and among Native Americans [106]. Gallstones are more common in North America, Europe, and Australia, and are less prevalent in Africa, India, China, Japan, Kashmir, and Egypt [67]. They are seen in all age groups, but the incidence increases with age, and about a quarter of women over 60 years will develop them [27, 90]. In most cases they do not cause symptoms, and only 10 % and 20 % will eventually become symptomatic within 5 years and 20 years of diagnosis [34, 98].

The etiology of gallstone disease is considered to be multifactorial, with roles for both genetic and environmental factors [7, 71]. Gallstones are four to ten times more frequent in older than younger subjects. Biliary cholesterol saturation increases with age, due to a decline in the activity of cholesterol 7α hydroxylase, the rate-limiting enzyme for bile acid synthesis [70]. Deoxycholic acid proportion in bile increases with age through enhanced 7α dehydroxylation of the primary bile acids by the intestinal bacteria [30]. In all populations of the world, regardless of overall gallstone prevalence, women during their fertile years are almost twice as likely as men to experience cholelithiasis. Increased levels of estrogens, as a result of pregnancy or hormone therapy, or the use of combined (estrogen-containing) forms of hormonal contraception, may increase cholesterol levels in bile and also decrease gallbladder movement, resulting in gallstone formation [101]. The highest incidence of cholelithiasis is observed in Indians Pima in Northern America, in citizens of the United States of Mexican origin, and the lowest in people of Bantu and Masai in Africa, where this disease practically does not occur [9, 58]. These racial differences in cholesterol gallstone prevalence provide strong evidence of the genetic predisposition to the disease. The next important risk factor is obesity, which raises the risk of cholesterol gallstones by increasing biliary secretion of cholesterol, as a result of an increase in 3-hydroxy-3-methylglutaryl-coenzyme A (HMGCoA) reductase activity. Rapid weight loss is associated with occurrence of sludge and gallstones in 10–25 % of patients in a few weeks of initiating the slimming procedures [113]. If a person loses weight too quickly, the liver secretes extra cholesterol; in addition there is rapid mobilization of cholesterol from adipose tissue stores. In fasting associated with severely fat-restricted diets, gallbladder contraction is reduced, and the accompanying gallbladder stasis favors gallstone formation. Western diet, i.e., increased intake of fat and refined carbohydrates and decrease in fiber content, is a potent risk factor for development of gallstones. People with diabetes generally have high levels of fatty acids called triglycerides. These fatty acids may increase the risk of gallstones. Gallbladder function is impaired in the presence of diabetic neuropathy, and regulation of hyperglycemia with insulin seems to raise the lithogenic index of bile. Other conditions which predispose to stone formation are sedentary behavior, with low physical activity, injuries of the spinal cord with autonomic nervous system dysfunction, and total parenteral nutrition [3].

5.3 Pathogenesis of Gallstone Disease

Gallstones can be divided on the basis of their chemical composition: pure cholesterol gallstones (which contain at least 90 % cholesterol), mixed (which contain varying proportions of cholesterol, bilirubin, and other substances such as calcium carbonate, calcium phosphate, and calcium palmitate), and pigment (they contain less than 20 % of cholesterol and the main compound is calcium bilirubinate) [100]. Following the epidemiologic studies in America and Europe, cholesterol and mixed concrements predominate; pigment stones are more common in Asian countries, and in Europe they can be found extremely rare [50, 51].

The main mechanisms of cholesterol lithogenesis include biliary cholesterol hypersecretion, supersaturation, and crystallization, stone formation and growth, bile stasis along with mucus hypersecretion, and gel formation within the gallbladder. Cholesterol is virtually insoluble in aqueous solution, but in bile it is made soluble by association with bile salts and phospholipids in the form of mixed micelles and vesicles. The formation of water-insoluble cholesterol crystals is due to a misbalance between the three aforementioned major lipids present in the bile: cholesterol, bile salts, and phospholipids. The ternary phase diagram by Admirand

and Small and later by Wang and Carey has clarified the importance of the relative amounts of bile salts and phospholipids needed to solubilize biliary cholesterol [1, 108, 110]. The cholesterol saturation index (CSI) or lithogenic index has its maximum normal value set at 1 for micellar-phase boundary. Above this boundary, bile becomes supersaturated and the CSI value is higher than 1. Unsaturated bile contains cholesterol that is solubilized in thermodynamically stable simple and mixed micelles. In supersaturated bile, phospholipids solubilize cholesterol into vesicles. Monohydrate crystals can precipitate from these cholesterol-enriched vesicles, become entrapped in gallbladder mucin gel together with bilirubinate (biliary sludge), and ultimately agglomerate into a macroscopic gallstone. However, supersaturated gallbladder bile with cholesterol crystals occurs frequently in healthy individuals, suggesting that microcrystals can be flushed into the duodenum during normal postprandial gallbladder contraction. In patients who form gallstones, pronucleating factors such as biliary glycoproteins and mucin, as well as impaired gallbladder, allow these microcrystals to be retained and to eventually grow over months or years into macroscopic gallstones [83].

Bile contains few proteins, which promote crystallization and gallstone formation, i.e., mucus glycoproteins – mucins. Gallbladder mucins are overproduced and act as pronucleating factors in gallstone disease. Mucins are found within the insoluble matrix of gallstones. In supersaturated model bile, they accelerate the nucleation of cholesterol monohydrate crystals. In animals fed a lithogenic diet, an increase in the synthesis and secretion of gallbladder mucins occurs before crystal formation, and, subsequently, crystals grow predominantly within a mucin gel that accumulates on the gallbladder wall. Likewise, in humans (e.g., in morbidly obese subjects) who develop gallstones during rapid weight loss, the concentration of mucins in gallbladder bile increases before the appearance of crystals.

Gallbladder hypomotility may be a key factor in the pathogenesis of cholelithiasis because it allows time for cholesterol microcrystals to precipitate from lithogenic bile that is supersaturated with cholesterol [73, 82, 92]. The impaired gallbladder motility observed in vivo in a subset of patients with gallstones correlates well with the decreased in vivo contractility of both gallbladder strips and isolated gallbladder smooth muscle cells [16, 112].

Despite the fact that cholesterol gallstones arise in the presence of lithogenic bile, supersaturation itself is not a sufficient factor for lithogenesis. Most of individuals with supersaturated bile do not develop lithiasis, because the time necessary for cholesterol crystallization (nucleation) and growth of the gallstone is longer than time of bile retention in the gallbladder. Risk of cholelithiasis increases in all conditions associated with gallbladder dysmotility such as pregnancy, obesity and rapid weight loss in obese patients, diabetes mellitus, and total parenteral nutrition. Acromegalics are at high risk for gallstone formation during treatment with the somatostatin analogue octreotide, which suppresses postprandial CCK release and gallbladder contractility [62]. A similar effect was achieved during studies on mice with genetic deletion of CCK-1 receptor – it evoked gallbladder stasis, increasing the risk of gallstone formation [109]. A primary role for gallbladder motility in gallstone formation is also supported by the observation that daily CCK injection during total parenteral nutrition or inclusion of dietary fat to enhance CCK release during rapid weight loss restores gallbladder contractility and can prevent gallstone formation [95].

It is still debatable whether gallbladder dysmotility is a primary factor in cholesterol gallstone disease or secondary to inflammation and excess cholesterol accumulation in gallbladder smooth muscle and whether perhaps other factors account for gallbladder hypomotility.

5.4 Telocytes in the Human Gallbladder

5.4.1 Interstitial Cajal-Like Cells and Telocytes

In recent years, our understanding of the physiology and regulatory mechanisms of smooth muscle tissue and the role of interstitium has been enhanced by the study of a population of newly described cells, the so-called interstitial Cajal-like cells (ICLCs). Such name was adopted, as these cells were thought to represent the phenotypic similarity (at least by means of immunohistochemistry, as they expressed c-Kit marker – CD117) to the archetypal, enteric interstitial cell of Cajal (ICC). Characteristic feature of ICLCs was the small, triangular-shaped body, and the presence of several very long prolongations. Multiple research teams have identified such cells in various tissues outside of the gut [20], including the pancreas [65], lungs [78], heart [37], skin [15], skeletal muscle [80], parotid gland [64], meninges and the choroid plexus [64], ureter, urethra, urinary bladder [36, 115], kidney [85], blood vessels [14], male [19, 35] and female reproductive organs [21, 22, 25, 43], mammary glands [39, 75, 79], and placenta [97]. During the last 5 years Prof. Popescu with his teammates examined such cells, and little by little, it became clear that the ultrastructure of ICLCs was different from that of ICCs and that the difference between these cells was not only semantic, as they have different ultrastructural and immunophenotypic patterns, and therefore these cells should be functionally distinct as well [77]. To distinguish ICLCs and to avoid confusion from other interstitial cell type, i.e., fibroblasts, fibrocytes, fibroblast-like cells, or mesenchymal cells, a new name was coined as telocytes (TCs) to replace ICLCs, according to the morphological characteristics through immunohistochemistry and electron microscopy [77]. Because in Greek affix "telos" means "goal," "end," and "fulfillment," the extremely long but thin prolongations of TCs were named as telopodes (Tps). The concept of TCs has been currently adopted by other laboratories. Telocytes were recently discovered in the wall of the human gallbladder by Hinescu et al. [45] and in bile ducts by Ahmadi et al. [2]. To date, there has been no direct evidence that TCs are directly involved in the regulation of gallbladder motility. However, a decrease in the density of TCs in the muscle layer of the gallbladder could be hypothetically related to bile stasis and thus might contribute to gallstone formation, because TCs are suggested to be involved in signaling processes [15, 37].

5.4.2 *Morphology of Telocytes*

TCs are proven to be different from fibroblasts and mesenchymal stem cells in terms of gene profile and proteomics and even in microRNA expression. They have principle morphological characteristics in contrast with other types of cells [66]. Currently, the most demonstrative and widespread methods for identification of TCs include transmission electron microscopy (TEM), immunohistochemistry, and immunofluorescence. TEM is considered the most accurate method for identifying telocytes [87].

General feature of telocyte is a small, oval-shaped cellular body, containing a nucleus and surrounded by a low amount of cytoplasm. The cellular body average dimensions are, as measured on EM images, 9.39 ± 3.26 μm (min = 6.31 μm; max = 16.42 μm). The nucleus occupies approximately 25 % of the cell volume and contains clusters of heterochromatin attached to the nuclear envelope. The perinuclear cytoplasm is rich in mitochondria (which occupy about 5 % of the cell body and are also particularly presented in podoms) and contains a small Golgi complex, as well as elements of rough and smooth endoplasmic reticulum and cytoskeletal elements (thin and intermediate filaments) [6, 21, 23]. The cell periphery is represented by a usual plasmalemma, with no (or thin and discontinuous) basal lamina and caveolae (about 2–3 % of cytoplasmatic volume; ~0.5 caveolae/μm of cell membrane length).

The shape of the telocytes is related to the number of their telopodes: piriform for one prolongation, spindle for two telopodes, triangular for three, stellate, etc. Presumably, their spatial appearance would be that of a polyhedron with a different number of vertices, depending on their telopode number [23, 33, 77].

TCs have a variable number of telopodes (Tps) (very long cellular extensions), which are probably the longest cellular prolongations in the human body. Tps are made by an alternation of dilated portions, named podoms (250–300 nm), containing mitochondria and endoplasmic reticulum and podomers (~80 nm) with thin segments. The main characteristics of Tps:

1. Number: 1–5, frequently only two to three telopodes are observed on a single section, depending on site and angle of section, since their 3D convolutions impede them to be observed at their full length in a 2D very thin section.
2. Length: tens – up to hundreds of μm, as measured on EM images. However, under favorable condition in cell cultures, their entire length can be captured.
3. Thickness: uneven caliber, mostly below 0.2 μm (resolving power of light microscopy), visible under electron microscopy, only 0.10 μm ± 0.05 μm (min = 0.03 μm; max = 0.24 μm).
4. Moniliform aspect, with many dilations along.
5. Presence of "Ca2+-release units" at the level of the dilations, accommodating mitochondria, elements of endoplasmic reticulum and caveolae.
6. Branching, with a dichotomous pattern.
7. Organization in a network – forming a labyrinthine system by tridimensional convolution and overlapping, communicating through gap junctions.

These characteristic features make Tps clearly different from neuronal dendrites, processes of antigen-presenting dendritic cells or fibroblasts and myofibroblasts [26, 77].

Popescu and Enciu experimentally demonstrated that TC morphology can be changed by aging or by modifying the redox balance of the cell culture environment. Oxidative stress impaired the ability of TCs to form Tps and the migration pathway length. Aging further aggravated this effect [31]. This should be considered in clinical studies, since the age of patients itself might be an independent factor influencing TC morphology.

Furthermore, TCs demonstrate specific direct (homocellular and heterocellular junctions) and/or indirect (chemical, paracrine/juxtacrine signaling, microvesicles and exosomes, sex hormone, and microRNAs) contacts with various surrounding cells [23, 26, 33].

Tps are connected to each other by homocellular junctions and appear to form a 3D network in the interstitial space at the border of smooth muscle cell bundles and contain cytoskeleton elements. Connection between TCs-exosomes-gap junctions-cytoskeleton seems to be the equivalent of the primitive nervous system [13, 22, 96, 114].

Heterocellular nanocontacts were frequently described between TCs and myocytes or TCs and immune cells. Their contacts with mast cells provide a reason to predict participating in immunoreactions (mastocyte-mediated immunoregulation/immunosurveillance). Moreover, various studies have revealed that this class of cells plays an ambiguous role in immune response. During the physical process, TCs can be activated to maintain homeostasis to induce proliferation, differentiation, and tissue regeneration. On the other hand, they initiate the tissue inflammation to induce pathogenesis under some challenges [111].

TCs also surrounded stem cell niches with telopodes and heterocellular contacts. In addition, they establish physical contacts with nerve endings, blood vessels, and different types of progenitor cells. Accumulating studies have shown that telocytes play an indisputable role in neo-angiogenesis. As TCs have cytoskeleton elements (myosin-14, periplakin), they could be responsible for detecting the smooth muscle cell stretch [18, 21, 38, 114].

TCs release at least three types of extracellular vesicles, exosomes (45 ± 8 nm), ectosomes (128 ± 28 nm), and multivesicular cargos (MVCs; 1 ± 0.4 μm), from their Tps and, occasionally, from the cell body [87]. Specifically, extracellular vesicles and/or exosomes are shed or released from Tps in uterine telocytes. Mediators, such as IL-6, vascular endothelial growth factor (VEGF) and nitric oxide, are secreted from TCs. Growth factors, including IL-6, VEGF, macrophage inflammatory protein 1α (MIP-1α), MIP-2, and monocyte chemoattractant protein 1 (MCP-1), are significantly expressed along with additional cytokines, including IL-2, IL-10, IL-13, and chemokines, such as growth-related oncogene/keratinocyte-derived chemokine (GRO-KC), in the secretome of cultured rodent cardiac TCs [17]. Besides, in a renal ischemia-inflammatory injury model, TCs activated nuclear factor kappa B (NF-kB) signaling pathway and upregulated the mRNA levels of pro-inflammatory cytokines such as IL-1 and TNF-α [17, 56].

Telocytes are also involved in the electrical modulation of excitable tissue, such as the smooth muscle of the gut and uterus. Assuredly, they could spontaneously initiate electric activity and play a role in modulation of glandular and immune activity [29]. According to literature, ion channels, such as T-type calcium and small-conductance calcium-activated potassium channels, are present in TCs [23].

5.4.3 Specific Features of the TCs in Gallbladder

The first detailed description of TC in gallbladder was presented by Hinescu et al. [45]. He followed the original visualization methods described by Santiago Ramón y Cajal [12].

5.4.3.1 Methylene Blue Staining

Methylene blue staining was the first step in identifying cell profiles in a stroma [12].

As shown in Fig. 5.1a, selectively stained cells in fresh tissue samples in the human gallbladder possess the characteristic features of TCs – positioning in interstitium, network-like appearance, and characteristic thin, very long, and moniliform cell prolongations. In specimens undergoing methylene blue vital staining followed by cryosectioning, Hinescu et al. approximated the count of TCs about from 100 to 110 cells per 1 mm^2. He also reported the dual distribution of TC in the human gallbladder: in the close vicinity of the epithelium and in the interstitial spaces between bundles of smooth muscle fibers. Sometimes "beads" of the moniliform cytoplasmic processes were visible (figures nr add).

5.4.3.2 Toluidine Blue Staining

In the specimens stained with toluidine blue, Hinescu et al. estimated TC localization, quantity, or some morphological details (e.g., cytoplasmic processes). In the human adult gallbladder, TCs are mainly placed near small vessels (Fig. 5.1b, c), in the subepithelial region of lamina propria and between smooth muscle bundles in muscularis. A combined analysis on semithin and ultrathin section enabled to determine the relative proportion of TC in the subepithelial and muscularis interstitium (about 7 %, and ~5 %, respectively). Hinescu et al. estimated that gallbladder TCs represent ~5.5 % of subepithelial cells and such cells were similar to those present in other organs as reported by other authors [10, 18, 42, 52, 60, 79, 86, 89], with characteristic cell shapes and with a long, thin process.

5.4.3.3 Electron Microscopy

Transmission electron microcopy (TEM) images obtained by Hinescu et al. [45] reveal the presence of two subpopulations of gallbladder TCs – located in the

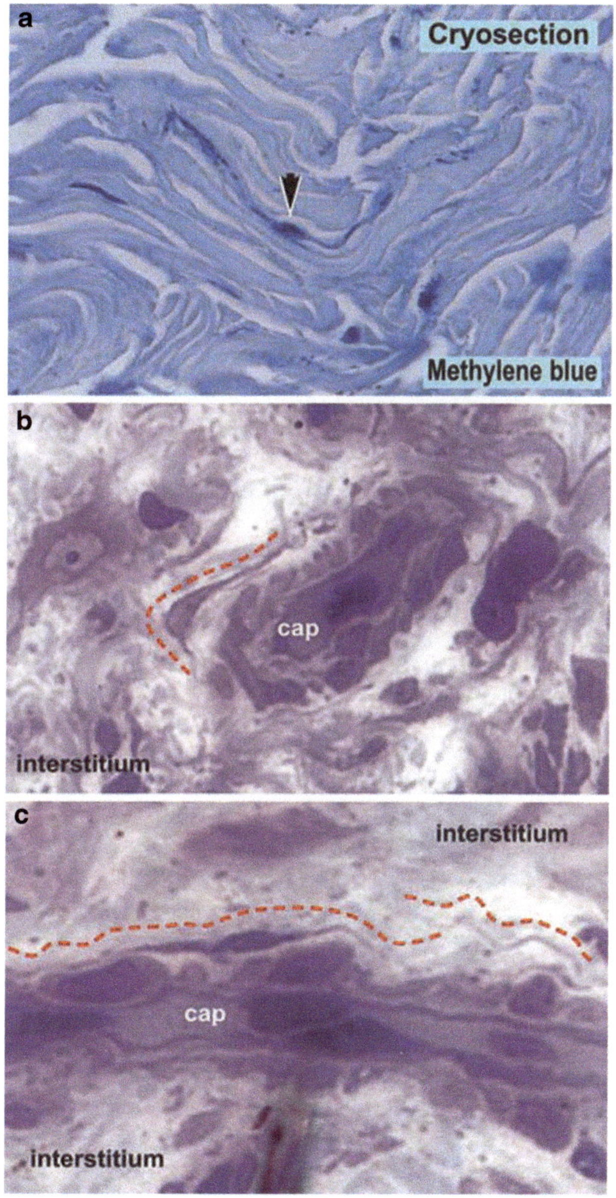

Fig. 5.1 Adult human gallbladder stained with methylene blue (**a**). Human gallbladder semithin sections stained with toluidine blue (nonconventional light microscopy). Interstitial cells having telocyte morphology, with thin cellular processes, are illustrated. Cellular processes embrace cross-sectioned (**b**) or longitudinally sectioned small diameter vessels (**c**). Long cellular processes running between smooth muscle bundles are suggestive for telocytes. Original magnification 40. *Red dot lines* follow the telocyte profile in order to make it more evident (From Hinescu et al. [45])

subepithelial region and between muscular bundles – with similar ultrastructural features: (1) location in the non-epithelial space, (2) close contacts with targets (nerve bundles and/or epithelia, and/or smooth muscle cells, and/or capillaries), and (3) characteristic cytoplasmic processes: (a)Number (1–3, frequently: 2–3); (b) Length (tens up to hundreds of nm); (c) Thickness (uneven caliber, <0.5 nm); (d) Aspect: moniliform, usually with mitochondria in dilations; (e) Presence of "Ca2+ release units"; (f) Branching: dichotomous pattern; (g) Organization in network—labyrinthic system: overlapping cytoplasmic processes; (4) Gap junctions: with smooth muscle cells or with each other; (5) Basal lamina: occasionally present; (6) Caveolae: 2–3 % of cytoplasmic volume; ~0.5 caveolae/μm of cell membrane length; (7) Mitochondria: 5–10 % of cytoplasmic volume; (8) Endoplasmic reticulum: about 1–2 %, either smooth or rough; (9) Cytoskeleton: intermediate and thin filaments, as well as microtubules, present; (10) Myosin thick filaments: undetectable. Most of these specific features of TCs are presented in Fig. 5.2. Such characteristic ultrastructural patterns of gallbladder TCs fulfill morphologic criteria proposed for TCs present in the other organs.

5.4.4 Immunohistochemical Profile of Telocytes

As any type of cells, TCs have their own immunohistochemical profile. Current studies demonstrate expressions or (co-)expression of different markers in gallbladder telocytes, which in the same time not peculiar for detection in a given organ. TCs are immunohistochemically positive for CD 34, CD117/c-Kit, plated-derived growth factor receptor alpha and beta (PDGFRα and PDGFRβ), VEGF, inducible nitric oxide synthase (iNOS), calveolin-1, vimentin, connexin 43, estrogen and progesterone receptors, CD44, desmin, nestin, and cadherin-11. In addition, they are immunohistochemically negative for procollagen-1, CD31/PECAM-1 (endothelial cells), α-SMA (myofibroblasts, pericytes, and vascular SMCs), CD11c (dendritic cells and macrophages), CD90/Thy-1 (fibroblasts), and, sometimes, c-kit/CD117 (mast cells). For instance, Tps of TCs might be positive for vimentin, a cytoskeleton protein [13, 21, 24, 25, 28, 76, 97, 105]. TCs are also negative for CD68 and other markers associated with immune functions (CD1a, CD62-P), suggesting a clear difference between TCs and macrophages [21] (Figs. 5.3 and 5.4).

5.4.5 Immunohistochemistry of the Gallbladder TCs

The density of the immunohistochemical staining differs according to the localization of TCs in the gallbladder wall. Immunohistochemistry with using of antibodies against CD117/c-kit revealed a high density of TC, mainly in the human adult gallbladder lamina propria and mast cells. CD117/c-kit expression was detected in cells with characteristic morphology susceptible to classification as TCs (one or more very long, thin processes, sometimes with a beads-on-string appearance, on spindle-shaped cell bodies). In the adult human gallbladder lamina propria, TCs expressed CD34 at high levels, as well as the PDGFRα. CD117/CD34 double immunohistochemical reaction is also positive: CD117 mainly stains the cell body, while CD34 preferentially stains the cell processes. Almost a majority of TCs, from lamina propria and muscularis, are negative for a-SMA. In contrast, constant positive immunostaining is

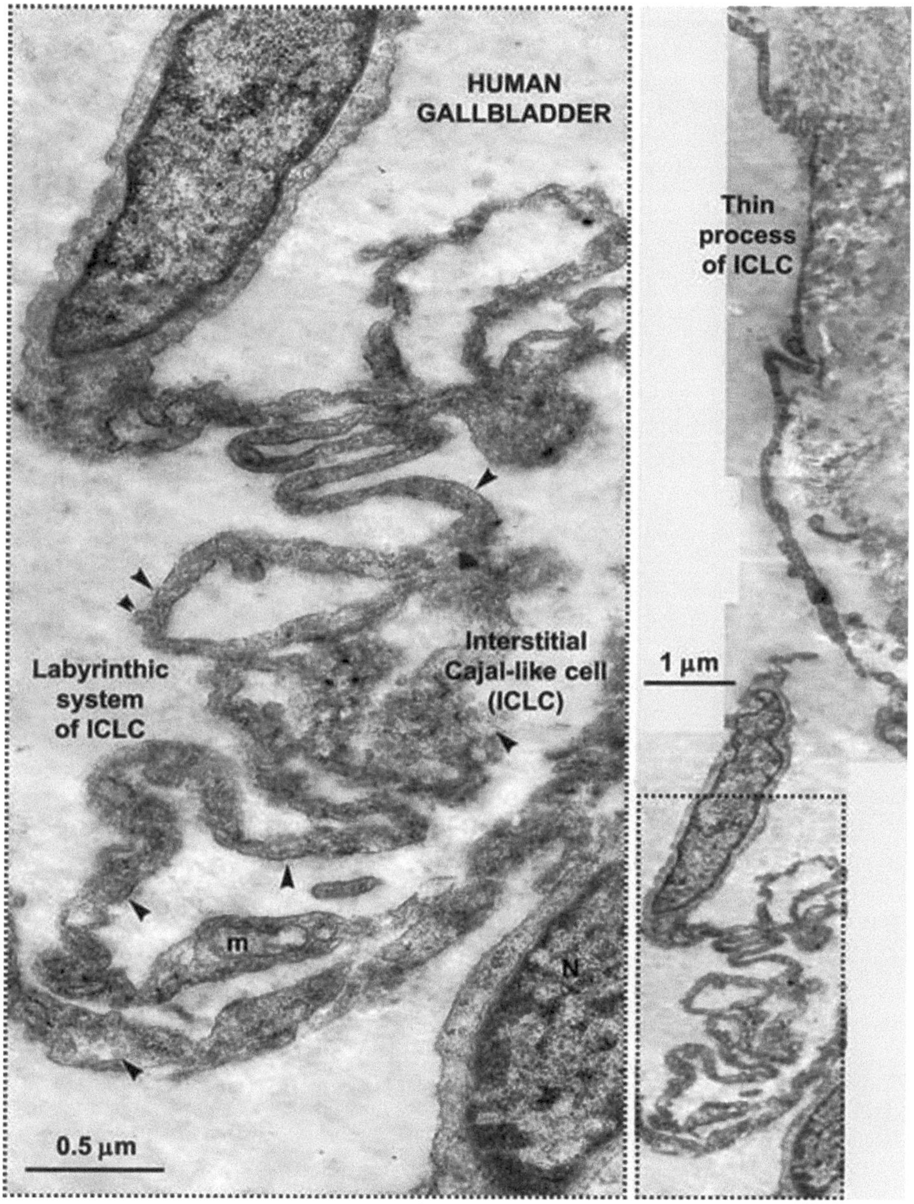

Fig. 5.2 Adult human gallbladder: *TEM* original magnification, 40,000. Photographic reconstruction illustrating the cell shape and a typical telocyte with long, thin, elaborate branching system of cell processes. *Left image* presents at a higher magnification area delimited by the thin border in the photographic reconstruction. *N* nucleus, *m* mitochondria, *arrowheads* = caveolae (From Hinescu et al. [45])

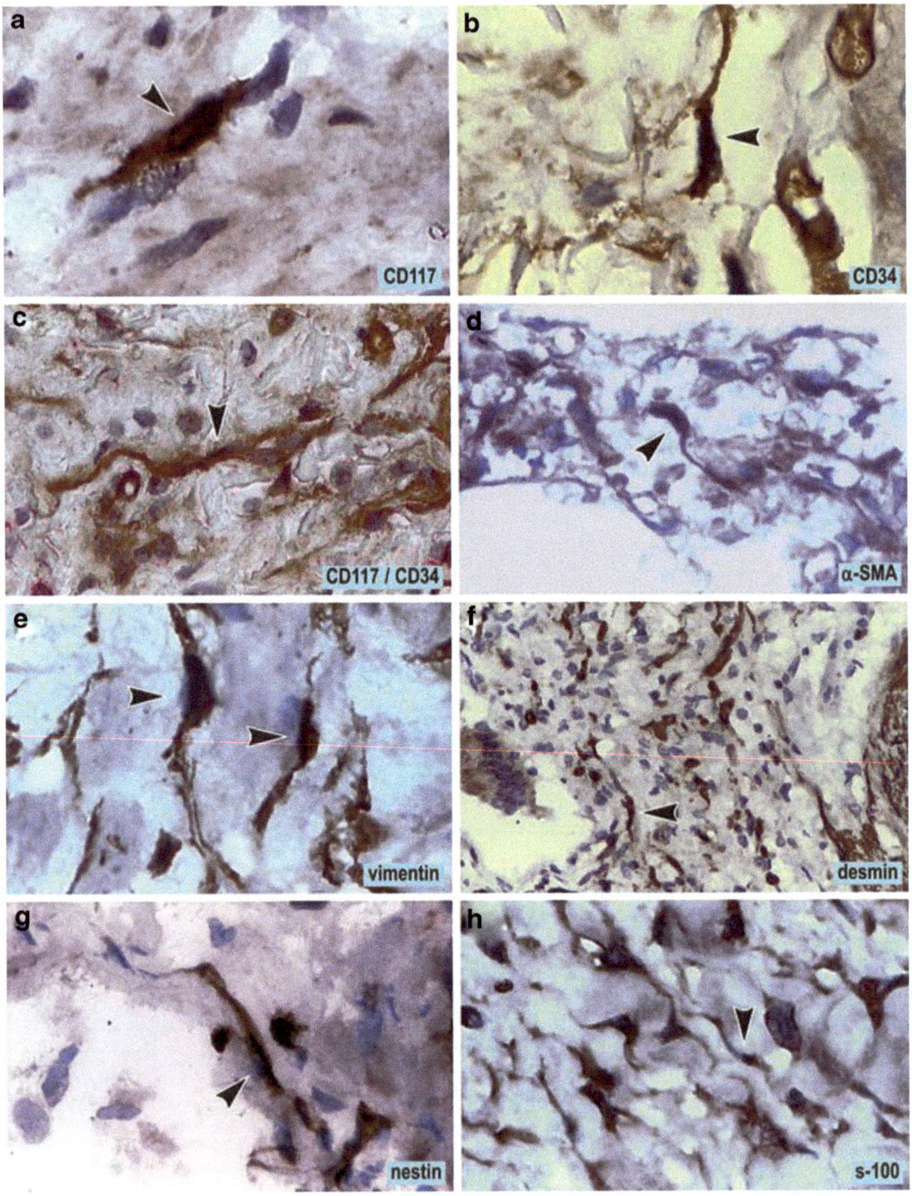

Fig. 5.3 Immunocytochemical reactions for CD117 (**a**), CD 34 (**b**), double staining for CD117/CD34 (**c**), smooth muscle actin (**d**), vimentin (**e**), desmin (**f**), nestin (**g**), S-100 (**h**). Mayer's hematoxylin counterstaining; original magnification A–E, G, H 40; F 20. *Arrows* indicate cells with a profile suggestive for telocytes (From Hinescu et al. [45])

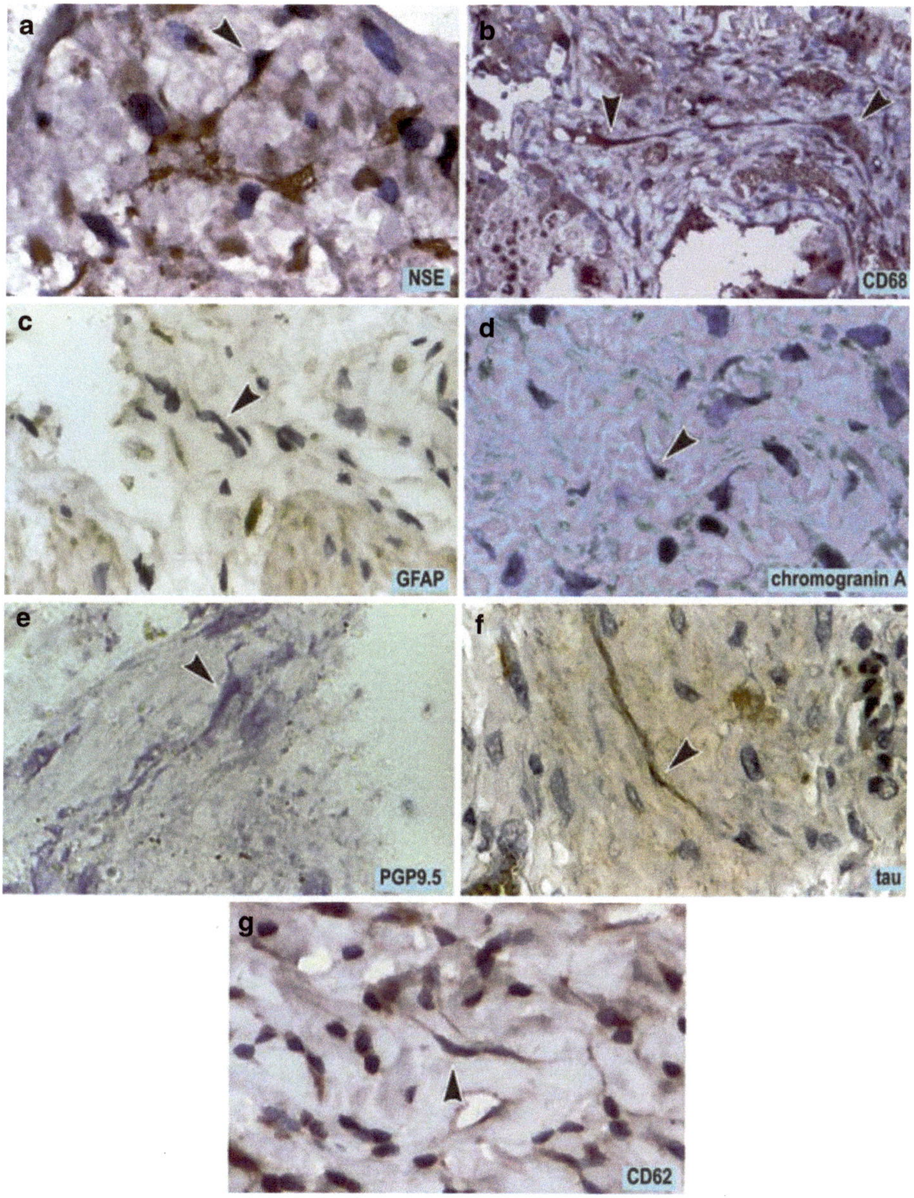

Fig. 5.4 Adult human gallbladder, immunostaining for: NSE (**a**), CD 68 (**b**), GFAP (**c**), chromogranin (**d**), PGP9.5 (**e**), tau protein (**f**), CD62P (**g**). Mayer's hematoxylin counterstaining; original magnification 40 (From Hinescu et al. [45])

observed for vimentin and nestin, and scattered cells were found to be positive for desmin. Some additional antigens showed inconstant or weak immunostaining for S-100, NSE, but clear negative immunostaining for CD68, GFAP, chromogranin A, and CD62 [45]. In contrast, tau protein was constantly positive. The observations from PGP9.5 immunostaining are disputable – in some specimens double labeling of TCs with characteristic morphology using PGP9.5/c-Kit antibodies reveals double-positive cells. These data support the distinction between ICLC and other cells present in interstitium or adjacent layers (fibroblasts, pericytes, macrophages, mast cells, smooth muscle cells, neuroendocrine cells, neural cells).

5.5 TCs in Gallstone Disease

Multiple factors are responsible for gallstone formation. Regardless of cholesterol supersaturation, hydrophobic bile salts, pronucleating proteins, and mucus hypersecretion with gel formation in the gallbladder as described previously [4, 83, 103, 107], gallbladder dysmotility may be a "triggering" event in the pathogenesis of cholesterol gallstones, providing the time necessary for the precipitation of cholesterol microcrystals from bile supersaturated with cholesterol and their subsequent growth to macroscopic stones [91, 103]. Gallbladder motility involves multiple regulatory mechanisms, including smooth muscle and enteric nervous circuit activity as well as, possibly, the recently described gallbladder interstitial cells – telocytes.

So far the published data on TCs in the human gallbladder remained limited. Hinescu et al. [45] (discussed widely above) reported that TCs in the human gallbladder typically appeared individually or in small clusters of two to three cells. The TCs were located in the *lamina propria*, with some very close to the epithelium, and in the connective tissue spaces between bundles of smooth muscle cells. TCs accounted for 5.5 % of the subepithelial cells in the gallbladder wall. Meanwhile, Ahmadi et al. [2] identified c-Kit-positive cells with characteristic morphology in the subepithelial and muscular layers of the gallbladder and extrahepatic bile ducts (which are denser than those of the gallbladder) running parallel to circular smooth muscle fibers, forming a cellular network. These authors did not find TCs in intrahepatic bile ducts. Further studies [59, 68, 69] brought more facts on the role of the gallbladder TCs.

5.5.1 Histopathological Findings in Patients with Cholelithiasis

As reported in studies by [59, 69], the histopathological evaluations showed chronic cholecystitis of varying intensity in patients with gallstones. The inflammation was assessed and designated as mild, moderate, or severe. In the study group, severe inflammation was predominant (Fig. 5.5). In contrast, only mild to intermediate inflammation was observed in the control group. The thickness of the gallbladder

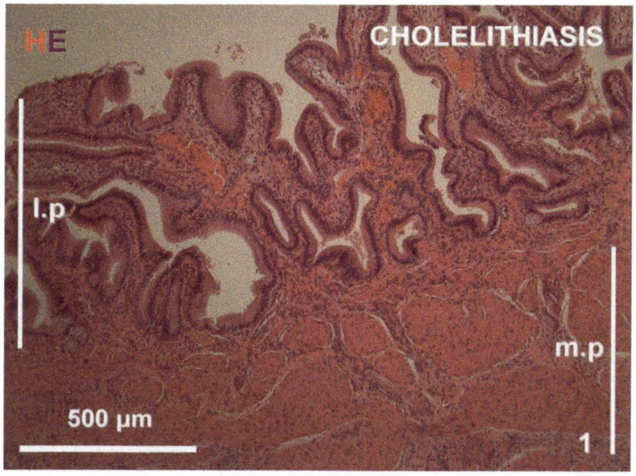

Fig. 5.5 Cross section of the gallbladder wall from the cholelithiatic group showing mild infiltration with inflammatory cells and a thickened muscular layer. H&E staining. *l.p.* lamina propria, *m.p.* muscularis propria (From Matyja et al. [59])

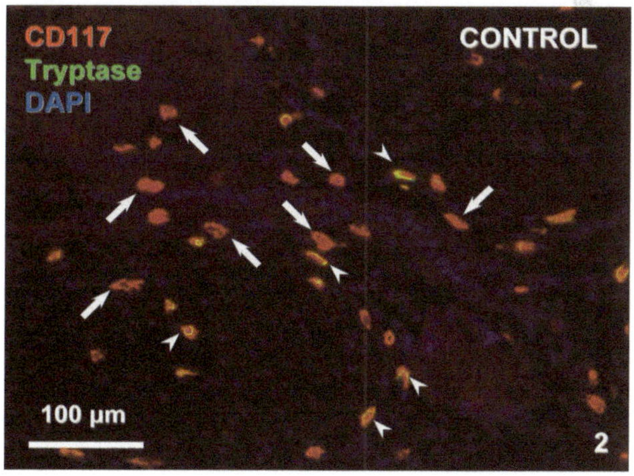

Fig. 5.6 Cross section of the gallbladder wall of a control patient stained for CD117 (*red*) and tryptase (*green*). The nuclei are counterstained with DAPI (*blue*). CD117-positive/tryptase-negative TCs (*arrows*), and CD117-positive/tryptase-positive mast cells (*arrowheads*) are indicated (From Matyja et al. [59])

muscularis propria was significantly increased in the study group compared with the control group (4,453 ± 597 vs. 3,610 ± 346 µm; $p < 0.01$).

The number of mast cells in the gallbladder wall was evaluated in specimens immunostained for tryptase. Mast cells were present in all layers of the gallbladder wall and predominantly localized to the *lamina propria* (Figs. 5.6 and 5.7). The total number of mast cells in the gallbladder wall was significantly higher in patients with gallstones than in control subjects (224 ± 41 vs. 161 ± 37, respectively; $p < 0.05$). In immunostained slides, c-Kit and tryptase double-positive mast cells were generally round or oval shaped, with a centrally located nucleus.

The c-Kit-positive/mast cell tryptase-negative cells were considered to be telocytes. We found them predominantly located in the corpus, but these cells were also observed in the gallbladder fundus and neck. TCs had a centrally located nucleus and were mostly fusiform in shape with small branches that were visible in some sections; however, sparse, round tryptase/c-Kit-positive cells were also

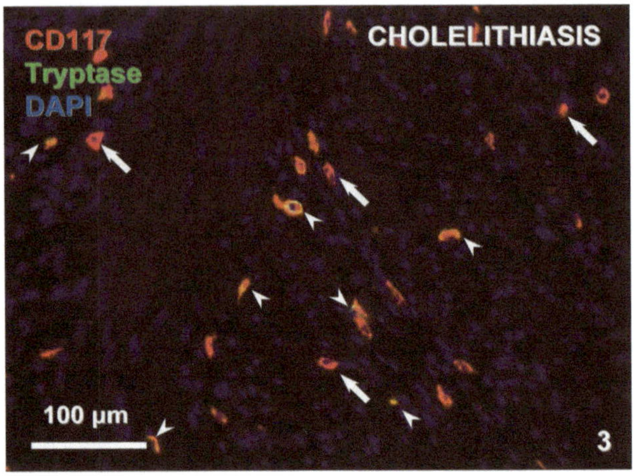

Fig. 5.7 Cross sections of the gallbladder wall of a cholelithiatic patient stained for CD117 (*red*) and tryptase (*green*). The nuclei are counterstained with DAPI (*blue*). CD117-positive/tryptase-negative TCs (*arrows*) and CD117-positive/tryptase-positive mast cells (*arrowheads*) are indicated (From Matyja et al. [59])

present. Numerous TCs were detected, mostly in the *muscularis propria*, and some TCs were observed in the connective tissue separating the smooth muscle bundles (Fig. 5.5).

The number of TCs in the gallbladder wall corpus was significantly lower in the study group than in the control group (3.03 ± 1.43 vs. 6.34 ± 1.66 cell/FOV in the *muscularis propria*; $p < 0.001$) (Figs. 5.6 and 5.7). CD-34-positive cells were visualized in the gallbladder; however, they appeared to be mainly of vascular origin. CD-34-positive cells were rarely encountered in the *muscularis propria*. These CD-34-positive cells were c-Kit negative concurrently.

5.6 On Possible Mechanisms of TC Impairment

In order to elucidate the possible mechanisms leading to TCs depletion in gallbladder, the consecutive studies [59] concentrated on the role of bile properties, including bile salt concentrations. Significant decrease in the mean concentrations of glycocholic and taurocholic acids in the bile from patients with cholelithiasis compared with the controls ($p < 0.02$ and $p < 0.05$, respectively) was reported (Fig. 5.8). Moreover, there was also a positive correlation between the TC count and the concentrations of glycocholic ($r = 0.45$, $p = 0.039$) and taurocholic ($r = 0.32$, $p = 0.05$) acids. No significant differences in the concentrations of the other bile acids examined (glycochenodeoxycholic, glycodeoxycholic, taurochenodeoxycholic, and taurodeoxycholic acid) were observed between the two groups of patients (Fig. 5.9). There were also no significant differences in the concentrations of cholesterol, bile salts, or phospholipids in the bile between the two groups (Fig. 5.10).

However, the calculated cholesterol saturation index (CSI) [108] was significantly higher in patients with cholecystolithiasis (1.23 ± 0.84) than in the controls

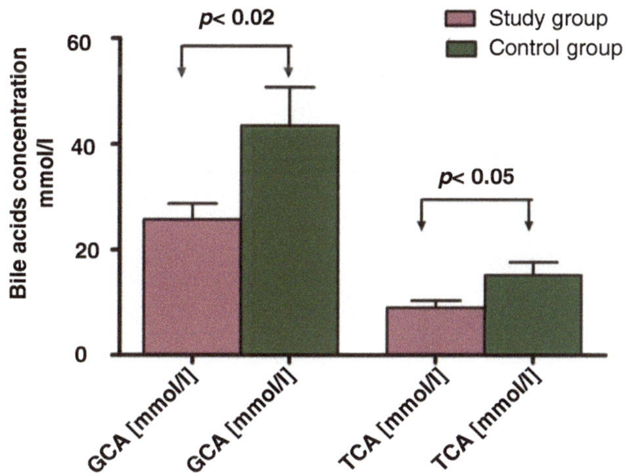

Fig. 5.8 Concentrations of glycocholic acid (*GCA*) and taurocholic acid (*TCA*) in gallbladder bile from patients with cholecystolithiasis ($n=24$) and controls ($n=25$). The data are expressed as the mean values ± SD (From Matyja et al. [59])

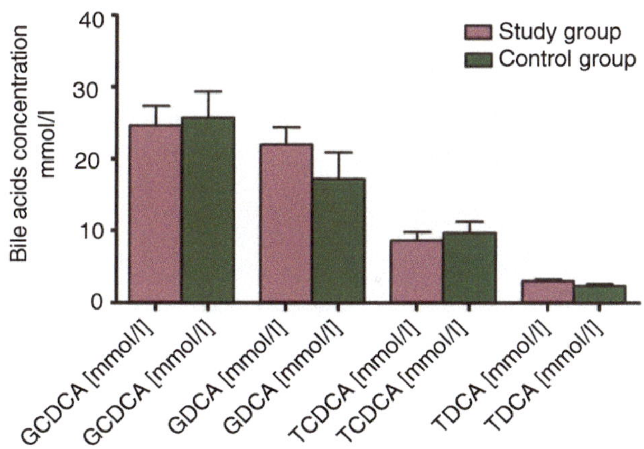

Fig. 5.9 Bile acid concentrations of glycochenodeoxycholic acid (*GCDCA*), glycodeoxycholic acid (*GDCA*), taurochenodeoxycholic acid (*TCDCA*), and taurodeoxycholic acid (*TDCA*) in gallbladder bile from patients with cholecystolithiasis ($n=24$) and controls ($n=25$). The data are expressed as the mean values ± SD (From Matyja et al. [59])

(0.78±0.33) ($p<0.05$). These results revealed also an important negative correlation ($r=-0.62$, $p=0.001$) between the CSI and the TC count in the gallbladder wall (Fig. 5.11a, b).

Even under physiological conditions, the cholesterol content in the aqueous solution of gallbladder bile is relatively high because cholesterol is incorporated into mixed micelles, together with bile salts and phospholipids. When the bile cholesterol levels increase or the secretion of solubilizing bile salt is diminished, the solution becomes supersaturated. In this case, excess cholesterol is stored in vesicles (spheres composed of cholesterol and phospholipids, with no bile salts), provided that enough phospholipid is available. When relatively low amounts of phospholipids are present, cholesterol crystal formation occurs, leading to gallstone formation [82, 84, 103]. Moreover, cholesterol crystallization is also promoted by hydrophobic bile salts (e.g., chenodeoxycholate and deoxycholate) and phospholipids with unsaturated acyl chains [63]. Human bile contains a mixture of both hydrophobic

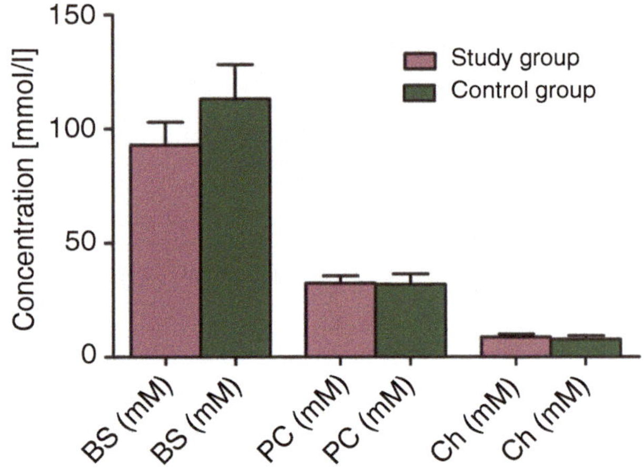

Fig. 5.10 Concentration of total bile acids (*BS*), phospholipids (*PC*), and cholesterol (*Ch*) in gallbladder bile from patients with cholecystolithiasis ($n = 24$) and controls ($n = 25$). The data are expressed as the mean values ± SD (From Matyja et al. [59])

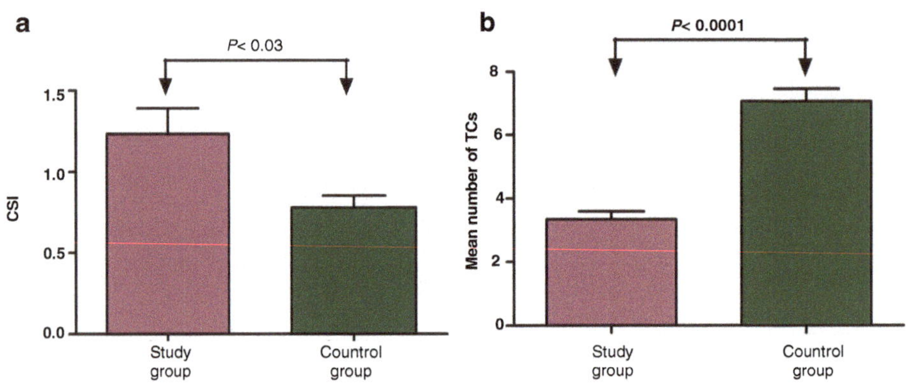

Fig. 5.11 Cholesterol saturation index (*CSI*) (**a** – *left part*) and the number of TCs (**b** – *right part*) in patients with cholecystolithiasis ($n = 24$) and controls ($n = 25$). The data are expressed as the mean values ± SD (From Matyja et al. [59])

(e.g., deoxycholate) and hydrophilic (e.g., ursodeoxycholic) acids, and thus, the final hydrophobicity of the bile salts depends on the relative contents of these components [44, 47, 72]. Therefore, the biliary bile salt composition affects human gallstone formation, which opens several therapeutic possibilities [48, 49, 55, 102]. Our results showed that there were no significant differences between the two groups of patients in the mean concentrations of total bile acids, phospholipids, or cholesterol in vesicular bile, except for the lower mean concentrations of glycocholic and taurocholic acid in the patient group. The lower amounts of these mostly hydrophilic bile acids were associated with the increased lithogenicity index (CSI) in the study group. Interestingly, in patients with gallstones, a significant positive correlation between the mean number of TCs and the concentrations of glycocholic and taurocholic acids was found. We conclude that glycocholic and taurocholic acids are somehow protective to TCs. However, it is unclear whether this is an artifact of the statistical analysis or whether TCs are indeed preserved by these acids, and the exact mechanism of this possible protective effect should be examined further, including in an experimental model.

We acknowledge that possible mechanisms underlying the destructive influence of bile on TCs remain speculative rather than empiric, as reports on the role of TCs in gallstone pathophysiology are sparse. Some insights were provided by Hu et al. [46] who demonstrated that the expression of c-Kit mRNA and protein in the gallbladder wall was significantly decreased in the gallbladders of guinea pigs fed a high-cholesterol diet. A study by Xu and Shaffer [112] reported that gallbladder hypomotility was impaired by the increased bile cholesterol level. A study by Lavoie [54] proved that excess cholesterol in the smooth muscle of the gallbladder attenuates the ability of the muscle to contract as a result of changes in signal transduction and ion channel activity, decoupled membrane receptor-ligand interactions, and disturbances in contractile protein activity. Moreover, in a subsequent study on guinea pigs fed a lithogenic diet, Lavoie et al. [53] reported cholesterol accumulation in gallbladder smooth muscles in the plasma membrane, especially membrane caveolae, leading to a decrease in membrane fluidity and a subsequent change in rhythmic electric activity. Hypertrophy was also detected in the *muscularis propria*, and the contractile response to an agonist was decreased. It is not yet known whether such mechanisms could influence TC activity as well. Furthermore, TCs seem not to be directly involved in the pacemaking mechanism but rather possibly act as modulators of the contractility, e.g., through the released exosomes/shedding vesicles [81]. Elements of the TC network are interacting with each other (homocellular connections) as well as with other cell types (heterocellular connections). As reported in the heart, the homocellular junctions occur at both podomeric and podomic levels, either side to side (presumably for exchanging information) or end to end (probably for relaying, passing on information). Heterocellular junctions are encountered between TCs and myocytes, fibroblasts, mast cells, macrophages, pericytes, endothelial, or Schwann cells. TCs are integrating such cellular types into a complex 3D network. These networks are providing both structural and functional support for long-distance signaling [5, 37, 74].

Another important mechanism underlying TC loss concerns the chronic inflammatory processes involving the gallbladder wall. Portincasa et al. [82] described impaired gallbladder motility caused by mild inflammation. Indeed, we observed inflammatory infiltration, predominantly localized in the *lamina propria*, in the gallbladders of patients with gallstones, which was associated with a significant increase in the mast cell count. However, as we reported previously, TC loss does not correlate with inflammatory grade or mast cell count [69].

Nevertheless, other indirect inflammatory effects on gallbladder TCs cannot be excluded. For example, apoptotic mechanisms leading to the loss of gallbladder TCs should be considered. The apoptosis might be caused by multiple mechanisms, including chronic inflammatory reactions [40], and it is possible that the reduced number of TCs in the inflamed gallbladder wall in patients with gallstones could be caused by an imbalance between the apoptosis and regeneration of TCs. Finally, the reduced number of TCs might be primarily caused by the trans-differentiation of TCs precursor cells, as reported by Torihashi et al. in the smooth muscles in the gut [99].

Furthermore, the recent studies on the role of unsaturated fatty acids (UFA) in bile phospholipids fraction in relationship to TCs performed by our team revealed that in patients with cholelithiasis, the levels of UFA differ significantly between healthy subjects and patients with cholelithiasis. Moreover, such changes are statistically related to the decreased TCs count in the gallbladder muscularis propria (unpublished).

Saturated fatty acid (SFA) concentrations in the phospholipid fraction of the bile in gallstone patients were slightly decreased as compared with control (47.7 % ± 3.0 vs. 50.7 % ± 7.5; $p = 0.07$, respectively). The monounsaturated fatty acid (MUFA) concentrations in the phospholipid fraction of the bile in gallstone patients remained unchanged between both groups (13.7 % ± 2.5 vs. 15.2 % ± 54.1; $p = 0.17$). Surprisingly however, the polyunsaturated fatty acid (PUFA) concentrations in the phospholipid fraction of the bile in gallstone patients were significantly higher in patient suffering from cholelithiasis compared with the control subjects (39.6 % ± 4.7 vs. 34.0 % ± 6.8; $p = 0.0046$). The omega-3 PUFA levels did not differ between both groups (3.68 % ± 1.1 for gallstone patients, 3.79 % ± 1.6 for controls, respectively, $p = 0.81$), whereas the omega-6 PUFA concentrations were significantly elevated in gallstone patients (35.8 % ± 4.3 for gallstone patients, 30.2 % ± 6.0 for controls, $p = 0.0015$).

Such increase originated mainly from the significant elevation of the C18:2(n-6) (linoleic acid) serum concentrations (28.8 % ± 4.8 vs. 22.8 % ± 4.9, respectively, for cholelithiasis and controls; $p = 0.0005$). In addition, the omega-6 PUFA to omega 3 PUFA ratio was significantly higher in gallstone patients compared to controls (11.0 ± 3.1 vs. 8.41 ± 2.3; $p = 0.0121$).

The study revealed also the moderate correlation of telocyte count in the gallbladder muscularis propria with the (1) total PUFA concentrations ($r = -0.543$; $p < 0.05$), (2) omega-6 PUFA levels ($r = -0.415$; $p < 0.05$), and (3) omega 6- PUFA to omega-3 PUFA ratio ($r = -0.559$; $p < 0.05$).

Thus, we conclude that the disrupted polyunsaturated fatty acid concentrations in the gallbladder bile, with elevation of omega-6 PUFA, constitute important factors influencing TC density in the gallbladder wall, being one of the possible pathophysiological components for the gallstone disease development.

5.7 Conclusions

The specific role of TCs in the gallstone disease reminds relatively incompletely described. There are still more questions rising than we have answers available. But whether the telocytes regulate the smooth muscle functions, or are the crucial element for their regeneration, or play a role in the cellular communication, or can act as the sensor elements, the importance of further investigation of TC ultrastructure and electrophysiological studies is indisputable. But despite the fact that many of pathophysiological mechanisms of gallstone formation still need to be elucidated, we can surely conclude that TCs are the important players in cholelithiasis and a possible target for the novel therapies.

Acknowledgments The authors acknowledge the involvement of many colleagues who have participated in our studies and contributed with helpful discussions.

References

1. Admirand WH, Small DM. The physicochemical basis of cholesterol gallstone formation in man. J Clin Invest. 1968;47:1043–52.
2. Ahmadi O, Nicholson Mde L, Gould ML, Mitchell A, Stringer MD. Interstitial cells of Cajal are present in human extrahepatic bile ducts. J Gastroenterol Hepatol. 2010;25:277–85.
3. Amigo L, Zanlugo S, Mendoza H. Risk factors and pathogenesis of cholesterol gallstones: state of the ART. Eur Rev Med Pharmacol Sci. 1999;3:241–6.
4. Apstein MD, Carey MC. Pathogenesis of cholesterol gallstones: a parsimonious hypothesis. Eur J Clin Invest. 1996;26:343–52.
5. Barile L, Gherghiceanu M, Popescu LM, Moccetti T, Vassalli G. Ultrastructural evidence of exosome secretion by progenitor cells in adult mouse myocardium and adult human cardiospheres. J Biomed Biotechnol. 2012; Article ID 354605, 2012. doi:10.1155/2012/354605.
6. Bei Y, Wang F, Yang C, Xiao J. Telocytes in regenerative medicine. J Cell Mol Med. 2015;19:1441–54.
7. Bennion LJ, Grundy SM. Risk factors for the development of cholelithiasis in man. N Engl J Med. 1978;299:1161–7.
8. Berci G. Historical overview of surgical treatment of biliary stone disease. In: MacFadyen BV, Arregui M, Eubanks S, Olsen DO, Peters JH, Soper NJ, Swanström LL, Wexner SD, editors. Laparoscopic surgery of the abdomen. New York: Springer; 2004. p. 139–42.
9. Biss K, Ho KJ, Mikkelson BS. Some unique biologic characteristics of the Masai of East Africa. N Engl J Med. 1971;284:694–9.
10. Bobryshev YV. Subset of cells immunopositive for neurokinin-1 receptor identified as arterial interstitial cells of Cajal in human large arteries. Cell Tissue Res. 2005;321:45–55.
11. Bouchier TA. Gallstones. Proc R Soc Med. 1977;70(9):597–9.
12. Cajal RS. Histologie du systeme nerveux de l'Homme et de Vertebres. Grand sympathique, Paris, 1911, Maloine.
13. Campeanu RA, Radu BM, Cretoiu SM, Banciu DD, Banciu A, Cretoiu D, Popescu LM. Near-infrared low-level laser stimulation of telocytes from human myometrium. Lasers Med Sci. 2014;29:1867–74.
14. Cantarero I, Luesma MJ, Junquera C. The primary cilium of telocytes in the vasculature: electron microscope imaging. J Cell Mol Med. 2011;15:2594–600.
15. Ceafalan L, Gherghiceanu M, Popescu LM, Simionescu O. Telocytes in human skin–are they involved in skin regeneration? J Cell Mol Med. 2012;16:1405–20.
16. Chen Q, Amaral J, Biancani P, Behar J. Excess membrane cholesterol alters human gallbladder muscle contractility and membrane fluidity. Gastroenterology. 1999;116:678–85.
17. Chi C, Jiang XJ, Su L, Shen ZJ, Yang XJ. In vitro morphology, viability and cytokine secretion of uterine telocyte-activated mouse peritoneal macrophages. J Cell Mol Med. 2015;19(12):2741–50.
18. Ciontea SM, Radu E, Regalia T, Ceafalan L, Cretoiu D, Gherghiceanu M, Braga RI, Malincenco M, Zagrean L, Hinescu ME, Popescu LM. C-kit immunopositive interstitial cells (Cajal-type) in human myometrium. J Cell Mol Med. 2005;9:407–20.
19. Corradi LS, Jesus MM, Fochi RA, Vilamaior PS, Justulin LA Jr, Góes RM, Felisbino SL, Taboga SR. Structural and ultrastructural evidence for telocytes in prostate stroma. J Cell Mol Med. 2013. In press: doi:10.1111/jcmm.12021.
20. Cretoiu D, Cretoiu SM, Simionescu AA, Popescu LM. Telocytes, a distinct type of cell among the stromal cells present in the lamina propria of jejunum. Histol Histopathol. 2012;27:1067–78.
21. Cretoiu SM, Cretoiu D, Marin A, Radu BM, Popescu LM. Telocytes: ultrastructural, immunohistochemical and electrophysiological characteristics in human myometrium. Reproduction. 2013;145:357–70.
22. Cretoiu SM, Cretoiu D, Popescu LM. Human myometrium – the ultrastructural 3D network of telocytes. J Cell Mol Med. 2012;16:2844–9.

23. Cretoiu SM, Popescu LM. Telocytes revisited. Biomol Concepts. 2014;5(5):353–69. doi:10.1515/bmc-2014-0029.
24. Cretoiu SM, Radu BM, Banciu A, Banciu DD, Cretoiu D, Ceafalan LC, Popescu LM. Isolated human uterine telocytes: immunocytochemistry and electrophysiology of T-type calcium channels. Histochem Cell Biol. 2015;143:83–94.
25. Cretoiu SM, Simionescu AA, Caravia L, Curici A, Cretoiu D, Popescu LM. Complex effects of imatinib on spontaneous and oxytocin-induced contractions in human non-pregnant myometrium. Acta Physiol Hung. 2011;98:329–38.
26. Dawidowicz J, Szotek S, Matysiak N, Mielańczyk L, Maksymowicz K. Electron microscopy of human fascia lata: focus on telocytes. J Cell Mol Med. 2015;19(10):2500–6. doi:10.1111/jcmm.
27. Dray X, Joly F, Reijasse D, Attar A, Alves A, Panis Y, Valleur P, Messing B. Incidence, risk factors, and complications of cholelithiasis in patients with home parenteral nutrition. J Am Coll Surg. 2007;204:13–21.
28. Duquette R, Shmygol A, Vaillant C, Mobasheri A, Pope M, Burdyga T, Wray S. Vimentin-positive, c-kit negative interstitial cells in human and rat uterus: a role in pacemaking? Biol Reprod. 2005;72:276–83.
29. Edelstein L, Smythies J. The role of telocytes in morphogenetic bioelectrical signaling: once more unto the breach. Front Mol Neurosci. 2014;7:41.
30. Einarsson K, Nilsell K, Leijd B, Angelin B. Influence of age on secretion of cholesterol and synthesis of bile acids by the liver. N Engl J Med. 1985;313:277–82.
31. Enciu AM, Popescu LM. Telopodes of telocytes are influenced in vitro by redox conditions and ageing. Mol Cell Biochem. 2015;410(1–2):165–74. doi:10.1007/s11010-015-2548-2.
32. Everhart JE, Khare M, Hill M, Maurer KR. Prevalence and ethnic differences in gallbladder. Gastroenterology. 1999;117(3):632.
33. Faussone-Pellegrini MS, Popescu LM. Telocytes. Biomol Concepts. 2011;2:481–9.
34. Friedman GD, Raviola CA, Fireman B. Prognosis of gallstones with mild or no symptoms: 25 years of follow-up in a health maintenance organization. J Clin Epidemiol. 1989; 42:127–36.
35. Gevaert T, De Vos R, Everaerts W, Libbrecht L, Van Der Aa F, van den Oord J, Roskams T, De Ridder D. Characterization of upper lamina propria interstitial cells in bladders from patients with neurogenic detrusor overactivity and bladder pain syndrome. J Cell Mol Med. 2011;15:2586–93.
36. Gevaert T, De Vos R, Van Der Aa F, Joniau S, van den Oord J, Roskams T, De Ridder D. Identification of telocytes in the upper lamina propria of the human urinary tract. J Cell Mol Med. 2012;16:2085–93.
37. Gherghiceanu M, Popescu LM. Cardiac telocytes – their junctions and functional implications. Cell Tissue Res. 2012;348:265–79.
38. Gherghiceanu M, Popescu LM. Heterocellular communication in the heart: electron tomography of telocyte-myocyte junctions. J Cell Mol Med. 2011;15:1005–11.
39. Gherghiceanu M, Popescu LM. Interstitial Cajal-like cells (ICLC) in human resting mammary gland stroma. Transmission electron microscope (TEM) identification. J Cell Mol Med. 2005;9:893–910.
40. Gibbons SJ, De Giorgio R, Faussone-Pellegrini MS, Garrity-Park MM, Miller SM, Schmalz PF, Young-Fadok TM, Larson DW, Dozois EJ, Camilleri M, Stanghellini V, Szurszewski JH, Farrugia G. Apoptotic cell death of human interstitial cells of Cajal. Neurogastroenterol Motil. 2009;21:85–93.
41. Gordon-Taylor G. On gallstones and their sufferers. Br J Surg. 1937;25:241–51.
42. Harhun MI, Pucovsky V, Povstyan OV, Gordienko DV, Bolton TB. Interstitial cells in the vasculature. J Cell Mol Med. 2005;9:232–43.
43. Hatta K, Huang ML, Weisel RD, Li RK. Culture of rat endometrial telocytes. J Cell Mol Med. 2012;16:1392–6.

44. Heuman DM. Quantitative estimation of the hydrophilic-hydrophobic balance of mixed bile salt solutions. J Lipid Res. 1989;30(5):719–30.
45. Hinescu ME, Ardeleanu C, Gherghiceanu M, Popescu LM. Interstitial Cajal-like cells in human gallbladder. J Mol Histol. 2007;38:275–84.
46. Hu WM, Luo HS, Ding XW, Wang L. Expression of c-kit messenger ribonucleic acid and c-kit protein in the gallbladders in guinea pigs of high cholesterol diet. Dig Dis Sci. 2009;54:1651–5.
47. Humbert L, Maubert MA, Wolf C, Duboc H, Mahé M, Farabos D, Seksik P, Mallet JM, Trugnan G, Masliah J, Rainteau D. Bile acid profiling in human biological samples: comparison of extraction procedures and application to normal and cholestatic patients. J Chromatogr B Analyt Technol Biomed Life Sci. 2012;899:135–45.
48. Jüngst D, Brenner G, Pratschke E, Paumgartner G. Low-dose ursodeoxycholic acid prolongs cholesterol nucleation time in gallbladder bile of patients with cholesterol gallstones. J Hepatol. 1989;8(1):1–6.
49. Kim JK, Cho SM, Kang SH, Kim E, Yi H, Yun ES, Lee DG, Cho HJ, Paik YH, Choi YK, Haam SJ, Shin HC, Lee DK. N-3 polyunsaturated fatty acid attenuates cholesterol gallstones by suppressing mucin production with a high cholesterol diet in mice. J Gastroenterol Hepatol. 2012;27(11):1745–51. doi:10.1111/j.1440-1746.2012.07227.x.
50. Kurtin WE, Schwesinger WH, Diehl AK. Age-related changes in the chemical composition of gallstones. Int J Surg Investig. 2000;2:299–307.
51. LaMont JT, Smith BF, Moore JR. Role of gallbladder mucin in pathophysiology of gallstones. Hepatology. 1984;4:51S–6.
52. Lang RJ, Klemm MF. Interstitial cell of Cajal-like cells in the upper urinary tract. J Cell Mol Med. 2005;9:543–56.
53. Lavoie B, Balemba OB, Nelson MT, Ward SM, Mawe GM. Morphological and physiological evidence for interstitial cell of Cajal-like cells in the guinea pig gallbladder. J Physiol. 2007;579:487–501.
54. Lavoie B, Nausch B, Zane EA, Leonard MR, Balemba OB, Bartoo AC, Wilcox R, Nelson MT, Carey MC, Mawe GM. Disruption of gallbladder smooth muscle function is an early feature in the development of cholesterol gallstone disease. Neurogastroenterol Motil. 2012;24(7):e313–24. doi:10.1111/j.1365-2982.2012.01935.
55. Lee DK, Jang SI. Can fish oil dissolve gallstones? J Gastroenterol Hepatol. 2012;27(11):1649–51. doi:10.1111/j.1440-1746.2012.07234.x.
56. Li L, Lin M, Li L, Wang R, Zhang C, Qi G, Xu M, Rong R, Zhu T. Renal telocytes contribute to the repair of ischemically injured renal tubules. J Cell Mol Med. 2014;18:1144–56.
57. Litynski GS. Profiles in laparoscopy: Mouret, Dubois, and Perissat: the laparoscopic breakthrough in Europe (1987–1988). JSLS. 1999;3(2):163–7.
58. Lopis S. The incidence cholelithiasis in the Bantu. Clin Proc Child Hosp Dist Columbia. 1947;3:338.
59. Matyja A, Gil K, Pasternak A, Sztefko K, Gajda M, Tomaszewski KA, Matyja M, Walocha JA, Kulig J, Thor P. Telocytes: new insight into the pathogenesis of gallstone disease. J Cell Mol Med. 2013;17(6):734–42.
60. McCloskey KD, Hollywood MA, Thornbury KD, Ward SM, McHale NG. Kit-like immunopositive cells in sheep mesenteric lymphatic vessels. Cell Tissue Res. 2002;310:77–84.
61. Morgenstern L. Carl Langenbuch and the first cholecystectomy. Surg Endosc. 1992;6(3):113–4.
62. Moschetta A, Stolk MF, Rehfeld JF, Portincasa P, Slee PH, Koppeschaar HP, Van Erpecum KJ, Vanberge-Henegouwen GP. Severe impairment of postprandial cholecystokinin release and gall-bladder emptying and high risk of gallstone formation in acromegalic patients during Sandostatin LAR. Aliment Pharmacol Ther. 2001;15:181–5.
63. Moschetta A, van Berge-Henegouwen GP, Portincasa P, Palasciano G, van Erpecum KJ. Cholesterol crystallization in model biles: effects of bile salt and phospholipid species composition. J Lipid Res. 2001;42(8):1273–81.

64. Nicolescu MI, Bucur A, Dinca O, Rusu MC, Popescu LM. Telocytes in parotid glands. Anat Rec (Hoboken). 2012;295:378–85.
65. Nicolescu MI, Popescu LM. Telocytes in the interstitium of human exocrine pancreas: ultrastructural evidence. Pancreas. 2012;41:949–56.
66. Niculite CM, Regalia TM, Gherghiceanu M, Huica R, Surcel M, Ursaciuc C, Leabu M, Popescu LM. Dynamics of telopodes (telocytes prolongations) in cell culture depends on extracellular matrix protein. Mol Cell Biochem. 2015;398:157–64.
67. Njeze GE. Gallstones. Niger J Surg. 2013;19(2):49–55. doi:10.4103/1117-6806.119236.
68. Pasternak A, Gajda M, Gil K, Matyja A, Tomaszewski KA, Walocha JA, Kulig J, Thor P. Evidence of interstitial Cajal-like cells in human gallbladder. Folia Histochem Cytobiol. 2012;50(4):581–5. doi:10.5603/19673.
69. Pasternak A, Gil K, Matyja A, Gajda M, Sztefko K, Walocha JA, Kulig J, Thor P. Loss of gallbladder interstitial Cajal-like cells in patients with cholelithiasis. Neurogastroenterol Motil. 2013;25(1):e17–24. doi:10.1111/nmo.12037. Epub 2012 Nov 1.
70. Paumgartner G, Gerok W, Bertolotti M, Bortolotti S, Menozzi D. Ageing and bile acid metabolism: studies on 7α hydroxylation of cholesterol in humans. In: Paumgartner G, Gerok W, editors. Trends in bile acid research. Lancaster: Kluwer Academic Publishers; 1989. p. 75–8.
71. Paumgartner G, Sauerbruch T. Gallstones: pathogenesis. Lancet. 1991;338:1117–21.
72. Pomare EW, Heaton KW. Bile salt metabolism in patients with gallstones in functioning gallbladders. Gut. 1973;14(11):885–90.
73. Pomeranz IS, Shaffer EA. Abnormal gallbladder emptying in a subgroup of patients with gallstones. Gastroenterology. 1985;88:787–91.
74. Popescu BO, Gherghiceanu M, Kostin S, Ceafalan L, Popescu LM. Telocytes in meninges and choroid plexus. Neurosci Lett. 2012;516:265–9.
75. Popescu LM, Andrei F, Hinescu ME. Snapshots of mammary gland interstitial cells: methylene-blue vital staining and c-kit immunopositivity. J Cell Mol Med. 2005;9:476–7.
76. Popescu LM, Ciontea SM, Cretoiu D. Interstitial Cajal-like cells in human uterus and fallopian tube. Ann N Y Acad Sci. 2007;1101:139–65.
77. Popescu LM, Faussone-Pellegrini MS. TELOCYTES – a case of serendipity: the winding way from interstitial cells of Cajal (ICC), via interstitial Cajal-like cells (ICLC) to TELOCYTES. J Cell Mol Med. 2010;14:729–40.
78. Popescu LM, Gherghiceanu M, Suciu LC, Manole CG, Hinescu ME. Telocytes and putative stem cells in the lungs: electron microscopy, electron tomography and laser scanning microscopy. Cell Tissue Res. 2011;345:391–403.
79. Popescu LM, Hinescu ME, Ionescu N, Ciontea SM, Cretoiu D, Ardeleanu C. Interstitial cells of Cajal in pancreas. J Cell Mol Med. 2005;9:169–90.
80. Popescu LM, Manole E, Serboiu CS, Manole CG, Suciu LC, Gherghiceanu M, Popescu BO. Identification of telocytes in skeletal muscle interstitium: implication for muscle regeneration. J Cell Mol Med. 2011;15(6):1379–92. doi:10.1111/j.1582-4934.2011.01330.x.
81. Popescu LM, Nicolescu MI. Telocytes and stem cells. In: dos Santos Goldenberg RC, Campos de Carvalho AC, editors. Resident stem cells and regenerative therapy. Oxford: Academic; 2013. p. 205–31. doi:10.1016/B978-0-12-416012-5.00011-6.
82. Portincasa P, Di Ciaula A, Vendemiale G, Palmieri V, Moschetta A, Vanberge-Henegouwen GP, Palasciano G. Gallbladder motility and cholesterol crystallization in bile from patients with pigment and cholesterol gallstones. Eur J Clin Invest. 2000;30:317–24.
83. Portincasa P, Moschetta A, Palasciano G. Cholesterol gallstone disease. Lancet. 2006; 368(9531):230–9.
84. Portincasa P, Moschetta A, van Erpecum KJ, Calamita G, Margari A, van Berge-Henegouwen GP, Palasciano G. Pathways of cholesterol crystallization in model bile and native bile. Dig Liver Dis. 2003;35(2):118–26.

85. Qi G, Lin M, Xu M, Manole CG, Wang X, Zhu T. Telocytes in the human kidney cortex. J Cell Mol Med. 2012;16:3116–22.
86. Radu E, Regalia T, Ceafalan L, Andrei F, Cretoiu D, Popescu LM. Cajal-type cells from human mammary gland stroma: phenotype characteristics in cell culture. J Cell Mol Med. 2005;9:748–52.
87. Roatesi I, Radu BM, Cretoiu D, Cretoiu SM. Uterine telocytes: a review of current knowledge. Biol Reprod. 2015;93(1):10, 1–13. doi:10.1095/biolreprod.114.125906.
88. Sandler RS, Everhart JE, Donowitz M, Adams E, Cronin K, Goodman C, Gemmen E, Shah S, Avdic A, Rubin R. The burden of selected digestive diseases in the United States. Gastroenterology. 2002;122:1500–11.
89. Sergeant GP, Thornbury KD, McHale NG, Hollywood MA. Interstitial cells of Cajal in the urethra. J Cell Mol Med. 2006;10:280–91.
90. Shaffer EA. Gallstone disease: epidemiology of gallbladder stone disease. Best Pract Res Clin Gastroenterol. 2006;20:981–96.
91. Shaffer EA. The role of the gallbladder in gallstone formation. In: Fisher MM, Goresky CA, Shaffer EA, Strasberg SM, editors. Gallstones. New York: Plenum; 1979. p. 223–49.
92. Sharma BC, Agarwal DK, Dhiman RK, Baijal SS, Choudhuri G, Saraswat VA. Bile lithogenicity and gallbladder emptying in patients with microlithiasis: effect of bile acid therapy. Gastroenterology. 1998;115:124–8.
93. Shehadi WH. The biliary system through the ages. Int Surg. 1979;64:63–78.
94. Shepherd JA. The contribution of Robert Lawson Tait to the development of abdominal surgery. Surg Annu. 1986;18:339–49.
95. Sitzmann JV, Pitt HA, Steinborn PA, Pasha ZR, Sanders RC. Cholecystokinin prevents parenteral nutrition induced biliary sludge in humans. Surg Gynecol Obstet. 1990; 170:25–31.
96. Smythies J, Edelstein L. Telocytes, exosomes, gap junctions and the cytoskeleton: the makings of a primitive nervous system? Front Cell Neurosci. 2013;7:278. doi:10.3389/fincel.2013.00278.
97. Suciu L, Popescu LM, Gherghiceanu M, Regalia T, Nicolescu MI, Hinescu ME, Faussone-Pellegrini MS. Telocytes in human term placenta: morphology and phenotype. Cells Tissues Organs. 2010;192:325–39.
98. Thistle JL, Cleary PA, Lachin JM, Tyor MP, Hersh T. The natural history of cholelithiasis: The National Cooperative Gallstone Study. Ann Intern Med. 1984;101:171–5.
99. Torihashi S, Nishi K, Tokutomi Y, Nishi T, Ward S, Sanders KM. Blockade of kit signaling induces transdifferentiation of interstitial cells of cajal to a smooth muscle phenotype. Gastroenterology. 1999;117:140–8.
100. Trotman BW. Pigment gallstone disease. Gastroenterol Clin North Am. 1991;20:111–26.
101. Valdivieso V, Covarrubias C, Siegel F, Cruz F. Pregnancy and cholelithiasis: pathogenesis and natural course of gallstones diagnosed in early puerperium. Hepatology. 1993;17(1):1–4.
102. van Erpecum KJ, Venneman NG, Portincasa P, Vanberge-Henegouwen GP. Review article: agents affecting gall-bladder motility – role in treatment and prevention of gallstones. Aliment Pharmacol Ther. 2000;14 Suppl 2:66–70.
103. van Erpecum KJ. Biliary lipids, water and cholesterol gallstones. Biol Cell. 2005;97(11): 815–22.
104. van Gulik TM. Langenbuch's cholecystectomy, once a remarkably controversial operation. Neth J Surg. 1986;38(5):138–41.
105. Vannucchi MG, Traini C, Guasti D, Giulio DP, Faussone-Pellegrini MS. Telocytes subtypes in human urinary bladder. J Cell Mol Med. 2014;18:2000–8.
106. Vázquez MC, Rigotti A, Zanlungo S. Molecular mechanisms underlying the link between nuclear receptor function and cholesterol gallstone formation. J Lipids. 2012;2012:547643. doi:10.1155/2012/547643. Epub 2011 Nov 1.
107. Venneman NG, van Erpecum KJ. Pathogenesis of gallstones. Gastroenterol Clin North Am. 2010;39(2):171–83.

108. Wang DQ, Carey MC. Characterization of crystallization pathways during cholesterol precipitation from human gallbladder biles: identical pathways to corresponding model biles with three predominating sequences. J Lipid Res. 1996;37:2539–49.
109. Wang DQ, Schmitz F, Kopin AS, Carey MC. Targeted disruption of the murine cholecystokinin-1 receptor promotes intestinal cholesterol absorption and susceptibility to cholesterol cholelithiasis. J Clin Invest. 2004;114:521–8.
110. Wang DQH, Carey MC. Complete mapping of crystallization pathways during cholesterol precipitation from model bile: influence of physical-chemical variables of pathophysiologic relevance and identification of a stable liquid crystalline state in cold, dilute and hydrophilic bile salt-containing systems. J Lipid Res. 1996;37:606–30.
111. Wang J, Ye L, Jin M, Wang X. Global analyses of chromosome 17 and 18 genes of lung telocytes compared with mesenchymal stem cells, fibroblasts, alveolar type II cells, airway epithelial cells, and lymphocytes. Biol Direct. 2015;10(1):9. doi:10.1186/s13062-015-0042-0.
112. Xu QW, Shaffer EA. The potential site of impaired gallbladder contractility in an animal model of cholesterol gallstone disease. Gastroenterology. 1996;110:251–7.
113. Yang H, Petersen GM, Roth MP, Schoenfield LJ, Marks JW. Risk factors for gallstone formation during rapid loss of weight. Dig Dis Sci. 1992;37:912–8.
114. Yang XJ, Yang J, Liu Z, Yang G, Shen ZJ. Telocytes damage in endometriosis-affected rat oviduct and potential impact on fertility. J Cell Mol Med. 2014. doi:10.1111/jcmm.12427.
115. Zheng Y, Zhu T, Lin M, Wu D, Wang X. Telocytes in the urinary system. J Transl Med. 2012;10:188. doi:10.1186/1479-5876-10-188.

Artur Pasternak is assistant professor at the Department of Anatomy at Jagiellonian University Medical College, Krakow, Poland, and surgeon in the 1st Department of General, Oncological, and GI Surgery. His main research interests focus on the interstitial Cajal-like cells/telocytes in the human biliary system. Dr. Pasternak got his Ph.D. degree *cum laude* in medicine at the Jagiellonian University in 2011, with the thesis on the role of the interstitial Cajal-like cells and bile lithogenecity in the pathogenesis of gallstone disease. He is actively engaged in the new research group on telocytes established in Krakow, Poland involving collaborations with the universities and clinical hospitals. Dr. Pasternak has published multiple scientific papers concerning etiopathogenesis of cholelithiasis within internationally well-recognized high-impacted journals and has written ten book chapters.

Krzysztof Gil is an associate professor at the Department of Pathophysiology at the Jagiellonian University Medical College, Krakow, Poland. His major research themes include basic and clinical neurogastroenterology issues, including brain-gut axis regulatory mechanisms, vagus nerve stimulation for the experimental obesity management, as well as research on the interstitial cells of Cajal and recently described cells – telocytes. These gastrointestinal regulatory cells are extremely responsive to neural, hormonal, and local influences. Because of this, alterations of gastric and intestinal motility are common manifestations of many disease processes. In his research he is implementing both morphological (immunohistochemical) and physiological techniques. He is involved in teaching dental, graduate, pharmacy, and medical students. He has written numerous book chapters for students and more than 60 scientific publications. He serves as the secretary editor of ***Folia Medica Cracoviensia*** journal. Since October 2014 he is also a head of the "International PhD studies in medical sciences" program at the Jagiellonian University Medical College, Faculty of Medicine.

Andrzej Matyja is a professor at the 1st Department of General, Oncological, and GI Surgery at Jagiellonian University Medical College, Krakow, Poland. He performs the whole range of complex general and oncological surgery procedures. Professor Matyja is a leading expert in giant abdominal wall hernia repair surgery. His current scientific research is focused on telocytes as new players in gallstone disease. He is also involved in studies regarding postoperative gastroparesis and function of the autonomic system in the internal and external innervation of the stomach. Professor Matyja has published over 100 original articles and chapters. In 2014 Professor Matyja was awarded with "*Gloria Medicinae*," the most prestigious medal of Polish Society of Medicine.

Chapter 6
Features of Telocytes in Agricultural Animals

Ping Yang, Nisar Ahmed, Shakeeb Ullah, Qiusheng Chen, and Yonghua Zheng

Abstract Telocytes have been identified in almost all kinds of organs and tissues in human, rat, mouse and other mammals, and they are considered to be important during tissue regeneration and repair, and they have potential roles in the pathogenesis of some diseases. However, there is still short of knowledge about telocytes in agricultural animals. In this chapter, we try to testify the existence of telocytes and demonstrate the morphological and ultrastructural characteristics of telocytes in agricultural animals. The results confirmed the existence of telocytes in the reproductive and gastrointestinal tract of agriculture animals, and they share the same morphology and ultrastructure with telocytes in other reported mammals. The results also promoted the study of telocytes and enhanced the researchers to pay more attentions on this new type of interstitial cells and it is helpful to create new treatments for the various reproductive disorders and gastrointestinal problems.

6.1 Introduction

Medical research focusing on human and rat and mouse are already considering telocyte histology and pathology, but in agricultural animal, research of telocyte is usually ignored. Nevertheless, recent studies of telocyte show a growing interest in the functional morphology of different organs and species and emphasize the consideration of the potential important role of this cell type in the agricultural animal. The aim of this chapter was to point out the little research performed in agricultural animal related to telocyte and give a resume of the importance of telocyte according to the study of the functional morphology in some agricultural animal.

Agricultural animals play a crucial role in the rural and agricultural economies of the world. They not only produce food directly but also provide major inputs to crop agri-

P. Yang • N. Ahmed • S. Ullah • Q. Chen
College of Veterinary Medicine, Nanjing Agricultural University,
Nanjing, People's Republic of China

Y. Zheng (✉)
Department of Respiratory Medicine, Xinhua Hospital, Jiaotong University, Shanghai, China
e-mail: zhengyonghua0118@hotmail.com

X. Wang, D. Cretoiu (eds.), *Telocytes*, Advances in Experimental Medicine and Biology, 913, DOI 10.1007/978-981-10-1061-3_6

culture. Actually, agricultural animals is a broader term including about 5,000 breeds comprising 40 edible or economic animal species, such as cattle, buffalo, horse, swine, sheep, goat, turtle, poultry, fishes, etc. Agricultural animals meet approximately 30 % of total human needs for food and agriculture in the form of eggs, milk, meat, fiber, fertilizer for crops, fuel, and draught power and also provide a very important cash source in many farming systems. Livestock are the only ready source of cash to buy inputs for crop production – seeds, pesticides, and fertilizers for many smallholder farmers.

For over 150 million years, poultry has been on the earth, dating back from the original wild jungle fowl to now that we include chickens, peafowl, turkeys, pigeons ducks, geese, pheasants, guinea fowl, and those in the list of species under the general class named poultry. Poultry provides human beings with companionship, fiber, and food in the form of eggs, meat, and feathers. Chicken and other poultry can be reared in almost all parts of the world and people all over the world like poultry products. In the last 20 years, the interest in poultry and poultry products has grown enormously due to easy availability and cost-effective values. Almost every country in the world has a poultry industry of some kind. In addition, turtles are one of the common species of reptiles and are widely distributed in world. This species is famous for its economic and pharmacologic value, and hence, they are subjected to extensive harvesting pressure.

Telocytes (TCs) are a novel interstitial (stromal) cell type described in many tissues and organs of humans and laboratory mammals (www.telocytes.com). A TC is characterized by a small cell body (9–15 μm) and a variable number (one to five) of extremely long and thin telopodes (Tps), with alternating regions of podomers (~80 nm) and podoms (250–300 nm). Tps are interconnected by homo- and heterocellular junctions and form three-dimensional networks. Moreover, Tps release three types of extracellular vesicles: exosomes, ectosomes, and multivesicular cargos, which are involved in paracrine signaling. Agricultural animals present significant economical and scientific notability; so, to determine the identificational and functional studies of TCs in agricultural animals is of great importance. LM Popescu (2010) established that TEM features alone can serve as the gold standard for identification of TCs [1]. In the level of light microscopy, by single immunohistochemistry, TCs do not express specific marker(s); at present, CD34 labeling remains the best available choice for the single immunohistochemical identification of TC; however, sometimes it may be confused with endothelial cells in general. But by double immunohistochemistry, CD34/PDGFRα or CD34/CD31 labeling can be an accurately oriented diagnosis [2]. Indeed, some researchers in agricultural animals had identified TCs by TEM. They had used the immunohistochemical staining and light microscopic techniques to localize and especially quantify the numbers of TCs in the various regions of different animals, but regarding the agricultural animals, these studies are regrettably low due to lack of the availability of easy choice-specific antibodies in the markets. Recently, only three papers have been published on telocytes in the agriculture animals (turtle and chicken), but a lot of research is waiting for publishing in other agriculture animals, such as, goat, sheep, etc., because the telocytes play a key role within the body of living organisms along with other cells.

Here, we discussed the novel work about the telocytes in agriculture animals, which is the first step to explore the importance of telocytes.

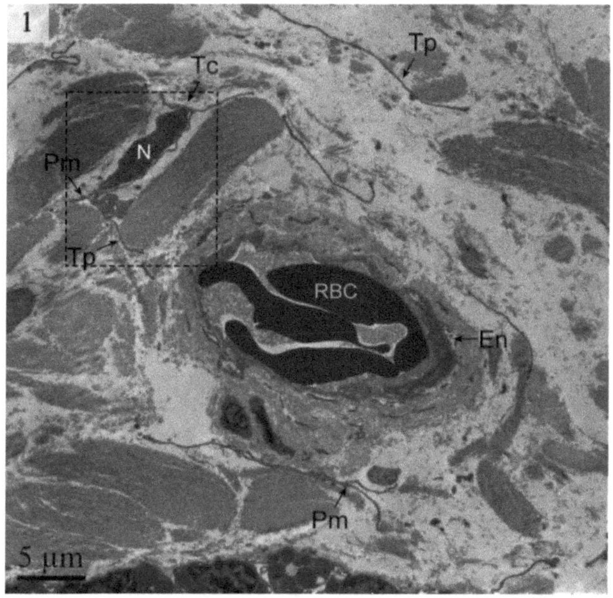

Fig. 6.1 Digitally colored TEM image of the turtle uterus. Long telopodes are present around the capillary. The podom and a telocyte are also visible. (*Tc* telocyte, *Tp* telopode, *RBC* red blood cell, *En* endothelium). The scale bar represents 5 μm

6.2 Telocytes in the Reproductive System of Agriculture Animals

Reproductive system is one of the most important systems within the living organisms. The reptilian and poultry oviduct, unlike mammals, is an important organ that performs major functions including sperm storage, egg formation, maturation, and transportation of unfertilized or fertilized eggs to the exterior of the body. The organ is composed of the following five segments or regions (from anterior to posterior): infundibulum, magnum, isthmus, shell gland (uterus), and vagina. The uterus is regarded as an important segment of the oviduct, which performs several important functions, such as salt and water absorption, egg calcification, egg transport, and egg pigmentation, while sperm storage also takes place in the uterus of the reptilian and poultry. Ullah et al. describe the ultrastructure characteristics of telocytes in the uterus of female adult soft-shelled turtles, *Pelodiscus sinensis*, through transmission electron microscopy (Figs. 6.1 and 6.2) [3]. Remarkably, this is the first study in reptiles, besides mammals. The general features of the telocytes, identified in the interstitial spaces of the uterus of soft-shelled turtle *Pelodiscus sinensis*, were similar to those already reported in mammalian species and are consistent with the diagnostic criteria of telocytes, explained by Popescu such as the existence of long, moniliform and convoluted telopodes. The telocytes makes homo- and heterocellular junctions by connecting with each other, and with different cell types by their long cell processes, these are able to carry signals over long distances. Conversely, local (paracrine) signaling of telocytes is achieved by shedding vesicles. The

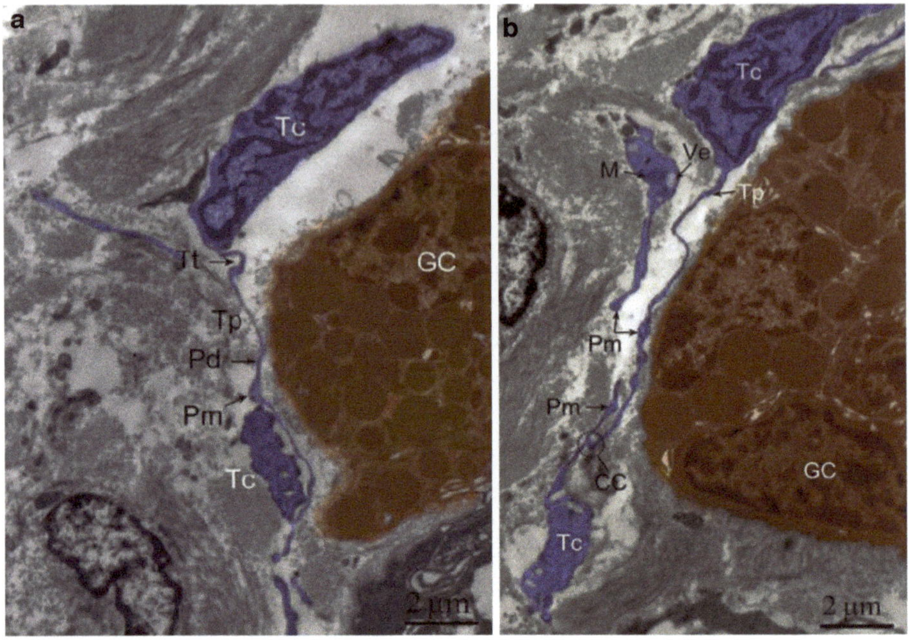

Fig. 6.2 Digitally colored TEM image of the turtle uterus. (**a**) At least 2 telocytes with their extensive telopodes are visible in the lamina propria near gland cells. Telopodes with long alternating podom and podomer are visible. (*Tt* tortuous telopode, *Tc* telocytes, *GC* gland cells, *Pm* podom, *Pd* podomer). The scale bar represents 2 μm. (**b**) Two telocytes are clearly visible and telocytes connect with telopodes from other telocytes. The encircled area shows the close connection between telopodes. Mitochondria and vesicles are also visible. (*TC* telocyte, *Tp* telopode, *GC* gland cell, *CC* close connection, *Ve* vesicle, *M* mitochondria.) The scale bar represents 2 μm

telocytes were embedded in the collagen fiber [4], which suggests that telocytes might be involved in homeostasis, remodeling, regeneration, and tissue repair. Telopodes are connected with each other by either end-to-end or side-to-side contacts but rarely by end-to-side connections in the uterus. These data provide morphological evidence for the presumption that telopodes might convey signal indication or have unique communication between telocytes. The telopodes also present in a close proximity to secretory glands and blood vessel in the lamina propria beneath the simple columnar epithelium. Furthermore, they are also observed in close connections with smooth muscle fibers formed by gap junctions and connective tissues. These findings suggest that telocytes could influence the timing of contractile activity in smooth muscle cells and indicate the key role of telocytes in uterine contraction. Further studies need to explore the potential bio-functions of telocytes in certain pathological conditions in the uterus and investigate the mechanisms of interaction between telocytes and other cells.

Testes are components of both the reproductive and endocrine systems. The main functions of the testis are the production of sperm (spermatogenesis) and androgens, primarily testosterone. The testis parenchyma consists of seminiferous tubules and interstitial tissue. The testis interstitial tissue (TIS) is a highly vascularized loose connective tissue that includes Leydig cells, blood vessels, leukocytes, and fibroblasts. Spermatogenesis occurs in the seminiferous tubules, which are composed of Sertoli cells (SCs) and maturing germ cells that are surrounded by one (e.g., in rats and mice) or more (e.g., humans and turtles) layers of peritubular cells. Recently, Yang et al. for the first time reported the information about the existence of telocytes in the testis through transmission electron microscopy and CD34 immunohistochemistry (Figs. 6.3, 6.4, and 6.5) [5]. This is the first time he used CD34 for the identification of telocytes in the agriculture animals.

In the turtle testis, several layers of peritubular cells were clearly present in the walls of the seminiferous tubules. Some peritubular cells presented the typical characteristics of smooth muscle cells, such as nuclei that exhibited many irregularities (i.e., concertina nuclei), abundant actin filaments with dense focal bodies, and caveolae. The peritubular cells located in the inner layer (near the seminiferous tubules) exhibited thicker processes and contained more actin filaments and dense material than the cells located in the outer layer. The outermost layer around peritubular cells was identified as a novel type of interstitial cell known as TCs according to the diagnostic criteria for TCs suggested by Popescu et al. [1]. However, in terms of morphology, these cells are easily confused with peritubular cells because both cell types

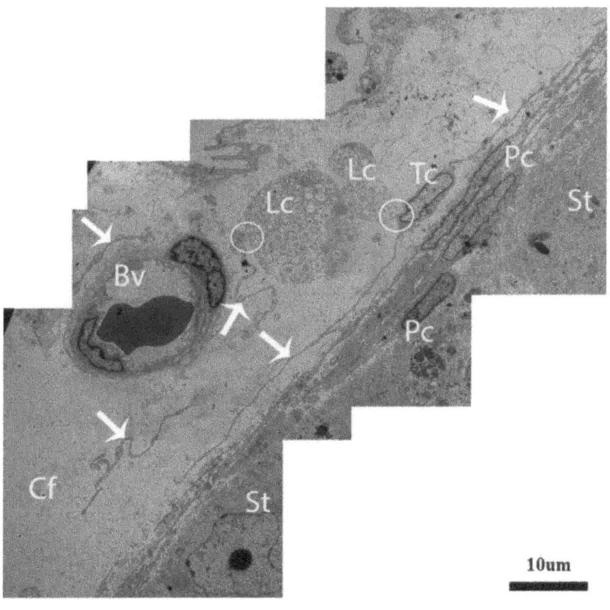

Fig. 6.3 TEM micrograph of a turtle testis. TCs with very long thin prolongations (telopodes, *arrows*) bordering several layers of the peritubular cells. The encircled area indicates the heterocellular junction (point contact) between the TCs and the Leydig cells. Several telopodes are present in close proximity to the blood vessels and Leydig cells within the interstitial tissue. *St* seminiferous tubules, *Tc* telocyte, *Pc* peritubular cell, *Lc* Leydig cell, *Cf* collagen fiber, *Bv* blood vessels. The scale bar is 10 μm

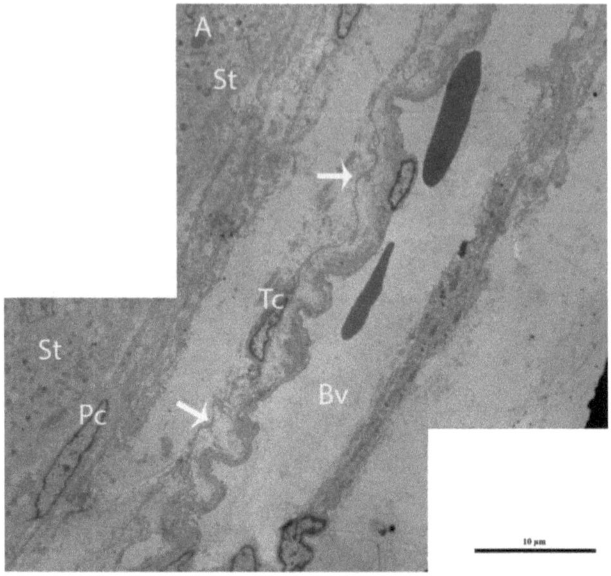

Fig. 6.4 TEM micrograph of a turtle testis. The TCs with extremely long processes (*arrow*) bordering the large blood vessel exist between the peritubular cells and the blood vessel within the interstitial tissue. *St* seminiferous tubules, *Pc* peritubular cell, *Tp* telopod, *Bv* blood vessels. The scale bar is 10 μm

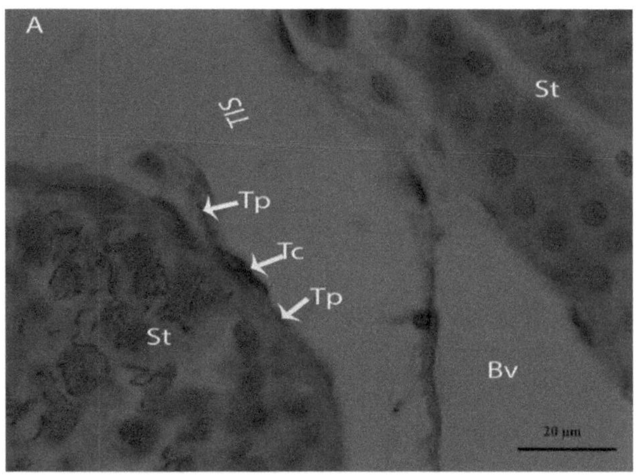

Fig. 6.5 Photomicrograph of a turtle testis. CD34-positive TCs with long thin telopodes extending from the cell body and bordering the seminiferous tubules are shown. *Tc* telocyte, *Tp* telopod, *St* seminiferous tubules, *TIS* testis interstitial tissue, *Bv* blood vessel. The scale bars are 20 μm

have long processes. A typical TC with two extremely long processes (Tps) near the peritubular cells. The observation of the TCs revealed that the cell body was small, the nuclei exhibited moderate heterochromatin at the periphery with a small amount of cytoplasm surrounding the nuclei, and the cytoplasm contained numerous mitochondria. Furthermore, the Tps have (cellular extensions) exhibited alternating thin segments (podomers) and dilated bead-like regions (podoms). The podoms accommodated the mitochondria and vesicles. Whereas the processes of the peritubular cells have thicker and greater numbers of spinelike short process. The appearances of the nuclei of the peritubular cells were similar to those of smooth muscle cells (i.e., concertina nuclei). The TCs shared obvious substantial connections with the

Leydig cells in the interstitial tissue along with few fibers around the blood vessels, which suggests that the TCs are indirectly involved in the secretion of the testosterone, rostenedione, and dehydroepiandrosterone within the testis of agriculture animals. The prolongations of the vascular TCs exhibited greater numbers of bands and secreted various numbers of vesicles into the extracellular compartment. TCs might act as cellular guides for immune cells via the blood stream, and this vesicle density seems to evidence cell-to-cell communication via the direct stimulation of target cells or receptor-mediated interactions. Furthermore, homocellular junctions, such as end-to-side membrane contacts (less than 13 nm) known as nanocontacts, were present between the Tps surrounding the blood vessels. Some uncommon junctions were also present between the Tps and cell bodies of the TCs. Electron microscopy also revealed octopus-like TCs with extremely twisted. These Tps formed a labyrinthine system with three-dimensional convolutions that overlapped each other and were also involved in the secretion of a large number of vesicles.

The endothelium and TCs were positively labeled for CD34. The immunostaining revealed numerous positive endothelial cells in the small and large blood vessels in the connective tissue between the seminiferous tubules of the testes. CD34 was expressed in the TC cell bodies and their cellular elongations, and the TCs were observed to be bipolar in shape with thin several prolongations that extended to surround the adjacent peritubular cells of the seminiferous tubule. The extremely long Tps were present in the close contact with the peritubular cells, Leydig cells, and blood vessel. Collectively, its morphological evidence suggested that the interactions between Sertoli cells, peritubular cells, Leydig cells, and germ cells are essential for the regulation of spermatogenesis in the agriculture animals.

6.3 Telocytes in the Gastrointestinal Tract (GIT) of Agriculture Animals

GIT is responsible for consuming and digesting food stuffs, absorbing nutrients, and expelling waste martial. The tract consists of the stomach and intestines and is divided into the upper and lower gastrointestinal tracts. By the broadest definition, the GIT includes all structures between the mouth and the anus. Yang et al. identified the existence of telocyte in chicken (Figs. 6.6 and 6.7) [6]. They found in chicken telocytes almost have similar characteristics like telocytes present in other organs. TCs have with a small body and a variable number of Tps, containing a nucleus, surrounded by a small amount of cytoplasm. Tps have a particular ultrastructural signature, consisting of an alternation of thin segments (podomers) and dilations (podoms). Each TC can have 1–5 Tps; the shape of the telocytes is according to the number of their telopodes: piriform for one prolongation, spindle for two telopodes, triangular for three, stellate, etc. The thin rim of cytoplasm surrounding the nucleus contains a small Golgi apparatus, some mitochondria, and few cisternae of the rough endoplasmic reticulum (RER) and smooth endoplasmic reticulum (SER). Telocytes seem to have strategic position in the ileum, in between blood capillaries and stem

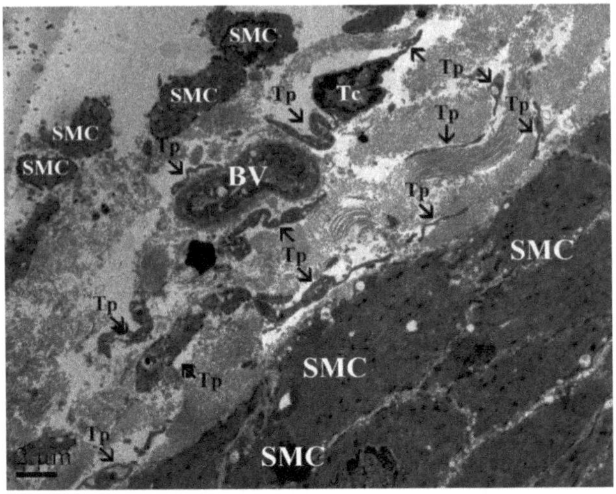

Fig. 6.6 TEM image of some telocytes along the submucosal surface of the circular muscle layer in the chicken ileum. TCs were arranged around the ganglia of the submucosal plexus. *Tc* telocytes, *Tp* telopods. Bars = 2 μm

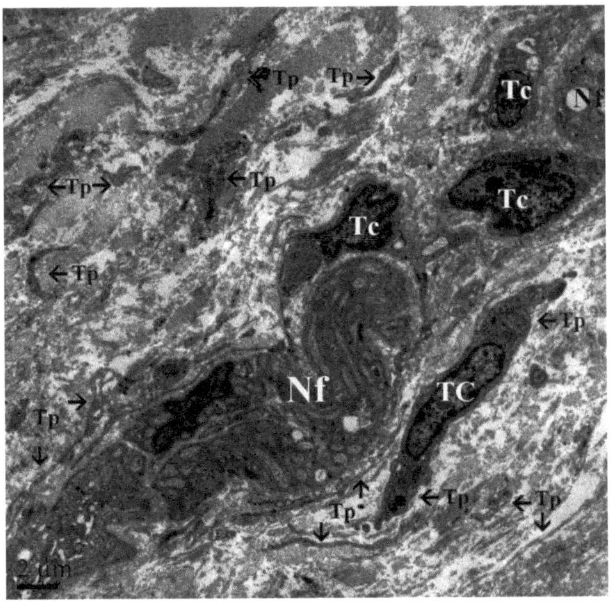

Fig. 6.7 TEM image of some telocytes along the submucosal surface of the circular muscle layer in the chicken ileum. TCs were arranged around the ganglia of the submucosal plexus. *Tc* telocytes, *Tp* telopods. Bars = 2 μm

cells, and are in close contact with nerve endings. Telocytes could be nurse cells integrating local (short-distance signals: direct contacts, exosomes, shed vesicles) and long-distance signals through the long TPs, because of their 3D network.

TCs were distributed within the longitudinal and circular muscle layers. The TCs of the longitudinal layer have distinct close contacts, not gap junctions, with adjacent SMC and with blood vessel. TCs were also present along the submucosal surface of the circular muscle layer, lamina propria of the ileum, and around the ganglia of the submucosal plexus. These findings suggest that the telocytes have a role in contraction of GIT, reparation of injured tissues during diseases, like in the skeletal muscles, but there is a great need to explore telocyte-specific biomarker to clarify the cell in the functional aspect and, in an easier way, network biomarkers to understand more about the interaction between proteins, genes, and signal pathways and dynamic network works to define and predict time-dependent telocyte function and morphological features. The ultrastructural evidence for the existence of telocytes in the muscularis and the lamina propria in the ileum of the chicken and telopodes connect with immune cells, smooth muscle cells, nerve fibers, and blood vessels. Further studies must explore the potential bio-functions of telocytes in certain pathological conditions in intestine and investigate the mechanisms of interaction between telocytes and other cells.

6.4 Conclusion

In this chapter, we have demonstrated the morphological and ultrastructural characteristics of telocytes in the few organs of reproductive and gastrointestinal tract of agricultural animals. This information not only enhanced the interest of researchers to conduct more studies on agricultural animals but is also helpful to treat the various reproductive disorders as well as gastrointestinal problems.

References

1. Popescu LM, Faussone-Pellegrini M-S. TELOCYTES – a case of serendipity: the winding way from interstitial cells of cajal (ICC), via interstitial cajal-like cells (ICLC) to TELOCYTES. J Cell Mol Med. 2010;14:729–40.
2. Cretoiu SM, Popescu LM. Telocytes revisited. Biomol Concepts. 2014;5:353–69.
3. Ullah S, Yang P, Zhang L, Zhang Q, Liu Y, Chen W, Waqas Y, Le Y, Chen B, Chen Q. Identification and characterization of telocytes in the uterus of the oviduct in the chinese soft-shelled turtle, pelodiscus sinensis: TEM evidence. J Cell Mol Med. 2014;18:2385–92.
4. Rusu M, Mirancea N, Mănoiu V, Valcu M, Nicolescu M, Păduraru D. Skin telocytes. Ann Anat Anat Anz. 2012;194:359–67.
5. Yang P, Ahmad N, Hunag Y, Ullah S, Zhang Q, Waqas Y, Liu Y, Li Q, Hu L, Chen Q. Telocytes: novel interstitial cells present in the testis parenchyma of the Chinese soft-shelled turtle pelodiscus sinensis. J Cell Mol Med. 2015;19:2888–99.
6. Yang P, Liu Y, Ahmed N, Ullah S, Liu Y, Chen Q. Ultrastructural identification of telocytes in the muscularis of chicken ileum. Exp Ther Med. 2015;10:2325–30.

Chapter 7
The Telocyte Subtypes

Maria-Giuliana Vannucchi and Maria-Simonetta Faussone-Pellegrini

Abstract Several cells are endowed in the interstitial space of the connective tissue; among them, a peculiar type has been recently described and named telocyte (TC). The increasing interest on this cell type has allowed identifying it in almost all the organs. All TCs have a proper ultrastructural feature that makes them undoubtedly recognizable under the transmission electron microscope (TEM). On the contrary, a complex often confusing picture comes out from the immunohistochemical investigations either due to the technical procedures used or, intriguingly, to the possibility that diverse subtypes of TC might exist.

Among the several markers used to label the TC, the most common are the CD34 and the PDGFRalpha, and, in many organs, the TC expresses both these markers. An exception is represented by the human urinary bladder where none of the TC, as recognized under the TEM, was double labelled. All the data indicate that TCs show immunohistochemical differences depending on the organ where they are located and/or the animal species.

On the basis of their ubiquitous distribution, TCs are unanimously considered organizers of the connective tissue because of their ability to form 3-D networks. Close to this common role, numerous other roles have been attributed to the TC. Indeed, each of the TC subtype likely plays an own organ-/tissue-specific role contributing to different aspects of physiological regulation in the various anatomical niches they occupy.

7.1 The Interstitial Cells

Several types of interstitial cells are present in the connective tissue so that, next to the fibroblasts/fibrocytes responsible for the matrix formation, there are mast cells, resident macrophages and, occasionally, lymphocytes, plasma cells, granulocytes

---

M.-G. Vannucchi, MD, PhD • M.-S. Faussone-Pellegrini (✉)
Department of Experimental and Clinical Medicine, Research Unit of Histology and Embryology, University of Florence, Viale Pieraccini 6, Florence 50134, Italy
e-mail: mariasimonetta.pellegrini@unifi.it

X. Wang, D. Cretoiu (eds.), *Telocytes*, Advances in Experimental Medicine and Biology, 913, DOI 10.1007/978-981-10-1061-3_7

and monocytes. Recently, another type of interstitial cell has been added to the previous ones. This cell type seems to be present in all organs so far examined and has been named telocyte (TC) [15, 21, 37]. Moreover, other interstitial cell types might be present in some organs. In the gut, for example, there are the so-called interstitial cells of Cajal (ICC), discovered by Cajal in 1889, responsible for pacemaker activity, and, according to some researchers, the so-called PDGFRα-positive cells are considered to be capable of neurotransmission [3, 25, 26, 32]. Interestingly, in the gut [44], liver [20, 47], uterus [14], heart [50] salivary glands [2] and urinary bladder [45], the PDGFRα-positive cells and the telocytes (identified by their CD34 positivity, Fig. 7.1) are the same cell type. In the urinary tract, many interstitial cell types have been described by immunohistochemistry, and several and confusing identities have been attributed to them: ICC [24], TC [21, 49], PDGFRα-positive cells [27, 28] and a lot of other various (inconsistent) names [41].

7.2 The Telocytes

The cells today called TC correspond to those formerly called interstitial Cajal-like cells (ICLC) [7, 11, 23, 34–36, 39]. This name, however, was quite ambiguous and was changed by the same authors in telocyte (TC) [37]. All the cells identified as TC have a proper and common ultrastructural feature that makes them easily and undoubtedly recognizable under the transmission electron microscope (TEM) [8, 10, 12, 15, 23, 29–31, 37, 42, 48]. TCs are surely not fibroblasts even if in some organs these cells, at variance with the TC typical ultrastructure and similarly to fibroblasts, are rich in rough endoplasmic reticulum (Fig. 7.2a). Likely, TC might

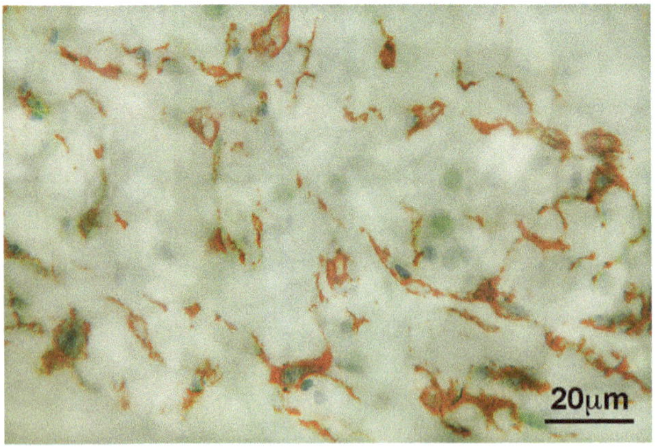

Fig. 7.1 Immunohistochemistry, light microscope. In *brown*, DAB revelation of the CD34 positivity. Telocytes are CD34-positive and form a 3-D network in the connective tissue. Human gastric submucosa. Bar = 20 μm

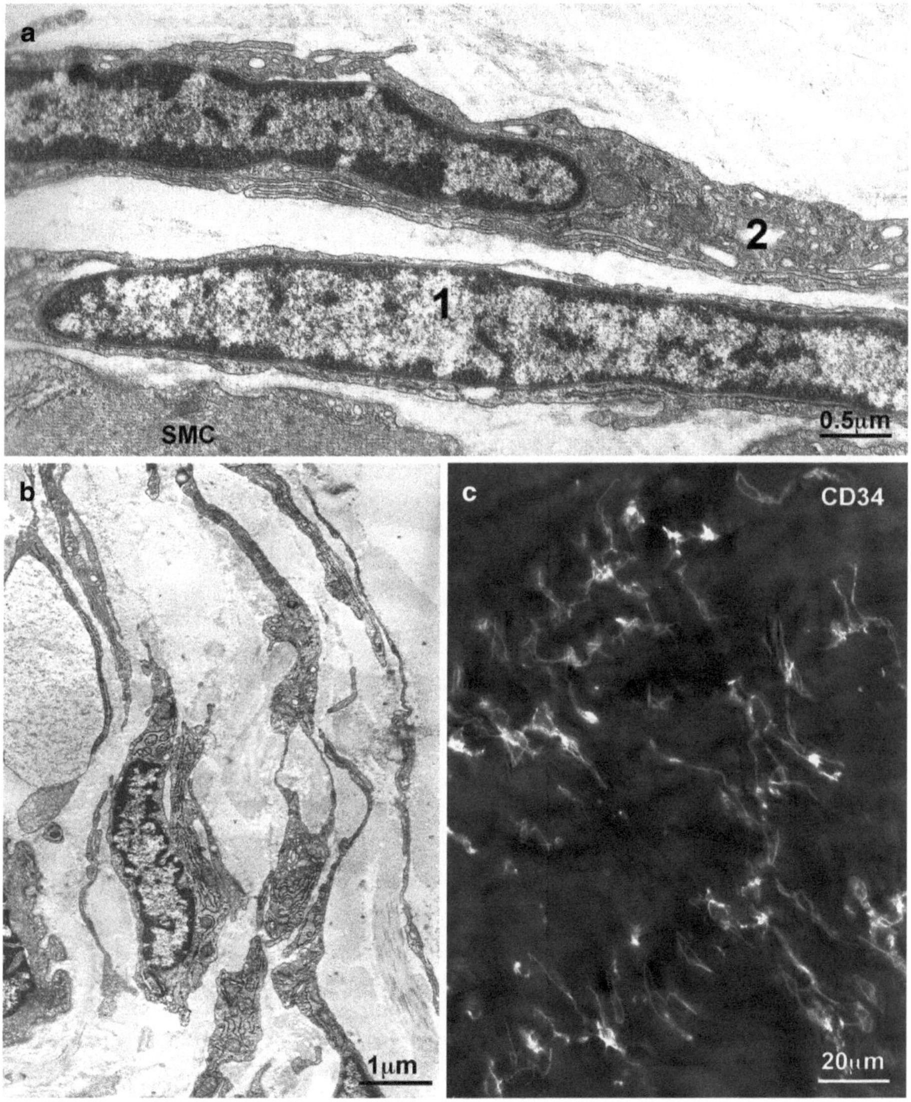

Fig. 7.2 (**a, b**) Transmission electron microscope. (**a**) Rabbit colonic submucosa. Two telocytes: a typical one (1) with a scarce perinuclear cytoplasm and an atypical one (2) with a perinuclear cytoplasm rich in mitochondria and endoplasmic reticulum. (**b**) Rabbit colonic submucosa. Several telocytes that with their long processes (telopodes) come in contact with each other and form a 3-D network. (**c**) Immunohistochemistry, fluorescence microscope. CD34-positive telocytes form a 3-D network. Bar: (**a**) = 0.5 μm, (**b**) = 1 μm, (**c**) = 20 μm

correspond to the cells called in the past and still called by some authors fibrocytes. Indeed, the morphology of these two cell types is very similar.

A discriminant element, especially under the light microscope, to make sure that we are looking at the TC is their organization in networks (Fig. 7.2b,c). Indeed, these cells are constantly organized in a 3-D network intermingled with resident and not resident connective tissue cells or in a 2-D network lining the connective border of various tissues and around blood vessels. The TC located around blood vessels presumably corresponds to the cells known as 'adventitial cells' and those located around enteric ganglia and nerve strands to those known as 'covering cells' [19].

7.3 TC Subtypes

A careful revision of our and literature data obtained in different animal species and in different organs of the same animal suggests that the cells identified as TC might differ from each other either in the ultrastructure or in the immunohistochemical labelling. Ultrastructural differences are the lesser evident; on the contrary, the immunohistochemical ones are prominent. Therefore, the possibility that the ultrastructural and immunohistochemical differences reported for these cells might correspond to diverse subtypes of TC is very likely and fascinating.

7.3.1 Gut and Urinary Tract

These two apparatus are the most extensively investigated for identifying the different types of interstitial cells. In particular, a controversy exists on whether in these two organs the TC and the so-called PDGFRα-positive cells are the same cell type. The PDGFRα-positive cells have been considered by some authors [3, 25–27] to correspond to the so-called fibroblast-like cell (FLC). In the human urinary bladder, however, the FLC ultrastructure resulted much more similar to that of fibroblasts than to that of TC [39]. Moreover, when the immune TEM was applied, these cells were CD34-negative [39]. These results would suggest that the TC and the PDGFRα cells are two distinct cell types. However, according to several authors, in the gut [34], liver [20, 47], uterus [14], heart [50] and salivary glands [2], the TC and the PDGFRα-positive cells are a unique cell type since in these organs, all the cells identified as TC were double labelled with CD34 and PDGFRα (Fig. 7.3a–c) and shared identical ultrastructural features. In the human urinary bladder, a picture more complex than in the other organs comes out. In particular, while TCs in the submucosa and detrusor were CD34-positive and PDGFRα-negative, those in the suburothelium were PDGFRα-positive and CD34-negative. To note, both these two populations formed 3-D networks and shared a

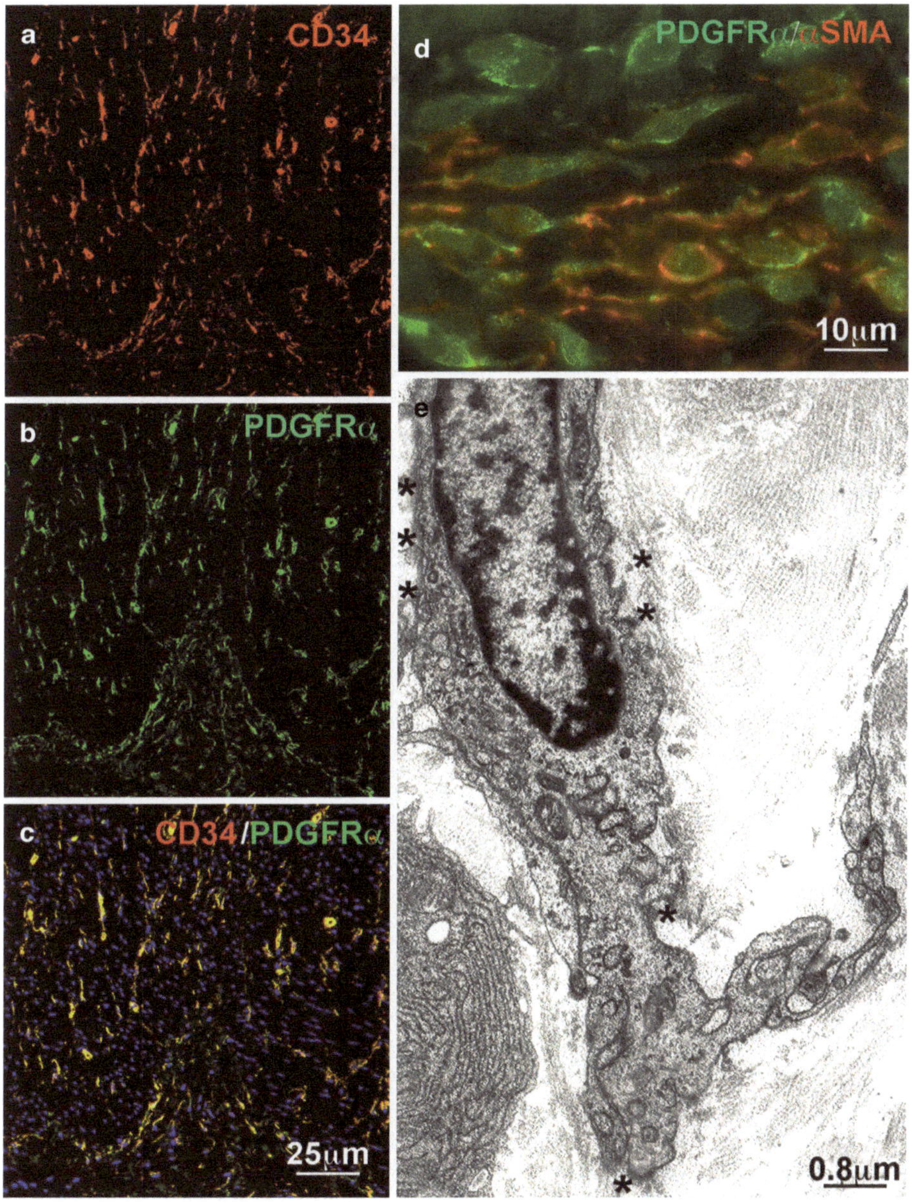

Fig. 7.3 (**a–d**) Immunohistochemistry, fluorescence microscope. (**a–c**) Telocytes labelling in the muscle coat and submucosa of the human small intestine. In *red*, CD34 immunoreactivity (**a**), in *green* PDGFRα immunoreactivity (**b**), in *yellow* double CD34/PDGFRα immunoreactivity (**c**). (**d**) Telocytes labelling in the suburothelial area of the human bladder. Some of the telocytes are PDGFRα-positive only (in *green*), whereas others are double PDGFRα (in *green*)- and αSMA (in *red*)-positive. (**e**) Transmission electron microscope, human bladder. A 'hybrid' telocyte possessing along its contour numerous attachment plaques (*asterisks*). Bar: (**a–c**) = 25 μm, (**d**) = 10 μm, (**e**) = 0.8 μm

typical TC ultrastructure [21, 22, 41, 45]. In the mouse bladder [28], PDGFRα-positive cells were seen throughout the entire bladder wall and, similarly to TC, had a spindle or stellate body and possessed multiple processes that contacted one another forming a loose 3-D network. Moreover, in the human urinary bladder, among the suburothelial PDGFRα-positive cells, some of those closer to the urothelium were also αSMA-positive (Fig. 7.3d) [45]. The presence of PDGFRα-positive/αSMA-positive TC is quite intriguing. Indeed, when these cells were observed under the TEM, they showed some peculiarities compared to the typical TC such as a larger body and cell processes possessing attachment plaques (Fig. 7.3e) similar to the fibronexus typical of the myofibroblasts. These TCs, therefore, were indicated as 'hybrid' TC [45]. In summary, on the basis of our experience, two TC subtypes can be identified under the TEM in the human urinary bladder: one with the characteristic small oval body and at least two long, thin, varicose processes and the other with a larger body containing several cisternae of the rough endoplasmic reticulum (RER) and a well-developed Golgi apparatus.

7.3.2 Placenta

In the chorial villi, TCs were seen to show different immunohistochemical profiles [42]: some of them were c-kit-positive; many were CD34-positive and co-expressed vimentin and caveolin-1. Moreover, double labelling for c-kit and CD34 disclosed three cell populations, one c-kit-positive, one CD34-positive and a third one c-kit/CD34-positive; despite that, all of them shared the same TC ultrastructural profile [42]. Finally, in cultured cells from the same placental villi, some were double-positive for c-kit and iNOS and others for c-kit and VEGF [42].

7.3.3 Striated Muscle

In the skeletal muscle, TC expressed c-kit-, caveolin-1 and CD34 and secreted VEGF [43]. Several phenotypes of TC have been described in the myocardium [9, 50]. In particular, in the mouse heart, most but not all of the TC were CD34-positive and some c-kit-positive, and those isolated from the mouse, rat and human heart showed a CD34-/PDGFRα-positivity [50].

7.3.4 Liver

In the mouse and human liver, TCs were CD34/PDGFRα-positive [20, 47].

7.3.5 Female Reproductive Tract

In mice and monkeys, the PDGFRα-positive cells formed extensive networks in the ovary, oviduct and uterus, and their distribution and density were similar between the two species [33]. However, significant differences in gene expression have been found among these organs and also among cells from different tissue regions within the same organ (e.g., myometrium vs. endometrium). Under the TEM, two different types of TC were seen in the *lamina propria* of mouse oviducts but in different periods of life: one during oestrum and the other during pregnancy (personal observations, Fig. 7.4a). Interestingly, since the TC of the oviduct expresses both oestrogen and progesterone receptors [11], it can be reasonably concluded that, in this organ, there are not two TC subtypes but a unique type that, under specific stimuli such as those linked to pregnancy, acquires different features. Finally, it has been reported that isolated human uterine telocytes are CD34-/PDGFRα-positive [14].

All above data indicate that TC shows several immunohistochemical differences depending on the organ and/or animal species. Beyond these differences, also technical

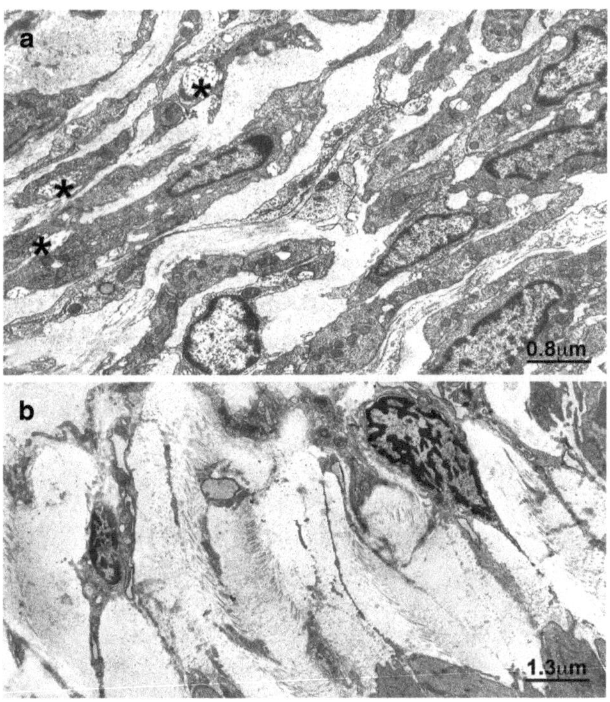

Fig. 7.4 (**a, b**) Transmission electron microscope. (**a**) Mouse oviduct during pregnancy. Telocytes are rich in organelles and embrace small blocks of new synthetized extracellular matrix (*asterisks*). (**b**) Submucosa of rabbit colon. The telocyte 3-D network embraces thick collagen bundles likely determining their orientation. Bar: (**a**) = 0.8 μm, (**b**) = 1.3 μm

factors might cause controversial results. The various laboratories often use different technical procedures for fixation, antibody unmasking and so on. Moreover and unpredictably, it happens sometimes that antibodies purchased by the same provider give controversial or no results when bought at different times. In the presence of such events, two main points could help in understanding whether or not the labelled cells are likely TC: if they form networks and if they are CD34-positive. Indeed, among all the immunohistochemical markers that label the TC, the CD34 seems to be that shared by most of them.

7.4 TC Roles

Many roles have been suggested for TC, presumably too many for each TC subtype. Nevertheless, it is likely that all TC subtypes play one common role. About it, it has to be borne in mind that TCs are unanimously considered organizers of the connective tissue by their ability to form 3-D networks, either alone or intermingled with other cell types, embracing and, probably, determining the orientation of the collagen and elastic fibres (Figs. 7.2b and 7.4b). Moreover, the TC, by forming a 3-D scaffold, should be able to follow organ distension and relaxation, to avoid anomalous organ deformation and to control blood vessels closure or rheology [4, 6, 13, 19, 23, 34, 38, 40, 44–46].

In our opinion, it is equally reasonable to conceive that each subtype plays an own organ-/tissue-specific role. Among the proposed organ-specific role, a TC net might participate in the control of muscle tone and motor activity [16]. In the gut, the mixed TC and ICC net has been correlated with the possibility that TC might favour the spreading of the neurotransmission signals directed to the ICC, which in turn are electrically coupled to the smooth muscle cells [34]. In favour of this hypothesis, it has been recently suggested that the 'smooth muscle cells are electrically coupled to both ICC and PDGFRα-positive cells (the cells we consider to correspond to the TC) forming an integrated unit called the SIP syncytium' [28]. More, it has been hypothesized that the 'hybrid TC' (Fig. 7.3d,e) present in the urinary bladder shares myofibroblast functions (i.e., participate to the sensory system described in the bladder) and also have the possibility, on demand, to differentiate in organ resident myofibroblasts [18, 45]. In the oviduct during pregnancy, since TCs are able to acquire features similar to those of the fibroblasts, they might produce the extracellular matrix (ECM) (Fig. 7.4a). This possibility is also suggestive for a close relationship between TC and fibroblasts [46]. However, at present whether TC might differentiate into fibroblasts and vice versa remains unanswered.

Among other fascinating roles, TCs have been proposed to be involved in the repair and regeneration of different tissues and organs, including the heart, lung, skeletal muscle, skin, meninges and choroid plexus, eye, liver, uterus and urinary system [5, 17–19, 36, 38, 46]. Moreover, it has also been demonstrated that TCs have a morphogenetic role, and, at present, the best example has been reported for the cardiac TC [1, 4, 36].

7.5 Conclusions

All what mentioned and discussed above induces us to conclude that many of the different interstitial cell types presently described likely correspond to a unique type of which several subtypes can be identified, each of them organ specific for morphology and role. Briefly, in our opinion, TCs are a heterogeneous population of cells that contributes to different aspects of physiological regulation in the various anatomical niches they occupy.

References

1. Albulescu R, Tanase C, Codrici E, Popescu DI, Cretoiu SM, Popescu LM. The secretome of myocardial telocytes modulates the activity of cardiac stem cells. J Cell Mol Med. 2015;19:1783–94.
2. Alunno A, Ibba-Manneschi L, Bistoni O, Rosa I, Caterbi S, Gerli R, Manetti M. Telocytes in minor salivary glands of primary Sjögren's syndrome: association with the extent of inflammation and ectopic lymphoid neogenesis. J Cell Mol Med. 2015;19:1689–96.
3. Baker SA, Hennig GW, Salter AK, Kurahashi M, Ward SM, Sanders KM. Distribution and Ca(2+) signalling of fibroblast-like (PDGFR(+)) cells in the murine gastric fundus. J Physiol. 2013;591(Pt 24):6193–208.
4. Bani D, Formigli L, Gherghiceanu M, Faussone-Pellegrini MS. Telocytes as supporting cells for myocardial tissue organization in developing and adult heart. J Cell Mol Med. 2010;14:2531–8.
5. Bei Y, Wang F, Yang C, Xiao J. Telocytes in regenerative medicine. J Cell Mol Med. 2015;19:1441–54.
6. Bosco C, Díaz E, Gutiérrez R, González J, Parra-Cordero M, Rodrigo R, Barja P. A putative role for telocytes in placental barrier impairment during preeclampsia. Med Hypotheses. 2015;84:72–7.
7. Briggs Boedtkjer D, Rumessen J, Baandrup U, Skov Mikkelsen M, Telinius N, Pilegaard H, Aalkjaer C, Hjortdal V. Identification of interstitial Cajal-like cells in the human thoracic duct. Cells Tissues Organs. 2013;197:145–58.
8. Cantarero Carmona I, Luesma Bartolomé MJ, Junquera Escribano C. Identification of telocytes in the lamina propria of rat duodenum: transmission electron microscopy. J Cell Mol Med. 2011;15:26–30.
9. Chang Y, Li C, Lu Z, Li H, Guo Z. Multiple immunophenotypes of cardiac telocytes. Exp Cell Res. 2015;338(2):239–44. doi:10.1016/j.yexcr.2015.08.012. Epub 2015 Aug 21.
10. Chen X, Zheng Y, Manole CG, Wang X, Wang Q. Telocytes in human oesophagus. J Cell Mol Med. 2013;7:1506–12.
11. Cretoiu SM, Cretoiu D, Suciu L, Popescu LM. Interstitial Cajal-like cells of human Fallopian tube express estrogen and progesterone receptors. J Mol Histol. 2009;40:387–94.
12. Cretoiu D, Cretoiu SM, Simionescu AA, Popescu LM. Telocytes, a distinct type of cell among the stromal cells present in the lamina propria of jejunum. Histol Histopathol. 2012;27:1067–78.
13. Cretoiu SM, Popescu LM. Telocytes revisited. Biomol Concepts. 2014;5:353–69.
14. Cretoiu SM, Radu BM, Banciu A, Banciu DD, Cretoiu D, Ceafalan LC, Popescu LM. Isolated human uterine telocytes: immunocytochemistry and electrophysiology of T-type calcium channels. Histochem Cell Biol. 2015;143:83–94.
15. Dawidowicz J, Szotek S, Matysiak N, Mielańczyk Ł, Maksymowicz K. Electron microscopy of human fascia lata: focus on telocytes. J Cell Mol Med. 2015. doi:10.1111/jcmm.12665.

16. Diaz-Flores L, Gutierrez R, Saez FJ, Diaz-Flores Jr L, Madrid JF. Telocytes in neuromuscular spindles. J Cell Mol Med. 2013;17:457–65.
17. Díaz-Flores L, Gutiérrez R, García MP, Sáez FJ, Díaz-Flores Jr L, Valladares F, Madrid JF. CD34+ stromal cells/fibroblasts/fibrocytes/telocytes as a tissue reserve and a principal source of mesenchymal cells. Location, morphology, function and role in pathology. Histol Histopathol. 2014;29:831–70. Review.
18. Díaz-Flores L, Gutiérrez R, García MP, González M, Sáez FJ, Aparicio F, Díaz-Flores Jr L, Madrid JF. Human resident CD34+ stromal cells/telocytes have progenitor capacity and are a source of αSMA+ cells during repair. Histol Histopathol. 2015;30:615–27.
19. Faussone-Pellegrini MS, Popescu LM. Telocytes. Biomol Concepts. 2011;2:481–9.
20. Fu S, Wang F, Cao Y, Huang Q, Xiao J, Yang C, Popescu LM. Telocytes in human liver fibrosis. J Cell Mol Med. 2015;19:676–83.
21. Gevaert T, De Vos R, Van Der Aa F, Joniau S, van den Oord J, Roskams T, De Ridder D. Identification of telocytes in the upper lamina propria of the human urinary tract. J Cell Mol Med. 2012;16:2085–93.
22. Gevaert T, Vanstreels E, Daelemans D, Franken J, Van Der Aa F, Roskams T, Ridder D. Identification of different phenotypes for the interstitial cells in the upper and deep lamina propria of the dome of the human bladder. J Urol. 2014;192:1555–63.
23. Hinescu ME, Popescu LM, Gherghiceanu M, Faussone-Pellegrini MS. Interstitial Cajal-like cells in rat mesentery: an ultrastructural and immunohistochemical approach. J Cell Mol Med. 2008;12:260–70.
24. Johnston L, Woolsey S, Cunningham RM, O'Kane H, Duggan B, Keane P, McCloskey KD. Morphological expression of KIT positive interstitial cells of Cajal in human bladder. J Urol. 2010;184:370–7.
25. Kurahashi M, Nakano Y, Hennig GW, Ward SM, Sanders KM. Platelet-derived growth factor receptor α-positive cells in the tunica muscularis of human colon. J Cell Mol Med. 2012;16:1397–404.
26. Kurahashi M, Nakano Y, Peri LE, Townsend JB, Ward SM, Sanders KM. A novel population of subepithelial platelet-derived growth factor receptor α-positive cells in the mouse and human colon. Am J Physiol Gastrointest Liver Physiol. 2013;304:G823–34.
27. Koh BH, Roy R, Hollywood MA, Thornbury KD, McHale NG, Sergeant GP, Hatton WJ, Ward SM, Sanders KM, Koh SD. Platelet-derived growth factor receptor-α cells in mouse urinary bladder: a new class of interstitial cells. J Cell Mol Med. 2012;16:691–700.
28. Lee H, Koh BH, Peri LE, Sanders KM, Koh SD. Purinergic inhibitory regulation of murine detrusor muscles mediated by PDGFRα+ interstitial cells. J Physiol. 2014;592:1283–93.
29. Li H, Zhang H, Yang L, Lu S, Ge J. Telocytes in mice bone marrow: electron microscope evidence. J Cell Mol Med. 2014;18:975–8.
30. Nicolescu MI, Bucur A, Dinca O, Rusu MC, Popescu LM. Telocytes in parotid glands. Anat Rec (Hoboken). 2012;295:378–85.
31. Nicolescu MI, Popescu LM. Telocytes in the interstitium of human exocrine pancreas: ultrastructural evidence. Pancreas. 2012;41:949–56.
32. Peri LE, Sanders KM, Mutafova-Yambolieva VN. Differential expression of genes related to purinergic signaling in smooth muscle cells, PDGFRα-positive cells, and interstitial cells of Cajal in the murine colon. Neurogastroenterol Motil. 2013;25:e609–20.
33. Peri LE, Koh BH, Ward GK, Bayguinov Y, Hwang SJ, Gould TW, Mullan CJ, Sanders KM, Ward SM. A novel class of interstitial cells in the mouse and monkey female reproductive tracts. Biol Reprod. 2015;92:102. doi:10.1095/biolreprod.114.124388.
34. Pieri L, Vannucchi MG, Faussone-Pellegrini MS. Histochemical and ultrastructural characteristics of an interstitial cell type different from ICC and resident in the muscle coat of human gut. J Cell Mol Med. 2008;12:1944–55.
35. Popescu LM, Ciontea SM, Cretoiu D. Interstitial Cajal-like cells in human uterus and fallopian tube. Ann N Y Acad Sci. 2007;1101:139–65. Review.

36. Popescu LM, Gherghiceanu M, Manole CG, Faussone-Pellegrini MS. Cardiac renewing: interstitial Cajal-like cells nurse cardiomyocyte progenitors in epicardial stem cell niches. J Cell Mol Med. 2009;13:866–86.
37. Popescu LM, Faussone-Pellegrini MS. TELOCYTES – a case of serendipity: the winding way from Interstitial Cells of Cajal (ICC), via Interstitial Cajal-Like Cells (ICLC) to TELOCYTES. J Cell Mol Med. 2010;14:729–40.
38. Popescu LM, Curici A, Wang E, Zhang H, Hu S, Gherghiceanu M. Telocytes putative stem cells in ageing human heart. J Cell Mol Med. 2015;19:31–45.
39. Rasmussen H, Rumessen JJ, Hansen A, Smedts F, Horn T. Ultrastructure of Cajal-like interstitial cells in the human detrusor. Cell Tissue Res. 2009;335:517–27.
40. Richter M, Kostin S. The failing human heart is characterized by decreased numbers of telocytes as result of apoptosis and altered extracellular matrix composition. J Cell Mol Med. 2015. doi:10.1111/jcmm.12664.
41. Rusu MC, Folescu R, Mănoiu VS, Didilescu AC. Suburothelial interstitial cells. Cells Tissues Organs. 2014;199:59–72.
42. Suciu L, Popescu LM, Gherghiceanu M, Regalia T, Nicolescu MI, Hinescu ME, Faussone-Pellegrini MS. Telocytes in human term placenta: morphology and phenotype. Cells Tissues Organs. 2010;92:325–39.
43. Suciu LC, Popescu BO, Kostin S, Popescu LM. Platelet-derived growth factor receptor-β-positive telocytes in skeletal muscle interstitium. J Cell Mol Med. 2012;16:701–7.
44. Vannucchi MG, Traini C, Manetti M, Ibba-Manneschi L, Faussone-Pellegrini MS. Telocytes express PDGFRα in the human gastrointestinal tract. J Cell Mol Med. 2013;17:1099–108.
45. Vannucchi MG, Traini C, Guasti D, Del Popolo G, Faussone-Pellegrini MS. Telocytes subtypes in human urinary bladder. J Cell Mol Med. 2014;18:2000–8.
46. Vannucchi, D Bani, M-S Faussone-Pellegrini. Telocytes contribute as cell progenitors and differentiation inductors in tissue regeneration. Curr Stem Cell Res Ther. 2016;11(5):383–9. [Epub ahead of print].
47. Wang F, Bei Y, Zhao Y, Song Y, Xiao J, Yang C. Telocytes in pregnancy-induced physiological liver growth. Cell Physiol Biochem. 2015;36:250–8.
48. Yang Y, Sun W, Wu SM, Xiao J, Kong X. Telocytes in human heart valves. J Cell Mol Med. 2014;18:759–65.
49. Zheng Y, Zhu T, Lin M, Wu D, Wang XJ. Telocytes in the urinary system. Transl Med. 2012;10:188.
50. Zhou Q, Wei L, Zhong C, Fu S, Bei Y, Huică RI, Wang F, Xiao J. Cardiac telocytes are double positive for CD34/PDGFR-α. J Cell Mol Med. 2015;19:2036–42.

Maria-Giuliana Vannucchi MGV is a specialist in neurology. She spent 2 years (1990–1991) at the University of Yale in the Department of Neurobiology working on brain ageing. Presently, she is an Associate Professor in Histology and works in the Histology and Embryology Unit, Department of Experimental and Clinical Medicine at the University of Florence.

MGV has been working for several years in the field of the neurotransmission either in the central nervous system or in the periphery at the level of the enteric nervous system. She also studied stromal cells formerly named interstitial Cajal-like cells (ICLC) and now telocytes (TC) in the gastrointestinal and urinary tract. All these topics have been studied by electron microscope examination and immunohistochemistry in normal and pathological conditions and during differentiation. Her results have been published in numerous and prestigious International Journals, most of which with IF. Hirsch index in 2015 is 26.

To date, MGV is author of 80 full papers and from 1995 till now has been asked as referee of 18 international journals.

Maria-Simonetta Faussone-Pellegrini born and still living in Florence, Italy, is Professor Emeritus of Histology at the Faculty of Medicine of the University of Florence, member of the Editorial Board of the Journal of Cellular and Molecular Medicine and of the Italian Journal of Anatomy and Embryology, and Honorary Foreign Member of the Romanian Academy of Medical Science. Her main researches deal on the Interstitial Cells of Cajal (ICC) that in 1977 she successfully proposed to be the intestinal pacemaker cells. She also studied enteric and cerebral neuronal and glial cells, gastrointestinal and urinary tract muscles, stromal cells formerly named Interstitial Cajal-like Cells (ICLC) and now telocytes (TC). All these topics have been studied by electron microscope examination and immunohistochemistry in normal and pathological conditions and during differentiation. Her publications are about 200, most of which in Journals with IF. Hirsch index in 2015 is 34.

Chapter 8
Telocytes in Cardiac Tissue Architecture and Development

Daniele Bani

Abstract The heart is a paradigm of organ provided with unique three-dimensional tissue architecture that is molded during complex organogenesis processes and is required for the heart's physiological function. The cardiac stroma plays a critical role in the formation and maintenance of the normal heart architecture, as well as of its changes occurring in cardiac diseases. Recent studies have shown that the cardiac stroma, including the epicardium, myocardial interstitium, and endocardium, contains typical telocytes: these cells establish complex spatial relationships with cardiomyocytes and cardiac stem cells suggestive for a regulatory role over three-dimensional organization of heart tissues. Telocytes appear early during prenatal heart development and represent a major stromal cell population in the adult heart. Numerous studies have highlighted that telocytes, through juxtacrine and paracrine mechanisms, can behave as nursing cells for cardiac muscle stem cells modulating their growth and differentiation. On these grounds, a possible role of telocytes in cardiac regeneration can be postulated: this hypothesis is supported by recent experimental findings that reduction of cardiac telocytes due to hypoxia may concur to explain the negligible regenerative ability of the post-infarcted heart, while grafting of telocytes in the injured myocardium improves adverse heart remodeling. The increasing knowledge on the properties of cardiac telocytes is orienting the research toward their role as key regulators of the three-dimensional architecture of the heart and new promising targets for cardiac regenerative medicine.

8.1 Introduction

The mechanisms presiding over the three-dimensional assembling of cells to give rise to differentiated tissues and organs are a major area of investigation in stem cell

D. Bani, MD
Department of Experimental & Clinical Medicine, Section of Anatomy & Histology, Research Unit of Histology & Embryology, University of Florence, viale G.Pieraccini, 6, 50139 Florence, Italy
e-mail: daniele.bani@unifi.it

X. Wang, D. Cretoiu (eds.), *Telocytes*, Advances in Experimental Medicine and Biology, 913, DOI 10.1007/978-981-10-1061-3_8

biology and have a pivotal importance for regenerative medicine. What we have learned from the numerous experimental and clinical studies on stem cell grafting for organ repair is that the probability of a successful outcome is substantially reduced when the host organ has a complex three-dimensional architecture and negligible when this architecture has been altered by disease. Stem cell-based treatment of the post-infarcted, failing heart is a typical example of the substantial inability of the regenerative approach to re-create a structurally complex tissue such as the myocardium [44].

The events and mechanisms responsible for the transition from undifferentiated aggregates of embryonic stem cells to well-defined organ precursors are basically characterized by the appearance of the mesenchyme and extracellular matrix (ECM) to give rise to the stroma. This is capable of forming a supporting frame required to integrate cells into functional assemblies and regulate their differentiation, thereby determining the proper cell composition and architecture of the organs [1, 12, 19, 20, 45], Hay et al. 1983, 2005]. For instance, during heart development, spatiotemporal deposition of fibronectin, a key molecule mediating cell-ECM interaction, influences the correct migration of myocardial precursor cells to form the primitive heart tube [42]. Moreover, evidence has been provided that spatially oriented changes in cell growth, migration, and differentiation, such as those involved in heart compartmentalization and valve formation, are influenced by mechanical forces originated by hemodynamic stresses in the beating primitive heart tube and mediated by the cardiac stroma [22].

8.2 Morpho-Functional Features of Cardiac Telocytes and Their Possible Roles in the Adult Heart

As stated above, the heart is a paradigm of organ provided with unique three-dimensional tissue architecture that is molded during complex organogenesis processes and is required for the heart's physiological function [10]. The cardiac stroma plays a critical role in the formation and maintenance of the normal heart architecture, as well as of its changes occurring in cardiac diseases [27].

Fibroblasts are considered the main cardiac stromal cell type: they are held responsible for the formation and renewal of ECM, but, being differentiated cells, in the normal adult heart, their proliferative capability is very low and barely sufficient for self-renewal [10]. Accordingly, like other differentiated cells, they grow very slowly and rapidly undergo senescence when cultured in vitro [8]. Morphologically, typical fibroblasts can be identified in the cardiac stroma by transmission electron microscopy. They appear as fusiform or stellate cells with euchromatic nuclei and cytoplasms containing numerous RER profiles and a well-developed Golgi apparatus. Their cell surface lacks a basal lamina and often forms grooves containing bundles of collagen microfibrils, indicating that these cells can preside over the spatial orientation of the newly formed ECM macromolecules. Cardiac fibroblasts can also

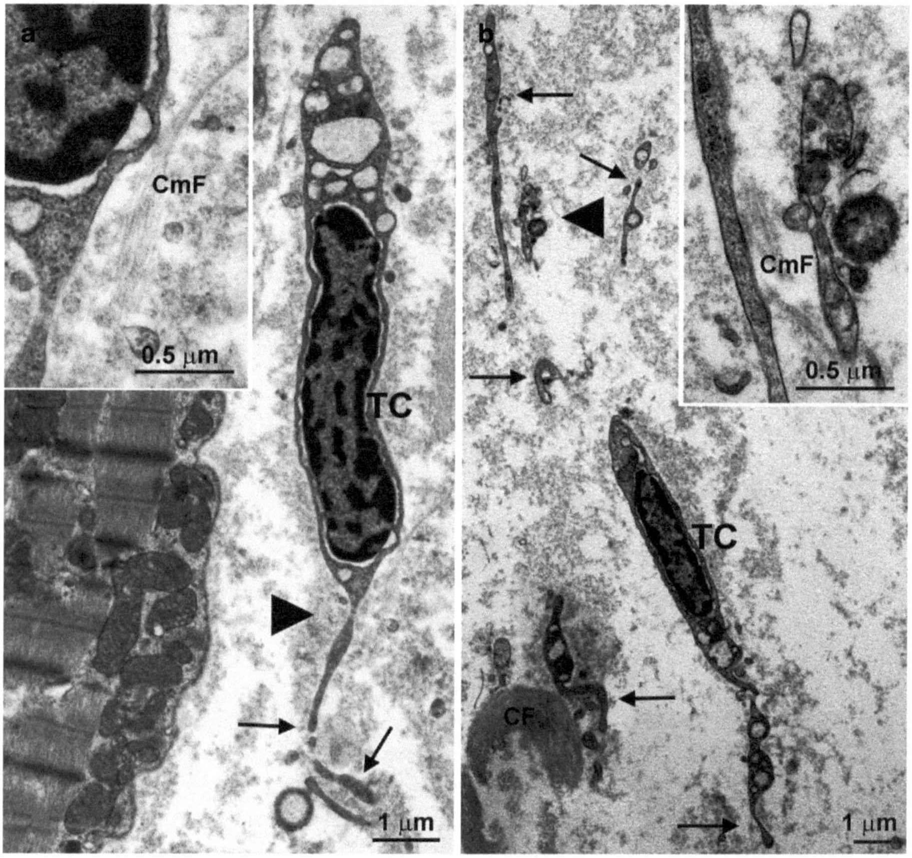

Fig. 8.1 Typical telocytes (TC) in the stroma of adult swine heart (**a**, **b**). These cells show spindle-like shape, heterochromatic nuclei, and scarce cytoplasms containing cisternae of rough endoplasmic reticulum (RER). They are provided with elongated telopodes (*arrows*). Collagen fibers (*CF*) and microfibrils (*CmF*) are adjacent to the telopodes (**b**). The insets show higher magnifications of the areas indicated by the arrowheads (Reproduced from Bani and Nistri [5], with permission)

regulate cardiomyocyte proliferation and growth during development through paracrine and juxtacrine signals [23, 35].

Owing to the pioneering studies of Professor L.M. Popescu and co-workers, the cardiac stroma has also been demonstrated to harbor telocytes [21, 25], which have been identified as a normal cell population in the adult heart [5, 26, 41]. In particular, cells with the typical features of telocytes have been detected in all the connective tissue components of the heart, including the myocardial stroma [11, 21, 25, 39], the endocardium and heart valves [15, 46], and the epicardium [16, 36, 37]. As detailed in another authoritative chapter of this book, the most characteristic and reliable hallmarks of cardiac telocytes are their ultrastructural features: a small, irregular cell body (average diameter of 10 μm), containing a nucleus with a peripheral heterochromatin rim, and a scarce cytoplasm with few organelles. The cell surface is characterized by very long, thin processes, or telopodes, whose number determines the shape of the cell body (spindle, triangular, or stellate), and lacks a basal lamina (Fig. 8.1). At variance with cardiac fibroblasts, which can be viewed as terminally differentiated cells, cardiac telocytes retain stemness features, as judged by the expression of the hematopoietic stem cell marker CD34, the protein encoded by the embryonic gene Nanog, and the myocardial stem cell markers sca-1 and c-kit [9, 50], as well as by the ability of these cells to be cultured and expanded in vitro [6, 34, 40, 48]. Moreover, cardiac telocytes have been found to establish complex interactions with cardiomyocytes through juxtacrine (i.e., direct cell-cell contacts) [17, 18] as well as paracrine (i.e., release of exosomes containing short-range growth factors and cytokines) [14] mechanisms: these findings suggest that the two cardiac cell types are functionally correlated and might represent a functional unit.

The putative relation between telocytes and fibroblasts coexisting in the cardiac stroma is an intriguing matter and an interesting subject for future investigation. During organ morphogenesis, ECM is produced and organized by mesenchymal cells which progressively differentiate into fibroblasts. Later on, telocytes appear in the stroma and sometimes represent the largest population at term development, while typical fibroblasts are scarce. Therefore, it could be hypothesized that telocytes might differentiate from preexisting fibroblasts and that both cell types might be involved in the synthesis/organization/maintenance of cardiac ECM. Possibly, fibroblasts could be chiefly held responsible for ECM production, while telocytes would preside over its spatial organization. This view is supported by the observations that, in diseases associated with a consistent loss of telocytes, such as systemic sclerosis, fibroblasts increase in number and produce a large amount of ECM but are unable to organize it correctly; thus, the three-dimensional architecture of the damaged tissue is subverted and prone to hypoxia and fibrosis [28, 29]. These findings clearly indicate that an important functional relationship exists between these two cell types, but the question whether fibroblasts might differentiate into telocytes, and/or vice versa, remains unanswered.

8.3 Morpho-functional Features of Telocytes in the Developing Heart

Heart development takes place through a complex spatial-temporal interplay between different cell types needed to the heart for attaining its final size, shape, histological structure, and function. During this process, muscle and stromal stem cells are mutually stimulated to proliferate, differentiate, and integrate into a three-dimensional functional assembly. In this context, stromal cells featuring telocytes have been described in the developing mouse heart, where they establish close spatial relationships with cardiac muscle cells [4, 13]. In particular, cells with the ultrastructural features of telocytes were found in the epicardium of developing

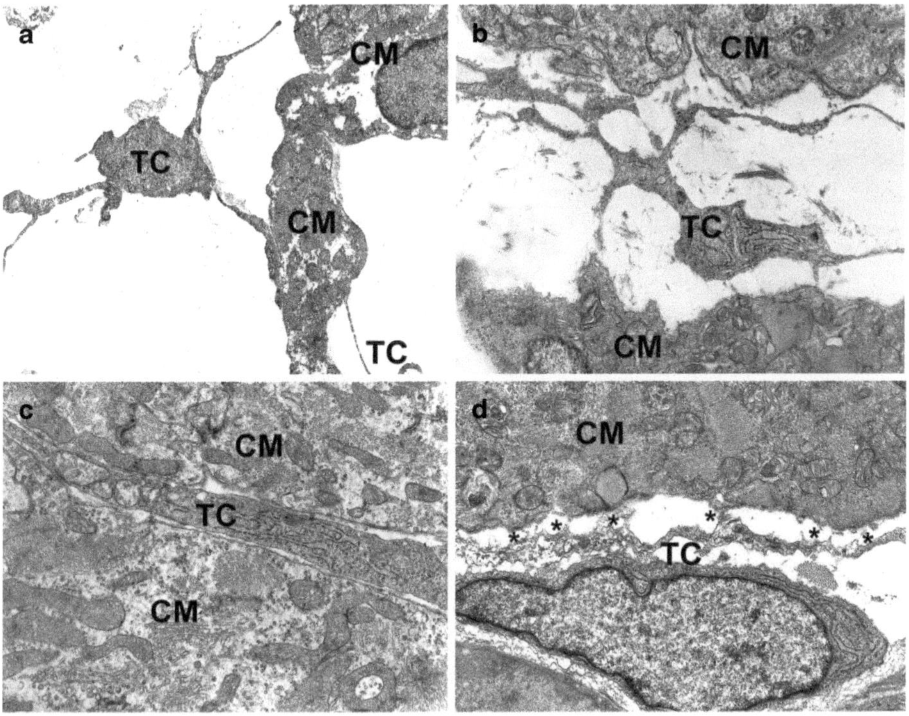

Fig. 8.2 Telocytes in the developing mouse heart. At midpregnancy (**a**, **b**), cells featuring telocytes are located in the wide space that separates the columns of immature cardiomyocytes. By their long, thin processes, the telocytes are in contact and border the cardiomyocytes nearby. At birth (**c**, **d**), telocytes acquire more differentiated features: (**c**) a telocyte process containing several rough endoplasmic cisternae is seen in the interstitial stroma between two cardiomyocytes, which appears reduced in size; (**d**) a telocyte process establishing numerous interactions (*asterisks*) with an adjacent cardiomyocyte, in the form of focal plasma membrane contacts. *CM* cardiomyocytes, *TC* telocytes (Reproduced from Faussone-Pellegrini and Bani [13], with permission]

mouse hearts since midpregnancy (day 14 of gestation) (Fig. 8.2). Telocytes progressively extended toward the myocardium where, with their cell bodies and telopodes, they created a framework co-oriented with the longitudinal and transverse axes of the cardiomyocytes [4]. These findings suggest that telocytes may be capable of molding the three-dimensional architecture of the developing heart and acting as nursing cells for muscle stem cells during their differentiation pathway. Indeed, in later (perinatal) developmental stages, telocytes appeared to mediate myocardial compaction from rudimental myocardial trabeculae and promote orthodox spatial organization of the ventricular wall [4]. The hypothesis of a morphogenetic role of cardiac telocytes has been further supported by the results of a study performed on in vitro reconstructed myocardium, in which telocytes were deemed necessary to the engineered tissue for acquisition of a proper three-dimensional architecture [49].

8.4 Telocytes as Key Players of Cardiac Regeneration

It is known that the heart behaves as a postmitotic organ, because mature cardiomyocytes have no regenerative ability. However, the adult heart retains a pool of cardiac stem cells, chiefly located in epicardial cardiogenic niches close to coronary artery branching, which have been credited for spontaneous, albeit limited, myocardial regeneration after injury [7, 43]. However, this phenomenon is not functionally relevant because the myocardial progenitors in the adult heart have an intrinsically low regenerative potential and are operating in an unfavorable environment, as that occurring in the diseased myocardium [24]. In principle, the persistence of a pool of cardiac stem cells in the adult heart provides the background for possible therapeutic approaches [33]. Therefore, studies aimed at identifying possible strategies to improve the recruitment, self-renewal, and in situ differentiation of resident cardiac progenitors are topical and clinically relevant. In the adult heart, besides forming a framework in the myocardial stroma, telocytes have been also found in the cardiogenic niches, where they form close contacts with the resident cardiac stem cells (Fig. 8.3), strongly suggestive for their function as supporting and nursing cells [16, 36]. Moreover, in vitro coculture studies have shown that telocytes secrete a wide range of cytokines and exosomes by which they exert paracrine modulation of cardiac stem cell growth and differentiation [2]. The postulated cooperative role of telocytes on myocardial histogenesis during heart development is closely related to the possibility that telocytes could also be involved in supporting myocardial regeneration occurring in the adult heart upon pathologic events, such as ischemia-induced infarction. This attractive hypothesis is based on multiple clues. First, like dermal telocytes in systemic sclerosis [29], cardiac telocytes have been shown to be prone to ischemic damage, and their framework in the heart stroma is rapidly disrupted during myocardial infarction [32]. The disappearance of cardiac telocytes may thus concur to explain the negligible regenerative ability of the post-infarcted heart. Second, a recent in vivo study in mice subjected to myocardial infarction has demonstrated substantial morpho-functional heart improvement upon grafting of induced pluripotent mesenchymal stem cells [32]. Of note, myocardial improvement coincided with the appearance of interconnecting telocytes, likely originated by in situ differentiation of the grafted mesenchymal precursors, in the stroma of the injured ventricular myocardium [32]. In fact, myocardial stem cells seem to require the close interaction with stromal cells for correct recruitment and commitment, a mechanism that may preside over the physiological turnover of the myocardium as well as disease-induced heart repair [3, 30]. The above data concur to suggest that cardiac telocytes are likely needed to stimulate and sustain the regenerative potential of the myocardium. This hypothesis deserves to be further investigated because it may lead to improve the current protocols of stem cell therapy for cardiac regeneration. In keeping with this concept, it has been recently shown that grafting of purified, cultured cardiac telocytes to rats subjected to myocardial infarction reduced infarct size and improved myocardial function. These effects were related to the reconstruction of a network of telocytes in the injured myocardium, which resulted in increased angiogenesis and reduced fibrosis [47].

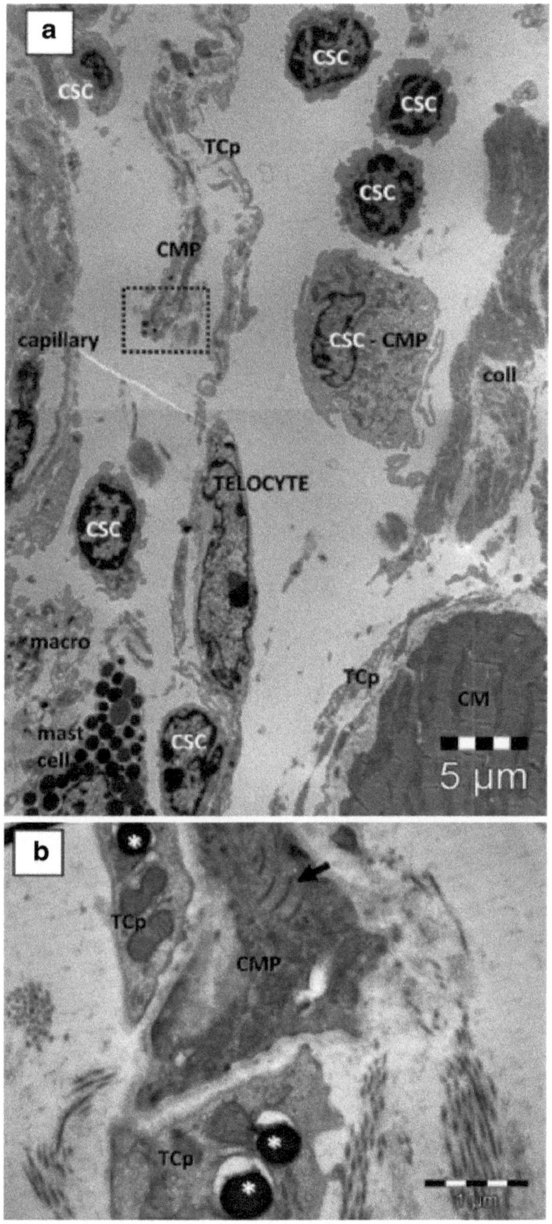

Fig. 8.3 Telocytes in a cardiogenic niche of mouse heart epicardium. (**a**) Putative cardiac stem cells (CSC), isolated or in small groups, cardiomyocyte progenitors (CMP), and cells with intermediate features (CSC-CMP) can be seen embedded within a loose extracellular matrix. (**b**) Details at higher magnification of the area marked in (**a**) with a *dotted square*. The CMP shows characteristic leptomeres (*arrow*), which precede the appearance of microfibrils, and is in close contact with telocyte processes (TCp) containing dense granules (*asterisk*) suggestive for paracrine signaling. *CM* cardiomyocyte; *macro*, macrophage; *coll*, collagen fibers (Reproduced from Gherghiceanu and Popescu [16], with permission)

An additional presumptive role of stromal cells in cardiac regeneration emerges from the observation that these cells can be reprogrammed in vitro to differentiate into cardiomyocytes [38]. However, whether stromal-to-muscle cell reprogramming might spontaneously occur in the adult heart, and whether telocytes might provide a contribution to such trans-differentiation, remains a matter of speculation and an area for future investigation.

8.5 Conclusions and Perspectives

The above notions provide support to the view that cardiac telocytes can play a key role in the regulation of heart morphogenesis and regeneration. In fact, they appear to possess the capability to shape a proper three-dimensional scaffold composed of their cell bodies and elongated processes and stimulate the growth and differentiation of muscle stem cells to give rise to the complex multicellular assembly constituting the heart. In this context, during embryonic development, mesenchymal cells have been found to extend thin filopodes interpreted as sensors for spatial information necessary for correct morphogenesis [31]. Such thin filopodes are very similar to telopodes, the typical processes of telocytes, which are conceivably provided with this sensing function. In pathological conditions, under the effects of local pro-inflammatory mediators, stromal cells are induced to differentiate into myofibroblasts, which are unable to generate a shaping framework but possess a fibrogenic phenotype needed for scarring, the typical emergency healing response to injury of the most evolved organisms aimed at preserving survival at the expense of function. Notwithstanding this, a large number of reports in the literature on regenerative medicine indicate that, under appropriate conditions, the original ability of stromal cells to orchestrate organ regeneration can be preserved or resumed. The increasing knowledge on the properties of cardiac telocytes is orienting the research toward their role as key regulators of the three-dimensional architecture of the heart and new promising targets for cardiac regenerative medicine.

References

1. Adams JC, Watt FM. Regulation of development and differentiation by the extracellular matrix. Development. 1993;117:1183–98.
2. Albulescu R, Tanase C, Codrici E, Popescu DI, Cretoiu SM, Popescu LM. The secretome of myocardial telocytes modulates the activity of cardiac stem cells. J Cell Mol Med. 2015;19:1783–94.
3. Ausoni S, Sartore S. The cardiovascular unit as a dynamic player in disease and regeneration. Trends Mol Med. 2009;15:543–52.
4. Bani D, Formigli L, Gherghiceanu M, Faussone-Pellegrini MS. Telocytes as supporting cells for myocardial tissue organization in developing and adult heart. J Cell Mol Med. 2010;14:2531–8.

5. Bani D, Nistri S. New insights into the morphogenic role of stromal cells and their relevance for regenerative medicine. Lessons from the heart. J Cell Mol Med. 2014;18:363–70.
6. Bei Y, Zhou Q, Fu S, Lv D, Chen P, Chen Y, Wang F, Xiao J. Cardiac telocytes and fibroblasts in primary culture: different morphologies and immunophenotypes. PLoS One. 2015;10(2):e0115991.
7. Beltrami AP, Barlucchi L, Torella D, Baker M, Limana F, Chimenti S, Kasahara H, Rota M, Musso E, Urbanek K, Leri A, Kajstura J, Nadal-Ginard B, Anversa P. Adult cardiac stem cells are multipotent and support myocardial regeneration. Cell. 2003;114:763–76.
8. Carlson S, Trial J, Soeller C, Entman ML. Cardiac mesenchymal stem cells contribute to scar formation after myocardial infarction. Cardiovasc Res. 2011;91:99–107.
9. Chang Y, Li C, Lu Z, Li H, Guo Z. Multiple immunophenotypes of cardiac Telocytes. Exp Cell Res. 2015. pii: S0014-4827(15)30073-2.
10. Crawford JR, Haudek SB, Cieslik KA, Trial J, Entman ML. Origin of developmental precursors dictates the pathophysiologic role of cardiac fibroblasts. J Cardiovasc Transl Res. 2012;5:749–59.
11. Cretoiu D, Hummel E, Zimmermann H, Gherghiceanu M, Popescu LM. Human cardiac telocytes: 3D imaging by FIB-SEM tomography. J Cell Mol Med. 2014;18:2157–64.
12. Doljanski F. The sculpturing role of fibroblast-like cells in morphogenesis. Perspect Biol Med. 2004;47:339–56.
13. Faussone-Pellegrini MS, Bani D. Relationships between telocytes and cardiomyocytes during pre- and post-natal life. J Cell Mol Med. 2010;14:1061–3.
14. Fertig ET, Gherghiceanu M, Popescu LM. Extracellular vesicles release by cardiac telocytes: electron microscopy and electron tomography. J Cell Mol Med. 2014;18:1938–43.
15. Gherghiceanu M, Manole CG, Popescu LM. Telocytes in endocardium: electron microscope evidence. J Cell Mol Med. 2010;14:2330–4.
16. Gherghiceanu M, Popescu LM. Cardiomyocyte precursors and telocytes in epicardial stem cell niche: electron microscope images. J Cell Mol Med. 2010;14:871–7.
17. Gherghiceanu M, Popescu LM. Heterocellular communication in the heart: electron tomography of telocyte-myocyte junctions. J Cell Mol Med. 2011;15:1005–11.
18. Gherghiceanu M, Popescu LM. Cardiac telocytes - their junctions and functional implications. Cell Tissue Res. 2012;348:265–79.
19. Hay ED. Cell and extracellular matrix: their organization and mutual dependence. Mol Cell Biol. 1983;2:509–48.
20. Hay ED. The mesenchymal cell, its role in the embryo, and the remarkable signaling mechanisms that create it. Dev Dyn. 2005;233:706–20.
21. Hinescu ME, Popescu LM. Interstitial Cajal-like cells (ICLC) in human atrial myocardium. J Cell Mol Med. 2005;9:972–5.
22. Hove JR, Köster RW, Forouhar AS, Acevedo-Bolton G, Fraser SE, Gharib M. Intracardiac fluid forces are an essential epigenetic factor for embryonic cardiogenesis. Nature. 2003; 421:172–7.
23. Ieda M, Tsuchihashi T, Ivey KN, Ross RS, Hong TT, Shaw RM, Srivastava D. Cardiac fibroblasts regulate myocardial proliferation through beta1 integrin signaling. Dev Cell. 2009; 16:233–44.
24. Kajstura J, Urbanek K, Rota M, Bearzi C, Hosoda T, Bolli R, Anversa P, Leri A. Cardiac stem cells and myocardial disease. J Mol Cell Cardiol. 2008;45:505–13.
25. Kostin S, Popescu LM. A distinct type of cell in myocardium: interstitial Cajal-like cells [ICLCs]. J Cell Mol Med. 2009;13:295–308.
26. Kostin S. Myocardial telocytes: a specific new cellular entity. J Cell Mol Med. 2010; 14:1917–21.
27. Lajiness JD, Conway SJ. The dynamic role of cardiac fibroblasts in development and disease. J Cardiovasc Transl Res. 2012;5:739–48.
28. Manetti M, Guiducci S, Ruffo M, et al. Evidence for progressive reduction and loss of telocytes in the dermal cellular network of systemic sclerosis. J Cell Mol Med. 2013;17:482–96.

29. Manetti M, Rosa I, Messerini L, Guiducci S, Matucci-Cerinic M, Ibba-Manneschi L. A loss of telocytes accompanies fibrosis of multiple organs in systemic sclerosis. J Cell Mol Med. 2014;18:253–62.
30. Mazhari R, Hare JM. Mechanisms of action of mesenchymal stem cells in cardiac repair: potential influences on the cardiac stem cell niche. Nat Clin Pract Cardiovasc Med. 2007;4 Suppl 1:S21–6.
31. McClay DR. The role of thin filopodia in motility and morphogenesis. Exp Cell Res. 1999;253:296–301.
32. Miao Q, Shim W, Tee N, Lim SY, Chung YY, Ja KP, Ooi TH, Tan G, Kong G, Wei H, Lim CH, Sin YK, Wong P. iPSC-derived human mesenchymal stem cells improve myocardial strain of infarcted myocardium. J Cell Mol Med. 2014;18:1644–54.
33. Müller P, Beltrami AP, Cesselli D, Pfeiffer P, Kazakov A, Böhm M. Myocardial regeneration by endogenous adult progenitor cells. J Mol Cell Cardiol. 2005;39:377–87.
34. Niculite CM, Regalia TM, Gherghiceanu M, Huica R, Surcel M, Ursaciuc C, Leabu M, Popescu LM. Dynamics of telopodes (telocyte prolongations) in cell culture depends on extracellular matrix protein. Mol Cell Biochem. 2015;398:157–64.
35. Noseda M, Schneider MD. Fibroblasts inform the heart: control of cardiomyocyte cycling and size by age-dependent paracrine signals. Dev Cell. 2009;16:161–2.
36. Popescu LM, Gherghiceanu M, Manole CG, Faussone-Pellegrini MS. Cardiac renewing: interstitial Cajal-like cells nurse cardiomyocyte progenitors in epicardial stem cell niches. J Cell Mol Med. 2009;13:866–86.
37. Popescu LM, Manole CG, Gherghiceanu M, Ardelean A, Nicolescu MI, Hinescu ME, Kostin S. Telocytes in human epicardium. J Cell Mol Med. 2010;14:2085–93.
38. Qian L, Huang Y, Spencer CI, Foley A, Vedantham V, Liu L, Conway SJ, Fu JD, Srivastava D. In vivo reprogramming of murine cardiac fibroblasts into induced cardiomyocytes. Nature. 2012;485:593–8.
39. Rusu MC, Pop F, Hostiuc S, Curca GC, Jianu AM, Paduraru D. Telocytes form networks in normal cardiac tissues. Histol Histopathol. 2012;27:807–16.
40. Suciu L, Nicolescu MI, Popescu LM. Cardiac telocytes: serial dynamic images in cell culture. J Cell Mol Med. 2010;14:2687–92.
41. Tao L, Wang H, Wang X, Kong X, Li X. Cardiac Telocytes. Curr Stem Cell Res Ther. 2016;11:404–9.
42. Trinh LA, Stainier DY. Fibronectin regulates epithelial organization during myocardial migration in zebrafish. Dev Cell. 2004;6:371–82.
43. Urbanek K, Torella D, Sheikh F, De Angelis A, Nurzynska D, Silvestri F, Beltrami CA, Bussani R, Beltrami AP, Quaini F, Bolli R, Leri A, Kajstura J, Anversa P. Myocardial regeneration by activation of multipotent cardiac stem cells in ischemic heart failure. Proc Natl Acad Sci U S A. 2005;102:8692–7.
44. Weil BR, Canty Jr JM. Stem cell stimulation of endogenous myocyte regeneration. Clin Sci (Lond). 2013;125:109–19.
45. Wessells NK. Tissue interactions and development. Menlo Park: The Benjamin/Cummings Publishing Company; 1977.
46. Yang Y, Sun W, Wu SM, Xiao J, Kong X. Telocytes in human heart valves. J Cell Mol Med. 2014;18:759–65.
47. Zhao B, Liao Z, Chen S, et al. Intramyocardial transplantation of cardiac telocytes decreases myocardial infarction and improves post-infarcted cardiac function in rats. J Cell Mol Med. 2014;18:780–9.
48. Zhou J, Zhang Y, Wen X, Cao J, Li D, Lin Q, Wang H, Liu Z, Duan C, Wu K, Wang C. Telocytes accompanying cardiomyocyte in primary culture: two- and three-dimensional culture environment. J Cell Mol Med. 2010;14:2641–5.
49. Zhou J, Wang Y, Zhu P, Sun H, Mou Y, Duan C, Yao A, Lv S, Wang C. Distribution and characteristics of telocytes as nurse cells in the architectural organization of engineered heart tissues. Sci China Life Sci. 2014;57:241–7.
50. Zhou Q, Wei L, Zhong C, Fu S, Bei Y, Huică RI, Wang F, Xiao J. Cardiac telocytes are double positive for CD34/PDGFR-α. J Cell Mol Med. 2015;19:2036–42.

Daniele Bani Born in Florence, Italy, September 20, 1961.

Degree in Medicine, with honors, University of Florence, 1986.

Fellowship from the Accademia Nazionale dei Lincei, 1987.

Fellowship from the Italian Association for Cancer Research (AIRC), 1989–1990.

Associate Professor of Histology, University of Florence Faculty of Medicine, November 1992. Full Professor of Histology, University of Florence Faculty of Medicine, November 2000 to date.

Main fields of expertise as a researcher: cell biology (cell culture, molecular biology, functional studies related to the biological effects of relaxin and of nitric oxide biology) and morphology (immunocytochemistry, morphometry, and electron microscopy).

Author or coauthor of 152 full-length papers in peer-reviewed journals, 16 articles on books, 1 international patent, and 85 presentations at international and Italian congresses. Invited speaker at international congresses.

Reviewer of international journals, including *Am J Physiol Cell Physiol*, *Am J Physiol Heart Circ Physiol*, *Biochem Pharmacol*, *Biochim Biophys Acta*, *Circulation*, *Circ Res*, *Diabetologia*, *Drug Discovery Today*, *Endocrinology*, *Exp Biol Med*, *FASEB Journal*, *Gynecol Obstet Invest*, *Hypertension*, *Int Immunopharmacol*, *It J Anat Embryol*, *J Appl Physiol*, *J Microsc*, *J Mol Cell Cardiol*, *J Pathol*, *J Pharmacol Exp Ther*, *J Sex Med*, *J Vasc Res*, *Med Sci Monit*, *Mol Cell Biochem*, *Mol Hum Reprod*, *Mol Ther*, *Nature Rev Cardiol*, *Pharmacol Ther*, *PLoS one*, *Regul Pept*, and *Reprod Biol Endocrinol*

Total impact factor of published articles, 594.9. H-index, 33

Chapter 9
The Potential Role of Telocytes for Tissue Engineering and Regenerative Medicine

Raymund E. Horch, Annika Weigand, Justus P. Beier, Andreas Arkudas, and Anja M. Boos

Abstract Despite recent advances in surgery, medicine and anaesthesiology as well as the development of microsurgical tissue transplantation, wear out of body parts remains a problem, and organ shortage does not allow to allocate enough donor organs for patients with vital diseases and conditions. The idea to create spare parts or spare organs from the patients own cells by combining engineering approaches to cellular and molecular medicine for th purpose of Tissue Engineering (TE) was fascinating when popularized in the early 1990ies. However clinically success was limited, mainly because of a lack in rapid vascularization of large scale TE replacement constructs useful for clinical purposes. The idea to utilize cells and cytokines to aid the human organism in gradually restoring lost tissue functions has drawn attention to the wider field of Regenerative Medicine (RM). Stem cells and putative stem cells, such as the recently discovered and meanwhile well described interstitial Telocytes, which are comprised of extremely long and thin prolongations named telopodes, may well become active players in the regenerative process. This article highlights the principles of TE and RM and the potential role of Telocytes with regard to tissue regeneration.

It was not until the end of the twentieth century that treating patients medically was based on the assumption that by doing so one could hope to reverse a pathologic process. The necessary loss of healthy tissue and/or tissue function was taken as an unavoidable drawback of this concept. Although complex plastic surgical reconstructions have been described already 600 before Christ in the Suśrutasamhitā Ayurveda Sanskrit text on medicine [1], surgery was mainly based on extirpative methods to cure diseases. Even the enormous advances in microsurgery and organ transplantation with their incredible achievements within the last decades depend

R.E. Horch (✉) • A. Weigand • J.P. Beier • A. Arkudas • A.M. Boos
Department of Plastic and Hand Surgery and Laboratory of Tissue Engineering and Regenerative Medicine, University Hospital Erlangen, Friedrich Alexander University Erlangen-Nuernberg FAU, Krankenhausstrasse 12, D-91054 Erlangen, Germany
e-mail: Raymund.Horch@uk-erlangen.de

X. Wang, D. Cretoiu (eds.), *Telocytes*, Advances in Experimental Medicine and Biology, 913, DOI 10.1007/978-981-10-1061-3_9

on donor sites to gain healthy tissue. This as well as donor organ shortage is a serious limitation toward a better cure of acute and chronic diseases without losing tissue function or hampering the patient's quality of life.

Attempts to develop prosthetic organs, which would be easily and universally available, such as the artificial heart, have met with limited success. Therefore a general use of synthetic tissues other than cochlear implants and prosthetic joints has not occurred up to now [2]. In contrast, distraction osteogenesis is an in vivo tissue engineering technique that gradually restores form by "coaching" bone to repair itself.

The prospect to culture replacement tissue and organs in the laboratory with the patient's own cells therefore seemed as a major breakthrough to overcome the current limitations of treatments. Pioneering inventions to serially culture cells under defined conditions and seeding them on appropriate scaffolds have not directly led to routine clinical therapies yet, mainly because the substitution of whole organs is a complex process and the multitude of influencing and interfering factors is not well enough understood. Nevertheless by deciphering the genetic code and by optimizing cell cultures and understanding ways to influence cell growth and modulate angiogenesis, tissue engineering (TE) has gained many insights that helped to better understand cellular and molecular mechanisms. It increased our knowledge of cellular pathophysiology and has expanded from the cellular to the biomolecular level. Today the detection of and in part the substitution of dysfunctional genes, molecules, cells, and organs have become clinically available in regenerative medicine (RM). In TE many approaches were hampered by the lack of vascularity of TE-generated replacement. Cells within larger constructs of clinically relevant size depend on sufficient nutrition from the very first moment they are implanted into the recipient organism. Our group has worked on the combination of TE with supermicrosurgical methods to generate intrinsic vascular supply to artificial constructs to overcome the remaining major problem with the lack of initial vascularization. We studied extrinsic and intrinsic vascularization strategies alone and in combination to accelerate vascularization (Figs. 9.1, 9.2, 9.3, and 9.4).

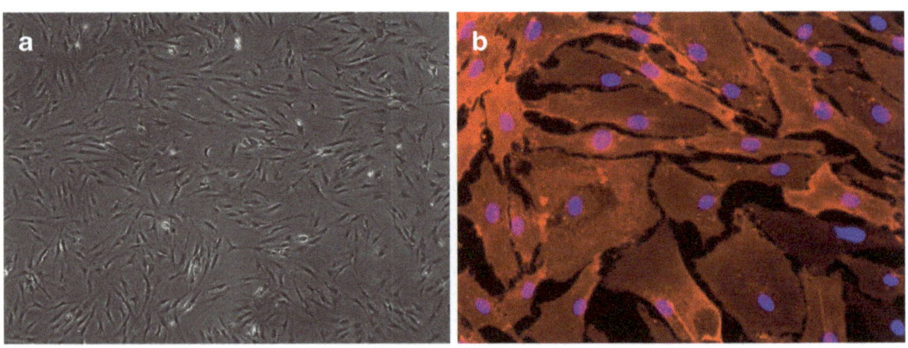

Fig. 9.1 (**a**) Adipose-derived stem cells (ADSC), (**b**) CD90 immunofluorescence staining of ADSC

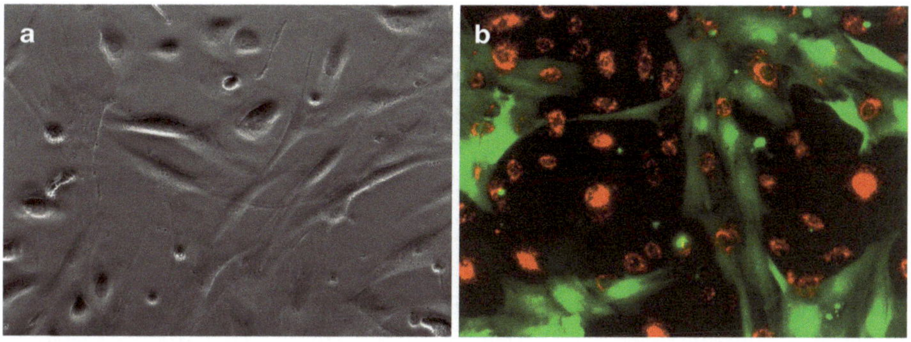

Fig. 9.2 (**a**) Coculture of adipose-derived stem cells (ADSC) and mammary epithelial cells (MEC). (**b**) Immunofluorescence microscopy of the coculture indicates possible fusion of ADSC and MEC (ADSC labeled in *green*, MEC in *red*)

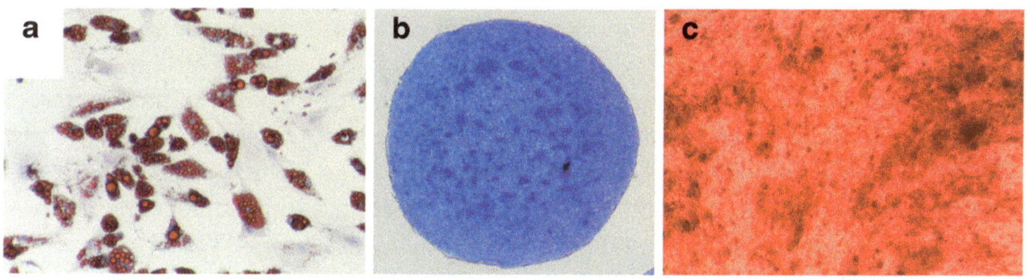

Fig. 9.3 Differentiation of human adipose-derived stem cells into the adipogenic (**a**), chondrogenic (**b**), and osteogenic (**c**) lineage

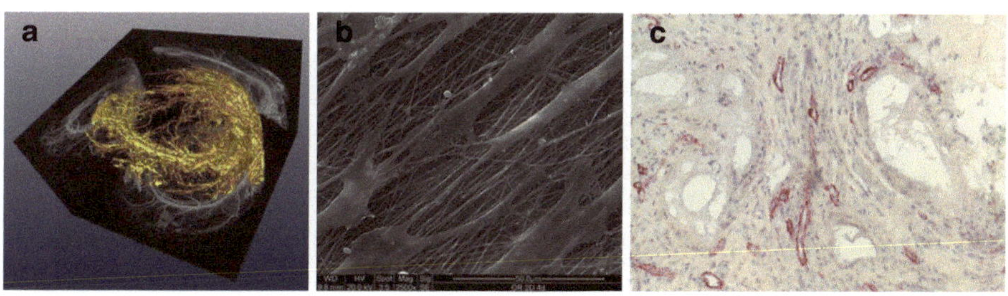

Fig. 9.4 (**a**) Isosurface rendering of micro-CT showing a dense vascular network emerging from the AV-loop for vascularization of a PCL-collagen-nanofiber scaffold (rat AV-loop model). (**b**) Primary rat myoblast seeded on electrospun PCL-collagen-nanofibers prior to rat AV-loop implantation procedure (magnification see data bar below). (**c**) Immunohistochemical staining of alpha smooth muscle actin in a tissue-engineered bone in the sheep model

The utilization of putative stem cell properties with extracellular matrix components, growth factors, and supporting matrices to achieve independently growing tissue is one way to deliver functional replacements. More insights are gained in cell-cell interaction and to manipulate cell behavior in the past years, but still the ideal cell source is not found so far. Possibly telocytes can play a role for tissue engineering and regenerative medicine purposes. According to the literature, it is more and more evident that they have a regenerative potential and act in cellular communication through their network-forming telopodes. Talking about tissue engineering, histological findings and in vitro experiments can be the first step, but we always should keep in mind the need for upscaling for later clinical application. In the upscaling process, vascularization of the engineered tissue is essential. Arteriovenous loop models are providing an ideal platform for preclinical testing of putative therapeutic concepts in regenerative medicine. The following book chapter should give an overview about what is known so far about the potential role of telocytes for tissue engineering and regenerative medicine.

Telocytes are identified as a new cell type among "classical" interstitial cells. The key feature that distinguishes telocytes from any other cell is their extremely long and thin prolongations named telopodes [3] (Fig. 9.5). Therefore, telocytes are shortly defined as "cells with telopodes" [4]. Telopodes are organized in a network – forming a labyrinthine system by tridimensional convolution and overlapping, communicating through gap junctions. Telocytes have "strategic" positioning in a tissue, in between blood capillaries and their specific target cells (e.g., smooth muscle cells, cardiomyocytes), and in close contact with nerve endings. To underline that telopodes could establish close contacts, like synapses, with immunoreactive cells, in various organs, telopodes have been called as stromal synapses [5]. Telocytes seem to be involved in intercellular signaling, taking into account the 3D network of telopodes and their strategic position. For that, two mechanisms could be considered. A paracrine and/or juxtacrine secretion of small signal molecules and shedding microvesicles, which play unique roles in horizontal transfer of important macromolecules among neighboring cells (e.g., proteins or RNAs, microRNA included). Such a mechanism, via shed vesicles, may serve to rapid phenotype adjustment in a variety of conditions [3, 6].

Telocytes are described in a broad range of tissues and organs so far. For example, they are found in the heart, and it is described that the number of telocytes decreases during myocardial infarction [7]. In the lung, they are supposed to stimulate and induce cell proliferation through the inhibition of apoptosis, possibly leading to reduced development of interstitial fibrosis in the lung [8]. In scleroderma, telocytes were reported with severe ultrastructural damages, reduced in number, and missing [9]. In patients with systemic sclerosis, the 3D network of telocytes is affected, for example, in the myocardium, lung, and gastric wall [10]. Authors are concluding that telocytes might be key players in the regulation of skin homeostasis, repair, and regeneration and a possible new target for the treatment of this disease. Telocytes seem to be important for liver regeneration in a murine model of partial hepatectomy. There they might influence proliferation of hepatocytes and/or the activation of stem/progenitor cells [11]. In patients with

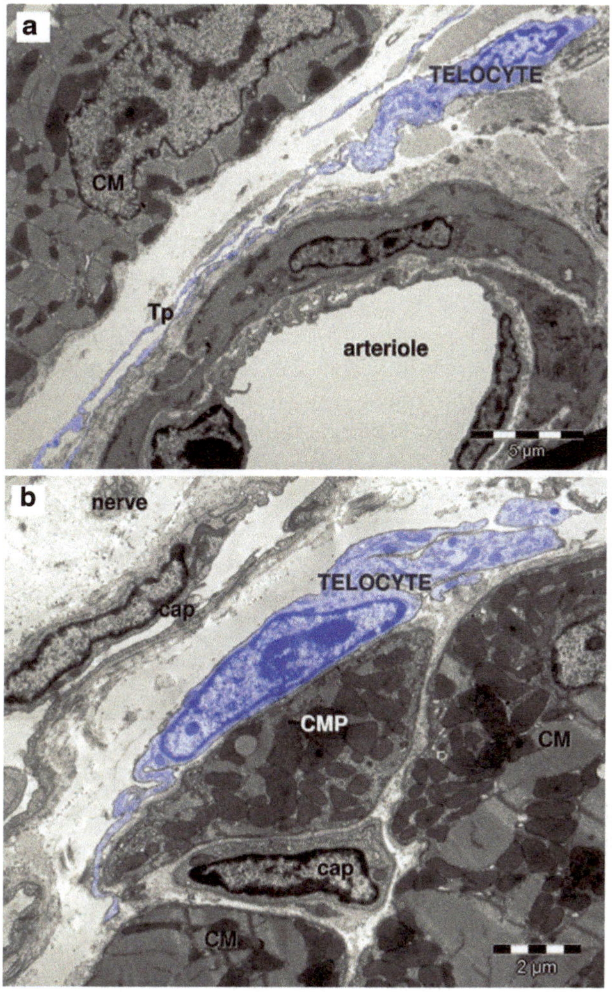

Fig. 9.5 Digitally colored transmission electron micrographs of mouse heart show telocytes (TC) in blue. (**a**) Cardiac structural unit composed by cardiomyocyte (CM), blood vessel, and TC. Telocyte extends its characteristic, very long and thin, process (Tp-telopode) between arteriole and cardiomyocyte. (**b**) Regenerative cardiac unit composed by cardiomyocyte progenitor (CMP), capillary (cap), nerve, and TC (Figure kindly provided by Prof. L.M.Popescu, reprinted with permission from: Polykandriotis et al. [33]; Copyright © 2010 The Authors Journal compilation © 2010 Foundation for Cellular and Molecular Medicine/Blackwell Publishing Ltd)

gallstone disease, telocytes are less described in the gallbladder wall [12, 13]. Furthermore, they are reported to be present in the urinary system and in the kidneys [14]. They are described as special cells that can adapt their morphology to the organ activity [15–17].

In tissue engineering approaches, cells are an integral part for generation of a functional tissue. Because regenerative potential is limited in humans, stem cell-based therapy remains a hope for regeneration in the future. So far a broad range of methods and cells (embryonic stem cells; induced pluripotent stem cells; hematopoietic stem cells; multipotent mesenchymal stromal cells derived from the blood, bone marrow, umbilical cord, and adipose tissue; etc.) have been investigated, but we are still waiting for the breakthrough in regenerative medicine. There are recent publications indicating that telocytes and stem cells are linked together [4, 18].

In the current literature, a potential power of regeneration is described for telocytes [19, 20]. Possibly, they play an important role for tissue repair and regeneration guiding putative stem and progenitor cells in stem cell niches in a spectrum of tissues and organs [21]. There are a lot of publications dealing with the role of telocytes in heart regeneration [22]. They can act as mediators for heart compaction from embryonic myocardial trabeculae during mouse heart development [23]. Transplantation of cardiac telocytes and mesenchymal stem cells decreases peripheral fibrosis of the border zone when directly injected into tissue damage zones. Rebuilding the cardiac telocyte network decreased infarction size and improved myocardial function [7]. Coming from ultrastructural studies, it has been shown that telocytes form a cardiac interstitial network which is involved in the structural support for overall information from the vascular system (endothelial cells and pericytes), nervous system (Schwann cells), immune system (macrophages, mast cells), interstitium (fibroblasts, extracellular matrix), stem cells/progenitor cells, and working cardiomyocytes [24]. This interstitial system coordinates possibly cardiac renewal, regeneration, and repair [24]. Telocytes act as "nursing" cardiomyocyte progenitors in epicardial stem cell niches and are therefore active players in cardiac renewing. In the newt heart, a supporting network of stromal cells is primarily developed possibly being telocytes because all ultrastructural criteria are fulfilled. Authors state that it would be an important goal for regenerative medicine to find some factors to stimulate telocytes, as autologous in situ cells.

Given these facts, telocytes may well become active players in the regenerative process. Since they can sometimes express stem cell markers like c-kit, Sca-1, and Oct 4 [25–27], they might possibly also have a role in angiogenesis and vascular stability during tissue repair. It has been shown that they express VEGF and PDGFR-β in skeletal muscle interstitium [28, 29]. Cells in the interstices have been studied by a growing number of research groups, and there is evidence that these cells may play a crucial role in tissue regeneration [30]. Telocytes have more and more gained international attention within the last years [3]. Ultrastructural techniques showed a complex interstitial cell population in various tissues changing during health and disease processes [31]. For example, telocytes were found with significantly higher numbers in the epicardium than in the myocardium, and they were distributed longitudinally and within the cross network of myocardium [32].

According to the current literature and the huge number of hinds for the regenerative potential of telocytes, there are great expectations for these cells at the moment from the tissue engineering and regenerative medicine community. Possibly further molecular and cellular studies of telocytes functions could develop new tools for regenerative reprogramming of lost organ functions [19, 33]. One can therefore rightly speculate that telocytes are possibly a novel target for therapeutic strategies, and further improvements in telocytes research will someday implement them a significant position in regenerative medicine. Further research will therefore focus on culturing telocytes under distinct conditions and to study the effect of cultured telocytes alone and together with other cells to possibly generate 3D replacement constructs.

References

1. Rana RE, Arora BS. History of plastic surgery in India. J Postgrad Med. 2002;48:76–8.
2. Jenkins DD, Yang GP, Lorenz HP, Longaker MT, Sylvester KG. Tissue engineering and regenerative medicine. Clin Plast Surg. 2003;30:581–8.
3. Popescu LM, Faussone-Pellegrini MS. TELOCYTES – a case of serendipity: the winding way from Interstitial Cells of Cajal (ICC), via Interstitial Cajal-Like Cells (ICLC) to TELOCYTES. J Cell Mol Med. 2010;14:729–40.
4. Popescu LM, Curici A, Wang E, Zhang H, Hu S, Gherghiceanu M. Telocytes and putative stem cells in ageing human heart. J Cell Mol Med. 2015;19:31–45.
5. Popescu LM, Gherghiceanu M, Cretoiu D, Radu E. The connective connection: interstitial cells of Cajal (ICC) and ICC-like cells establish synapses with immunoreactive cells. Electron microscope study in situ. J Cell Mol Med. 2005;9:714–30.
6. Cocucci E, Racchetti G, Meldolesi J. Shedding microvesicles: artefacts no more. Trends Cell Biol. 2009;19:43–51.
7. Zhao B, Chen S, Liu J, Yuan Z, Qi X, Qin J, Zheng X, Shen X, Yu Y, Qnin TJ, Chan JY, Cai D. Cardiac telocytes were decreased during myocardial infarction and their therapeutic effects for ischaemic heart in rat. J Cell Mol Med. 2013;17:123–33.
8. Zheng Y, Cretoiu D, Yan G, Cretoiu SM, Popescu LM, Fang H, Wang X. Protein profiling of human lung telocytes and microvascular endothelial cells using iTRAQ quantitative proteomics. J Cell Mol Med. 2014;18:1035–59.
9. Manetti M, Guiducci S, Ruffo M, Rosa I, Faussone-Pellegrini MS, Matucci-Cerinic M, Ibba-Manneschi L. Evidence for progressive reduction and loss of telocytes in the dermal cellular network of systemic sclerosis. J Cell Mol Med. 2013;17:482–96.
10. Manetti M, Rosa I, Messerini L, Guiducci S, Matucci-Cerinic M, Ibba-Manneschi L. A loss of telocytes accompanies fibrosis of multiple organs in systemic sclerosis. J Cell Mol Med. 2014;18:253–62.
11. Wang F, Song Y, Bei Y, Zhao Y, Xiao J, Yang C. Telocytes in liver regeneration: possible roles. J Cell Mol Med. 2014;18:1720–6.
12. Franks I. Gallbladder: loss of interstitial Cajal-like cells in the gallbladder might contribute to gallstone formation. Nat Rev Gastroenterol Hepatol. 2012;9:689.
13. Pasternak A, Gil K, Matyja A, Gajda M, Sztefko K, Walocha JA, Kulig J, Thor P. Loss of gallbladder interstitial Cajal-like cells in patients with cholelithiasis. Neurogastroenterol Motil: Off J Eur Gastrointest Motil Soc. 2013;25:e17–24.
14. Qi G, Lin M, Xu M, Manole CG, Wang X, Zhu T. Telocytes in the human kidney cortex. J Cell Mol Med. 2012;16:3116–22.
15. Zheng Y, Zhu T, Lin M, Wu D, Wang X. Telocytes in the urinary system. J Transl Med. 2012;10:188.
16. Gevaert T, De Vos R, Van Der Aa F, Joniau S, van den Oord J, Roskams T, De Ridder D. Identification of telocytes in the upper lamina propria of the human urinary tract. J Cell Mol Med. 2012;16:2085–93.
17. Vannucchi MG, Traini C, Guasti D, Del Popolo G, Faussone-Pellegrini MS. Telocytes subtypes in human urinary bladder. J Cell Mol Med. 2014;18:2000–8.

18. Popescu LM, Gherghiceanu M, Suciu LC, Manole CG, Hinescu ME. Telocytes and putative stem cells in the lungs: electron microscopy, electron tomography and laser scanning microscopy. Cell Tissue Res. 2011;345:391–403.
19. Horch RE, Kneser U, Polykandriotis E, Schmidt VJ, Sun J, Arkudas A. Tissue engineering and regenerative medicine - where do we stand? J Cell Mol Med. 2012;16:1157–65.
20. Ceafalan LC, Popescu BO, Hinescu ME. Cellular players in skeletal muscle regeneration. BioMed Res Int. 2014;2014:957014.
21. Haines DD, Juhasz B, Tosaki A. Management of multicellular senescence and oxidative stress. J Cell Mol Med. 2013;17:936–57.
22. Bani D, Nistri S. New insights into the morphogenic role of stromal cells and their relevance for regenerative medicine. Lessons from the heart. J Cell Mol Med. 2014;18:363–70.
23. Bani D, Formigli L, Gherghiceanu M, Faussone-Pellegrini MS. Telocytes as supporting cells for myocardial tissue organization in developing and adult heart. J Cell Mol Med. 2010;14:2531–8.
24. Gherghiceanu M, Popescu LM. Cardiac telocytes – their junctions and functional implications. Cell Tissue Res. 2012;348:265–79.
25. Suciu L, Popescu LM, Gherghiceanu M, Regalia T, Nicolescu MI, Hinescu ME, Faussone-Pellegrini MS. Telocytes in human term placenta: morphology and phenotype. Cells Tissues Organs. 2010;192:325–39.
26. Bojin FM, Gavriliuc OI, Cristea MI, Tanasie G, Tatu CS, Panaitescu C, Paunescu V. Telocytes within human skeletal muscle stem cell niche. J Cell Mol Med. 2011;15:2269–72.
27. Galiger C, Kostin S, Golec A, Ahlbrecht K, Becker S, Gherghiceanu M, Popescu LM, Morty RE, Seeger W, Voswinckel R. Phenotypical and ultrastructural features of Oct4-positive cells in the adult mouse lung. J Cell Mol Med. 2014;18:1321–33.
28. Popescu LM, Manole E, Serboiu CS, Manole CG, Suciu LC, Gherghiceanu M, Popescu BO. Identification of telocytes in skeletal muscle interstitium: implication for muscle regeneration. J Cell Mol Med. 2011;15:1379–92.
29. Suciu LC, Popescu BO, Kostin S, Popescu LM. Platelet-derived growth factor receptor-beta-positive telocytes in skeletal muscle interstitium. J Cell Mol Med. 2012;16:701–7.
30. Zheng Y, Zhang M, Qian M, Wang L, Cismasiu VB, Bai C, Popescu LM, Wang X. Genetic comparison of mouse lung telocytes with mesenchymal stem cells and fibroblasts. J Cell Mol Med. 2013;17:567–77.
31. Cretoiu SM, Popescu LM. Telocytes revisited. Biomol Concepts. 2014;5:353–69.
32. Zhao B, Liao Z, Chen S, Yuan Z, Yilin C, Lee KK, Qi X, Shen X, Zheng X, Quinn T, Cai D. Intramyocardial transplantation of cardiac telocytes decreases myocardial infarction and improves post-infarcted cardiac function in rats. J Cell Mol Med. 2014;18:780–9.
33. Polykandriotis E, Popescu LM, Horch RE. Regenerative medicine: then and now – an update of recent history into future possibilities. J Cell Mol Med. 2010;14:2350–8.

Raymund E. Horch Univ.-Prof. Dr. Dr. h. c. Prof. h. c. Raymund E. Horch, MD, is currently acting the President of the German Society of Plastic Reconstructive and Aesthetic Surgeons (DGPRÄC). As a Director he heads the Department of Plastic and Hand Surgery and the Laboratory for Tissue Engineering and Regenerative Medicine at the University Hospital Erlangen, Friedrich Alexander University Erlangen-Nuernberg FAU. He is at the same time also acting as the head/spokesman of all Surgical Departments and Divisions of the Erlangen University Hospital and as the spokesman of the convention of Executive University Plastic Surgeons in Germany.

His clinical focus lies on Reconstructive, Aesthetic and Hand Surgery and Breast Surgery. Within the reconstruction of defects after tumor surgery with complex microsurgical tissue transplants he has developed modern interdisciplinary concepts. In his research Prof Horch was intensely invovled in creating replacement tissues by means of Tissue Engineering and Regenerative Medicine from the very beginning.

His scientific work includes several hundreds of publications, currently more than 380 PubMed articles, numerous book sections and editor and coeditor of several text books on clinical and research topics (Tissue Engineering, Burns, Reconstruction, Wound Healing, Vacuum Therapy, and Microsurgery). In addition to numerous invitations as a "visiting professor" at renowned international universities his scientific work was honored with a variety of prestigious awards and honors worldwide.

Chapter 10
Presence of Telocytes in a Non-innervated Organ: The Placenta

Cleofina Bosco and Eugenia Díaz

Abstract This chapter discusses the relationship between failure in placentation and the subsequent alterations in the normal structure of the placenta. Interstitial Cajal-like cells (ICLC) were observed for the first time in the human placenta in 2007 and later were named telocytes. Strong evidence confirms that in the placental chorionic villi, TC are located strategically between the smooth muscle cells (SMC) of the fetal blood vessel wall and the stromal myofibroblasts. As the placenta is a non-innervated organ and considering the strategic position of telocytes in chorionic villi, it has been postulated that their function would be related to signal transduction mechanisms involved in the regulation of the blood flow in the fetal vessels, as well as in the shortening/lengthening of the chorionic villi providing the necessary rhythmicity to the process of maternal/fetal metabolic exchange. In this context, telocytes represent part of a functional triad: "SMC of fetal blood vessel-telocyte-myofibroblast." This triad takes part in the regulation of fetal growth and development via transport of nutrients and gases. This chapter also discusses the alterations in the metabolic maternal–fetal exchange, leading to intrauterine growth retardation and preeclampsia. Additionally, the apoptosis undergoing in the preeclamptic hypoxic placenta affects all the chorionic villi cells, including telocytes and myofibroblast, and not only trophoblast, as it has been so far considered. In consequence, we proposed that apoptosis affects the triad structure and alters the placental function, subsequently affecting the normal fetal growth and development.

C. Bosco (✉)
Laboratorio de Placenta y Desarrollo Fetal, Anatomy and Developmental Biology Program, Institute of Biomedical Sciences, Faculty of Medicine, University of Chile, Independencia 1027, Casilla, Santiago 7 70079, Chile
e-mail: cbosco@med.uchile.cl

E. Díaz
Anatomy and Developmental Biology Program, Institute of Biomedical Sciences, Faculty of Medicine, University of Chile, Santiago, Chile

X. Wang, D. Cretoiu (eds.), *Telocytes*, Advances in Experimental Medicine and Biology, 913, DOI 10.1007/978-981-10-1061-3_10

Abbreviations

CTB	Cytotrophoblast
EVT	Extravillous trophoblast
ICC	Interstitial cells of Cajal
ICLC	Interstitial Cajal-like cells
IUGR	Intrauterine growth restriction
NO	Nitric oxide
PE	Preeclampsia
ROS	Reactive oxygen species
SMC	Smooth muscle cells
STB	Syncytiotrophoblast
TC	Telocytes
TP	Telopode
UL-vWF	Ultra-large vWB multimers
VEGF	Vascular endothelial growth factor

10.1 Telocytes

Ramón y Cajal, the eminent Spanish neuropathologist of the nineteenth century, discovered a particular cell type in the gut, which he named "interstitial neuron." In the early 1970s, electron microscopy studies established that this interstitial cell type did not correspond to neurons and, consequently, were renamed "interstitial cells of Cajal" (ICC) [1]. It is now widely accepted that in the gut, the ICC are pacemaker cells involved in the regulation of gastrointestinal motility [2, 3]. ICC were also found outside the gastrointestinal tract and were named "interstitial Cajal-like cells" (ICLC). An example of this is the mammary gland [4], where ICLC were found in the intralobular stroma and were identified by methylene blue vital staining and c-kit immunoreactivity [5]. Later, ICLC were also found in many other human organs such as in the fallopian tube [6], gallbladder [7], placenta [8, 9], etc. Due to all these confusing names, Popescu and Faussone-Pellegrini [10] proposed the term telocytes (TC) for these cells.

The most striking ultrastructural feature of TC is that they possess few and long cytoplasmic projections emerging from an elongated cell body, with small perinuclear cytoplasm and oval nucleus, presenting one or more nucleoli [10]. Cell-to-cell communication between TC occurs through the cytoplasmic's very long moniliform prolongations called telopodes (TP), frequently two to three per cell. The TP present thin fibrillar-like segments called podomers, and dilated cistern-like regions denoted as podoms. Mitochondria and endoplasmic reticulum cisternae are accumulated inside podoms [8, 9]. In addition, homocellular junctions correspond to TC communication established between telopodes, and heterocellular junctions to those established with other cell types. Additionally, these junctions could also be established with the surrounding connective tissue (stromal synapses), and metabolite signaling can be released via microvesicles or exosomes [9]. Furthermore, the expression of c-kit/CD117 is commonly accepted as a specific marker for TC, while vimentin expression is considered as an alternative or supplementary marker [8].

10.2 Placental Telocytes

For the first time in 2007, Suciu et al. [8] reported that TC were present in the extra-embryonic mesoderm of the villi from the human term placenta. Placental TC were immunohistochemically positive for the following markers: vimentin, c-kit, CD34, VEGF, caveolin-1 and iNOS, αSMC, neuron-specific enolase, S-100, and nestin [9]. The light microscopy study of Suciu et al. [8] on semithin sections of human term placenta demonstrated that TC were easily identifiable as cells with long cellular processes (TP) that surround blood vessels, or extend into the connective tissue beneath the trophoblast, or were interposed between arterioles and the trophoblast basement membrane, in both large, peripheral stem villi and small stem villi. The authors also proposed that the presence of TP suggests a juxta- and/or paracrine activity with immunoreactive cells. These TP were connected by gap junctions forming a network extending into the placental stroma between SMC layers of large blood vessels and myofibroblasts [8, 9, 11]. The former authors considered "that these cells may be involved in tissue remodelling and, since the placenta is a non-innervated organ, they may also intervene in blood flow regulation." Regarding their disposition, myofibroblast are arranged in parallel to the longitudinal axis of the anchoring and stem villi [12, 13]. Even if no distinct junctions between TC and myofibroblasts are apparent, the close contact between them suggests some sort of communication. Based on the placental lack of innervation and the strategic position of TC in placental villi [8, 9], Bosco et al. [11] have proposed to consider TC as placental pacemakers. However, in order to avoid any possible misinterpretation, it seems advisable to reserve the pacemaker denomination for the gastrointestinal ICC [3] and consider placental TC related to cellular signaling [9, 14].

Studies from Nicolescu and Popescu [15] and Bosco et al. [16] in the pancreas, an organ where there is no presence of muscle layers, have suggested that TC may constitute the pancreatic pacemaker, considering the strategic position of these cells between neurons and pancreatic exocrine ducts, as well as between neurons and blood vessels, thus providing rhythmicity to the exocrine ducts and flow regulation in the pancreatic blood vessels.

10.3 Placental Telocytes Origin

It has been demonstrated that normal placental development ensures normal fetal development and growth [17]. All the events involved are highly related to each other and are susceptible to the effects of intra- and extraplacental environmental factors, including the nutritional status of the mother and nutrient intake during pregnancy [18, 19].

As we previously described, the first fetal–placental villi develops as trophoblast sprouts, an epithelial tissue subsequently invaded by extraembryonic mesoderm, forming the secondary villi which are then transformed, by vasculogenesis, into tertiary villi [19]. Due to the fact that TC have mesodermal origin, which has been immunohistochemically evidenced by the presence of vimentin-positive cells [9, 14] have postulated that they probably differentiated from mesenchymal cells of the mesoderm present in the secondary and tertiary villi.

10.4 Placenta

The placenta is the site where the physiological exchange between the mother and the fetus occurs. Placental function is to transfers nutrients to the fetus, excrete waste products into the maternal blood, and modify maternal metabolism at different stages of pregnancy, via hormones [20]. Therefore, it is reasonable to assume that the placenta functions as a lung, gastrointestinal tract, kidney, liver, as well as an endocrine organ for the fetus [20, 21]. Additionally, the placenta constitutes an immunological barrier between the fetus and the mother and produces and secretes different hormones [21], as well as a variety of cytokines and signaling molecules [22]. The events that characterize normal placental growth are considered to be important determinants of fetal growth and development. All of these events are likely interrelated and susceptible to the effects of many environmental factors, including maternal food intake, both before and during pregnancy [18].

The blastocyst is an embryonic structure constituted by an inner cell mass, the embryoblast, formed at 4–5 days of gestation, surrounded by the external trophoblast. The cells of the trophoblast layer differentiate into two layers, an inner cytotrophoblast or epithelial stem cells (CTB) and an outer syncytiotrophoblast (STB), both of which contribute to the formation of the villi and ultimately the placenta [19]. The stem villi, which represent the central branches of the villous trees, are characterized by a loose connective tissue or stroma in which the fetal arteries and veins are embedded [19].

Human placenta development depends critically on the differentiation of CTB [23], process for which two differentiation pathways exist. In one, CTB remains in the fetal compartment and fuse themselves to form multinucleate STB that cover the floating chorionic villi. These villi, which are in direct contact with the maternal blood in the intervillous space, perform the nutrient and gas exchange for the fetus [24, 25]. In the other pathway, a subset of CTB in anchoring chorionic villi aggregate into cell columns that attach to the uterine wall [26]. From there, CTB differentiated to extravillous trophoblasts (EVT) ulteriorly invading the uterine arteries walls [23–25]. As a result, these cells replace the endothelial and muscular linings of uterine spiral arteries, a process that initiates maternal blood flow to the placenta and greatly enlarges the diameter and low resistance of maternal spiral arteries, thereby increasing blood flow to the placenta and allowing an adequate supply of oxygen and nutrients to the growing fetus [26- 28]. Insufficient EVT invasion contributes to the development of preeclampsia (PE), which often results in fetal intrauterine growth restriction (IUGR), maternal hypertension, and proteinuria [28]. This insufficient EVT invasion is associated with a significant reduction in the uteroplacental blood flow, developing placental hypoxia which in turn results in further elevated oxidative stress and apoptosis, as observed in PE placentas [29–32].

10.5 Placenta Oxidative-Nitrosative Stress

Pregnancy itself is a condition of increased susceptibility to oxidative stress arising from the increased metabolic activity in placental mitochondria and reduced scavenging power of antioxidants [33]. The production of reactive oxygen species (ROS) within the fetal–placental unit induces cellular damage by acting on protein and lipids. ROS form as a natural product of the normal metabolism of oxygen and generate potentially superoxide anion, hydrogen peroxide, and hydroxyl radical [34].

Although the cause of PE still remains unknown, it has been proposed that enhanced oxidative stress is a basic component of this condition that could provide the connection between abnormal placentation and the maternal syndrome [35, 36]. Oxidative stress occurs when there is an imbalance between the production of ROS and the ability of the biological system to readily detoxify these reactive intermediates or easily repair the resulting damage [37, 38].

The placenta also produces nitric oxide (NO) [39], giving rise to another local source of free radicals, likely contributing to endothelial dysfunction [30]. In a placental environment where superoxide anion and NO are present in abundance, the interaction of the two chemical species inevitably will yield peroxynitrite, a potent prooxidant which also increases the oxidative stress (in this condition defined as nitrosative stress) observed in PE placenta [40]. As the placenta is a non-innervated organ [14], its blood flow must be regulated by autocrine/paracrine factors produced in the organ such as NO [34]. A decrease in NO availability could adversely affect placental blood flow regulation, which could, in turn, account for the abnormal fetal development [30].

10.6 Myofibroblast Placental Function

The myofibroblast are cells located intermediately between fibroblasts and smooth muscle cells. They exhibit an important cytoplasmic microfilamentous apparatus such as bundles of actin microfilaments with associated contractile proteins such as non-muscle myosin [41–44]. Hence, these cells are considered to be specialized contractile fibroblast with an important role in establishing tension. By scanning electron microscopy, it has been demonstrated that the stellate form of myofibroblast-like cells in placental villi and their tendency to establish a three-dimensional networks in this organ [13, 45, 46] require contractile structures to generate the necessary force for blood propulsion [47].

It has been suggested that the contraction of myofibroblast adjusts the blood flow in fetal vessels and increases the turgor, imparting mechanical stability to the villous tree in the maternal bloodstream [12]. It has also been proposed that contraction of longitudinally arranged myofibroblasts within anchoring villi may influence the length and width of the intervillous space, thus regulating maternal intervillous blood pressure [48, 49]. The stimulus for this type of contraction has not yet been established because the placenta is a non-innervated organ, but Suciu et al. [9, 14] have proposed that TC function would be related to mechanisms of signal transduction to myofibroblast involved in the regulation of the blood flow in fetal vessels, as well as in the shortening/lengthening of the chorionic villi. Furthermore, Suciu et al. [8, 9] have also demonstrated that TC were connected by gap junctions in a network extending into the stromal and large blood vessels of the placental myofibroblasts. These authors concluded that the close contact between these cells suggests some sort of communication or coupling. Additionally, it was also demonstrated that TC [9] and myofibroblasts [48] generate NO, which, in turn, may modulate the tone of perivascular contractile sheets.

10.7 Placenta and Telocyte Functions

The placenta provides an excellent model for understanding the relationship between hypoxia, organogenesis (organ development), and angiogenesis (blood vessel development). In order to elucidate the role of hypoxia in the regulation of cellular and organ functions, it is necessary to understand placental development and some placental pathologies.

It is well known that defects in the processes of embryonic implantation and fetal placentation can result in the condition of PE, which occurs in 5–10 % of pregnancies and is responsible for diseases of pregnancy like spontaneous abortion, preterm birth, and IUGR [50]. PE is a multisystem disorder which is a major cause of maternal morbidity and mortality worldwide. The cardinal features of PE are hypertension and proteinuria, clinical signs which are manifested after 20 weeks gestation in women who were not previously known to be hypertensive. Other signs and symptoms include edema and headache, and in severe cases, the condition is associated with seizures (eclampsia), liver and kidney dysfunction, as well as clotting abnormalities [29]. In PE pregnancies, the reduction of the utero-placental perfusion pressure and the ensuing placental ischemia/hypoxia during late pregnancy may be caused by inadequate EVT invasion of the uterine spiral arteries in the first trimester of pregnancy [27]. Placental ischemia/hypoxia may trigger the release of placental factors that initiate a cascade of cellular and molecular events leading to apoptosis of TC [11] and endothelial and vascular smooth muscle cell dysfunction [19], thereby increasing vascular resistance and arterial pressure [23, 51–53].

Placental function is to exchange nutrients to the fetus, excrete waste products into the maternal blood, and modify maternal metabolism at different stages of pregnancy via hormones. Therefore, it is plausible to say that the placenta functions in a similar way as the lungs, gastrointestinal tract, kidneys, liver, as well as an endocrine organ for the fetus [19, 21]. In this context, it is important to emphasize that in PE the apoptosis of placental TC [11] affects placental functions related to the regulation of fetal blood flow and the intraplacental blood volume.

Considering that the placenta functions as a lung for the fetus, some authors have postulated in human and mouse lung that besides the conventional role of mechanical support for the TC network, TC's main role in the lung would be related to intercellular communication, proliferation, differentiation, as well as growth of the stem cells and repair mechanisms in injured tissue [54, 55]. The authors based this assumption on the existence, in the human lung, of TC near the stem cell niches in the lung [54], and the existence of TC and their TP in close vicinity of blood capillary and/or nerve fibers [55]. Additionally, Popescu et al. [54] observed TC in the mouse respiratory tree, located in the interstitial space of terminal and respiratory bronchioles, as well as in alveolar ducts in which TPs were connected with alveolar epithelial cells and the vicinity of small blood vessels. This crucial relationship has been also demonstrated in the placental villi, an organ that exchanges gases between mother and fetus. TC and their TP were described in the interstitial space between trophoblast epithelial cells and blood capillary [8, 11]. Taking into account that in PE the TC suffer apoptosis [11, 14], this condition consequently will affect the gases exchange process.

Considering that the placenta also functions as gut due to its absorbing metabolites functions, it is worthwhile to point out that Vannucchi et al. [56] described the presence of TC in the mucosa, submucosa, and muscle coat of the gastrointestinal tract. The authors described that TC formed a three-dimensional network in the submucosa and in the interstitium between muscle layers, and an almost continuous layer at the submucosal borders of muscularis mucosae and circular muscle layer. Moreover, TC encircled muscle bundles, nerve structures, blood vessels, funds of gastric glands, and intestinal crypts. Additionally, Milia et al. [57] found in the normal gut that TC were observed in all the ileal wall layers, from the mucosa to the subserosa, and according to the different disposition in the wall layers, they form a network. On the contrary, in the gut from Crohn's disease, characterized by derangement of the normal disposition of the intestinal walls, these authors observed that TC have disappeared. Milia et al. [57] concluded that due to the 3-D network of TP and their strategic position between immune cells, smooth muscle cells, blood and lymphatic vessels, and nerve endings, the loss of TC might have important pathophysiological implications, contributing to the disorder of the intestinal wall architecture, gut dysmotility, and impaired immune surveillance. In this context, it seems reasonable to assume that placental TC apoptosis observed in PE [11] would affect the regulation of fetal blood flow, the intraplacental blood volume, as well as the shortening/lengthening of the chorionic villi.

In another line of evidence, PE is characterized by a maternal hypercoagulable state and intravascular coagulation, microthromboses in several organs, and impairment of the uteroplacental circulation [58]. The thromboresistance of the placental endothelium is maintained as long as natural anticoagulant pathways are functionally present in the endothelial plasma membrane [59, 60]. The main anticoagulant pathway in the placenta is mediated by thrombomodulin (TM), an endothelial cell-surface glycoprotein [61]. In this regard, Bosco et al.[32] found immunohistochemical expression of TM in the stromal cells of placental villi of PE, and these authors concluded that the presence of TM in stromal cells may suggest a role in preserving the function of these cells in villous contractility and modulation of the intervillous space affecting both maternal and fetal–placental circulation. This is further supported by the lack of IURG in these PE cases. These authors further postulate that TM-positive placental stromal cells correspond to TC and myofibroblasts (Fig.10.1a) and that their functions and communication between them favor the metabolic exchange and protect the organ against the hypercoagulable state. The lack of autonomic innervations in the placenta implicates that the blood flow must be regulated by humoral mechanisms and by autocrine/paracrine factors produced in the organ [33]. The main vasoactive agent, NO, is secreted by the endothelium and by stromal placental TC and myofibroblasts, a fact that further supports this idea [9, 12]. Additionally, Kroll and Waltenberger [62] found that vascular endothelial growth factor (VEGF) enhances the activity of endothelial NOS (eNOS) and inducible NOS (iNOS) in endothelial cells via KDR receptor. VEGF can induce the production of NO in the placenta, but an excess of NO and superoxide anion induces the formation of an excess of peroxynitrite [63], thus increasing the nitrosative stress. Therefore, the increased VEGF expression observed by Parra-Cordero et al. [64] and Bosco et al. [65] in the stromal TC and myofibroblasts of PE placentas (Fig. 10.1b) suggests an increase in the NO production and subsequently an increase in peroxynitrite formation (Fig. 10.1c) and finally an increase of the apoptotic programmed cell death, evidenced by a greater expression of the proapoptotic protein BAX in TC and myofibroblast (Fig. 10.1d). The apoptosis in TC causes loss of their TP [11, 66] and hence of the intercellular connections between the myofibroblasts and smooth muscle cells from the blood vessels.

Bosco et al. [65] have also suggested that the oxidative and nitrosative stress developed in PE placentas is followed by a significant increase of maternal plasma vWF levels. Additionally, Dong et al. [67] and Li et al. [68] demonstrated that ultra-large vWF multimers (UL-vWF), unlike plasma forms of vWF secreted by stimulated endothelial cells, are anchored to the endothelial surface as extraordinarily long string-like structures capable of binding platelets. These UL-vWF multimeric strings were rapidly cleaved in the presence of the normal plasma metalloprotease ADAMTS-13, limiting thrombus growth [68–70]. These multimers are also more reactive with platelet receptors in the presence of high fluid shear stress [69]. Furthermore, Lancellotti et al. [71] found that peroxynitrite oxidizes the vWF-cleaving metalloprotease ADAMTS-13, thus contributing to the prothrombotic effects. Additionally, Myatt et al. [40] and Bosco et al. [62] also reported the presence of nitrotyrosine residues in the stromal cell of PE placentas. Taking all these

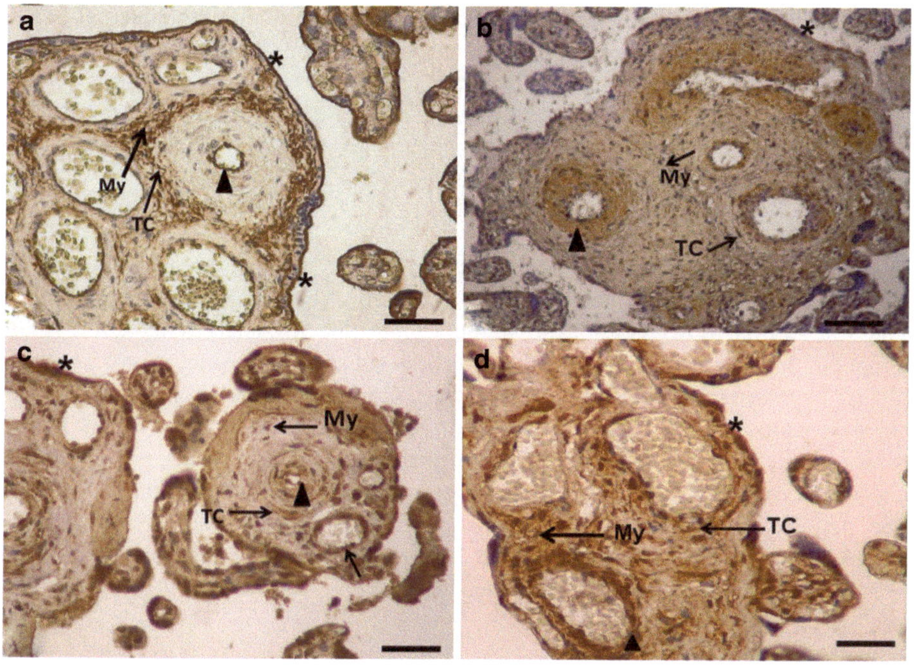

Fig. 10.1 Placentas of preeclamptic women. (**a**) Anti-thrombomodulin monoclonal antibody expression in PE placental villi. An intense immunohistochemical labeling in the apical zone of the syncytiotrophoblast (*), endothelium (*arrow head*), telocytes (*TC*), and myofibroblast (*My*) is displayed. (**b**) Antihuman VEGF165 monoclonal antibody labeling. VEGF immunostaining was moderate in telocytes (*TC*), myofibroblast (*My*), and syncytiotrophoblast (*) and intense in the muscular layer of the arteries (*arrow head*) and veins of the stem villi. (**c**) Anti-nitrotyrosine residues rabbit polyclonal antibody labeling. The labeling was intense in telocytes (*TC*), myofibroblast (*My*), syncytiotrophoblast (*), and the muscular layer of the arterioles (*arrow head*) and the veins of the stem villi. (**d**) Anti-BAX rabbit polyclonal antibody labeling. The mark for the proapoptotic protein BAX was intense in the syncytiotrophoblast (*), telocytes (*TC*), myofibroblast (*My*), and the muscular layer (*arrow head*) of the fetal blood of the stem villi. (**a**) Calibration bar 50 μm. Antibody working dilution=1:50. (**b**) Calibration bar 70 μm. Antibody working dilution=1:50. (**c**) Calibration bars 60 μm. Antibody working dilution=1:100. (**d**) Calibration bars 60 μm. Antibody working dilution=1:100. Original magnification: 400× in (**a**, **c**) and (**d**); 200× in (**b**)

evidences into account, we consider it plausible to postulate that the maternal hypercoagulable state due to the impairment of ADAMTS-13 alters the normal longitudinal contraction/relaxation of the villi, a mechanism through which TC modulate the rhythmicity of the process.

In a model of cardiac hypoxia, Kostin [66] found that increased interstitial fibrosis and fibrillar collagen lead to TC cell death via apoptosis, with shrinkage and shortening of telopodes, and Bosco et al. [11] found a considerable amount of collagen deposits located near the telopodes of TC in PE placentas. Apoptosis of placental TC from the chorionic villi leads to alteration of the TP and therefore to a possible loss of the synaptic-like connections that relate TC with themselves, with myofibroblasts and with smooth muscle cells from fetal blood vessels. All these events point to a decrease of the contact surface of the villi with the maternal blood in the intervillous space, with a consequent reduction of maternal metabolic contribution to the fetus.

10.8 Final Remarks

Finally, we postulate that SMC of fetal blood vessels, TC, and myofibroblast act as a triad and that the coordinate function of TC contribute to the normal placental function. Apoptosis of placental TC from the chorionic villi leads to alteration of the telopodes and therefore to a possible loss of the synaptic-like connections. The loss of these synapses would lead to loss of the TC triad function probably related to signal transduction mechanisms involved in the regulation of the fetal vessels blood flow, as well as in the shortening/lengthening of the chorionic villi, providing the necessary rhythmicity to the process of maternal/fetal metabolic exchange.

Acknowledgments This chapter was supported by Grant 1090245 Fondo Nacional de Desarrollo Cientifico y Tecnologico (FONDECYT). The authors wish to thank Miss Paulina González Díaz for her kind assistance with language editing.

Conflict of Interest The authors confirm that this article content has no conflict of interest.

References

1. Faussone-Pellegrini MS, Thuneberg L. Guide to the identification of interstitial cells of Cajal. Microsc Res Tech. 1999;47:248–66.
2. Der T, Bercik P, Donnelly G, Jackson T, Berezin I, Collins SM, Huizinga JD. Interstitial cells of cajal and inflammation-induced motor dysfunction in the mouse small intestine. Gastroenterology. 2000;119:1590–9.
3. Sanders KM, Koh SD, Ward SM. Interstitial cells of cajal as pacemakers in the gastrointestinal tract. Annu Rev Physiol. 2006;68:307–43.
4. Gherghiceanu M, Popescu LM. Interstitial Cajal-like cells (ICLC) in human resting mammary gland stroma. Transmission electron microscope (TEM) identification. J Cell Mol Med. 2005;9:893–910.
5. Popescu LM, Andrei F, Hinescu ME. Snapshots of mammary gland interstitial cells: methylene blue vital staining and c-kit immunopositivity. J Cell Mol Med. 2005;9:476–7.
6. Popescu LM, Ciontea SM, Cretoiu D, Hinescu ME, Radu E, Ionescu N, Ceausu M, Gherghiceanu M, Braga RI, Vasilescu F, Zagrean L, Ardeleanu C. Novel type of interstitial cell (Cajal like) in human fallopian tube. J Cell Mol Med. 2005;9:479–523.
7. Hinescu ME, Ardeleanu C, Gherghiceanu M, Popescu LM. Interstitial Cajal-like cells in human gallbladder. J Mol Histol. 2007;38:275–84.
8. Suciu L, Popescu LM, Gherghiceanu M. Human placenta: de visu demonstration of interstitial Cajal-like cells. J Cell Mol Med. 2007;11:590–07.
9. Suciu L, Popescu LM, Gherghiceanu M, Regalia T, Nicolescu MI, Hinescu ME, Faussone-Pellegrini MS. Telocytes in human term placenta: morphology and phenotype. Cells Tissues Organs. 2010;192:325–39.

10. Popescu LM, Faussone-Pellegrini MS. TELOCYTES – a case of serendipity: the winding way from interstitial cells of Cajal (ICC), via interstitial Cajal-like cells (ICLC) to TELOCYTES. J Cell Mol Med. 2010;14:729–40.
11. Bosco C, Díaz E, Gutiérrez R, González J, Parra-Cordero M, Rodrigo R, Barja P. A putative role for telocytes in placental barrier impairment during preeclampsia. Med Hypotheses. 2015;84:72–7.
12. Graf R, Langer JU, Schonfelder G, Oney T, Hartel- Schenk S, Reutter W, Schmidt HH. The extravascular contractile system in the human placenta. Morphological and immunocytochemical investigations. Anat Embryol (Berl). 1994;190:541–8.
13. Kohnen G, Kertschanska S, Demir R, Kaufmann P. Placental villous stroma as a model system for myofibroblast differentiation. Histochem Cell Biol. 1996;105:415–29.
14. Bosco C, Díaz E, Gutiérrez R, González J, Parra-Cordero M, Rodrigo R, Barja P. Placental hypoxia developed during preeclampsia induces telocytes apoptosis in chorionic villi affecting the maternal-fetus metabolic exchange. Curr Stem Cell Res Ther. 2016;11:420–5.
15. Nicolescu MI, Popescu LM. Telocytes in the interstitium of human exocrine pancreas: ultrastructural evidence. Pancreas. 2012;41:949–56.
16. Bosco C, Díaz E, Gutiérrez R, González J, Pérez J. Ganglionar nervous cells and telocytes in the pancreas of Octodon degus. Auton Neurosci Basic Clin. 2013;177:224–30.
17. Bosco C, Díaz E. Placental hypoxia and foetal development versus alcohol exposure in pregnancy. Alcohol Alcohol. 2012;47:109–17.
18. Rosso P. Placental growth, development, and function in relation to maternal nutrition. Fed Proc. 1980;39:250–4.
19. Bosco C. Alcohol and xenobiotics in placenta damage. In: Preedy VR, Watson RR, editors. Comprehensive handbook of alcohol related pathology, vol. 2. London: Elsevier Science Press; 2005. p. 921–35.
20. Cross CJ. Placental function in development and disease. Reprod Fertil Dev. 2006;18:71–6.
21. Bosco C. The cytotrophoblast: the first human amine precursor uptake and decarboxylation (APUD) cell. Med Sci Res. 1995;23:205–7.
22. Pollheimer J, Knöfler M. Signalling pathways regulating the invasive differentiation of human trophoblasts: a review. Placenta. 2005;26:S21–30.
23. Zhou Y, Damsky CH, Chiu K, Roberts JM, Fisher SJ. Preeclampsia is associated with abnormal expression of adhesion molecules by invasive cytotrophoblasts. J Clin Invest. 1993;91:950–60.
24. Zhou Y, Fisher SJ, Janatpour M, Genbacev O, Dejana E, Wheelock M, Damsky CH. Human cytotrophoblasts adopt a vascular phenotype as they differentiate. A strategy for successful endovascular invasion? J Clin Invest. 1997;99:2139–51.
25. Zhou Y, Damsky CH, Fisher SJ. Preeclampsia is associated with failure of human cytotrophoblasts to mimic a vascular adhesion phenotype. One cause of defective endovascular invasion in this syndrome? J Clin Invest. 1997;99:2152–64.
26. Genbacev O, Joslin R, Damsky CH, Polliotti BM, Fisher SJ. Hypoxia alters early gestation human cytotrophoblast differentiation/invasion in vitro and models the placental defects that occur in preeclampsia. J Clin Invest. 1996;97:540–50.
27. Zhou Y, Genbacev O, Damsky CH, Fisher SJ. Oxygen regulates human cytotrophoblast differentiation and invasion: implications for endovascular invasion in normal pregnancy and in pre-eclampsia. J Reprod Immunol. 1998;39:197–213.
28. Redman CW, Sargent IL. Placental debris, oxidative stress and pre-eclampsia. Placenta. 2000;21:597–602.
29. Redman CWC. Current topic: pre-eclampsia and the placenta. Placenta. 1991;112:301–8.
30. Myatt L, Kossenjans W, Sahay R, Eis A, Brockman D. Oxidative stress causes vascular dysfunction in the placenta. J Matern Fetal Med. 2000;9:79–82.
31. Allaire AD, Ballenger KA, Wells SR, McMahon MJ, Lessey BA. Placental apoptosis in preeclampsia. Obstet Gynecol. 2000;96:271–6.
32. Bosco C, Parra M, Barja P, Rodrigo R, Fernández V, Suarez M, Muñoz H. Increased immunohistochemical expression of thrombomodulin at placental perivascular myofibroblast in severe preeclampsia (PE). Histol Histopathol. 2005;4:1045–55.

33. Myatt L. Control of vascular resistance in the human placenta. Placenta. 1992;13:329–41.
34. Myatt L, Cui X. Oxidative stress in the placenta. Histochem Cell Biol. 2004;122:369–82.
35. Roberts JM, Hubel CA. Is oxidative stress the link in the two-stage model of pre-eclampsia? Lancet. 1999;354:788–9.
36. Bowen RS, Moodley J, Dutton MF, Theron AJ. Oxidative stress in pre-eclampsia. Acta Obstet Gynecol Scand. 2001;80:719–25.
37. Takagi Y, Nikaido T, Toki T, Kita N, Kanai M, Ashida T, Ohira S, Konishi I. Levels of oxidative stress and redox-related molecules in the placenta in preeclampsia and fetal growth restriction. Virchows Arch. 2004;444:49–55.
38. Rodrigo R, Parra M, Bosco C, Fernández V, Barja P, Guajardo J, Messina R. Pathophysiological basis for the prophylaxis of preeclampsia through early supplementation with antioxidant vitamins. Pharmacol Ther. 2005;107:177–97.
39. Dotsch J, Hogen N, Nyul Z, Hanze J, Knerr I, Kirschbaum M, Rascher W. Increase of endothelial nitric oxide synthase and endothelin-1 mRNA expression in human placenta during gestation. Eur J Obstet Gynecol Reprod Biol. 2001;97:163–7.
40. Myatt L, Rosenfield RB, Eis AL, Brockman DE, Geer I, Lyall F. Nitrotyrosine residues in placenta. Evidence of peroxynitrite formation and action. Hypertension. 1996;28:488–93.
41. Gabbiani G, Hirschel BJ, Ryan GB, Statkov PR, Majno G. Granulation tissue as a contractile organ: a study of structure and function. J Exp Med. 1972;135:719–34.
42. Iwasaki H, Isayama T, Ichiki T, Kikuchi M. Intermediate filaments of myofibroblasts. Immunochemical and immunocytochemical analyses. Pathol Res Pract. 1987;182:248–54.
43. Fuchs E, Cleveland DW. A structural scaffolding of intermediate filaments in health and disease. Science. 1998;279:514–9.
44. Desmoulière A, Chaponnier C, Gabbiani G. Tissue repair, contraction, and the myofibroblast. Wound Repair Regen. 2005;13:7–12.
45. Castellucci M, Kaufmann P. Evolution of the stroma in human chorionic villi throughout pregnancy. Bibl Anat. 1982;3:40–5.
46. Castellucci M, Kaufmann P. A three-dimensional study of the normal human placental villous core: II. Stromal architecture. Placenta. 1982;3:269–5.
47. Feller AC, Schneider H, Schmidt D, Parwaresch MR. Myofibroblast as a major cellular constituent of villous stroma in human placenta. Placenta. 1985;6:405–15.
48. Graf R, Schonfelder G, Muhlberger M, Gutsmann M. The perivascular contractile sheath of human placental stem villi: its isolation and characterization. Placenta. 1995;16:57–66.
49. Farley AE, Graham CH, Smith GN. Contractile properties of human placental anchoring villi. Am J Physiol Regul Integr Comp Physiol. 2004;287:680–5.
50. Fryer BH, Simon MC. Hypoxia, HIF and the placenta. Cell Cycle. 2006;5:495–8.
51. Di Federico E, Genbacev O, Fisher SJ. Preeclampsia is associated with widespread apoptosis of placental cytotrophoblasts within the uterine wall. Am J Pathol. 1999;155:293–301.
52. Many A, Hubel CA, Fisher SJ, Roberts JM, Zhou Y. Invasive cytotrophoblasts manifest evidence of oxidative stress in preeclampsia. Am J Pathol. 2000;156:321–31.
53. Khalil RA, Granger JP. Vascular mechanisms of increased arterial pressure in preeclampsia: lessons from animal models. Am J Physiol Regul Integr Comp Physiol. 2002;283:R29–45.
54. Popescu LM, Gherghiceanu M, Suciu LC, Manole CG, Hinescu ME. Telocytes and putative stem cells in lungs: electron microscopy, electron tomography and laser scanning microscopy. Cell Tissue Res. 2011;345:391–403.
55. Zheng Y, Li H, Manole CG, Sun A, Ge J, Wang X. Telocytes in trachea and lungs. J Cell Mol Med. 2011;15:2262–8.
56. Vannucchi MG, Traini C, Manetti M, Ibba-Manneschi L, Faussone-Pellegrini MS. Telocytes express PDGFRα in the human gastrointestinal tract. J Cell Mol Med. 2013;17:1099–2108.
57. Milia AF, Ruffo M, Manetti M, Rosa I, Conte D, Fazi M, Messerini L, Ibba-Manneschi L. Telocytes in Crohn's disease. J Cell Mol Med. 2013;17:1525–36.
58. Kanfer A, Bruch JF, Nguyen G, He CJ, Delarue F, Flahault A, Nessmann C, Uzan S. Increased placental antifibrinolytic potential and fibrin deposits in pregnancy-induced hypertension and preeclampsia. Lab Invest. 1996;74:253–8.

59. Labarrere CA, Faulk WP. Microvascular perturbations in human allografts: analogies in preeclamptic placentae. Am J Reprod Immunol. 1992;27:109–16.
60. Labarrere CA, Faulk WP. Fetal stem vessel endothelial changes in placentae from normal and abnormal pregnancies. Am J Reprod Immunol. 1992;27:97–100.
61. Dittman WA, Majerus PW. Structure and function of thrombomodulin: a natural anticoagulant. Blood. 1990;75:329–36.
62. Kroll J, Waltenberger J. VEGF-A induces expression of eNOS and iNOS in endothelial cells via VEGF receptor-2 (KDR). Biochem Biophys Res Commun. 1998;252:743–6.
63. Webster RP, Roberts VH, Myatt L. Protein nitration in placenta functional significance. Placenta. 2008;29:985–94.
64. Parra-Cordero M, Bosco C, González J, Gutiérrez R, Barja P, Rodrigo R. Immunohistochemical expression of von Willebrand factor in the preeclamptic placenta. J Mol Histol. 2011;42:459–65.
65. Bosco C, González J, Gutiérrez R, Parra-Cordero M, Barja P, Rodrigo R. Oxidative damage to preeclamptic placenta: immunohistochemical expression of VEGF, nitrotyrosine residues and von Willebrand factor. J Matern Fetal Neonatal Med. 2012;25:2339–45.
66. Kostin S. Myocardial telocytes: a specific new cellular entity. J Cell Mol Med. 2010;14:1917–21.
67. Dong J, Moake J, Nolasco L, Bernardo A, Arceneaux W, Shrimpton CN, Schade AJ, McIntire LV, Fujikawa K, López JA. ADAMTS-13 rapidly cleaves newly secreted ultralarge von Willebrand factor multimers on the endothelial surface under flowing conditions. Blood. 2002;100:4033–9.
68. Li Y, Choi H, Zhou Z, Nolasco L, Pownall HJ, Voorberg J, Moake JL, Dong JF. Covalent regulation of ULVWF string formation and elongation on endothelial cells under flow conditions. J Thromb Haemost. 2008;6:1135–43.
69. López JA, Dong JF. Shear stress and the role of high molecular weight von Willebrand factor multimers in thrombus formation. Blood Coagul Fibrinolysis. 2005;16:S11–6.
70. Furlan M. Proteolytic cleavage of von Willebrand factor by ADAMTS-13 prevents uninvited clumping of blood platelets. J Thromb Haemost. 2004;2:1505–9.
71. Lancellotti S, De Filippis V, Pozzi N, Peyvandi F, Palla R, Rocca B, Rutella S, Pitocco D, Mannucci PM, De Cristofaro R. Formation of methionine sulfoxide by peroxynitrite at position 1606 of von Willebrand factor inhibits its cleavage by ADAMTS-13: a new prothrombotic mechanism in diseases associated with oxidative stress. Free Radic Biol Med. 2010;48:446–56.

Chapter 11
Telocytes in Exocrine Glands Stroma

Mihnea Ioan Nicolescu

Abstract Stroma is viewed as the supportive framework of a predominant epithelial organ, comprising mostly of connective tissue, blood vessels and nerves. Since the discovery of telocytes one decade ago (Popescu and Faussone-Pellegrini J Cell Mol Med 2010;14(4):729–40), their presence was proven in several exocrine gland stromata, including major and minor salivary glands, mammary glands as well as exocrine pancreas.

Telocytes have been found in a close connection with acinar and ductal structures but also with their stromal neighbours – nerves, blood vessels or other connective elements, either cells or collagen fibres.

The approaches used to reveal the telocytes' location were immunohistochemistry and electron microscopy.

11.1 Salivary Glands

Saliva is produced by major and minor salivary glands. The first are responsible for about 90 % of the total salivary secretion. In humans, major/main salivary glands are the parotid, submandibular and sublingual pairs. The differences between their saliva are due to the differences in their cellular repertoire, namely, exclusively serous acini in parotids and mixed in the other two, with a mucous predominance in the sublinguals. Intriguingly, the telocytes' location did not show any significant difference between the major salivary glands. Their existence was suggested by immunohistochemistry and proven by electron microscopy, the most reliable technique for the direct identification of a cell or group of cells in a given tissue.

M.I. Nicolescu, MD, DMD, OMS, PhD, MSc
Histology and Cytology Department, "Carol Davila" University of Medicine and Pharmacy, 8 Eroilor Sanitari, RO-050474 Bucharest, Romania

"Victor Babeş" National Institute of Pathology, Bucharest, Romania
e-mail: mihnea.nicolescu@umf.ro

X. Wang, D. Cretoiu (eds.), *Telocytes*, Advances in Experimental Medicine and Biology, 913, DOI 10.1007/978-981-10-1061-3_11

In parotid glands, vimentin-positive cells were making stromal networks surrounding the acini, while others resided in septa (Fig. 11.1). Solitary CD117+ with long, moniliform processes were identified within the intraparotid large septa, around ducts and blood vessels. Although a defining trait of myoepithelial cells, smooth muscle α-actin also showed positive results on interacinar cells, outside their basal laminae, that sent long prolongations between and around parotid acini [5].

In adult submandibular glands (Fig. 11.2), CD117 exhibited positivity around ducts (panel A) and nerves (B and D panels) or in scarce interstitial location (panel C). Only excretory ducts presented surrounding CD117+ cells, while no positivity was encountered around striated or intercalary ducts in foetal samples (Fig. 11.3). In major salivary glands, CD117 marked preferentially the actual body of telocytes, while telopodes (the long, peculiar, moniliform prolongations of telocytes) were positive for vimentin and smooth muscle α-actin. Both structures showed selective positivity for CD34.

Interlobular telocytes were identified under electron microscopy in rat and human parotid samples (Fig. 11.4), with their long moniliform telopodes (Figs. 11.4 and 11.5). Interacinar telocytes confirmed the location suggested by

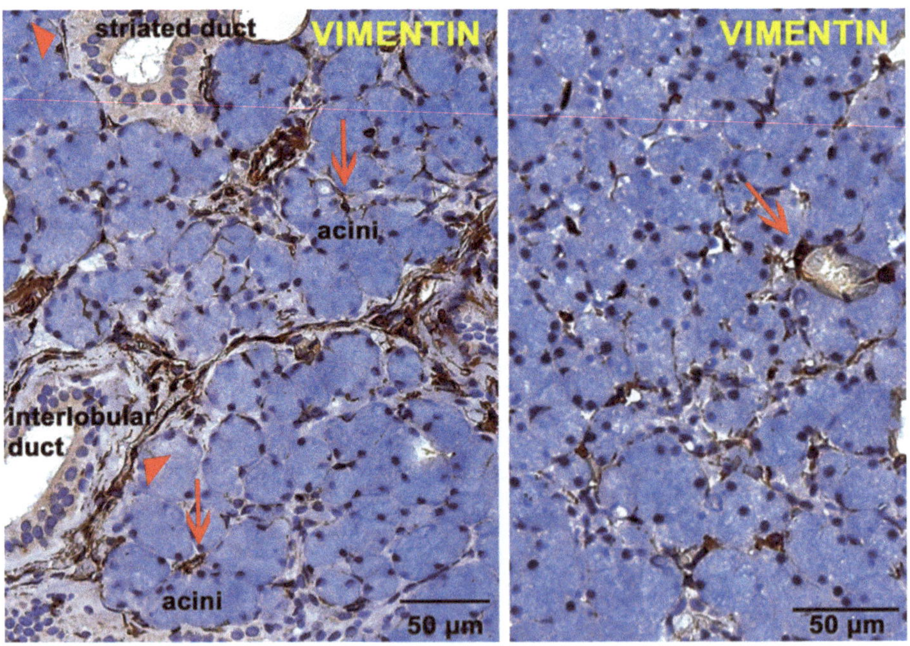

Fig. 11.1 Human parotid gland: vimentin immune staining. Spindle-shaped vimentin-positive cells with long prolongations are identified in septa, where they contribute to stromal networks. Periductal positive cells (*arrowhead*) are mostly uni-/bipolar, while peri-/interacinar ones (*arrows*) often have a multipolar appearance sending prolongations in the periphery of the neighbouring acini (Reproduced with permission from Nicolescu [5])

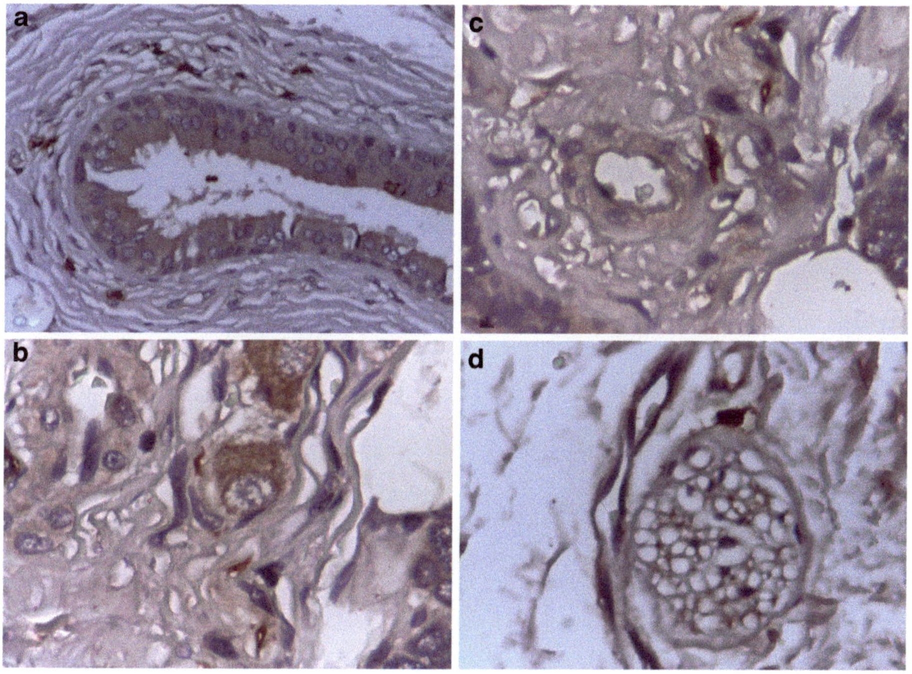

Fig. 11.2 Submandibular gland (human, adult): CD117 immunohistochemistry. (**a**) Positive cells around ducts. Ob. 20×. (**b**) Positive cellular processes around vegetative neural cells. Ob. 40×. (**c**) Positive interstitial cells. Ob. 40×. (**d**) Positive perineural cells. Ob. 40× (Reproduced with permission from Nicolescu [3])

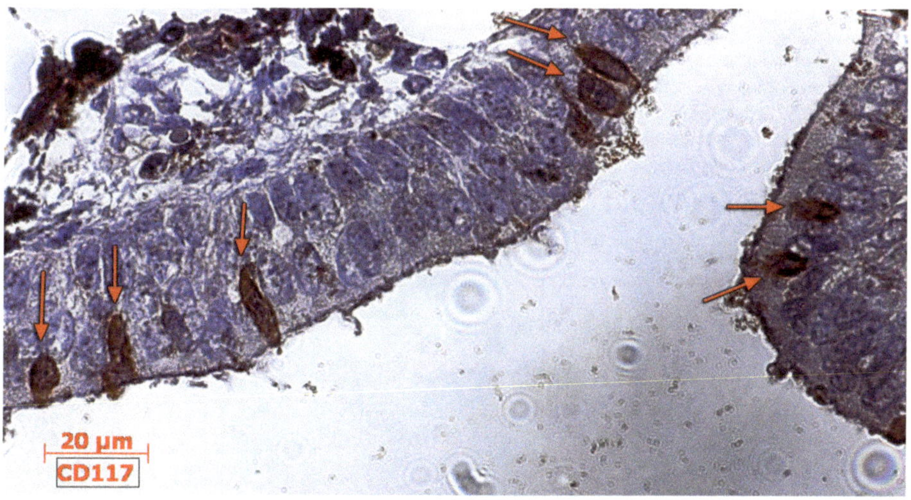

Fig. 11.3 Submandibular gland (human, foetal, 20 weeks): CD117 immunohistochemistry. *Arrows* indicate positive apical duct cells (Reproduced with permission from Nicolescu [3])

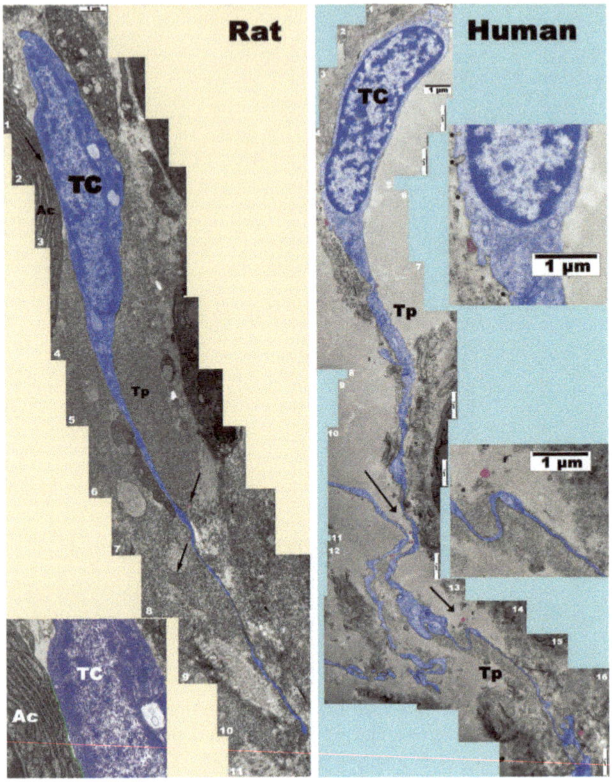

Fig. 11.4 *Left* panel – rat parotid gland: two-dimensional sequenced concatenation from 11 serial electron micrographs. *Right* panel – human parotid gland: two-dimensional sequenced concatenation from 16 serial electron micrographs. Stromal telocytes (*TC*, digitally colourized in *blue*) in rat (*left* panel) and human (*right* panel) parotid interstitium send off long telopodes (*Tp*). *Arrows* and *insets* indicate shedding microvesicles (digitally colourized in *purple*). *Left inset* shows (*bright green*) the stromal synaptic line between a telocyte and an acinar cell. *Right insets* prove why several sequenced electron micrographs are needed to observe a whole TC/Tp and why they were overlooked so far. *Asterisks* indicate TC caveolae. *Ac* acinar cell, *TC* telocyte, *Tp* telopode (Reproduced with permission from Nicolescu [5])

immunohistochemistry (Fig. 11.5). The long trajectory of telopodes includes very convoluted areas we consider to be elongation reserves, possibly needed for morphological reconfiguration of the cellular processes involved in intercellular signalling (Fig. 11.6).

Telocytes' close connection with neurovascular bundles (Fig. 11.7) presumably may support a role in regulation of vascular tonicity, by direct influence or via local neural pathways. Nevertheless, they might also influence the salivary flow and composition, by local interactions, e.g. with myoepithelial cells (Fig. 11.8). Heterocellular relations between telocytes and mast and plasma cells (Fig. 11.9) further emphasize the complexity of their interactions in salivary gland stroma.

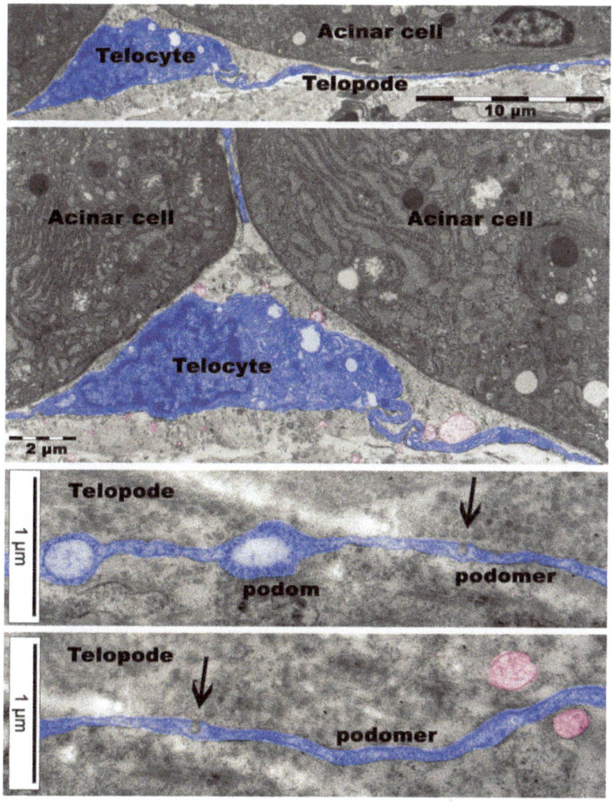

Fig. 11.5 Rat parotid gland: electron micrographs. A typical telocyte cell body with an interacinar localization is shown in the upper panels, digitally colourized in *blue*. Shed microvesicles were digitally colourized in *purple*. Note the moniliform aspect of telopodes (digitally colourized in *blue*). *Arrows* indicate caveolae. Note the alternation of thin segments (*podomeres*) and thicker portions (*podoms*) (Reproduced with permission from Nicolescu [5])

More recent studies showed, on minor salivary glands, the impairment of telocytes in autoimmune pathologies such as Sjogren's syndrome [1].

11.2 Mammary Glands

Telocytes have been previously reported as "interstitial Cajal-like cells" (ICLCs) in several organs [7], including human resting mammary gland stroma [2].

Toluidine blue-stained semithin sections showed long cellular processes [8], which were to be later named telopodes. Their ultrastructural details were thoroughly reviewed under transmission electron microscopy. Telopodes present an alternation of thin, long segments – podomeres – and dilations accommodating mitochondria and calcium-releasing units, podoms [10].

Telocytes and various cell types (plasma cells, lymphocytes, fibroblasts, mast cells and macrophages) presented multi-contact stromal synapses of variable length (up to several micrometres), comprising of close plasmalemmal apposition segments, with the cell membranes only 10–20 nm apart, as detailed in [2].

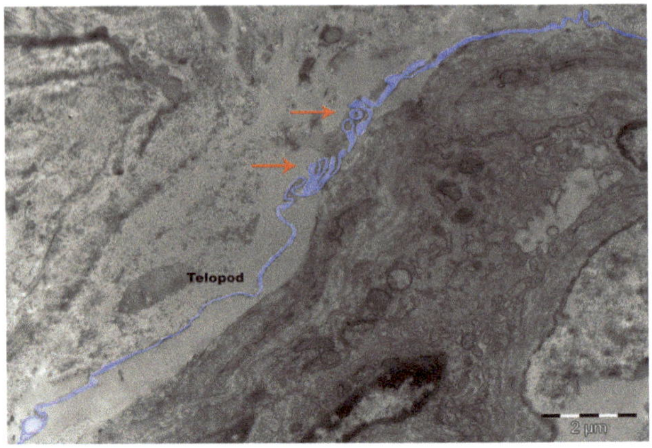

Fig. 11.6 Human submandibular gland: transmission electron microscopy. A telopode (digitally colourized in *blue*) shows seriated elongation reserve areas (*red arrows*) along its thin segments (podomeres) (Reproduced with permission from Nicolescu [3])

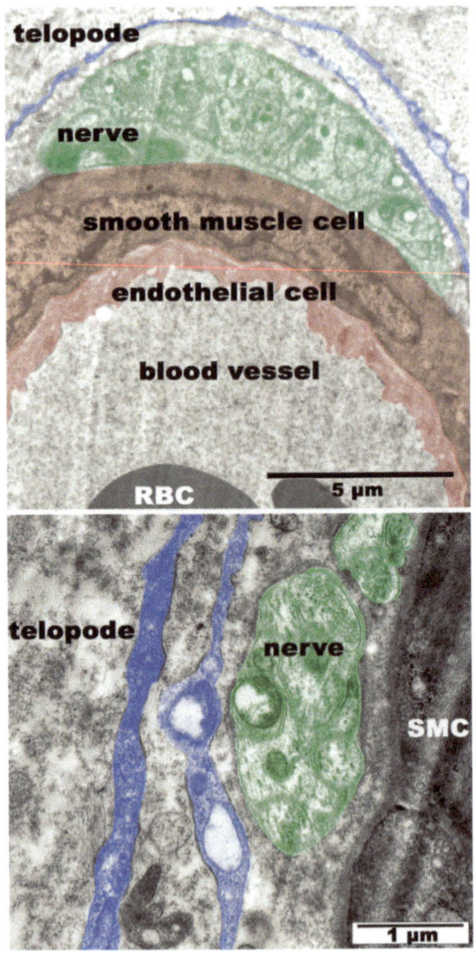

Fig. 11.7 Rat parotid gland: electron micrographs. Two telopodes (digitally colourized in *blue*) are bordering nerves (digitally colourized in *green*) and an arteriolar wall – digitally colourized in *brown* (muscular layer) and *red* (endothelium). *Lower panel* shows a cholinergic nerve between an arteriole and a telopode. *RBC* red blood cell, *SMC* smooth muscle cell (Reproduced with permission from Nicolescu [5])

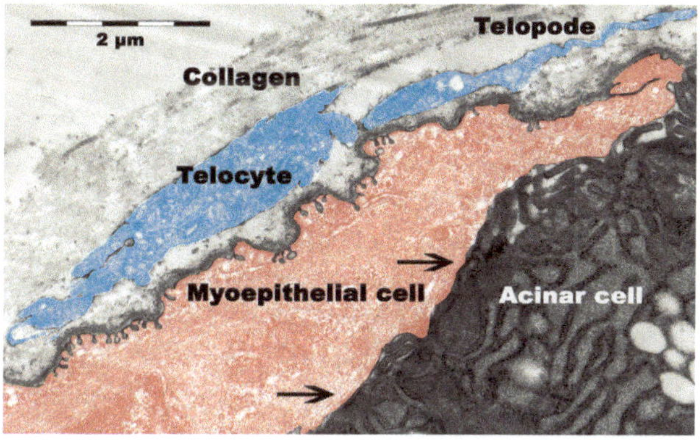

Fig. 11.8 Human parotid gland: electron micrograph. A telocyte (digitally colourized in *blue*) is present between collagen fibres and a myoepithelial cell (digitally colourized in *red*). *Arrows* indicate junction points between the myoepithelial cell and a parotid acinar cell (Reproduced with permission from Nicolescu [5])

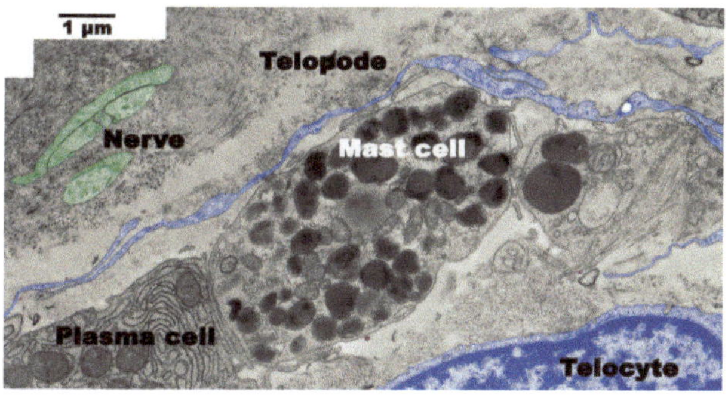

Fig. 11.9 Human parotid gland: electron micrograph. A mast cell surrounded by plasma cells establishes a multi-contact stromal synapse with a telopode (digitally colourized in *blue*). Shed microvesicles were digitally colourized in *purple*. A nearby nerve is digitally colourized in *green* (Reproduced with permission from Nicolescu [5])

Networks of CD34+ stromal cells surrounding microvessels and excretory units were reported in the human mammary gland stroma. Furthermore, $CD34^{+}/CD10^{\pm}/CD117^{-}/vimentin^{-}$ cells of inter-/intralobular mammary gland stroma showed signs of mesenchymal potency [6].

11.3 Exocrine Pancreas

As the case of other organs [7], too, pancreatic telocytes were initially described as ICLCs [9]. In human pancreas, telocytes and their telopodes form an extensive lattice network, especially around serous acini (Fig. 11.10). This network is based on homocellular contacts (between the same type of cells – here, between the long podomeres of telocytes), as they may be seen in Fig. 11.11. All the same, heterocellular connections are also present between telocytes and mast cells (Figs. 11.11b and 11.12a), macrophages (Fig. 11.12b) and stellate cells (Fig. 11.13a).

Branching pattern of telopodes might enhance the cellular telocytes signalling in the interstitial microenvironment. However, less periductal telocytes were encountered in exocrine pancreas compared to major salivary glands, where telocytes are more frequent around ducts than between acinar units [5]. As a true interconnecting hub, telocytes establish relations with all functional elements of the pancreas – acinar

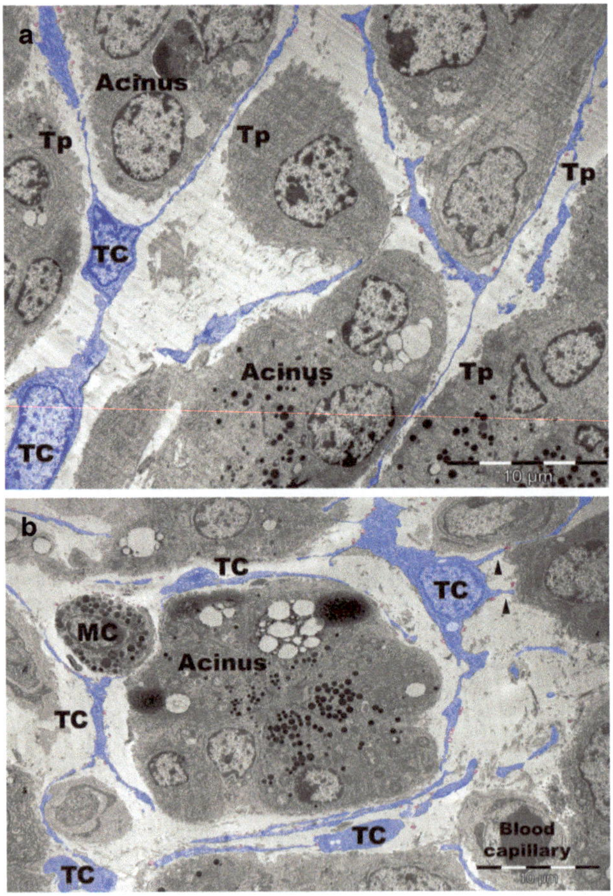

Fig. 11.10 Overview of telocytes' network in the pancreatic interstitium. Human exocrine pancreas: transmission electron microscopy. Telocytes (TC) were digitally colourized in *blue*, shed microvesicles in *purple*. (**a**) General topography of acini and the interstitial TC. Note the small TC cell body, with a nucleus and several very long and thin telopodes (Tp). (**b**) One human pancreatic acinus is circumvented by long Tp belonging to different TC. Arrowheads indicate two Tp in close contact with neighbour pancreatic acinar cells. Blood capillary and a mast cell (MC) are also present (Reproduced with permission from Nicolescu and Popescu [4])

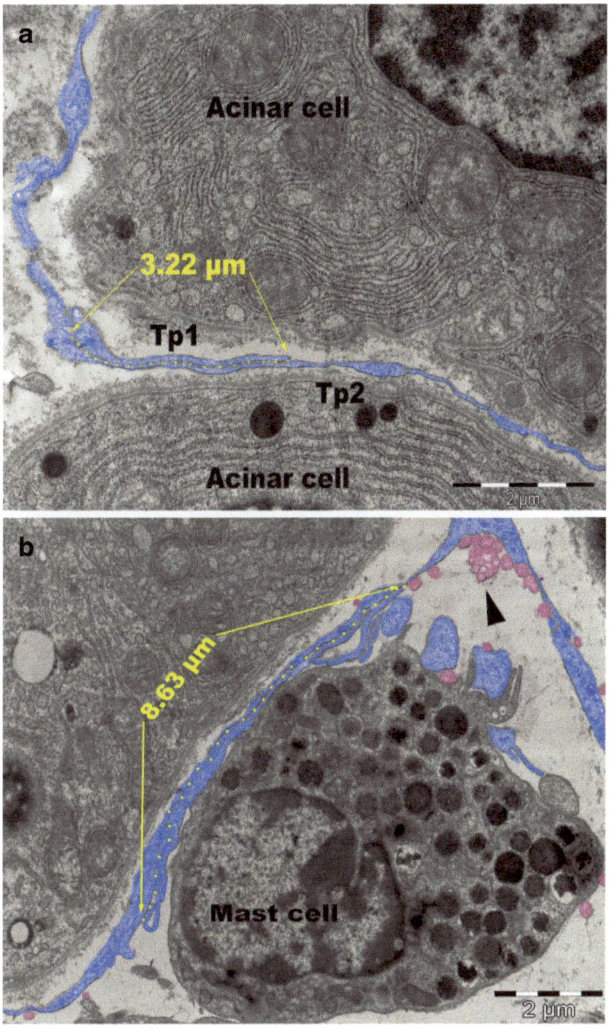

Fig. 11.11 Interpodomeric connections. Human exocrine pancreas: transmission electron microscopy. Telocytes digitally colourized in *blue*, shed microvesicles in *purple*. (**a**) Note the long (3.22 µm, *dotted yellow line*) linear plain contact between podomeres of two telopodes (*Tp1*, *Tp2*) in their intra-acinar trajectory. (**b**) Several Tps located between an acinar cell and a mast cell, bordering both of them. They establish a close intertelopodic convoluted plain contact (*dotted yellow line*) 8.63 µm long. A multivesicular body (*arrowhead*) is present at a Tp bifurcation (Reproduced with permission from Nicolescu and Popescu [4])

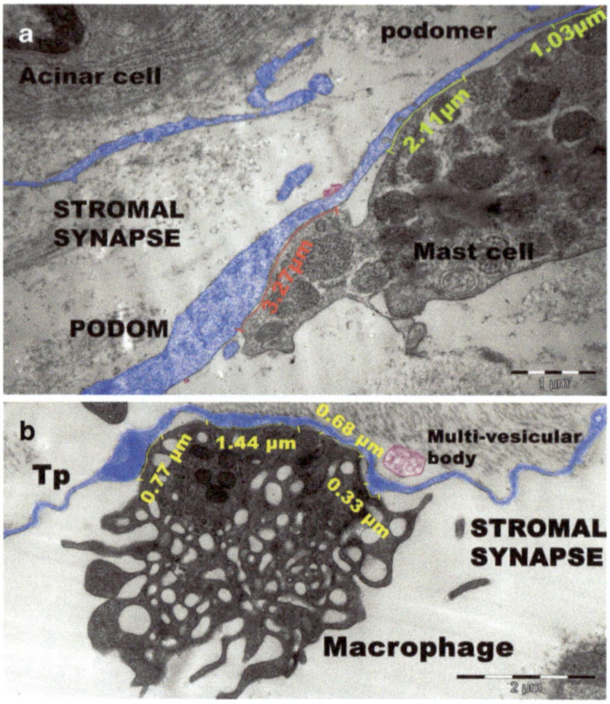

Fig. 11.12 Stromal synapses. Human exocrine pancreas: transmission electron microscopy. Telocytes digitally colourized in *blue*, shed microvesicles in *purple*. (**a**) Multi-contact (*bright yellow*) and "kiss-and-run" (*bright red*) stromal synapses between a telopode and a mast cell. Note that the contacts are established at both podomic and podomeric levels. (**b**) Multi-contact stromal synapse between a long telopode (*Tp*) and a macrophage. Note also a multivesicular body (Reproduced with permission from Nicolescu and Popescu [4])

cells, ducts, blood vessels and nerves (Figs. 11.13, 11.14, and 11.15). Telocytes release microvesicles to communicate with other cells, probably as alternative to stromal synapses for a telecrine/remote noncontact cellular cross talking. Thus, they act not as simply inert stromal elements but as active participants in regulating specific microenvironment [4].

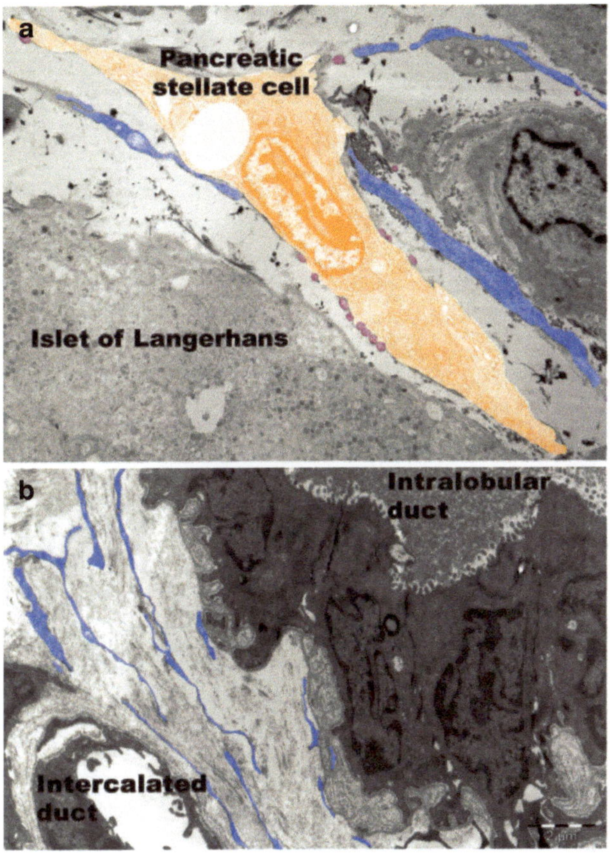

Fig. 11.13 Telocytes near stellate cells and pancreatic ducts: transmission electron microscopy. Telocytes and telopodes digitally colourized in *blue*, shed microvesicles in *purple*. (**a**) Human endocrine pancreas. Pancreatic stellate cell (digitally colourized in *brown*) in close relation to several telopodes. Note also abundant microvesicles. (**b**) Rat exocrine pancreas. Parallel telopodes bordering cells from the walls of pancreatic intercalated and intralobular ducts (Reproduced with permission from Nicolescu and Popescu [4])

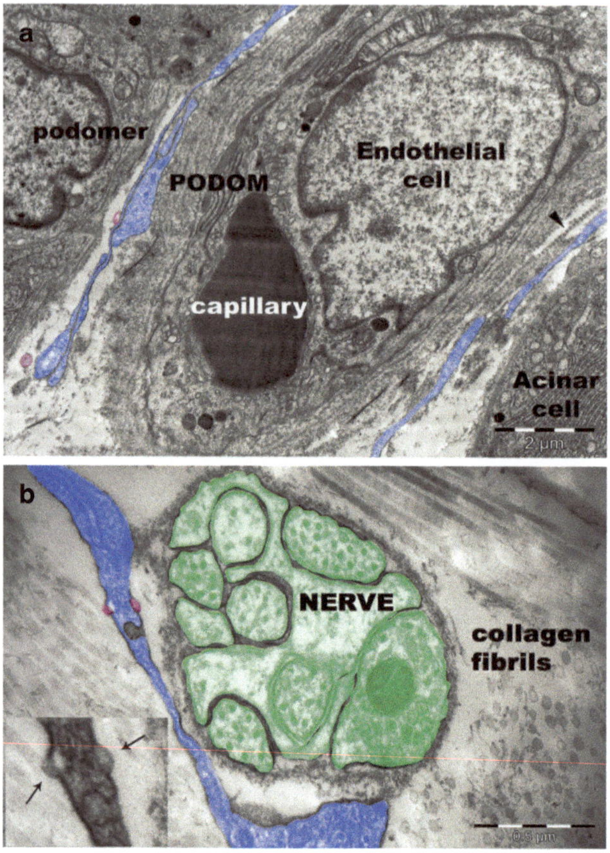

Fig. 11.14 Telocytes near blood vessels, nerves and collagen fibres: transmission electron microscopy. Telocytes and telopodes digitally colourized in *blue*, shed microvesicles in *purple*. (**a**) Human exocrine pancreas. Several telopodes are passing very close to endothelial cells. Note that telopodes establish contacts with other telopodes (podomere-podomere and podomere-podom). Note the similar diameter of podomere segments and collagen fibrils (*arrowhead*). (**b**) Rat exocrine pancreas. Telopode neighbouring a nerve (digitally colourized in *green*) in pancreatic interstitium. *Inset* shows nascent shed microvesicles (*arrows*). Also present transversal- and cross-cut collagen fibrils (Reproduced with permission from Nicolescu and Popescu [4])

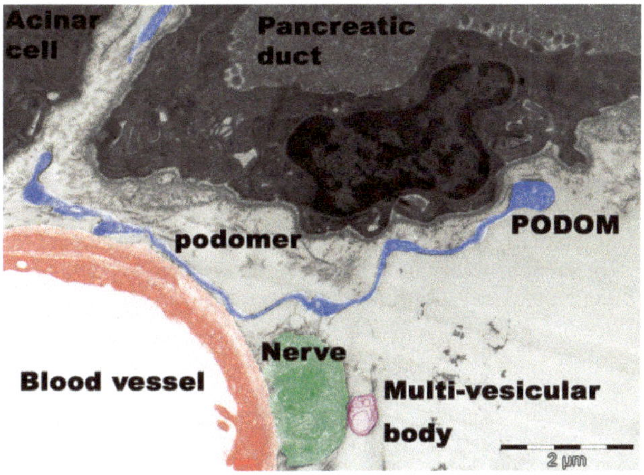

Fig. 11.15 Integrative role of telocytes. Rat exocrine pancreas: transmission electron microscopy. Telocytes digitally colourized in *blue*. Note the same telopode bordering blood vessel (digitally colourized in *red*), nerve (digitally colourized in *green*) and pancreatic acinar and ductal cells. Also note a multivesicular body (digitally colourized in *purple*) (Reproduced with permission from Nicolescu and Popescu [4])

References

1. Alunno A, et al. Telocytes in minor salivary glands of primary Sjögren's syndrome: association with the extent of inflammation and ectopic lymphoid neogenesis. J Cell Mol Med. 2015;19(7):1689–96.
2. Gherghiceanu M, Popescu LM. Interstitial Cajal-like cells (ICLC) in human resting mammary gland stroma. Transmission electron microscope (TEM) identification. J Cell Mol Med. 2005;9(4):893–910.
3. Nicolescu MI. Telocytes in pancreas and salivary glands [PhD Thesis]. Bucharest: Carol Davila University of Medicine and Pharmacy; 2012.
4. Nicolescu MI, Popescu LM. Telocytes in the interstitium of human exocrine pancreas: ultrastructural evidence. Pancreas. 2012;41(6):949–56.
5. Nicolescu MI, et al. Telocytes in parotid glands. Anat Rec. 2012;295(3):378–85.
6. Petre N, et al. Telocytes of the mammary gland stroma. Folia Morphol. 2016;75(2):224–31.
7. Popescu LM, Faussone-Pellegrini M-S. TELOCYTES – a case of serendipity: the winding way from interstitial cells of Cajal (ICC), via interstitial Cajal-like cells (ICLC) to TELOCYTES. J Cell Mol Med. 2010;14(4):729–40.
8. Popescu LM, Andrei F, et al. Snapshots of mammary gland interstitial cells: methylene-blue vital staining and c-kit immunopositivity. J Cell Mol Med. 2005;9(2):476–7.
9. Popescu LM, Hinescu ME, et al. Interstitial cells of Cajal in pancreas. J Cell Mol Med. 2005;9(1):169–90.
10. Suciu L, et al. Cardiac telocytes: serial dynamic images in cell culture. J Cell Mol Med. 2010;14(11):2687–92.

Mihnea Ioan Nicolescu I am currently an Assistant Professor of Histology and Cytology in the School of Dental Medicine in Bucharest, Romania. I have a double degree in Dental Medicine (2003) and Medicine (2006). After a 5-year resident training, I was granted the specialist degree in Oral and Maxillofacial Surgery (2009). I defended my PhD thesis in 2012, on the topic of telocytes' presence in the pancreas and salivary glands. Recently, I graduated (2015) my master's degree at King's College London (MSc in Regenerative Dentistry), with a thesis on the effect of neural mediators on salispheres.

My current publication record consists of 2 book chapters and 20 peer-reviewed publications, with more than 600 citations, resulting in a Hirsch index of 10. I submitted over 60 meeting abstracts at national/international scientific meetings (most of them as principal author) and served as a reviewer for 10 international ISI scientific journals.

Chapter 12
Telocyte Behaviour During Inflammation, Repair and Tumour Stroma Formation

L. Díaz-Flores, R. Gutiérrez, M. González-Gómez, L. Díaz-Flores Jr, F. Valladares, N. Rancel, F.J. Sáez, and J.F. Madrid

Abstract In this chapter, we outline the role of human CD34+ stromal cells/telocytes (CD34+ SC/TCs) as progenitor cells during repair. The in vivo activation phenomena of CD34+ SC/TCs in this process include increased size; separation from the neighbouring structures (mainly of the vascular walls); association with inflammatory cells, predominantly macrophages; development of the organelles of synthesis (rough endoplasmic reticulum and Golgi apparatus); cell proliferation with presence of mitosis and high proliferative index (transit-amplifying cells); and fibroblastic and myofibroblastic differentiation. A procedure to study these tissue-resident cells, comparison of their behaviour in vivo and in vitro and different behaviour depending on location, time, type of injury (including tumour stroma) and greater or lesser proximity to the injury are also considered.

12.1 Introduction

Currently, the morphologic, ultrastructural, histochemical and functional characteristics of telocytes (TCs) are well known (for revision, see www.telocytes.com). These stromal cells described by Popescu and his group [25, 51] are characterized by a small cell body and extremely long, thin moniliform processes (telopodes). Immunohistochemically, they are identified by double positive labelling for CD34

L. Díaz-Flores (✉) • R. Gutiérrez • M. González-Gómez • L. Díaz-Flores Jr • F. Valladares • N. Rancel
Department of Anatomy, Pathology, Histology and Radiology, Faculty of Medicine, University of La Laguna, Tenerife, Spain
e-mail: kayto54@gmail.com

F.J. Sáez
Department of Cell Biology and Histology, UFI11/44, School of Medicine and Dentistry, University of the Basque Country, UPV/EHU, Leioa, Spain

J.F. Madrid
Department of Cell Biology and Histology, School of Medicine, Regional Campus of International Excellence, "Campus Mare Nostrum" University of Murcia, Murcia, Spain

X. Wang, D. Cretoiu (eds.), *Telocytes*, Advances in Experimental Medicine and Biology, 913, DOI 10.1007/978-981-10-1061-3_12

and vimentin and by CD34 and PDGFR-α. Cell-surface glycoprotein CD34 expression in telocytes plays a significant role, and the use of this marker is important for several reasons: (a) it enables TC identification as a subpopulation of stromal cells (CD34+ SC/TCs), (b) it is present in progenitor cells and its potential functions include enhancing proliferation and blocking differentiation of stem or progenitor cells (maintenance of the undifferentiated progenitor/stem cell phenotype) and (c) it may vary with physiological and pathological processes, or during in vitro passages (it is a sometimes-on-sometimes-off molecule), and CD34+ SC/TCs may therefore be a source of other cells in which CD34 expression is absent. Thus, the role of CD34+ SC/TCs as progenitor cells during tissue repair requires particular attention and is the subject of this chapter.

12.2 Role of CD34+ SC/TCs in Tissue Regeneration/Repair

CD34+ SC/TCs can act in the regeneration and repair of tissues in two ways: as interstitial cells located in stem cell niches and as progenitor cells themselves.

As interstitial cells located in stem cell niches (networking between stem cells) (for revision, see [2, 20, 63]), CD34+ SC/TCs are integrated with stem cells and other components of the niche (blood vessels, extracellular matrix) [2, 6, 11, 13, 26, 27, 42, 43, 52–55, 64]. Thus, telocytes have been described in stem cell niches of different anatomical regions, including the skin; respiratory, gastrointestinal and urinary tracts; liver; pancreas; uterus; skeletal muscle; heart and vascular system; nervous system; meninges; choroid plexus; and ocular structures [13, 16, 27, 52, 53, 64, 66]. CD34+ SC/TCs in the niches carry out several functions (homeostatic control, support, nurse, regulation, guide, signal generation and induction) and are therefore in "unperturbed physiological conditions" [20]. Indeed, although the term "quiescent" may be used to describe this stage of these cells, it is not entirely accurate, since they perform numerous functions [14, 18, 22, 51].

When CD34+ SC/TCs act as progenitor cells, they enter into an activated stage, with changes in distribution, relationship and morphology, subsequent proliferation and the possibility of conserving CD34 positivity or differentiating into other cell types after losing CD34 expression [17, 20, 21, 63].

12.3 Selection of Human Tissues as Substrates to Study CD34+ SC/TC Behaviour During Human In Vivo Repair

The adipose tissue and enteric and gallbladder walls are excellent anatomic sites to study in vivo human CD34+ SC/TC behaviour during inflammatory-repair processes. Indeed, an advantage of using adipose tissue substrate is the existence of numerous studies of its stromal cells in culture (see below). The enteric wall is interesting because previous studies by different authors have confirmed that CD34+ SC/TCs are telocytes in this location [25, 45, 50, 62]. The gallbladder wall is important because intense angiogenesis and widespread fibroblastic proliferation occur in acute cholecystitis, whereas the polymorphonuclear infiltrate is scarce. Therefore, combined observations of these tissues affected by inflammatory-repair processes give an overview of CD34+ SC/TC behaviour in these conditions, which also enables comparison with observations by other authors of these cells in culture.

12.4 Importance of Resident CD34+ SC/TC Studies in Adipose Tissue Affected by Inflammatory-Repair Processes to Compare In Vivo and In Vitro Behaviour in These Cells

In vitro models to study the capacity of tissue CD34+ SC/TCs as progenitor cells include isolation and subsequent 2D and 3D cultures. Tissue-resident CD34+ SC/TCs may be isolated from several tissues, although most studies have been performed in the adipose tissue for the following reasons: (a) it is abundant, (b) it is easily obtainable in large quantities with no significant side effects and (c) its freshly isolated vascular fraction contains CD34+ SC/TCs in far greater quantities than in other tissues, including the bone marrow. In studies in culture, the common term for these cells is adipose-derived stromal cells or adipose stromal/stem/progenitor cells (ASCs), which are primarily CD34+, form part of the heterogenous population of the stromal vascular fraction, adhere to culture plastic and may proliferate and differentiate into adipocytes, osteoblasts, chondrocytes and myocytes [15, 24, 28–32, 47, 61, 69, 70]. Therefore, cells described as CD34+ SC/TCs in the adipose tissue are consistent with ASCs in the freshly isolated vascular fraction and in culture, since ASCs are found in the CD34+ population and are negative for pericytic, endothelial or other haematopoietic markers [5, 7–9, 35, 37, 40, 44, 46, 58, 60, 61, 67, 68]. Thus, the extensive literature on ASCs in culture can also be considered for the tissue-resident (native) CD34+ SC/TCs. According to the objective of this chapter on the role of CD34+ SC/TCs during repair, we also outline our contributions on the human adipose tissue during in vivo activation of CD34+ SC/TCs and compare the results with those of other authors on ASCs in culture.

12.5 A Procedure to Study Tissue-Resident CD34+ SC/TC Behaviour as Progenitor Cells In Vivo

An interesting procedure to study the in vivo behaviour of CD34+ SC/TCs as progenitor cells is to use surgically obtained tissue affected by inflammatory-repair processes (e.g. in periappendicitis, cholecystitis, complicated diverticulitis and localized peritoneal abscesses). The morphological findings of each repair stage (inflammatory, proliferative and remodelling) enable human tissue selection for the follow-up of tissue-resident CD34+ SC/TC behaviour at different evolutionary times [20, 21]. The inflammatory stage is characterized by vasodilation, oedema, leukocyte margination and interstitial infiltration of leukocytes, with a predominance of neutrophils. Angiogenesis with neovascularization; recruitment of inflammatory cells, predominantly macrophages; and stromal cell proliferation, with granulation tissue formation, occur in the proliferative stage. Numerous stromal cells with a tendency to be arranged parallel to each other are observed in early remodelling. Subsequently, cell numbers decrease and extracellular matrix increases.

12.6 CD34+ SC/TCs as Adult Progenitor Cells and Their Derived Committed Precursors as Transit-Amplifying Cells

Telocytes are a subpopulation of quiescent, slow-cycling resident cells, with proliferative potentiality and with the capacity to self-renew and to originate transit-amplifying cells, which may differentiate into αSMA+ cells and, in all likelihood, into other functionally mature cells. Therefore, telocytes may be considered as adult/somatic stem/progenitor cells and their proliferating derivatives as transit-amplifying cells or committed precursor cells, which, with more rapid though limited proliferation, increase the number of differentiated cells produced by each telocyte.

12.7 Morphologic Characteristics During the Transition from Quiescent CD34+ SC/TCs to Transit-Amplifying Cells (Initial Stage of CD34+ SC/TC Activation)

The changes in quiescent CD34+ SC/TCs (Fig. 12.1a) are evident during repair around vessels. Early morphologic phenomena of CD34+ SC/TC activation include increased size (becoming plumb cells, which augment their somatic volume and shorten their telopodes), loss of their arrangement with separation from the neighbouring structure (Fig. 12.1b) (from the vascular wall and the interstitial fibrils), association with inflammatory cells (neutrophils, macrophages, mast cells or lymphocytes, predominantly macrophages (Fig. 12.2), which contribute to the dissociation and detachment of the CD34+ SC/TCs), location in a new medium (in oedematous spaces) and structural modifications (the nucleus increased in size with prominent nucleoli, numerous intracytoplasmic ribosomes, either singly or in aggregates, as well as some profiles of the rough endoplasmic reticulum) (Fig. 12.3). During this initial stage of activation, transitional cell formed between quiescent and activated CD34+ SC/TCs are observed by light and electron microscopy. These

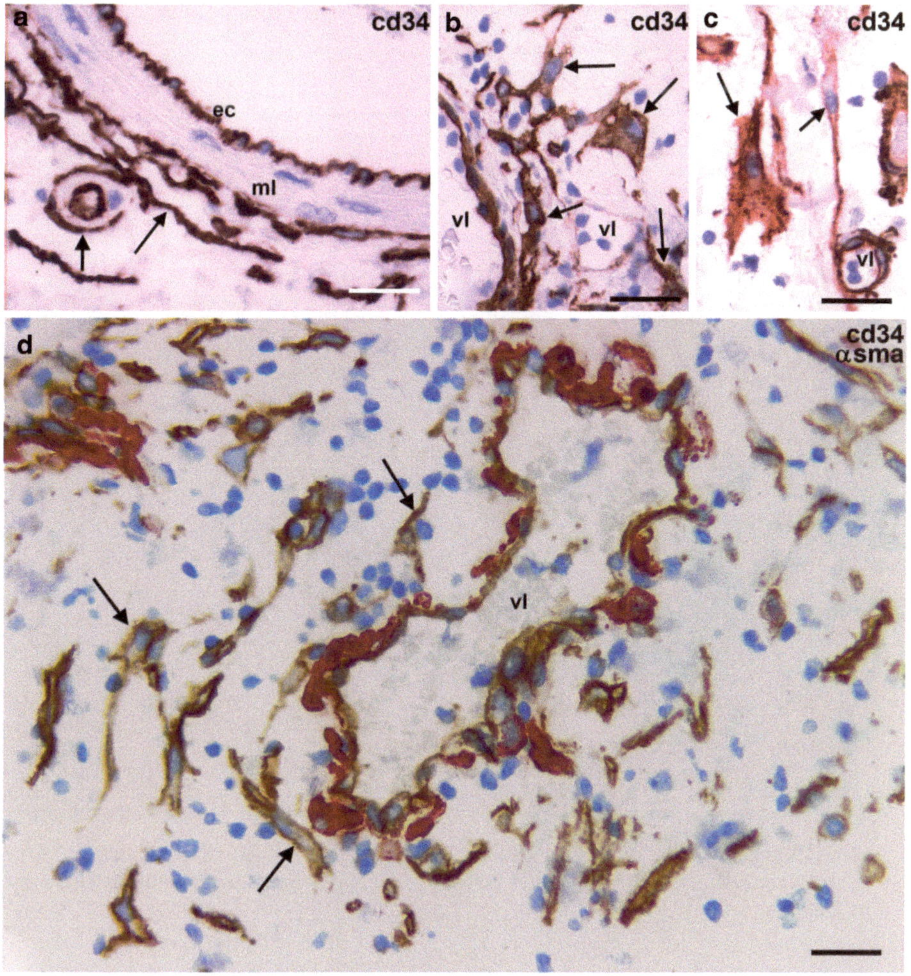

Fig. 12.1 (**a**) CD34+ SC/TCs (*arrows*) around vessels. One medium-sized and one small vessel are observed. Endothelial cells (in the intima) and SC/TCs (in the adventitia), both stained with CD34, sandwich the CD34-unstained media layer. (**b–d**) Activated CD34+ SC/TCs (*arrows*) during the inflammatory stage of repair are observed in oedematous spaces around different sized vessels. Most CD34+ SC/TCs increase in size and appear as large, plump, stellate or ovoid cells. The nuclei are also increased in size and show prominent nucleoli. Some cells conserve their telopodes. (**a–c**) CD34 immunoperoxidase labelling (using *brown* – **a** and **b** or *red* – **c** -visible label) with haematoxylin counterstain. (**d**) Section double stained with anti-CD34 (*brown*) and anti-αSMA (*red*). Bar: (**a**) 20 μm; (**b, c**) 10 μm; (**d**) 25 μm. (**a–c**) (Reproduced from Díaz-Flores et al. (2015), with permission of Histol. Histopathol)

transitional cell forms show an increased somatic region while conserving more or less evident telopodes (Fig. 12.3b). All these phenomena can be well demonstrated during the inflammatory stage of repair (in the first 24 h after tissue injury), in which vascular dilation of venules and capillaries, opening of intercellular contacts between

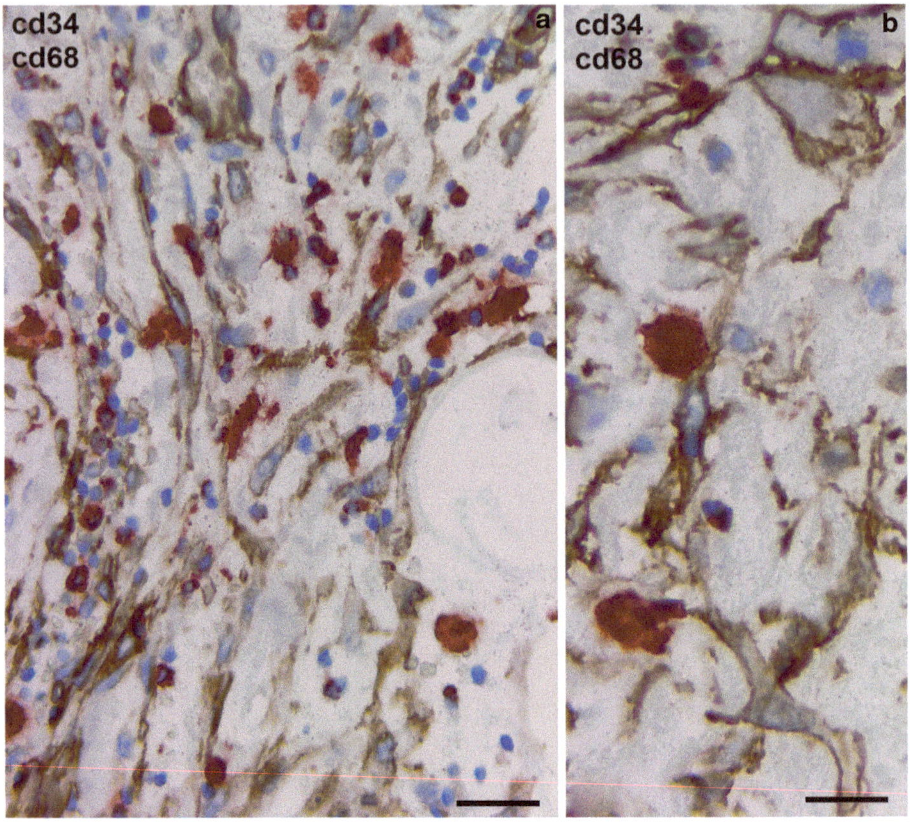

Fig. 12.2 Association between activated CD34+ SC/TCs (*brown*) and macrophages (*red*) during granulation tissue formation is demonstrated in a section double stained with anti-CD34 and anti-CD68. Bar: 25 μm

endothelial cells, increased vascular permeability, leukocyte margination and early diapedesis occur. All the activated stromal cells in this stage express CD34, and expression of αSMA is only observed in pericytes and smooth muscle cells.

12.8 Transit-Amplifying Cells Derived from CD34+ SC/TCs (Proliferative Stage of CD34+ SC/TCs and Derived Cells)

Transit-amplifying cells continue to express CD34 and coincide with the end of the inflammatory stage and with most of the proliferative stage of repair. Therefore, they become evident during granulation tissue formation, and their presence is associated with angiogenesis (proteolytic degradation of the vascular basal lamina, dissociation between endothelial cells and pericytes, endothelial cell migration and proliferation with budding, formation of the new vessel lumen, recruitment of pericytes around the

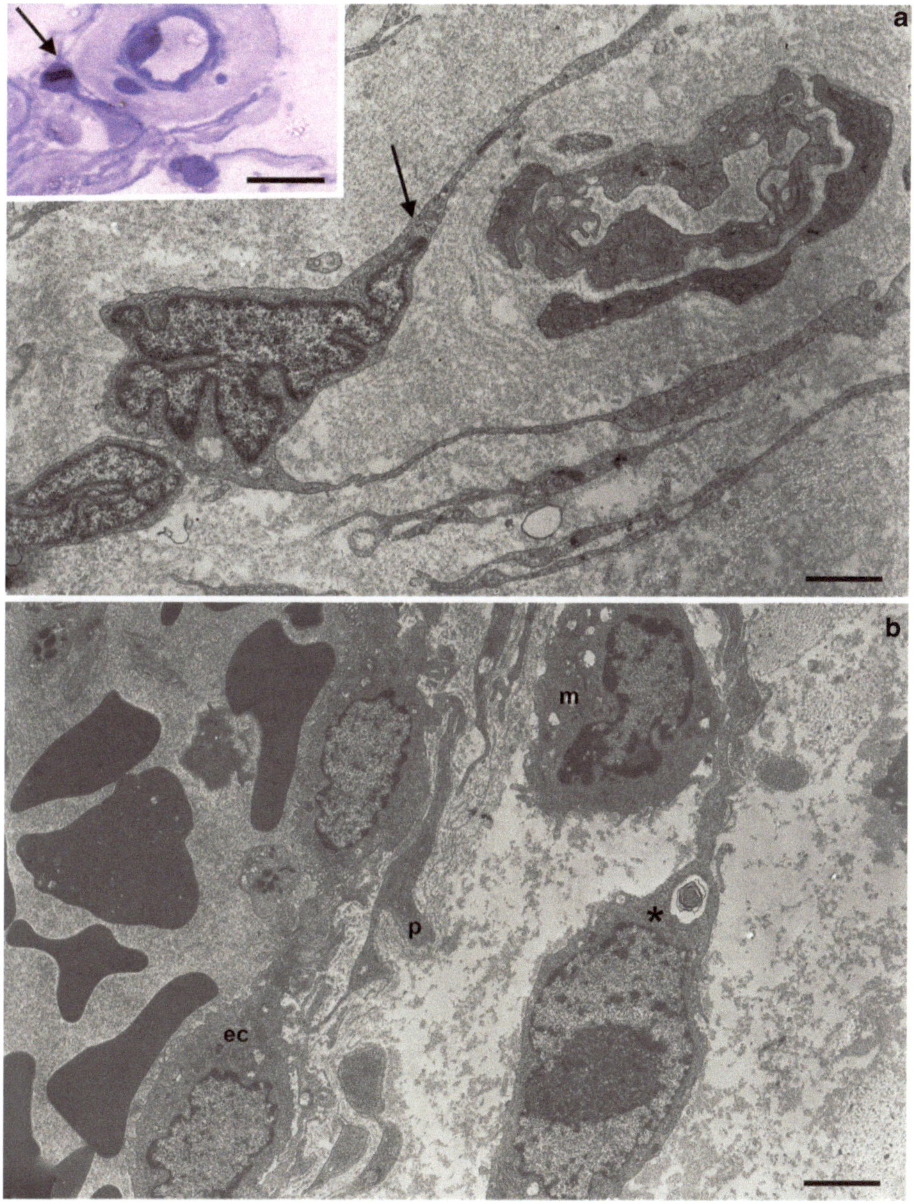

Fig. 12.3 Ultrastructural characteristics of quiescent (**a**) and early activated (**b**) telocytes. (**a**) Telocytes (*arrows*) around small vessels are demonstrated in semithin and ultrathin sections (Insert of **a**, vessel in semithin section) (**b**) In the first stages of repair through granulation tissue, an activated telocyte (*asterisk*) separates from the vessel wall (v) (*ec* endothelium, *p* pericyte, *m* macrophages). Uranyl acetate and lead citrate. Bar: 1 μm. *Insert*, toluidine blue, Bar: 25 μm (Reproduced from Díaz-Flores et al. (2015), with permission of Histol. Histopathol)

new vessels and formation of the new basal lamina). As the name suggests, transit-amplifying cells derived from CD34+ SC/TCs are proliferating cells, showing mitoses (Fig. 12.4a, b) and a high proliferative index (demonstrated in sections double stained with anti-Ki-67, detected exclusively within the nucleus and anti-CD34 detected in the

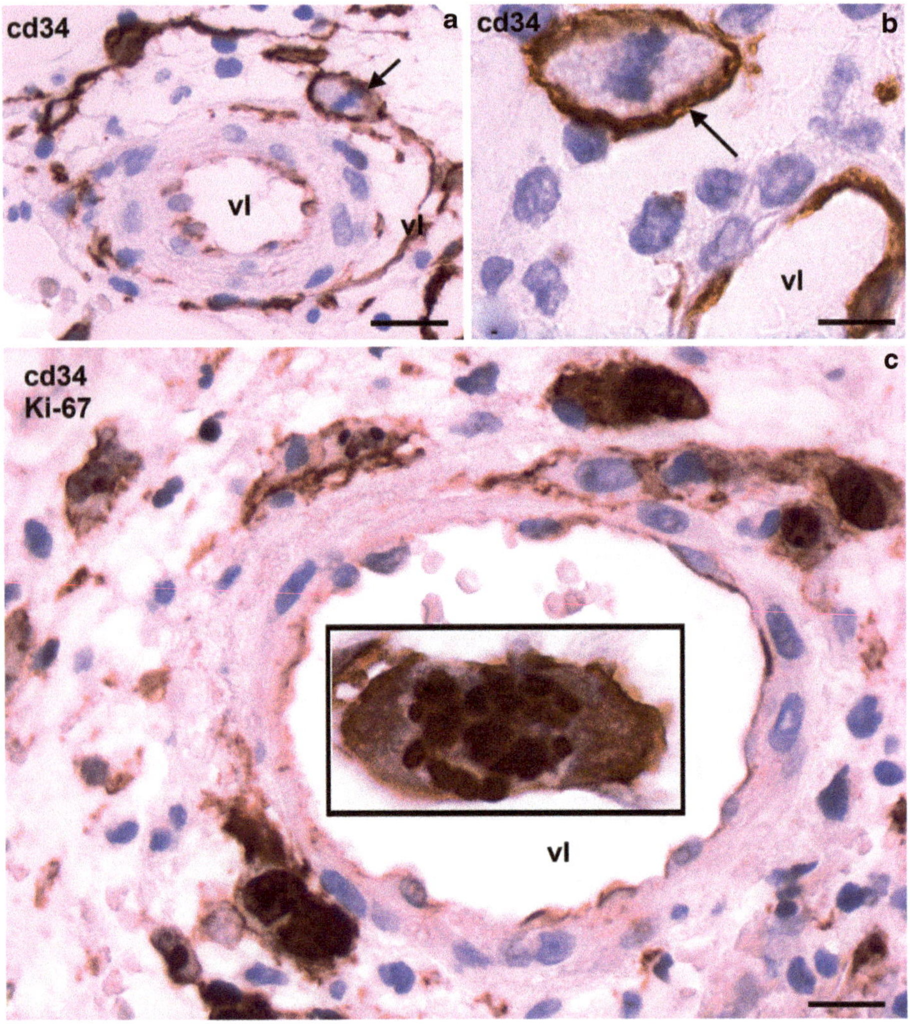

Fig. 12.4 (**a**, **b**) Activated CD34+ SC/TCs in mitosis (*arrows*) around different size vessels during the proliferative stage. The cells in mitosis show rounded morphology with peripheral expression of CD34 and chromosomes in the CD34-unstained central zone. (**c**) High proliferative index in CD34+ SC/TCs is demonstrated in a section double stained with anti-CD34 (peripheral expression) and anti-Ki-67 (expression in nuclei). In the insert, one activated CD34+ SC/TC in mitosis, showing anti-Ki-67-stained chromosomes. *vl* vessel lumen. Insert, 6 μm. Bar: (**a**) 20 μm, (**b**) 10 μm; (**c**) 15 μm; Insert, 6 μm (Reproduced from Díaz-Flores et al. (2015), with permission of Histol. Histopathol)

peripheral cytoplasm and cell membrane) (Fig. 12.4c). Therefore, a higher number of CD34+ SC/TC-derived cells occur in this proliferative stage. Although these cells continue to express CD34 during the early steps of this stage, co-expression of αSMA is observed in some cells. Likewise, neighbouring stromal cells express CD34 or αSMA with different intensity (Fig. 12.5). In successive steps of this stage, the number of stromal cells that express CD34 gradually diminishes, while that of αSMA-expressing cells increases. Finally, most stromal cells express αSMA. Stromal cells

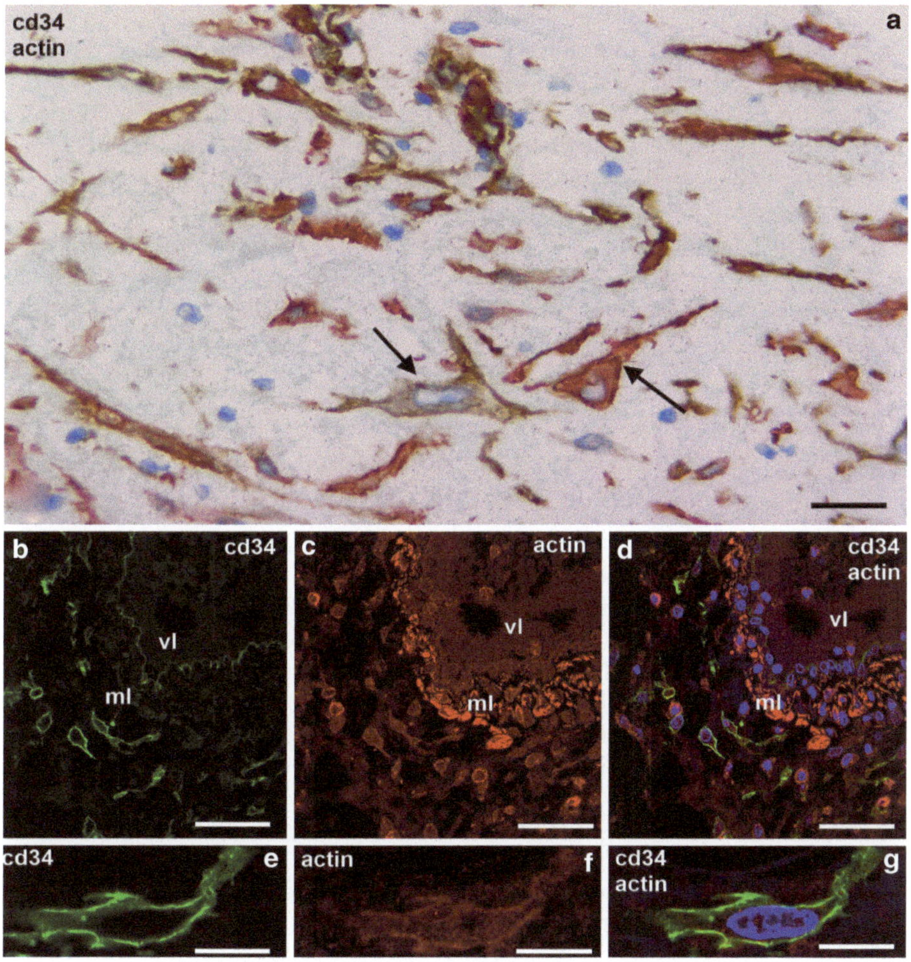

Fig. 12.5 (**a**) Activated CD34+ SC/TCs during differentiation into αSMA+ stromal cells. (**a**) A zone in which neighbouring CD34+ (*brown*) and αSMA+ (*red*) stromal cells (*arrows*), with different labelling intensity, is demonstrated in a section double stained with anti-CD34 and anti-αSMA. (**b**, **d**) Co-localization of CD34 and actin (*arrows*) in stromal cells around a vessel, demonstrated by double immunofluorescence. Nuclei are counterstained with DAPI. (**e**–**f**) Co-localization of CD34 and αSMA in a stromal cell. *vl* vessel lumen, *ml* media layer. (**a**) 25 μm; (**b**–**d**) 20 μm; **e**–**g**) 8 μm. (**b**–**g**) (Reproduced from Díaz-Flores et al. (2015), with permission of Histol. Histopathol)

expressing αSMA show ultrastructural characteristics typical of myofibroblasts (abundant organelles of synthesis and stress filaments). Increased organelles of synthesis are observed (activated fibroblasts) in cells without expression of αSMA.

12.9 Comparison of the Behaviour of Activated CD34+ SC/TCs In Vivo and In Culture

Numerous studies of the freshly isolated vascular fraction and their adipose stromal/stem progenitor cells (ASCs) and the culture of these cells support the concept that the CD34+ stromal population of the isolated stromal vascular fraction is the main progenitor with the ability to differentiate into multiple lineages, including adipocytes, osteoblasts, chondrocytes and myocytes [15, 24, 28–32, 47, 61, 69, 70]. Cultured CD34+ stromal cells lose CD34 expression in successive passages and gain expression of other markers, including αSMA [5, 7–9, 15, 24, 28–32, 35, 38–41, 44, 46, 47, 58, 60, 61, 68–70]. This behaviour of stromal cells in culture concurs with that of CD34+ SC/TCs in vivo during inflammatory-repair processes.

12.10 Different Behaviour of Activated CD34+ SC/TCs Depending on Location

During inflammatory-repair processes, CD34+ SC/TCs around vessels and in the interstitium of the connective tissue (e.g. in the submucosa and subserosa of the gastrointestinal tract or in the dermis) can undergo all the steps outlined above. However, CD34+ SC/TCs in other locations (e.g. in the endoneurium or between smooth muscle cells of the gastrointestinal muscular propria) may activate and undergo changes in number and morphology but retain CD34 positivity (non-published observations).

12.11 Importance of the Observation of CD34+ SC/TCs in Early Stages of Inflammatory-Repair Processes to Understand Their Role and Behaviour

The activation of CD34+ SC/TCs and their main changes occur over a short period in the initial stages of inflammation and repair, which explains why stromal cells co-expressing CD34 and αSMA are not detected in later stages (e.g. in tumour stroma or in remodelling stages of repair).

12.12 Different Behaviour of Activated CD34+ SC/TCs Depending on the Type of Injury

Depending on the type of injury, activated CD34+ SC/TCs may follow all the stages until loss of CD34 and gain of αSMA (wound scarring and stroma formation in some tumours – see below) or conserve CD34 expression (e.g. the stroma and capsule of some tumours, such as myopericytoma, melanocytic nevus, neurofibromas and lipomas, edematous and myxomatous processes or tissues around silicon prostheses).

12.13 Different Behaviour of Activated CD34+ SC/TCs Depending on Greater or Lesser Proximity to Injury

As mentioned above, the activation of CD34+ SC/TCs may occur with or without transformation to αSMA+ cells, in relation to the type of process [17, 19] and to the tissue location. This different form of activation is also in relation to proximity to the injury [20, 21]. Thus, in the zones with greater proximity, CD34+ SC/TCs can differentiate into αSMA+ cells, while in zones relatively far from injury, the activated cells retain CD34 positivity.

12.14 Possible Participation of Cells Other Than Resident CD34+ SC/TCs in the Origin of αSMA+

CD34+ circulating fibrocytes, which are a subset of circulating monocyte-like cells [10], may contribute to CD34+ SC/TCs in the tissues and may enter the injured tissues and also contribute to αSMA+ stromal cells [1, 57]. Although an important role of tissue-resident CD34+ SC/TCs has been demonstrated in the origin of αSMA+ stromal cells during inflammatory-repair processes, this does not exclude a contribution of CD34+ circulating fibrocytes.

12.15 CD34+ SC/TCs in the Tumour Stroma

Carcinoma-associated fibroblasts (CAFs) and myofibroblasts play an important role in tumour behaviour, formation and progression [59]. Differences in the stromal cell characteristics have been reported [3, 4, 12, 23, 33, 34, 36, 48, 49, 56, 65]. Indeed, as occurs during inflammatory-repair processes, different behaviour of activated CD34+ SC/TCs can be observed depending on the type of tumour. Thus, activated CD34+ SC/TCs may follow all the stages until loss of CD34 and gain of αSMA (myofibroblasts transformation) or conserve CD34 expression. Interestingly, many malignant invasive epithelial tumours lose CD34 expression and gain αSMA expression [3, 4, 23, 34, 48, 49, 56, 65], although this is not always the case (e.g. two third of lobular carcinoma of the breast – [23]).

References

1. Abe R, Donnelly SC, Peng T, Bucala R, Metz CN. Peripheral blood fibrocytes: differentiation pathway and migration to wound sites. J Immunol. 2001;166:7556–62.
2. Bei Y, Wang F, Yang C, Xiao J. Telocytes in regenerative medicine. J Cell Mol Med. 2015;19:1441–54.
3. Barth PJ, Ramaswamy A, Moll R. CD34(+) fibrocytes in normal cervical stroma, cervical intraepithelial neoplasia III, and invasive squamous cell carcinoma of the cervix uteri. Virch Arch. 2002;441:564–8.
4. Barth PJ, Moll R, Ramaswamy A. Stromal remodeling and SPARC (secreted protein acid rich in cysteine) expression in invasive ductal carcinomas of the breast. Virchows Arch. 2005;446:532–6.
5. Bassi G, Pacelli L, Carusone R, Zanoncello J, Krampera M. Adipose-derived stromal cells (ASCs). Transfus Apher Sci. 2012;47:193–8.
6. Bojin FM, Gavriliuc OI, Cristea MI, Tanasie G, Tatu CS, Panaitescu C, Paunescu V. Telocytes within human skeletal muscle stem cell niche. J Cell Mol Med. 2011;15:2269–72.
7. Boquest AC, Shahdadfar A, Fronsdal K, Sigurjonsson O, Tunheim SH, Collas P, Brinchmann JE. Isolation and transcription profiling of purified uncultured human stromal stem cells: alteration of gene expression after in vitro cell culture. Mol Biol Cell. 2005;16:1131–41.
8. Bourin P, Bunnell BA, Casteilla L, Dominici M, Katz AJ, March KL, Redl H, Rubin JP, Yoshimura K, Gimble JM. Stromal cells from the adipose tissue-derived stromal vascular fraction and culture expanded adipose tissue-derived stromal/stem cells: a joint statement of the International Federation for Adipose Therapeutics and Science (IFATS) and the International Society for Cellular Therapy (ISCT). Cytotherapy. 2013;15:641–8.
9. Braun J, Kurtz A, Barutcu N, Bodo J, Thiel A, Dong J. Concerted regulation of CD34 and CD105 accompanies mesenchymal stromal cell derivation from human adventitial stromal cell. Stem Cells Dev. 2013;22:815–27.
10. Bucala R, Spiegel LA, Chesney J, Hogan M, Cerami A. Circulating fibrocytes define a new leukocyte subpopulation that mediates tissue repair. Mol Med. 1994;1:71–81.
11. Campeanu RA, Radu BM, Cretoiu SM, Banciu DD, Banciu A, Cretoiu D, Popescu LM. Near-infrared low-level laser stimulation of telocytes from human myometrium. Lasers Med Sci. 2014;29:1867–74.
12. Catteau X, Simon P, Vanhaeverbeek M, Noel JC. Variable stromal periductular expression of CD34 and smooth muscle actin (SMA) in intraductal carcinoma of the breast. PLoS One. 2013;8:e57773.
13. Ceafalan L, Gherghiceanu M, Popescu LM, Simionescu O. Telocytes in human skin – are they involved in skin regeneration? J Cell Mol Med. 2012;16:1405–20.
14. Cretoiu SM, Popescu LM. Telocytes revisited. Biomol Concepts. 2014;5:353–69.
15. Daher SR, Johnstone BH, Phinney DG, March KL. Adipose stromal/stem cells: basic and translational advances: the IFATS collection. Stem Cells. 2008;26:2664–5.
16. Díaz-Flores L, Gutiérrez R, Sáez FJ, Díaz-Flores Jr L, Madrid JF. Telocytes in neuromuscular spindles. J Cell Mol Med. 2013;17:457–65.
17. Díaz-Flores L, Gutiérrez R, García MP, Sáez FJ, Díaz-Flores Jr L, Valladares F, Madrid JF. CD34+ stromal cells/fibroblasts/fibrocytes/telocytes as a tissue reserve and a principal source of mesenchymal cells. Location, morphology, function and role in pathology. Histol Histopathol. 2014;29:831–70.

18. Díaz-Flores L, Gutiérrez R, García MP, Sáez FJ, Aparicio F, Díaz-Flores Jr L, Madrid JF. Uptake and intracytoplasmic storage of pigmented particles by human CD34+ stromal cells/telocytes: endocytic property of telocytes. J Cell Mol Med. 2014;18:2478–87.
19. Díaz-Flores L, Gutiérrez R, García MP, Álvarez-Argüelles H, Díaz-Flores Jr L, López-Campos D. CD34-positive fibroblasts in Reinke's edema. Laryngoscope. 2014;124:E73–80.
20. Díaz-Flores L, Gutiérrez R, García MP, González M, Sáez FJ, Aparicio F, Díaz-Flores Jr L, Madrid JF. Human resident CD34+ stromal cells/telocytes have progenitor capacity and are a source of αSMA+ cells during repair. Histol Histopathol. 2015;30:615–27.
21. Díaz-Flores L, Gutiérrez R, Lizartza K, Goméz M, García MP, Sáez FJ, Díaz-Flores Jr L, Madrid JF. Behavior of in situ human native adipose tissue CD34+ stromal/progenitor cells during different stages of repair. Tissue-resident CD34+ stromal cells as a source of myofibroblasts. Anat Rec (Hoboken). 2015;298:917–30.
22. Díaz-Flores L, Gutiérrez R, García MP, González M, Díaz-Flores L Jr, Madrid JF. Telocytes as a source of progenitor cells in regeneration and repair through granulation tissue. Curr Stem Cell Res Ther. 2016;11(5):395–403.
23. Ebrahimsade S, Westhoff CC, Barth PJ. CD34+ fibrocytes are preserved in most invasive lobular carcinomas of the breast. Pathol Res Pract. 2007;203:695–8.
24. Erickson GR, Gimble JM, Franklin DM, Rice HE, Awad H, Guilak F. Chondrogenic potential of adipose tissue-derived stromal cells in vitro and in vivo. Biochem Biophys Res Commun. 2002;290:763–9.
25. Faussone-Pellegrini MS, Popescu LM. Telocytes Biomol Concepts. 2011;2:481–9.
26. Gherghiceanu M, Popescu LM. Cardiomyocyte precursors and telocytes in epicardial stem cell niche: electron microscope images. J Cell Mol Med. 2010;14:871–7.
27. Gherghiceanu M, Popescu LM. Cardiac telocytes – their junctions and functional implications. Cell Tissue Res. 2012;348:265–79.
28. Gimble J, Guilak F. Adipose-derived adult stem cells: isolation, characterization, and differentiation potential. Cytotherapy. 2003;5:362–9.
29. Gimble JM, Katz AJ, Bunnell BA. Adipose-derived stem cells for regenerative medicine. Circ Res. 2007;100:1249–60.
30. Halvorsen YC, Wilkison WO, Gimble JM. Adipose-derived stromal cells—their utility and potential in bone formation. Int J Obes Relat Metab Disord. 2000;24 Suppl 4:S41–4.
31. Hicok KC, Du Laney TV, Zhou YS, Halvorsen YD, Hitt DC, Cooper LF, Gimble JM. Human adipose-derived adult stem cells produce osteoid in vivo. Tissue Eng. 2004;10:371–80.
32. Justesen J, Pedersen SB, Stenderup K, Kassem M. Subcutaneous adipocytes can differentiate into bone-forming cells in vitro and in vivo. Tissue Eng. 2004;10:381–91.
33. Kacar A, Arikok AT, Kokenek Unal TD, Onder E, Hucumenoglu S, Alper M. Stromal expression of CD34, alpha-smooth muscle actin and CD26/DPPIV in squamous cell carcinoma of the skin: a comparative immunohistochemical study. Pathol Oncol Res. 2012;18:25–31.
34. Kuroda N, Jin YL, Hamauzu T, Toi M, Miyazaki E, Hiroi M, Moriki T, Enzan H. Consistent lack of CD34-positive stromal cells in the stroma of malignant breast lesions. Histol Histopathol. 2005;20:707–12.
35. Li H, Zimmerlin L, Marra KG, Donnenberg VS, Donnenberg AD, Rubin JP. Adipogenic potential of adipose stem cell subpopulations. Plast Reconstr Surg. 2011;128:663–72.
36. Li Q, Huang W, Zhou X. Expression of CD34, alpha-smooth muscle actin and transforming growth factor-beta1 in squamous intraepithelial lesions and squamous cell carcinoma of the cervix. J Int Med Res. 2009;37:446–54.
37. Lin CS, Xin ZC, Deng CH, Ning H, Lin G, Lue TF. Defining adipose tissue-derived stem cells in tissue and in culture. Histol Histopathol. 2010;25:807–15.
38. Lin CS, Ning H, Lin G, Lue TF. Is CD34 truly a negative marker for mesenchymal stromal cells? Cytotherapy. 2012;14:1159–63.
39. Lin CS, Xin ZC, Dai J, Lue TF. Commonly used mesenchymal stem cell markers and tracking labels: limitations and challenges. Histol Histopathol. 2013;28:1109–16.
40. Lin G, Garcia M, Ning H, Banie L, Guo YL, Lue TF, Lin CS. Defining stem and progenitor cells within adipose tissue. Stem Cells Dev. 2008;17:1053–63.

41. Lin G, Xu RH. Progresses and challenges in optimization of human pluripotent stem cell culture. Curr Stem Cell Res Ther. 2010;5:207–14.
42. Luesma MJ, Gherghiceanu M, Popescu LM. Telocytes and stem cells in limbus and uvea of mouse eye. J Cell Mol Med. 2013;17:1016–24.
43. Manetti M, Guiducci S, Ruffo M, Rosa I, Faussone-Pellegrini MS, Matucci-Cerinic M, Ibba-Manneschi L. Evidence for progressive reduction and loss of telocytes in the dermal cellular network of systemic sclerosis. J Cell Mol Med. 2013;17:482–96.
44. Maumus M, Peyrafitte JA, D'Angelo R, Fournier-Wirth C, Bouloumie A, Casteilla L, Sengenes C, Bourin P. Native human adipose stromal cells: localization, morphology and phenotype. Int J Obes (Lond). 2011;35:1141–53.
45. Milia AF, Ruffo M, Manetti M, Rosa I, Conte D, Fazi M, Messerini L, Ibba-Manneschi L. Telocytes in Crohn's disease. J Cell Mol Med. 2013;17:1525–36.
46. Mitchell JB, McIntosh K, Zvonic S, Garrett S, Floyd ZE, Kloster A, Di Halvorsen Y, Storms RW, Goh B, Kilroy G, Wu X, Gimble JM. Immunophenotype of human adipose-derived cells: temporal changes in stromal-associated and stem cell-associated markers. Stem Cells. 2006;24:376–85.
47. Mizuno H, Zuk PA, Zhu M, Lorenz HP, Benhaim P, Hedrick MH. Myogenic differentiation by human processed lipoaspirate cells. Plast Reconstr Surg. 2002;109:199–209.
48. Nakayama H, Enzan H, Miyazaki E, Kuroda N, Naruse K, Kiyoku H, Toi M, Hiroi M. CD34 positive stromal cells in gastric adenocarcinomas. J Clin Pathol. 2001;54:846–8.
49. Nimphius W, Moll R, Olbert P, Ramaswamy A, Barth PJ. CD34+ fibrocytes in chronic cystitis and noninvasive and invasive urothelial carcinomas of the urinary bladder. Virchows Arch. 2007;450:179–85.
50. Pieri L, Vannucchi MG, Faussone-Pellegrini MS. Histochemical and ultrastructural characteristics of an interstitial cell type different from ICC and resident in the muscle coat of human gut. J Cell Mol Med. 2008;12:1944–55.
51. Popescu LM, Faussone-Pellegrini MS. TELOCYTES – a case of serendipity: the winding way from interstitial cells of Cajal (ICC), via Interstitial Cajal-Like Cells ICLC) to TELOCYTES. J Cell Mol Med. 2010;14:729–40.
52. Popescu LM, Gherghiceanu M, Suciu LC, Manole CG, Hinescu ME. Telocytes and putative stem cells in the lungs: electron microscopy, electron tomography and laser scanning microscopy. Cell Tissue Res. 2011;345:391–403.
53. Popescu LM, Manole E, Serboiu CS, Manole CG, Suciu LC, Gherghiceanu M, Popescu BO. Identification of telocytes in skeletal muscle interstitium: implication for muscle regeneration. J Cell Mol Med. 2011;15:1379–92.
54. Popescu BO, Gherghiceanu M, Kostin S, Ceafalan L, Popescu LM. Telocytes in meninges and choroid plexus. Neurosci Lett. 2012;516:265–9.
55. Popescu LM, Nicolescu MI. Chapter 11 – Telocytes and stem cells. In: Goldenberg RCS, de Carvalho ACC, editors. Resident stem cells regenerative therapy. Elsevier Inc/Academic Press; Brazil. 2013. pp. 205–231.
56. Ramaswamy A, Moll R, Barth PJ. CD34+ fibrocytes in tubular carcinomas and radial scars of the breast. Virchows Arch. 2003;443:536–40.
57. Reilkoff RA, Bucala R, Herzog EL. Fibrocytes: emerging effector cells in chronic inflammation. Nat Rev Immunol. 2011;11:427–35.
58. Rodeheffer MS, Birsoy K, Friedman JM. Identification of white adipocyte progenitor cells in vivo. Cell. 2008;135:240–9.
59. San Martin R, Barron DA, Tuxhorn JA, Ressler SJ, Hayward SW, Shen X, Laucirica R, Wheeler TM, Gutierrez C, Ayala GE, Ittmann M, Rowley DR. Recruitment of CD34(+) fibroblasts in tumor-associated reactive stroma: the reactive microvasculature hypothesis. Am J Pathol. 2014;184:1860–70.
60. Sengenes C, Lolmede K, Zakaroff-Girard A, Busse R, Bouloumie A. Preadipocytes in the human subcutaneous adipose tissue display distinct features from the adult mesenchymal and hematopoietic stem cells. J Cell Physiol. 2005;205:114–22.

61. Suga H, Matsumoto D, Eto H, Inoue K, Aoi N, Kato H, Araki J, Yoshimura K. Functional implications of CD34 expression in human adipose-derived stem/progenitor cells. Stem Cells Dev. 2009;18:1201–10.
62. Vannucchi MG, Traini C, Manetti M, Ibba-Manneschi L, Faussone-Pellegrini MS. Telocytes express PDGFRα in the human gastrointestinal tract. J Cell Mol Med. 2013;17:1099–108.
63. Vannucchi MG, Bani D, Faussone-Pellegrini MS. Telocytes contribute as cell progenitors and differentiation inductors in tissue regeneration. Curr Stem Cell Res Ther. 2016;11(5):383–9.
64. Wang F, Song Y, Bei Y, Zhao Y, Xiao J, Yang C. Telocytes in liver regeneration: possible roles. J Cell Mol Med. 2014;18:1720–6.
65. Wessel C, Westhoff CC, Nowak K, Moll I, Barth PJ. CD34(+) fibrocytes in melanocytic nevi and malignant melanomas of the skin. Virchows Arch. 2008;453:485–9.
66. Xiao J, Wang F, Liu Z, Yang C. Telocytes in liver: electron microscopic and immunofluorescent evidence. J Cell Mol Med. 2013;17:1537–42.
67. Zimmerlin L, Donnenberg VS, Pfeifer ME, Meyer EM, Peault B, Rubin JP, Donnenberg AD. Stromal vascular progenitors in adult human adipose tissue. Cytometry A. 2010;77:22–30.
68. Zimmerlin L, Donnenberg VS, Rubin JP, Donnenberg AD. Mesenchymal markers on human adipose stem/progenitor cells. Cytometry A. 2013;83:134–40.
69. Zuk PA, Zhu M, Ashjian P, De Ugarte DA, Huang JI, Mizuno H, Alfonso ZC, Fraser JK, Benhaim P, Hedrick MH. Human adipose tissue is a source of multipotent stem cells. Mol Biol Cell. 2002;13:4279–95.
70. Zuk PA, Zhu M, Mizuno H, Huang J, Futrell JW, Katz AJ, Benhaim P, Lorenz HP, Hedrick MH. Multilineage cells from human adipose tissue: implications for cell-based therapies. Tissue Eng. 2001;7:211–28.

Chapter 13
Paracrine Signaling in the Prostatic Stroma: A Novel Role for the Telocytes Revealed in Rodents' Ventral Prostate

Bruno D.A. Sanches, Lara S. Corradi, Patricia S.L. Vilamaior, and Sebastião R. Taboga

Abstract The telocytes have recently been described in the prostate gland. In mature gland, they exist in close association with the acini and their telopodes form networks whose functions remain unclear. In this chapter, our group gives a brief introduction to telocytes and explores the history that led to such a concept and then discusses hypotheses and presents new evidences about the roles exerted by telocytes in the prostate. First is given emphasis on the role that these cells possibly play in paracrine signaling employed in the differentiation of smooth muscle periacinar are then discussed other roles potentially performed by telocytes in the prostate, such as the organizational, where these cells would act in order to delimit stromal

B.D.A. Sanches
Department of Structural and Functional Biology, State University of Campinas – UNICAMP, Bertrand Russel Av. s/n, Campinas, São Paulo, Brazil

Laboratory of Microscopy and Microanalysis, Department of Biology, Univ. Estadual Paulista – UNESP, Cristóvão Colombo St., 2265, Jardim Nazareth, São José do Rio Preto, São Paulo, Brazil

L.S. Corradi • P.S.L. Vilamaior
Campus de Araguaína – Escola de Medicina Veterinária e Zootecnia, Fund. Univ. Federal do Tocantins, BR 153 Km112, Araguaína, TO, Brazil

Laboratory of Microscopy and Microanalysis, Department of Biology, Univ. Estadual Paulista – UNESP, Cristóvão Colombo St., 2265, Jardim Nazareth, São José do Rio Preto, São Paulo, Brazil

S.R. Taboga (✉)
Department of Structural and Functional Biology, State University of Campinas – UNICAMP, Bertrand Russel Av. s/n, Campinas, São Paulo, Brazil

Laboratory of Microscopy and Microanalysis, Department of Biology, Univ. Estadual Paulista – UNESP, Cristóvão Colombo St., 2265, Jardim Nazareth, São José do Rio Preto, São Paulo, Brazil
e-mail: taboga@ibilce.unesp.br

X. Wang, D. Cretoiu (eds.), *Telocytes*, Advances in Experimental Medicine and Biology, 913, DOI 10.1007/978-981-10-1061-3_13

microenvironments, thereby assisting the differentiation of the prostatic anatomical components. In addition, the pacemaker function of smooth muscle cells contraction, as evidenced by the presence of caveolae and gap-type junction and, finally, the role of telocytes in prostate remodeling and the possible action as adult progenitor cells. Generally speaking, the chapter reaffirms the existence of telocytes as distinct cells of other stromal cells and the importance of this new cell type for normal metabolism and prostate development.

13.1 A Brief Introduction on the Prostate Telocytes

The epithelial-mesenchymal interactions since the early twentieth century are considered the core of the formation of various organs [37], among which are the lungs, kidneys, mammary glands, and prostate [4, 29, 35]. Several paracrine molecular pathways have been implicated in prostate development; such interactions initially are set between projections of the epithelium of the urogenital sinus (UGS) and the epithelium of the urogenital sinus (UGE) [9, 32, 42]. Such epithelial projections (prostatic buddings) reach the prostate inductive mesenchymes surrounding the periurethral musculature, among which the most studied is the ventral mensenchymal pad (VMP), which is responsible for inducing the differentiation of the ventral lobe of the prostate in rodents [39, 43, 44]. In these mesenchymes prostatic buddings undergo branching morphogenesis and suffer a progressive canalization in the proximal-distal direction and, at the end of the first month of postnatal life, functional acini are verified in the prostate in rodents [26, 32].

The presence of interstitial Cajal-like cells was detected in the prostate in intimate association with gland smooth muscle cells [34]; at the same time, a series of publications confirmed the existence of interstitial Cajal-like cells in other organs such as the mammary glands [11], pancreas [23], uterus [24], and others [22]. The Cajal cells were first described by Cajal [28], who proposed in the early twentieth century that such cells would be interstitial neurons; such a classification was defied for more detailed studies that have shown that these cells were truly interstitial cells of mesenchymal origin [10, 41]. Such cells would have a function in a smooth muscle contraction, acting as a pacemaker [14, 41], and immunostaining for c-kit has become the main identification method of this cell type [16]. The c-kit is a receptor which binds to stem cell factor (SCF) and generates a calcium mobilization by dependent pathways of src family kinase and PI3-kinase [46], which contributes to the activity of interstitial Cajal cells as pacemaker of smooth muscle contraction.

Evidences were accumulated to consider interstitial cells of Cajal (ICCs) a heterogeneous population of cells [30], with cells more alike to smooth muscle cells and more fibroblast-like cells [15]. The same fibroblast-like cells remained present in c-kit mutated mice. The c-kit marker is a factor considered essential for the differentiation of ICCs: the fibroblast-like cell subpopulation carries a smaller amount of caveolae, has gap junctions, and is usually CD34 positive [12, 38]. These cells were also supposed from performing a pacemaker function in the smooth muscle cells, only by inhibitory pathways of electrical stimulation via sk3, unlike the ICCs that would act in a stimulating manner via c-kit [13, 47].

In mutant mice presenting problems in contractility of smooth muscle cells, CD34-positive fibroblast-like cells in the intestine increased concentration in comparison to other ICCs that decreased their number; thus, it was proposed that the CD34-positive fibroblast-like cells were associated with regeneration of the other ICCs, so as to be an adult progenitor cell [48]; however, such cells also possess a pacemaker activity, and to consider them only immature ICCs is an incomplete assumption [38]. The existence of a distinct cell type of ICCs in the gut is widely accepted; however, the typical marker of these cells is controversial, CD34 immunostaining is not a consensus, and the PDGFR is also considered a marker [33], albeit many of the CD34-positive cells are also positive PDGFR [2].

Given the existence of a distinct cell type of ICCs and ambiguous interpretations or vague terms like fibroblast-like cells or CD34-positive interstitial Cajal-like cells, a new term has been coined to characterize this cell type, which has a role that extends beyond the auxiliary function in the contraction of smooth muscle cells (such cells were denominated as telocytes [22], and their thin and long cytoplasmic processes were termed telopods). With the characterization of this new cell type, distinct from the ICCs, a number of new studies have been conducted in this field, bringing light to an enormous amount of data on this cell type [20] (Nicolescu et al. 2012). New roles beyond the pacemaker activity were proposed for such cells, for example, a central role in organizing bodies' tissue compartmentalization and defining stromal microenvironments [31]. Other functions have been proposed and they vary from tissue to tissue such as, for example, acting on immune function in the duodenum [5] in regenerating the cardiac muscle [17] or direct action on cell differentiation of smooth muscle cells in the prostate, as proposed in this chapter (Figs. 13.1, 13.2, 13.3, 13.4, 13.5, 13.6, 13.7, and 13.8).

Such cells are present in the heart [17, 31], in the lungs [50], in the pancreas [19], in the parotid glands [20], in the uterus [7, 8], in the kidneys [27], and in the intestine [5], and more recently, they have been described in the prostate by our research group [6]. Changes in the structure of telocytes in pathological conditions have been identified for the dermis [18] and lung [50]. Such evidence may indicate that these cells may show, in general, an important organizational role in the maintenance of several organs. In terms of the prostate gland, ultrastructural evidence and histochemical approaches point out that telocytes could present an important role in the maintenance of the prostate architecture and they could be associated with the homeostasis of stroma and contribute to the compartmentalization of prostatic stroma microenvironments. This possible function of prostate telocytes could imply them with the investigation of pathological conditions in the prostate or during the prostate involution following chemical castration [6].

Prostate telocytes are morphologically similar to telocytes of the myocardium, kidneys, and other organs [6]. Such cells are distributed in the interacinar region that involves the smooth muscle layers of the acini, but unlike what is found in the myocardium,

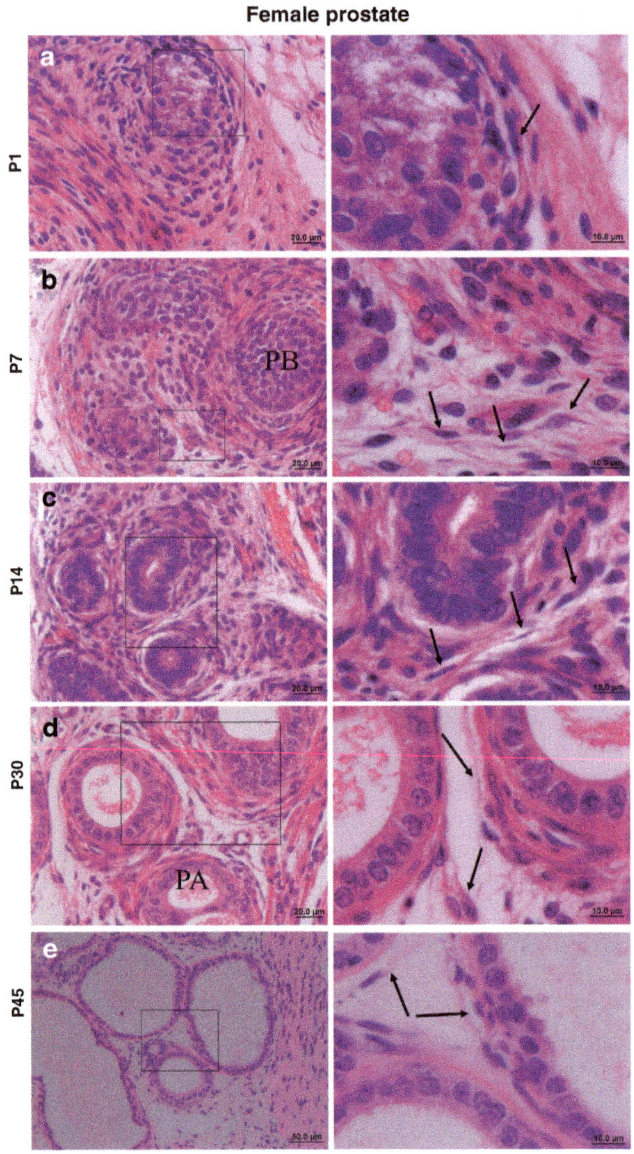

Fig. 13.1 Histology of female prostate development in the Mongolian gerbil. Histological sections of the developing prostatic complex were stained with hematoxylin-eosin to cover the early postnatal development period of the gland. (**a**) Cells with flattened nuclei are observed in the periphery of the inductive prostatic mesenchyme at P1. (**b**) At P7, fusiform cell bodies with small cytoplasmic projections are observed in interductal region in the developing. (**c**) At the end of the second week of development, spindle-shaped cell bodies with thin and longer projections are arranged on the periphery of the acini and juxtaposed to the developing smooth muscle. (**d**, **e**) Cell bodies with thin and long projections exhibiting the characteristic morphology of telocytes occupy the periductal region forming a network of projections (telopods) involving the prostatic acini. *Arrows* (telocytes in differentiation), *inserts* (details of regions where there are possible telocytes), *PB* prostatic buddings, *PA* prostatic acini

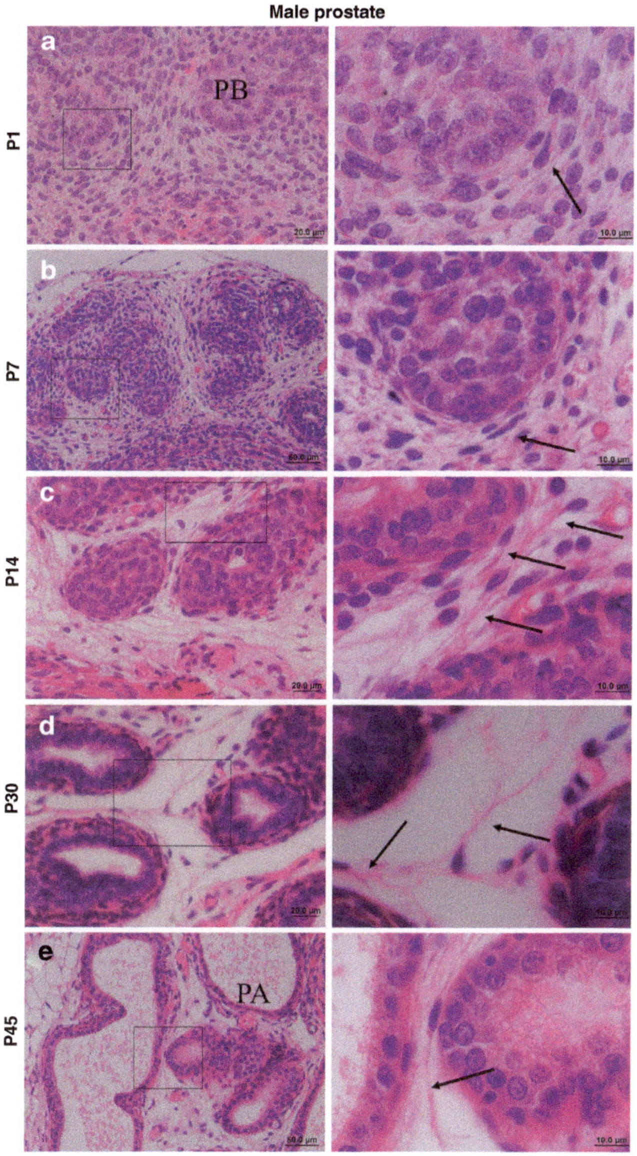

Fig. 13.2 Histology of male prostatic complex development in Mongolian gerbil. Histological sections of the developing prostatic complex were stained with hematoxylin-eosin to cover the early postnatal development period of the gland. (**a**) Cells with flattened nuclei are observed in the mesenchyme inductive prostate at P1. (**b**) At P7, fusiform cell bodies are verified in the vicinity of prostatic buds. (**c**) At P14, spindle-shaped cell bodies with thin and longer projections in the interductal region closely associated with the developing periductal smooth muscle. (**d**, **e**) As seen in females, fusiform cell bodies with thin and long projections, with the characteristic morphology of telocytes, occupy the periductal region forming a network of cytoplasmic projections (telopods) involving the prostatic acini. *Arrows* (telocytes in differentiation), *inserts* (details of regions where there are possible telocytes), *PB* prostatic buddings, *PA* prostatic acini

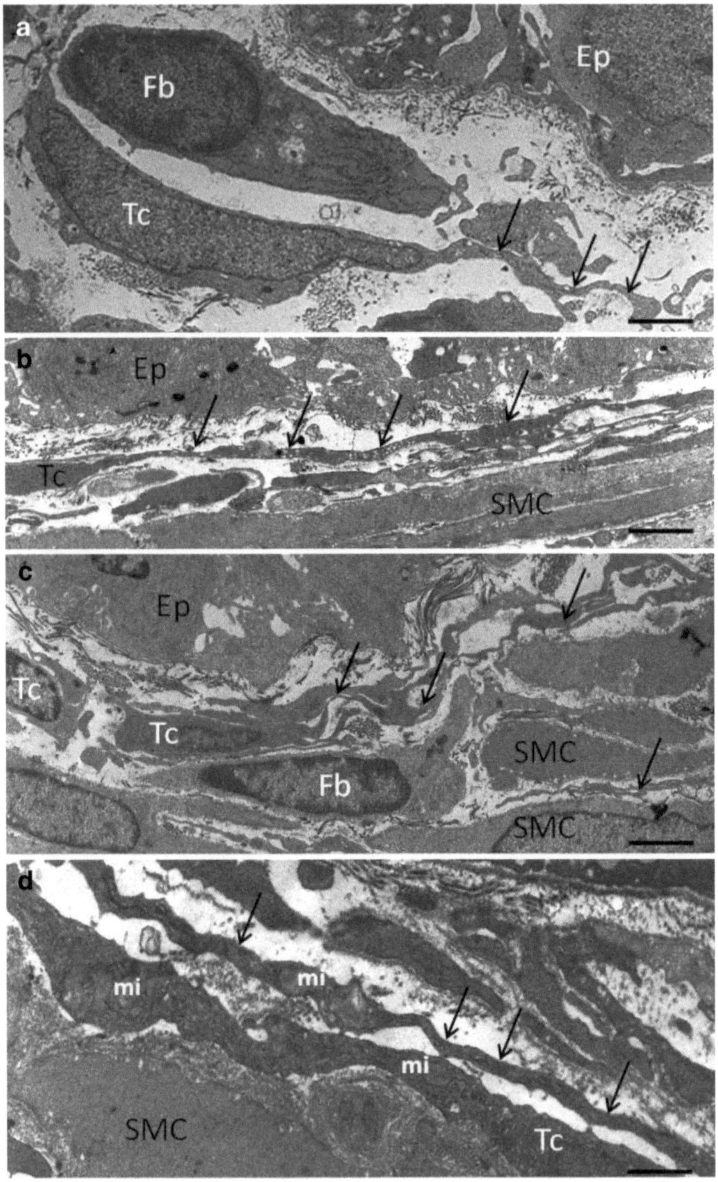

Fig. 13.3 Ultrastructural aspects of telocytes in the ventral prostate of Mongolian gerbil. Detailed description of the ultrastructure of telocytes in the prostate was made by Corradi and collaborators [6]. The figure (**a**) illustrates a few synthetic cytoplasmic organelles in the telocytes (*Tc*) in comparison with a large endoplasmic reticulum in the stromal fibroblast (*Fb*). The figure (**b**) illustrates an abundant cytoplasmic prolongaments (podomers) in the basis of the epithelium, with caveolae. Figure (**c**) reveals an extensive network of cytoplasmic podomers interposed to fibroblasts (*Fb*) and smooth muscle cells (*SMC*) in the basis of secretor epithelium (*Ep*). Figure (**d**) illustrates the telocytes podomers with several mitochondria (*mi*). In all figures, the podomers and telocyte prolongaments are illustrated by *black arrows*. Scale bars: Fig. (**a**)= 0.8 μm; Fig. (**b**) = 1.2 μm; Fig. (**c**) = 1.2 μm; Fig. (**d**) = 0.5 μm

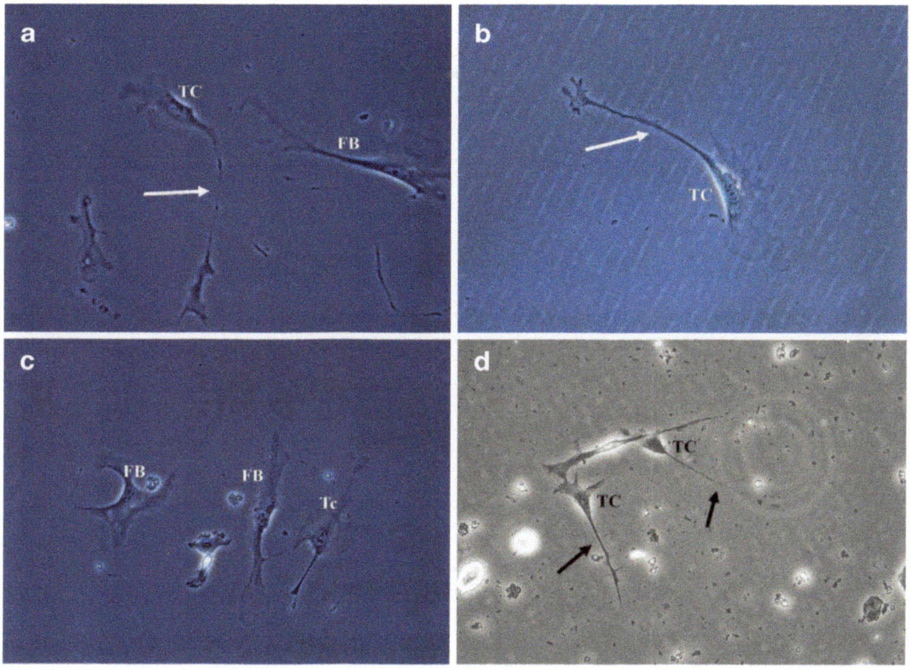

Fig. 13.4 Light microscopy showing telocytes and fibroblasts (**a–d**) in mice primary prostatic tissue culture after 72 h after the primary culture establishment. The telocytes have cell bodies with fusiform aspect and long and thin cytoplasmic projections (telopods) with moniliforme format, with dilatation (podoms) and thin segments (podomeres). Fibroblasts have shorter and thicker projections. *Arrow* (Telopods), *TC* telocytes, *FB* fibroblasts, scale bar (20 μm)

telocytes are not intermingled in the muscle (Figs. 13.1, 13.2, and 13.3) and telocytes in other organs; they form a network of thin and long cytoplasmic projections (telopods) which surround the acini (Figs. 13.3, 13.4, and 13.5). Similar to that seen in the myocardium [17, 31], telocytes are completely differentiated only in the postnatal period coinciding with the morphological and functional differentiation of the gland, and they are still present in the mature organ as described by Corradi and coworkers [6]. These cells are present in females (Fig. 13.1) and in the male prostate (Fig. 13.2) and display similar pattern of differentiation in the gland. The ultrastructural aspects of the telocytes in the prostate gland were detailed in previous work [6], and some aspects of the ultrastructure are here illustrated, such as few synthetic cytoplasmic organelles (Fig. 13.3a), abundant cytoplasmic prolongaments with caveolae (Fig. 13.3b), and extensive network of cytoplasmic prolongaments interposed to the fibroblasts and smooth muscle cells at the basis of secretor epithelium (Fig. 13.3c), and in the telocytes, podomers may be observed in several mitochondria (Fig. 13.3d).

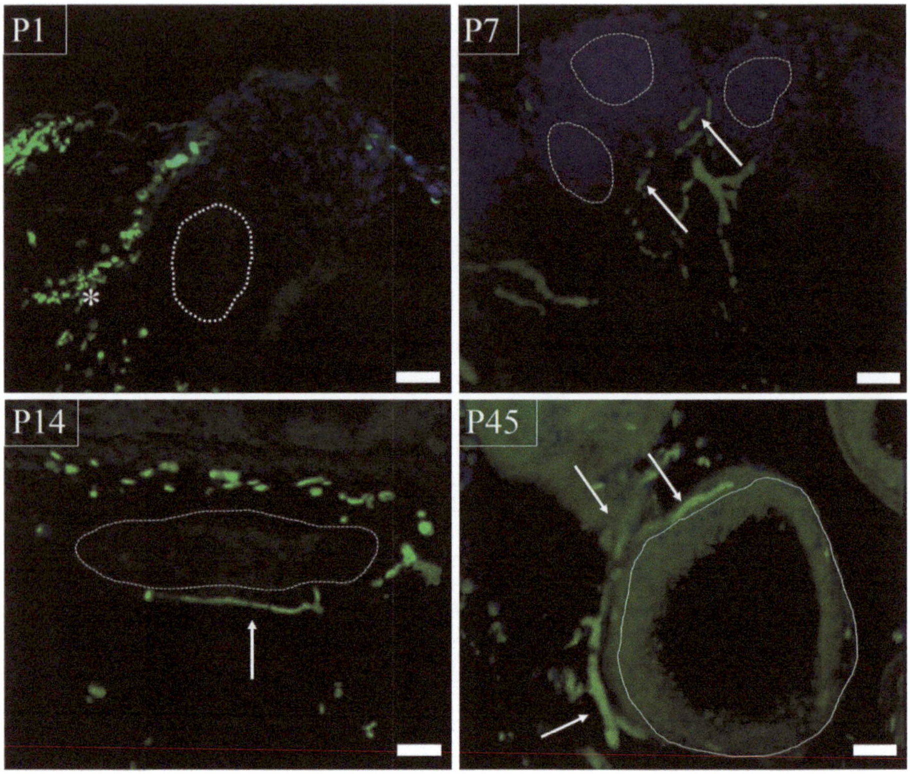

Fig. 13.5 Immunofluorescence assays for CD34 showing the progression of immunostaining for this factor during postnatal development. At P1, CD34-positive cells are found in peripheral portions of the prostate inductive mesenchyme. At P7, CD34-positive cells are found in the vicinity of prostatic buds. Already at P14, CD34-positive cells can be seen in periductal region with the typical telocyte morphology. At P45, such cells surrounding the acini form a network telopods. *Asterisk* (CD34-positive cells in the periphery of the prostate inductive mesenchyme – VMP), *arrow* (CD34-positive cells), scale bar (50 μm)

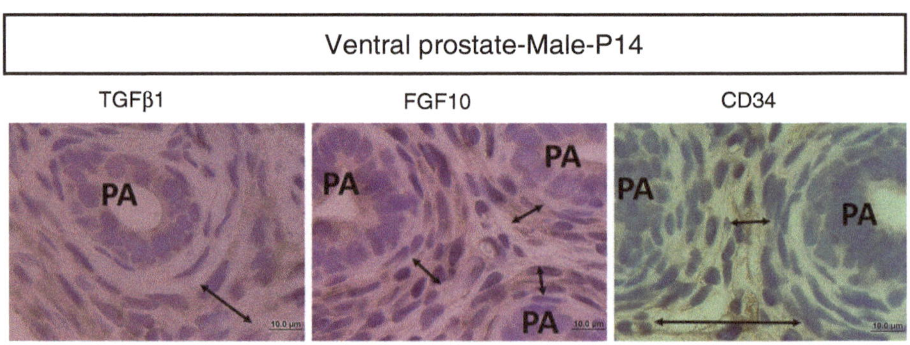

Fig. 13.6 Immunohistochemical assays for TGFβ1 and FGF10 in CD34. In interacinar region in a male at P14, where immunomarking for TGFβ1 in the center of the interacinar region can be verified. While in the periductal region, marking for FGF10 is observed. CD34 is colocated with TGFβ1, which indicates a possible secretion of TGFβ1 by telocytes. *PA* prostatic acini, *arrow* (immunomarked region)

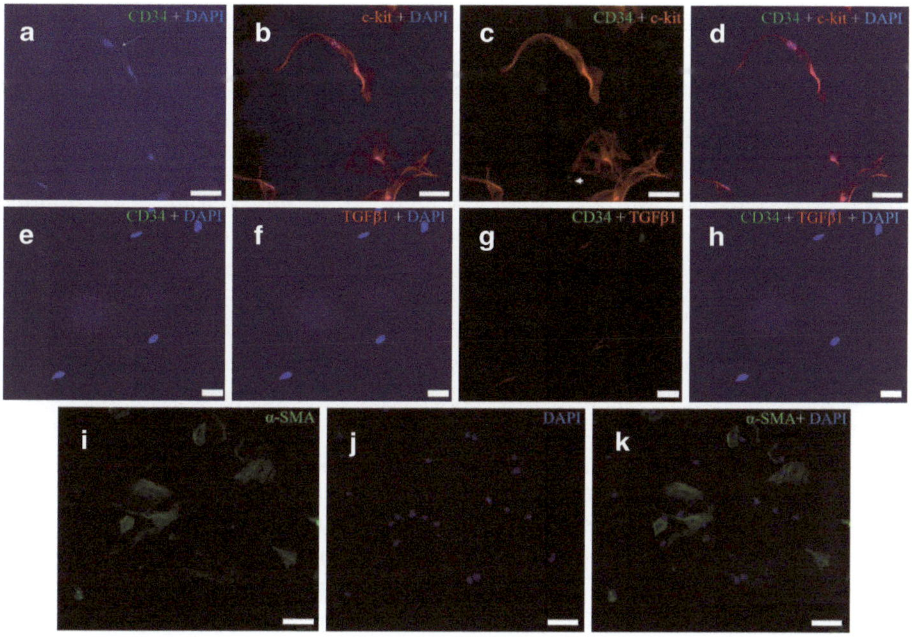

Fig. 13.7 Double-staining immunofluorescence assays for CD34/c-kit and CD34/TGFβ1 and immunofluorescence assay for α-SMA in telocytes on mice prostate primary culture. (**a**–**d**) There is a small colocalization marking for CD34 and c-kit in some telocytes, and there are c-Kit-negative telocytes (*arrow head*) and c-kit-positive and CD34-negative cells constituting other interstitial cells. The telocytes have thin and long projections (telopods) that are CD34 positive. (**e**–**h**) Telocytes show colocalization for CD34 and for TGFβ1 marking. (**i**–**k**) Some fusiform cell bodies with long and thin projections (telopods) showed positive staining for α-SMA. Bar (100 μm)

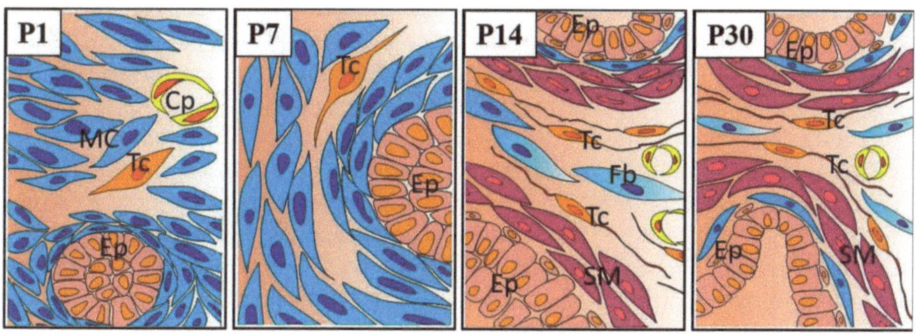

Fig. 13.8 Schematic diagram depicting the developing prostate telocytes and their role in prostate development. At P1, telocytes are not on the periphery of prostatic buds, and they do not possess their characteristic long and thin projections. From P7, the developing telocytes have already reached the periductal region; at P14, they have strong association with periductal smooth muscle; and from P30 onwards, they form a network of telopods involving the periductal smooth muscle. *MC* mesenchymal cells, *Tc* telocytes, *Cp* capillaries, *Ep* prostatic epithelium, *SM* smooth muscle

13.2 The Multiple Functions of Prostate Telocytes

13.2.1 The Telocytes and the Development of Ducts, Acini, and Smooth Muscle Layers

The prostate telocytes as it was verified in myocardium telocytes [17, 31] are fully differentiated only in the postnatal period, when they acquire their characteristic morphology with cell bodies of fusiform aspect with long and thin cytoplasmic projections (telopods) of the moniliform aspect because of their switching between dilations (podoms) and thin regions (podomers). The differentiation of prostate telocytes occurs simultaneously with the very differentiation of histological compartments of the gland, which occurs markedly in the postnatal period. At birth, solid epithelial cords (prostatic buds), undifferentiated and originating from the urogenital sinus, grow to reach their prostatic inductive mesenchymes. As the most studied of all, VMP responds by inducing the ventral lobe of the prostate in rodents [25, 32, 39, 42, 44]. In such, mesenchymes undergo epithelial-mesenchymal interactions that trigger the gradual differentiation of epithelial buds in the secretory portion of prostate from the most distal portion and the conductive portion from more proximal regions. At the same time, the inductive mesenchymes undergo progressive differentiation giving rise to the periductal musculature, which surrounds the acini and prostatic ducts, accounting for the gland contractility and the loose connective tissue of the prostatic stroma [26, 40].

Fibroblasts are the main cell population in the prostatic stroma but to this will be joined the nerve cells and mesenchymal stem cells, besides the telocytes, first described in 2013 in the prostate [6]. During the development of the prostate gland, considering the days for development of the gland as "P," at P14 (14 days in the postnatal development), both the male and female prostate in the Mongolian gerbil possess developing telocytes involving the periductal smooth muscle cells (Fig. 13.5); at this period the periductal smooth muscle layer undergoes a progressive cellular differentiation. At P30 (30 days in the postnatal development), telocytes form a complex network of telopods closely associated with the smooth muscle layers.

Immunohistochemical assays for TGFβ1, FGF10, and CD34 were performed, and the center of interacinar region in a male at P14 was TGFβ1 positive and colocalizes with the marking for CD34, while they were opposite to the immunomarking for FGF10 (Fig. 13.6). This immunostaining pattern suggests that telocytes secrete TGFβ1; this factor promotes the differentiation of smooth muscle cells of ducts and prostatic acini [21], besides inhibiting the expression of FGF10, which reduces the epithelial proliferative activity in the developing prostate and stimulates its morphological differentiation [26, 45]. This finding is confirmed by immunofluorescence assays for mice primary cultures of telocytes (Fig. 13.7a–k), in which TGFβ1 production is also verified (Fig. 13.7e–h).

Nevertheless, prostate telocytes also show positive staining for α-SMA (Fig. 13.7i–k), which can be considered an indication that they have a common progenitor cell with the smooth muscle cells of the gland, as previously proposed for ICCs [49], or they possibility are a reservoir of immature cells capable of differentiating into smooth muscle cells in cases of injury to smooth muscle, as it was suggested for telocytes in the myocardium [2]. There are c-kit positive cells presenting a small colocalization for CD34 marking and cells that express exclusively CD34 (Fig. 13.7a–d). However, this colocalization can also occur due to overlapping cells [47]. The telocytes are distinct from ICCs, in which the basic marker is the c- kit [16], and do not necessarily express such factor, so that the CD34 is the basic factor that distinguishes telocytes from ICCs. The telocytes also have a characteristic morphology with CD34-positive telopods, and the other c-kit-positive cells lack such cytoplasmic projections.

In general terms, our studies on prostatic stroma, under several steps of normal and pathological conditions, confirm the existence of telocytes as distinct cells from the interstitial cells of Cajal (ICCs). This new cell type displays in the prostate a role that changes throughout the development: initially they perform a direct role in the differentiation of smooth muscle cells and an indirect role in reducing the concomitant proliferative activity in the prostate epithelium (Fig. 13.8). In the adult period, prostate telocytes exhibit an auxiliary role in the smooth muscle cell contraction process, as well as a possible role in the replacement in other smooth muscle cell contraction auxiliary cells or even in the replacement of smooth muscle cells. Finally, the CD34 can be regarded as a more suitable marker for the detection of such cells in the prostatic stroma than c-kit.

13.2.2 Prostate Stromal Telocytes and Tissular Prostatic Compartmentalization

The differentiation of prostatic compartments depends on finely regulated epithelial-mesenchymal interactions coordinated by concentration gradients of various self-regulating paracrine factors. For example, the FGF10 secreted from mesenchymal cells promotes epithelial proliferation, but its increase leads to activation of inhibitory pathways involving Shh and Bmp4; these, in turn, have an inhibitory effect on epithelial proliferation [26, 44]. The TGFβ1 operates as well so as to directly inhibit the production of FGF10 by mesenchymal cells in the periductal region to reduce the proliferation of prostate epithelium and, at the same time, promote the differentiation of smooth muscle cells [21]; thus, the development of prostatic histological compartments takes place in a highly integrated and regulated manner over time. The telocytes involve the smooth muscle layer of the prostatic ducts and acini and may perform a role in compartmentalization of paracrine factors, so as to establish delimitations between the smooth muscle layer and surrounded connective tissue of the gland, which would contribute to the individualization of acini. The involvement of telocytes in tissue structure also has been suggested for myocardium telocytes [1].

13.2.3 Prostate Telocytes as Smooth Muscle Cell Pacemakers

As previously proposed to fibroblast-like cells CD34 present in the intestine [13, 47] and later to myometrium telocytes [7] and placenta [3], prostate telocytes can exert an auxiliary function in contractility of the ducts and acini smooth muscles cells in prostate stroma. In studies with the myometrium, such cells showed no excitatory activity on the smooth muscle cells, but they can perform a pacemaker activity [7], possibly by inhibitory pathways. The evidence for such a role in prostate telocyte is, as in the myometrium, the existence of caveolae and gap junctions between their membranes and membranes of smooth muscle cells (demonstrated by [6]).

13.2.4 Telocytes and the Prostate Stroma Remodeling

As it had been previously suggested for interstitial Cajal-like cells [48] and also designed for the telocytes concept, such cells exert a possible role in tissue remodeling or regeneration. The myocardial α-smooth muscle actin expression [2], as seen in prostate telocytes (Fig. 13.7i–k), could indicate a role in the remodeling of prostatic smooth muscle cells or a common origin with the smooth muscle cells. Furthermore, the expression of CD34 by telocytes (Fig. 13.5), a factor typical of undifferentiated cells and hematopoietic cells [36], supports the hypothesis that these cells would serve as a reservoir of cells active in stroma remodeling processes.

Funding The São Paulo Research Foundation (FAPESP), grant numbers 2013/15939-0 and 2013/16443-9, and the CNPq Brazilian National Research and Development Council (CNPq), grant number 301596/2011-5.

References

1. Bani D, Formigli L, Gherghiceanu M, et al. Telocytes as supporting cells for myocardial tissue organization in developing and adult heart. J Cell Mol Med. 2010;14(10):2531–8.
2. Bei Y, Zhou Q, Fu S, et al. Cardiac telocytes and fibroblasts in primary culture: different morphologies and immunophenotypes. PLoS One. 2015;10(2):e0115991.
3. Bosco C, Díaz E, Gutiérrez R, et al. A putative role for telocytes in placental barrier impairment during preeclampsia. Med Hypotheses. 2014;84:72–7.
4. Botchkarev VA, Kishimoto J. Molecular control of epithelial–mesenchymal interactions during hair follicle cycling. J Investig Dermatol Symp Proc. 2003;8:46–55. doi:10.1046/j.1523-1747.2003.12171.x.
5. Carmona IC, Bartolomé MJL, Escribano CJ. Identification of telocytes in the lamina propria of rat duodenum: transmission electron microscopy. J Cell Mol Med. 2011;15(1):26–30. doi:10.1111/j.1582-4934.2010.01207.x.
6. Corradi LS, Jesus MM, Fochi RA, et al. Structural and ultrastructural evidence for telocytes in prostate stroma. J Cell Mol Med. 2013;17:398–406.

7. Creţoiu SM, Creţoiu D, Marin A, et al. Telocytes: ultrastructural, immunohistochemical and electrophysiological characteristics in human myometrium. Reproduction. 2013;145(4): 357–70.
8. Creţoiu SM, Creţoiu D, Popescu LM. Human myometrium – the ultrastructural 3D network of telocytes. J Cell Mol Med. 2012;16(11):2844–9. doi:10.1111/j.1582-4934.2012.01651.x.
9. Cunha GR. The role of androgens in the epithelio-mesenchymal interactions involved in prostatic morphogenesis in embryonic mice. Anat Rec. 1973;1:87–96.
10. Faussone Pellegrini MS, Cortesini C, Romagnoli P. Ultrastructure of the tunica muscularis of the cardial portion of the human esophagus and stomach, with special reference to the so-called Cajal's interstitial cells. Arch Ital Anat Embriol. 1977;82:157.
11. Gherghiceanu M, Popescu LM. Interstitial Cajal-like cells (ICLC) in human resting mammary gland stroma. Transmission electron microscope (TEM) identification. J Cell Mol Med. 2005;9(4):893–910.
12. Horiguchi K, Komuro T. Ultrastructural observations of fibroblast-like cells forming gap junctions in the W/W(nu) mouse small intestine. J Auton Nerv Syst. 2000;80(3):142–7.
13. Iino S, Nojyo Y. Immunohistochemical demonstration of c-Kit-negative fibroblast-like cells in murine gastrointestinal musculature. Arch Histol Cytol. 2009;72(2):107–15.
14. Keith A. A new theory of the causation of enterostasis. Lancet. 1915;2:371–5.
15. Komuro T, Seki K, Horiguchi K. Ultrastructural characterization of the interstitial cells of Cajal. Arch Histol Cytol. 1999;62(4):295–316.
16. Komuro T, Zhou DS. Anti-c-kit protein immunoreactive cells corresponding to the interstitial cells of Cajal in the guinea-pig small intestine. J Auton Nerv Syst. 1996;61(2):169–74.
17. Kostin S. Myocardial telocytes: a specific new cellular entity. J Cell Mol Med. 2010;14:1917–21.
18. Manetti M, Guiducci S, Ruffo M, et al. Evidence for progressive reduction and loss of telocytes in the dermal cellular network of systemic sclerosis. J Cell Mol Med. 2013;17:482–96.
19. Nicolescu MI, Popescu LM. Telocytes in the interstitium of human exocrine pancreas: ultrastructural evidence. Pancreas. 2012;41(6):949–56. doi:10.1097/MPA.0b013e31823fbded.
20. Nicolescu MI, Bucur A, Dinca O, et al. Telocytes in parotid glands. Ann N Y Acad Sci. 2011;295(3):378–85. doi:10.1002/ar.21540.
21. Niu Y, Wang J, Shang Z, et al. Increased CK5/CK8-positive intermediate cells with stromal smooth muscle cell atrophy in the mice lacking prostate epithelial androgen receptor. PLoS One. 2011;6(7):e20202. doi:10.1371/journal.pone.0020202.
22. Popescu LM, Faussone-Pellegrini M-S. Telocytes – a case of serendipity: the winding from Interstitial Cells of Cajal (ICC), via Interstitial Cajal-Like Cells (ICLC) to Telocytes. J Cell Mol Med. 2010;14(4):729–40.
23. Popescu L, Hinescu M, Radu E, et al. CD117/c-kit positive interstitial (Cajal-like) cells in human pancreas. J Cell Mol Med. 2005;9:738–9.
24. Popescu LM, Ciontea SM, Creţoiu D. Interstitial Cajal-like cells in human uterus and fallopian tube. Ann NY Acad Sci. 2007;1101:139–65.
25. Prins GS, Birch L. The developmental pattern of androgen receptor expression in rat prostate lobes is altered after neonatal exposure to estrogen. Endocrinology. 1995;136:1303–14.
26. Prins GS, Putz O. Molecular signaling pathways that regulate prostate gland development. Differentiation. 2008;76(6):641–59. doi:10.1111/j.1432-0436.2008.00277.x.
27. Qi G, Lin M, Xu M, et al. Telocytes in the human kidney cortex. J Cell Mol Med. 2012;16(12):3116–22. doi:10.1111/j.1582-4934.2012.01582.x.
28. Ramon y Cajal S. Histologie du Systeme Nerveux de L'Homme et des Vertebres. Volume 2. Paris: A. Maloine; 1911.
29. Ribatti D, Santoiemma M. Epithelial-mesenchymal interactions: a fundamental developmental biology mechanism. Int J Dev Biol. 2014;58:303–6. doi:10.1387/ijdb.140143dr.
30. Rumessen JJ. Identification of interstitial cells of Cajal. Significance for studies of human small intestine and colon. Dan Med Bull. 1994;41(3):275–93.

31. Rusu MC, Pop F, Hostiuc S, et al. Telocytes form networks in normal cardiac tissues. Histol Histopathol. 2012;27:807–16.
32. Sanches BDA, Biancardi MF, Santos FCA, et al. Budding process during the organogenesis of the ventral prostatic lobe in mongolian gerbil. Microsc Res Tech. 2014;77:458–66.
33. Sanders KM, Ward SM, Koh SD. Interstitial cells: regulators of smooth muscle function. Physiol Rev. 2014;94(3):859–907. doi:10.1152/physrev.00037.2013.
34. Shafik A, Shafik I, El-Sibai O. Identification of c-kit-positive cells in the human prostate: the interstitial cells of Cajal. Arch Androl. 2005;51(5):345–51.
35. Shannon JM, Hyatt BA. Epithelial-mesenchymal interactions in the developing lung. Annu Rev Physiol. 2004;66:625–45.
36. Sidney LE, Branch MJ, Dunphy SE, et al. Concise review: evidence for CD34 as a common. Marker for diverse progenitors. Stem Cells. 2014;32:1380–9.
37. Spemann H, Schotté O. ¨Uber xeneplastische Transplantation als Mittel zur Analyse der embryonalen Induction. Naturwissenschaften. 1932;20:463–7.
38. Streutker CJ, Huizinga JD, Driman DK, et al. Interstitial cells of Cajal in health and disease. Part I: normal ICC structure and function with associated motility disorders. Histopathology. 2007;50:176–89. doi:10.1111/j.1365-2559.2006.02493.x.
39. Thomson AA, Marker PC. Branching morphogenesis in the prostate gland and seminal vesicles. Differentiation. 2006;7:382–92.
40. Thomson AA, Timms BG, Barton L, et al. The role of smooth muscle in regulating prostatic induction. Development. 2002;129(8):1905–12.
41. Thuneberg L. Interstitial cells of Cajal: intestinal pacemaker cells? Adv Anat Embryol Cell Biol. 1982;71:1–130.
42. Timms BG. Prostate development: a historical perspective. Differentiation. 2008;76:565–77.
43. Timms BG, Lee CW, Aumüller G, et al. Instructive induction of prostate growth and differentiation by a defined urogenital sinus mesenchyme. Microsc Res Tech. 1995;30:319–32.
44. Timms BG, Hofkamp LE. Prostate development and growth in benign prostatic hyperplasia. Differentiation. 2011;82:173–83.
45. Tomlinson DC, Grindley JC, Thomson AA. Regulation of Fgf10 gene expression in the prostate: identification of transforming growth factor-β1 and promoter elements. Endocrinology. 2004;145(4):1988–95.
46. Ueda S, Mizuki M, Ikeda H, et al. Critical roles of c-Kit tyrosine residues 567 and 719 in stem cell factor–induced chemotaxis: contribution of src family kinase and PI3-kinase on calcium mobilization and cell migration. Blood. 2002;99(9):3342–9.
47. Vanderwinden JM, Rumessen JJ, de Kerchove d'Exaerde Jr A, et al. Kit-negative fibroblast-like cells expressing SK3, a Ca2+ -activated K+ channel, in the gut musculature in health and disease. Cell Tissue Res. 2002;310(3):349–58.
48. Wang XY, Albertí E, White EJ, et al. Igf1r/CD34 immature ICC are putative adult progenitor cells, identified ultrastructurally as fibroblast-like ICC in Ws/Ws rat colon. J Cell Mol Med. 2009;13(9):3528–40.
49. Ward SM, Sanders KM. Physiology and pathophysiology of the interstitial cell of Cajal: from bench to bedside. I. Functional development and plasticity of interstitial cells of Cajal networks. Am J Physiol Gastrointest Liver Physiol. 2001;281(3):G602–11.
50. Zheng Y, Bai CE, Wang X. Potential significance of telocytes in the pathogenesis of lung diseases. Expert Rev Respir Med. 2012;6(1):45–9. doi:10.1586/ers.11.91.

Chapter 14
Primary Extragastrointestinal Stromal Tumours in the Hepatobiliary Tree and Telocytes

Somanath Padhi and Hemanta Kumar Nayak

Abstract The first decade of the twenty-first century witnessed the presence and light microscopic, immunophenotypic, and ultrastructural characterization of interstitial Cajal-like cells (coined as 'telocytes') in virtually every extragastrointestinal site of the human body by Laurentiu M. Popescu and his co-workers. Not surprisingly, stromal tumours, immunophenotypically similar to that of *telocytes* [CD117 (c-KIT) +/CD34 +], have also been sporadically reported outside the tubular gut (so-called extragastrointestinal stromal tumours, EGISTs), including the gall bladder, liver, and pancreas. A meticulous literature search from January 2000 to November 2015 have found 9 such case reports of EGISTs in the gall bladder, 16 in the liver, and 31 occurring in the pancreas. The site wise mean age at presentation for these tumours were reported to be 62.2 ± 16.6, 50.9 ± 20.1, and 55.3 ± 14.3 years, respectively. Six of nine EGISTs in the gall bladder were associated with gallstones. On pathological evaluation, these tumours exhibited prominent spindled cell morphology and consistent expression of CD117/c-KIT and CD34 on immunohistochemistry and variable expression of vimentin and α-smooth muscle actin. The biological behaviour of hepatic and pancreatic lesions was favourable compared to that in the gall bladder, following definitive surgery with or without imatinib therapy. While the exact pathophysiologic role played by *telocytes* in various organs is yet to be fully elucidated, there seems to be a direct link between these enigmatic stromal cells and pathogenesis of gallstones and origin of EGISTs, and a hope for targeted therapies. Furthermore, the possible role of *telocytes* in hepatic regeneration and liver fibrosis opens a new dimension for futuristic research.

S. Padhi, MD (✉)
Assistant Professor, Department of Pathology and Lab Medicine,
All India Institute of Medical Sciences,
Bhubaneswar, India
e-mail: somanath.padhi@gmail.com

H.K. Nayak, MD, DM
Department of Gastroenterology, Sanjay Gandhi Post Graduate Institute of Medical Sciences,
Lucknow, India
e-mail: drhemantanayak@gmail.com

X. Wang, D. Cretoiu (eds.), *Telocytes*, Advances in Experimental Medicine and Biology, 913, DOI 10.1007/978-981-10-1061-3_14

14.1 Introduction

Gastrointestinal stromal tumours (GISTs), first coined by Mazur and Clark in 1983, are the most common primary, nonepithelial, mesenchymal tumours of the tubular gastrointestinal tract which arise from or show the morphologic, immunophenotypic characteristics of the interstitial cells of Cajal (ICC). These are the results of gain-of-function mutation of c-KIT (up to 90 %) or platelet-derived growth factor receptor alpha (PDGFR-α, 5–7 %) proto-oncogenes, leading to ligand-independent activation of tyrosine kinase protein receptors in ICC and concomitant downstream activation of signal transduction pathways [1, 2]. Recently, discovery on gastrointestinal stromal tumour 1 (DOG 1, also called TMEM16A/FLJ10261/ORAOV2/anoctamin 1), a calcium-regulated chloride channel protein, and protein kinase C theta (PKC-θ) (a signalling molecule in T-cell activation) have been shown to be the most specific diagnostic biomarker of GISTs, especially in c-KIT/CD117 and PDGFR-α-negative tumours [3–6]. The most common sites of the origin of GISTs are the stomach (40–70 %), small intestine (20–40 %), and rarely the oesophagus, colon, and rectum (less than 10 %) [1, 7]. A subset of these GISTs arise from extragastrointestinal sites (so-called extragastrointestinal stromal tumours, EGISTs); most common being the soft tissues of the retroperitoneum, mesentery, and omentum [8, 9].

The first decade of the twenty-first century witnessed the presence and light microscopic, immunophenotypic, as well as ultrastructural characterization of interstitial Cajal-like cells (ICLC) in virtually every extragastrointestinal anatomic site of the human body by the pioneering work of Laurentiu M. Popescu and his coworkers as well as other researchers around the world [10–17]. These cells, which Popescu et al. earlier coined as ICLC, were renamed as 'telocytes' in 2010 by Popescu and Faussone-Pellegrini by virtue of their specific ultrastructural characteristics ('telopodes') [10]. Not surprisingly, EGISTs, immunophenotypically similar to *telocytes*, have been sporadically reported outside the tubular gut such as the pancreas [18–20], gall bladder (GB) [21–29], liver [30–45], and other rare anatomical sites [46–48]. With this, the well-established dogma that 'GISTs arise exclusively in the tubular gut' has been overpassed!

In our previous paper [20], we reviewed all reported cases of pancreatic EGISTs till 2013 ($n = 19$) and highlighted the pancreatic *telocytes*. Considering the fact that both the liver and GB are being embryologically related to the tubular gut, it seems prudent that GISTs may arise primarily from these sites (hepatobiliary EGISTs). Since the first report of EGIST in the GB (benign) by Ortiz-Hidalgo et al. [21] in the year 2000, and in the liver by Hu et al. [30] in 2003, there have been altogether 25 reported cases of EGISTs from these sites [21–45]. This chapter summarizes all the reported cases of hepatobiliary EGISTs in regard to clinical characteristic, diagnostic modalities used, pathology, immunohistochemical and molecular/genetic characteristics, management, and outcome. Furthermore, *telocytes* in the hepatobiliary tree are briefly highlighted.

14.2 Molecular Biology of GIST and EGIST (Fig. 14.1)

14.2.1 c-KIT

Molecular biology of GIST and EGIST is similar. These are characterized by their genetic expression of c-KIT and immunohistochemical staining of CD117, which occurs in 85–95 % of all GISTs [7, 9, 49]. c-KIT is a 145-KD transmembrane tyrosine kinase which serves as a receptor for stem cell factor. The binding of stem cell receptor to KIT results in homodimerization of its receptor with the activation of tyrosine kinase and concomitant activation of downstream intracellular signal transduction pathways, most notably RAS-RAF-MAPK and P13K-AKT-mTOR pathways. This results in modification of several cellular functions, which includes adhesion, migration, differentiation, and cellular proliferation with decrease in cellular apoptosis. These oncogenic potentials would ultimately lead to neoplasia. The mutation of the KIT proto-oncogene tends to cluster in four exons, namely, exon 9 (extracellular domain), exon 11 (intracellular juxtamembrane domain), exon 13 (split kinase domain), and exon 17 (kinase activation loop).

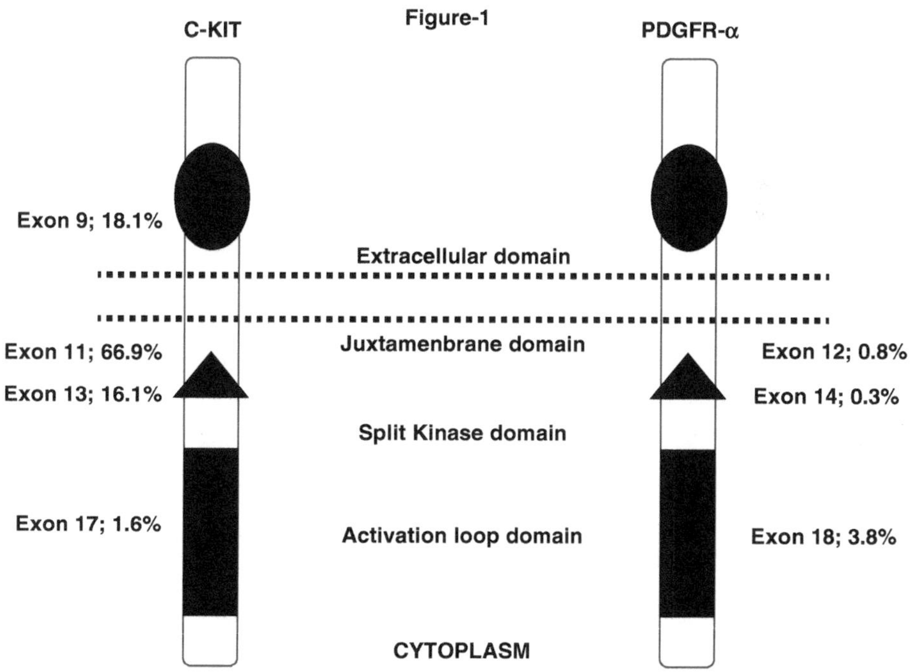

Fig. 14.1 Schematic representation of c-KIT and platelet-derived growth factor receptor alpha (*PDGFR-α*) molecule with location and frequency of mutation in gastrointestinal/extragastrointestinal stromal tumours

Exon 11 mutations, which encode for juxtamembrane domain, are the most common mutated regions of KIT. They account for 70 % of all the tumours and do not appear to be associated with any specific location, size, or clinical outcome [2]. In-frame deletions of one or more codons in exon 11 of KIT are the most common mutations, accounting for 60–70 %. The majority of these mutations involve the proximal part of KIT exon 11 between codons Gln550 and Glu561 [1]. Deletion of Trp557 and Lys558 in exon 11 codon, which is the most common simple deletion in GISTs, is associated with poorer clinical outcome with more aggressive metastatic behaviour [50]. Missense point mutation in KIT exon 11 is the next most common type of mutation, occurring in 20–30 % of GISTs. They involve almost exclusively three codons, Trp557, Val559, and Val560, in the proximal part and Leu576 in the distal part of exon 11. GIST with missense mutation at these regions seems to have better prognosis in gastric but not in small intestinal tumours.

Exon 9 mutations are the second most often involved region which entails mutations of the extracellular domain. These account for 10 % of tumours and are most commonly associated with GISTs of the small bowel with a known aggressive clinical behaviour. Primary mutation of exon 13 (split kinase domain) and exon 17 (loop) is rare, accounting for <1 % of the cases. Exon 13 involves missense mutations resulting in substitution of glutamate for lysine with a more malignant potential [2, 3]. Secondary cytogenetic changes occur in GISTs and likely cause tumour progression. These include loss of chromosome 14, 22q, 1p, 9p, and 11p and gains of chromosomes 5p, 20q, 8q, and 17q with resultant loss of tumour suppressor genes and gain of proto-oncogenes, respectively [51, 52].

14.2.2 PDGFR-α

In up to 5 % of cases of c-KIT negative, GISTs/EGISTs harbour a mutation in the gene encoding for PDGFR-α which most commonly involves the cytoplasmic (intracellular) domain such as exon 18 (activation loop domain) (3.8 %), followed by exon 12 and exon 14 (juxtamembrane domain).

14.3 EGISTs in the Gall Bladder and Liver

The clinical presentation and imaging modalities used for their characterization, management, pathology, and follow-up data of reported cases of hepatobiliary EGISTs are summarized in Tables 14.1 and 14.2, and a comparative review of different parameters among EGISTs in the pancreas, liver, and GB is presented in Table 14.3.

Table 14.1 Clinicopathological characteristics of all reported cases of primary extragastrointestinal stromal tumours in the gall bladder ($n=9$) (2000–2015) [21–29]

Year [ref]	Age (years)/ gender	C/F	Imaging (USG/CT)	Intervention	Size (max.)	Cytology	Mitosis	IHC		Outcome
								Positive	Negative	
2000 [21]	69/F	Pain	Mixed gallstones	Lap. chol.	2.4 cm (body)	Spindled	Nil	CD117, CD34, vimentin	CK, S-100, α-SMA, desmin	–
2002 [22]	34/F	Pain	Gallstones, polypoid mass	Open chol.	1.5 cm (neck)	Mixed	28/10 HPF	CD117, vimentin	CK, CD-34, myoglobin, α-SMA, desmin	Alive and disease-free, local recurrence, 6 months
2004 [23]	72/F	Pain, fever	Dilated and thickened gall bladder (?empyema)	Open chol.	6 cm (neck and body)	Spindled	>20/50 HPF	CD117, vimentin	CD34, S-100, desmin, α-SMA	Liver metastasis and death, 9 months, c-KIT negative
2004 [24]	79/F	Pain, fever	Gallstones, thickened walls? Empyema	Open chol.	4 cm	Mixed	Numerous	CD117, desmin, α-SMA	S-100, EMA, CK	Chemotherapy, death, 15 months
2005 [25]	68/F	Pain, mass	Mass invading right lobe of the liver	Surgery[a]	6 cm	Spindled[b]	Numerous, >5/10 HPF	CD117, vimentin, myoglobin, P53, NSE (focal)	CD34, S-100, desmin, α-SMA, CK, chromogranin, myosin	Alive, no evidence of recurrence, >12 months, c-KIT+, p53+
2009 [26]	48/F	Pain	Stones, polypoid mass	USG-FNA, [c]lap. chol.	2 cm (body)	Mixed	1/10HPF	CD117, vimentin	α-SMA, S-100, desmin, CD34, CD31	Death (5 days)? pulmonary embolism

(continued)

Table 14.1 (continued)

Year [ref]	Age (years)/ gender	C/F	Imaging (USG/CT)	Intervention	Size (max.)	Cytology	Mitosis	IHC Positive	IHC Negative	Outcome
2011 [27]	72/F	Pain	Gallstones, dilated common bile duct	Roux-en-Y HJ	(neck/body)[e]	Spindled	50/50HPF	PDGFR-α	CD117	Mutation in codon 824 in exon 18 of PDGFR-α gene +, c-KIT negative
2012 [28]	77/F	Pain, fever	Gallstone, mass (CT scan)	Open chol.[d] Billroth II GJ+ celiac LN	7 cm	Mixed (Ep > Sp)	>50/50HPF	CD117, vimentin, α-SMA (focally)	CD34, S-100, CK, CD30, ALK	Liver metastasis, imatinib, death due to GJ-stomy bleed – 1.5 months
2012 [29]	41/F	Pain	Not described	Chol + Roux-en-Y HJ + LN	(Body)[e]	Spindled	High	CD117	CD34, PDGFR-α	(

F female, *HJ* hepaticojejunostomy, *GJ* gastrojejunostomy, *LN* lymphadenectomy, *Ep* epithelioid morphology, *Sp* spindled cell, *SMA* smooth muscle actin, *NSE* neuron-specific enolase, *PDGFR* platelet-derived growth factor receptor, *CK* cytokeratin, *EMA* epithelial membrane antigen, *ALK* anaplastic large cell lymphoma kinase

[a]Transileal-vein portal vein thromboembolism (TIPE) followed by open cholecystectomy and extended right hepatectomy

[b]RACKET-shaped cells, cross striation on PTAH staining, the first case of malignant GIST in the GB with rhabdomyosarcomatous differentiation

[c]Ultrasound-guided fine-needle aspiration

[d]Positive

[e]Data not available

Table 14.2 Clinicopathological characteristics of all reported cases of primary hepatic extragastrointestinal stromal tumours (2000–2015) ($n=14/16$)[g] [30–45]

Year [ref]	Age, gender	Presentation	Location	Intervention	Size (max.)	Cytology	Mitosis	IHC Positive	IHC Negative	Medical therapy	Outcome, follow-up
2003 [30]	79/F	Pain, breathlessness	Rt lobe	Lobectomy	15 cm, satellite nodule[a]	Spindle, giant cells	4/10 HPF	CD117, CD34, vimentin	CD117, CD34, CK (negative at metastatic site)	No	Alive, lymph node metastasis (16 months). *The first case of primary hepatic GIST (malignant).* EM[c] study[a]
2003 [31]	60/M	Pain, weight loss	Right lobe	Biopsy	–	–	–	CD117	–	Imatinib	Death, lung metastasis (3 months)
2006 [32]	37/M	Mass in the liver	Rt lobe (V)	Segmental resection	18 cm	Spindle	20/50HPF	CD117, vimentin	CK, α-SMA, S-100, CD31, CD34, f VIII	Imatinib	Alive, lung metastasis, 36 months *The first case with lung metastasis successfully treated with imatinib*
2009 [33]	30/M	Abdominal fullness	Bilateral, multiple	Left hepatectomy + BDJ[d]	–	Mixed	75/50HPF	CD117, CD34	S-100, α-SMA	Imatinib	Alive, scar metastasis, 108 months *Mutation at exon 11 of c-KIT gene; negative, distinct c-KIT + (exon 11) GIST in stomach*

(continued)

Table 14.2 (continued)

Year [ref]	Age, gender	Presentation	Location	Intervention	Size (max.)	Cytology	Mitosis	IHC		Medical therapy	Outcome, follow-up
								Positive	Negative		
2009 [34]	17/M	Asymptomatic, HBV carrier	Rt lobe	USG[b] Bx, RFA[e]	5.1 cm	Mixed	Low	CD117, CD34, vimentin	S-100, α-SMA, desmin, CK	No	Alive, 3 months
2010 [35]	71/F	Symptomatic	Rt lobe	Biopsy	4.3 cm	Spindle	–	CD117, CD34, PDGFR-α, vimentin	S-100, desmin	No	Alive. *c-KIT gene mutation at exon 9, 11, 17 +; PDGFR-α negative*
2010 [36]	70/M	Loss of appetite, past h/o gastric carcinoma	Left lobe	Resection	20 cm	Epithelioid	1/50 HPF	PKC-θ, CD34, α-SMA	CD117, HMB-45, S-100, EMA	No	Favourable. *The first report of hepatic GIST with PDGFR-α mutation positive (V561D, exon 12)*
2011 [37]	28/F	Epigastric pain	Left lobe, (III–IV)	Resection	5 cm	Spindle	Low	DOG 1, CD117, CD34	–	Imatinib	Alive
2012 [38]	53/M	Pain (HBV carrier, gastric ulcer)	Right lobe	USG-FNB	20 cm	Spindle	Low	DOG 1 CD117, CD34, α-SMA, vimentin	CK, S-100	No	–
2012 [39]	59/M	Abdominal mass, (h/o testicular ca)	Falciform ligament	Resection	10.8 cm	Spindle	<5/50HPF	CD117, α-SMA	S-100, CK, CD34, vimentin	No	*The first report of GIST arising from hepatic falciform ligament*

2013 [40]	69/F	Epigastric discomfort	Segment IV	Partial resection	9.0 cm	Mixed	1–2/50HPF; 87/50HPF in recurrent tumour	CD117, vimentin, SMA, desmin	DOG 1, CD34, S-100, CD31, factor VIII, D2-40	No	Alive, recurrence, 108 months *Mutation of c-KIT and PDGFR-α not identified*, skeinoid fibre, *EM study; neuro secretory granules+*
2014 [41]	56/M	Asymptomatic	Right lobe	Resection	10 cm	Spindle	<5/50HPF	CD117, α-SMA	CD34, S-100, CK19	No	Alive, 12 months
2015 [44]	60/F	Asymptomatic	Right lobe	ECHRA[f]	Large	Spindle	–	CD117, CD34	–	–	Alive, *the first case treated with ECHRA* *Mutation of exon 11 of c-KIT+*
2015 [45]	67/F	Fatigue	Right lobe	Resection	7.4 cm	Mixed	8/50 HPF	CD117, CD34	Desmin, SMA, CK19, HMB45, AFP	Imatinib, sunitinib	Alive, recurrence, bone metastasis, 48 months. *Mutation of exon 11 of c-KIT gene +*

SMA smooth muscle actin, *NSE* neuron-specific enolase, *PDGFR* platelet-derived growth factor receptor, *CK* cytokeratin, *EMA* epithelial membrane antigen, *ALK* anaplastic large cell lymphoma kinase, *AFP* alpha-fetoprotein, *DOG 1* discovery on gastrointestinal stromal tumour 1

[a]Positive

[b]Data not available

[c]Electron microscopy

[d]Bile duct-jejunum anastomosis

[e]Ultrasound-guided biopsy, radiofrequency ablation

[f]Extracorporeal hepatic resection and autotransplantation

[g]In two of 16 cases [42, 43], full data were not accessible, and hence not presented in the Table

Table 14.3 Comparative review of all reported cases of primary extragastrointestinal stromal tumours (EGIST) in the gall bladder, liver, and pancreas (2000–2015)

Parameters	Gall bladder [21–29] (*n*=9)	Liver [30–45] (*n*=14/16)[a]	Pancreas [18–20] (*n*=31)
Male/female (*n*)	Nil/9	8/6	16/15
Mean age (years) (mean ± SD)	62.2±16.65	50.9±20.10	55.3±14.3
≤50/>50 years (*n*) (%)	3/6 (33.3/66.7)	4/10 (28/72)	8/23 (26/74)
Symptomatic (*n*) (%)	9 (100)	11 (78.5)	25 (80.6)
Localization (by imaging and/or gross examination) (*n*) (%)	6 (neck/body) (66.7)	10; right lobe (71.4) 2; left lobe (14.4) 1; both lobes (7.1) 1; falciform ligament (7.1)	Head; 11 (36.0), Tail; 7 (22.5), body + tail; 7 (22.5), head + body; 3 (10.0), uncinate process; 1 (3.0), whole pancreas; 1 (3.0), unspecified; 1 (3.0)
Mean size (cm) (range)	4.5 (0.7–7.5)	11.3 (5–20) (*n*=11/16)	11.4 (2.4–35) (*n*=30/31)
>10 cm (*n*) (%)	0	5/11 (45.5)	12 (40.0)
≤10 cm (*n*) (%)	9 (100)	6/11 (54.5)	18 (60.0)
Microscopy (n) (%)	–	*In 3; cytology not described*	*In 4; cytology not described*
Spindle cell	5/9 (55.5)	8/13 (61.5)	25/27 (92.6)
Spindled and epithelioid cells	3/9 (33.3)	4/13 (30.8)	1/27 (3.7)
Pure epithelioid cell	–	1/13 (7.7)	1/27 (3.7)
Mitosis	Numerous to –high in 8, sparse in 1	Variable (low to high)	<5/50 HPF; 12 5–10/50 HPF; 8 >10/50 HPF; 5 Not specified; 6
Immunohistochemistry (+) (n, %)	–	(*In 2, not described*)	–
CD117/c-KIT (*n*) (%)	7/9	13/14 (92.8)	30/31 (96.8)
CD34	1/9	9/14 (64.3)	22/31 (71.0)
DOG 1	*Not used*	2	2/31 (in rest not used)
PKC-θ	*Not used*	1	*Not used*
PDGFR-α	1/9	1	*Not used*
Myoglobin	1/9	–	–
Vimentin	5/9	6	4/6
A-SMA	1/9	5	–
Management (n, %)		*In 14; data available*	
Surgery	8/9	6/14	16/31(51.6)
Surgery + imatinib (Im) ± sunitinib (Su)	1/9 (Im)	4/14 (Im)	13/31(41.9)
Supportive ± imatinib	–	4/14	2/31 (6.5)
Outcome, n (%)	*Reported in 6*	*Reported in 12*	*Reported in 29*
Favourable	2/6	11/12 (91.6)	27/29 (93)

Table 14.3 (continued)

Unfavourable/aggressive	4/6	1/12 (8.4)	2/29 (7)
Duration of follow-up	3–108 months	3–108 months	5 days to 66 months
Mutation analysis done (n)	3/9 (2; c-KIT negative, 1; c-KIT+, 1; PDGFR-α+)	6/16 (3; c-KIT +, 1; PDGFR-α +, 1; negative for both c-KIT and PDGFR-α, 1; c-KIT negative lesion in the liver, but distinct positive lesion in the stomach, on follow-up)	3/31 (c-KIT+)

[a]In two of 16 cases, full data were not available

14.3.1 Clinical Characteristics

EGISTs in the GB ($n=9$) were reported exclusively in females, whereas an equal gender predilection was reported both at hepatic and pancreatic sites. The mean ages at diagnosis/presentation for these tumours were 62.2 ± 16.6 (GB), 50.9 ± 20.1 (liver), and 55.6 ± 13.6 years (pancreas) ($P>0.05$), and a higher proportion of these patients (60–75 %) were older than 50 years. All patients with GB-EGISTs presented with pain in the right upper quadrant with or without fever, palpable lump, and leucocytosis which was suggestive of acute cholecystitis or empyema, thus requiring urgent intervention. Similarly, a higher proportion (75–80 %) of hepatic and pancreatic tumours presented with non-specific symptoms and signs such as malaise, loss of appetite, epigastric pain, or weight loss. Among those with hepatic lesions, two were asymptomatic hepatitis B virus carrier, one had a past history of treated testicular cancer [39], and two had prior history of gastrectomy (one for carcinoma stomach [36], one for chronic gastric ulcer [38]). Notable was the fact that one of the cases of hepatic EGIST, on follow-up, also developed a distinct c-KIT positive gastric GIST [33] (Tables 14.1, 14.2, and 14.3).

14.3.2 Imaging Modalities

Routine abdominal ultrasound (USG) was used most often to characterize the lesions in the GB which demonstrated dilatation and thickening of walls with or without stones (six of nine cases). In addition, polypoid mass lesions were observed in two. Furthermore, computerized tomogram (CT) scan, demonstrated heterogeneously enhancing masses in two patients, suggestive of malignancy. On imaging, 6/9 (66.7 %) of EGISTs were localized to the neck and/or body of the GB with size ranging from 0.7 to 7.5 cm (mean = 4.5 cm). On abdominal CT scan supplemented with colour Doppler imaging, hepatic lesions showed a predilection for the right lobe

(10/16, 62.5 %); left lobe localization was observed in two patients, whereas bilateral multiple lesions were reported in one patient. In another patient, the lesion was reported to originate from the hepatic falciform ligament with contiguous involvement of the left hepatic lobe [39], and the presence of satellite nodules suggestive of hepatocellular carcinoma was reported in one patient [30]. The mean size of these lesions ($n = 11/16$) was 11.3 cm (range, 5–20 cm), and all these lesions occurred in noncirrhotic livers.

14.3.3 Management

Following radiological diagnosis, all patients with GB lesions underwent either open or laparoscopic cholecystectomy followed by anastomosis procedures, and in one patient, ultrasound-guided fine-needle aspiration cytology preceded definitive surgery [26]. Only one of the nine patients received postoperative imatinib mesylate (Gleevec), following histopathological diagnosis of EGIST [28] (Table 14.1). Ten of 14 patients with hepatic tumours (where data available) underwent elective resection of mass lesions (hepatectomy), of which four also received postoperative imatinib, and the remaining four patients were managed symptomatically with or without imatinib therapy (Table 14.2). In contrast, pancreatic lesions were managed by surgery alone (R0 resection) in 16, surgery and neoadjuvant imatinib in 13, and the remaining two cases were managed symptomatically (Table 14.3).

14.3.4 Biological Behaviour and Outcome

The biological behaviour of GB-EGISTs (data from seven patients) was aggressive (locoregional recurrence or metastasis) in three, whereas death occurred in four patients (two due to postoperative complications, two due to tumour) (duration of follow-up, 3–108 months) (Table 14.1). Similarly, among 12 patients with hepatic EGIST (where data available), six behaved aggressively, though only one patient died of lung metastasis [31] (duration of follow-up, 3–108 months) (Table 14.2). In comparison, 27/29 (93 %) patients with pancreatic EGISTs had a favourable outcome (Table 14.3).

14.3.5 Gross and Microscopic Pathology

Six of nine (66.7 %) GB-EGISTs originated from the neck and/or the body with size ranging from 0.7 to 7.5 cm (mean = 4.5 cm). In six cases, these lesions were associated with gallstones. These tumours were solid, tan to grey to purplish in colour with foci of haemorrhage, ulceration, and yellowish areas, though macroscopic evidence of necrosis was evident in two cases. In comparison, hepatic EGISTs were solid with size ranging from 5 to 20 cm (mean = 11.3 cm, $P<0.05$) (50 % of these were >10 cm in diameter) and showed a predilection for the right lobe. The sizes of pancreatic EGISTs [mean, 11.4 cm (range, 2.4–35 cm)] were comparable to that of hepatic lesions, and higher proportion of these (18/30, 60 %) were ≤10 cm. Fourteen of 31 (45 %) occurred in the head and/or body, whereas similar proportion involved the body and/or tail of the pancreas (Table 14.3). This variation in sizes may be attributed to the anatomic dimension of the GB and liver as well as to the anatomic location; the pancreas being a retroperitoneal organ, the lesions attained bigger size before being symptomatic.

On microscopic evaluation, EGISTs showed variable cellularity with predominant spindle cells in fascicles or storiform pattern at all three anatomic sites, though mixture of spindle and epithelioid cells were described variably. Furihata et al. [25] described a case of GB-EGIST with mixed spindled to epithelioid cells and racquet-shaped cells with cytoplasmic cross striation, highlighted on phosphotungstic acid haematoxylin (PTAH) staining, suggestive of rhabdomyosarcomatous differentiation. Hu et al. [30] described a spindle cell EGIST in the liver with giant cells at the hilar lymph node metastatic site; Yamamato et al. [36] described a case of pure epithelioid EGIST in the liver masquerading as a perivascular epithelioid cell tumour (PEComa). Hayashi et al. [40] reported a case of EGIST in the liver with extensive myxoid change, microcysts, skeinoid fibres, and a prominent haemangiopericytoma-like pattern. Among GB lesions, the mitotic figures were reported to be numerous to high in eight, whereas those in hepatic lesions were reported as variable (low to high). The mitotic figures in pancreatic tumours were as follows: ≤10/50 high-power field in (HPF) 20/31 (64.5 %), >10/50 HPF in 5/31 (16.1 %), and not specified in the remaining six cases. Thus, applying Fletcher's risk stratification criteria [53], the tumours were categorized as high-risk category for GB lesions (high mitosis, any size), and high to intermediate risk for hepatic and pancreatic lesions.

14.3.6 Immunohistochemical and Molecular Characteristics

On immunohistochemistry (IHC), hepatobiliary and pancreatic EGISTs were consistently positive for CD117/c-KIT (up to 93 %), making this as the single most reliable marker for these tumours. Furthermore, DOG 1 positivity was noted in two cases, each of hepatic and pancreatic EGISTs, whereas PKC-θ and PDGFR-α expression was reported in one each of hepatic EGIST. These tumours also showed variable expression of CD34 (60–70 %), vimentin, α-smooth muscle actin (SMA), neuron-specific enolase (NSE), and very rarely for desmin and myoglobin. These tumours were uniformly negative for cytokeratin, epithelial membrane antigen, CD30, alpha-fetoprotein, HMB-45, CD31, factor VIII, and D2-40, thus differentiating these unusual tumours from solitary fibrous tumour, epithelial malignant tumours (primary and metastatic), PEComas, vascular tumours, as well as rare primary sarcomas such as leiomyosarcoma, rhabdomyosarcoma, etc.

Mutation analysis studies were reported in three cases of GB tumours [23, 25, 27], six cases of hepatic tumours [33, 35, 36, 40, 44, 45], and only three cases of pancreatic tumours [18–20]. c-KIT-positive (exon 9 and 11) mutation was reported in one case of GB-EGIST [25], and c-KIT negative/PDGFR-α positive (codon 824 in exon 18) was reported in another [27]. Among hepatic lesions, two were c-KIT positive (exon 9, 11, 17) [44, 45], one was c-KIT positive/PDGFR-α negative [35], one was PDGFR-α positive (V561D, exon 12) [36], and one was negative for both c-KIT and PDGFR-α [40]. In one case, a distinct c-KIT-positive (exon 11) GIST occurred in the stomach during postoperative follow-up of primary hepatic tumour (c-KIT negative) [33] (Table 14.2).

14.4 Telocytes: The Enigmatic Interstitial Stromal Cells

The 2010, landmark studies by Laurentiu M. Popescu and Faussone-Pellegrini led to ultrastructural and immunophenotypic characterization of the peculiar, enigmatic, interstitial stromal cells in various extragastrointestinal, anatomic sites of the human body. These cells, earlier presumed to be ICLC by several researchers, were coined as *telocytes* [10]. These cells were characterized by the presence of

Fig. 14.2 Telocytes in extragastrointestinal sites: ultrastructural and immunophenotypic characteristics. (**a**) Rat exocrine pancreas. Nonconventional light microscopy, objective 100×. Tissue fixed with glutaraldehyde and postfixed in OsO4. Thin section of Epon-embedded material (~1 μm) was stained with toluidine blue; *cap* capillary, *ven* venule. At least four *telocytes* (*TC*) are present in the interstitium among acini (a). Note the cell bodies of TC and the emerging prolongations – *telopodes* (*dashed lines*). The length of the *telopodes* is very impressive – tens of micrometres (!); they are very thin (less than 0.5 μm) (Reproduced with permission from Popescu and Faussone-Pellegrini [10]). (**b**) Digitally coloured electron microscopic image of a *telocyte* in rat myometrium: *telocyte* (*blue*), smooth muscle cells (Sienna brown); *N* nuclei. Note three long, moniliform processes that encircle bundles of cross-cut smooth muscle cells. Original magnification ×6800. Inset: human pregnant myometrium. Primary confluent culture (day 8) showing a *telocyte* with at least prolongations with several 'beads' along *telopodes* (Reproduced with permission from Popescu and Faussone-Pellegrini [10]). (c) Immunohistochemistry: *telocytes* in human exocrine pancreas. Paraffin-embedded pancreas sections were incubated with polyclonal antibodies against CD117/c-KIT. Nuclei were counterstained with Mayer's haematoxylin. *Telocytes* (*arrows*) with fusiform body can be seen, having typical long, moniliform cytoplasmic processes that 'touch' the acini. Original magnification 100×, oil immersion (Reproduced with permission from Popescu and Faussone-Pellegrini [10]). (**d**) Human exocrine pancreas. Positive immunostaining of *telocytes* for CD34 (*arrows*), counterstained with Mayer's haematoxylin, 40× (Reproduced with permission from Popescu and Faussone-Pellegrini [10]). (**e**) Transmission electron microscope images showing the ultrastructure of liver (mice). *Telocytes* (*TCs*) with *telopodes* (*Tps*) in the perisinusoidal space of Disse (*D*) between endothelial cells (*E*) and hepatocytes (*H*). Note the upper *telopode* (*Tp*) which is more than 20 μm long. These cells were distinct from conventional perisinusoidal *Ito cells* and showed positivity for CD117/CD34, CD117/vimentin, and PDGFR-α/CD34 by double immunofluorescence labelling (Reproduced with permission from Xiao et al. [54])

specific cell characteristics termed as 'telopodes' [two to five, extremely long (ten to hundreds of μm) and thin (less than 0.2 μm) cell body prolongations with a moniliform aspect (many dilations along), and many caveolae] (Fig. 14.2a, b). These features were unequivocally distinct from the prototype ICC, as well as from other interstitial stromal cells such as fibroblasts, fibroblast-like stromal cells, or any other mesenchymal cells. Immunophenotypically, by using IHC and double immunofluorescence, these cells were shown to be positive for CD117/CD34 and CD117/vimentin (Fig. 14.2c, d).

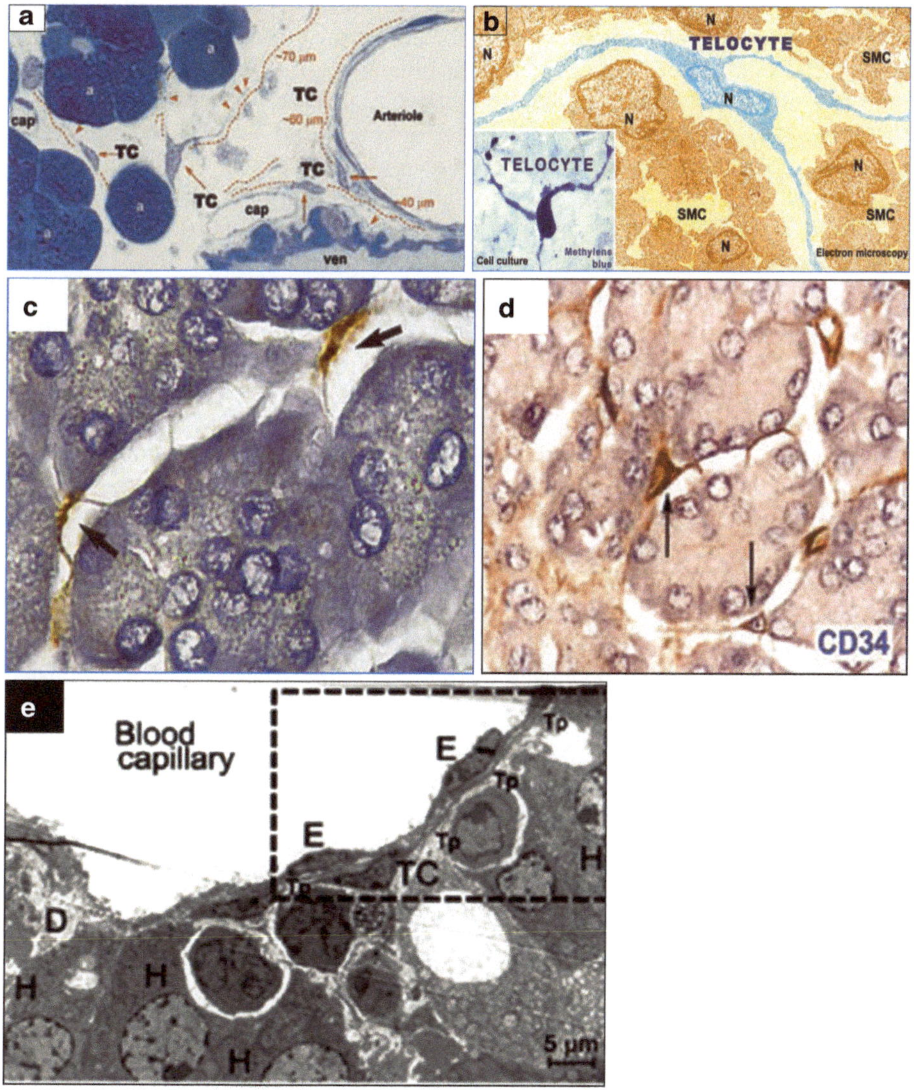

14.5 Telocytes in Hepatobiliary Tree: Possible Role and Future Directions (Table 14.4) [14–17, 54–61]

Hinescu et al. [14] in the year 2007 gave a detailed light microscopic (conventional sections stained with methylene blue, Epon-embedded semi-thin sections stained with toluidine blue) and transmission electron microscopic (TEM) evidence of the presence of *telocytes* in normal human GB. Three years later, Ahmadi et al. demonstrated the presence of similar cells in human GB and extrahepatic biliary ducts, by using double immunofluorescence confocal microscopy as well as TEM, but these cells were not demonstrated in the intrahepatic biliary ductules [15]. Huang et al. [55] in their guinea pig experiments demonstrated the presence of pleomorphic to spindle-shaped *telocytes* in the extrahepatic biliary tree, maximum at the distal common bile duct, thus suggesting the possible role of these cells in maintenance of function of the sphincter of Oddi. Pasternak et al. [12, 56] have convincingly demonstrated, quantitatively, the presence of *telocytes* in the wall of human GB and highlighted their role in GB motility and pathogenesis of gallstone disease. Furthermore, the role of *telocytes* in human cholecystitis and gallstone disease was further corroborated with similar observation by Matyja et al. in 2013 [57].

Rusu and colleagues in 2011 gave the histological evidence of hepatic, portal, and periportal *telocytes* in their experiments on human cadavers [17]. By using routine light microscopy (haematoxylin and eosin and Giemsa-stained sections) as well as IHC, CD117+ *telocytes* were demonstrated in the portal space, portal aspect of the limiting plate, and the periphery of the hepatic lobule, and these researchers postulated the possible role of these cells in idiopathic portal hypertension. However, more convincing evidence of hepatic telocytes came only recently from the study by the Chinese researchers in their mouse experiments where *telocytes* were demonstrated in the perisinusoidal space of Disse. These CD117+/CD34+, CD117+/PDGFR-α+, CD117+/vimentin + *telocytes* were found to be distinct from the conventional *Ito* cells (Fig. 14.2e) [54]. Furthermore, studies from the recent past have also postulated the role of *telocytes* as a part of hepatic stem cells involved in hepatic regeneration programme, and a decrease *telocyte* number has also been implicated in alteration in hepatocyte-matrix interaction and pathogenesis of hepatic fibrosis [58, 60, 61].

14.6 Conclusion

Although the presence of *telocytes* has been well established in virtually every anatomic organ outside the tubular gut, their exact pathophysiologic role is yet to be fully elucidated. In the hepatobiliary tree, there seems to be a direct link between *telocytes* and the gall bladder motility, bile flow, and gallstone pathogenesis. The origin of CD117/c-KIT-positive stromal tumours in hepatobiliary tree adds a new dimension to *telocytes* pathobiology and a hope for targeted therapy with tyrosine

Table 14.4 Evidence for the presence of 'telocytes' [previously interstitial Cajal-like cells (ICLC)] in hepatobiliary tree

Author, year, reference	Study subjects (*n*)	Tissues studied	Methodology	Results	Remark/hypothesis
Hinescu et al. (2007) [14]	Adult human (27; 11 males, 16 females)	Histologically normal GB remote from any pathology	LM ($n=27$) and TEM ($n=17$). Fresh (cryostat sections)- methylene blue; fixed Epon-embedded semi-thin sections- Toluidine blue, and TEM	*Telocytes* in subepithelium (7 %); muscularis layer (5 %), TEM; characteristic features met	Possible role in pace making, intercellular signalling, stone formation
Ahmadi et al. (2010) [15]	Human (16; males-5, females-11)	GB (fundus, body, neck), EHBD, IHBD	Double IF confocal microscopy (c-KIT, mast cell tryptase), TEM (GB)	*Telocytes* in subepithelial and muscular layer of the GB and EHBD; most prominent in muscular layer of EHBD parallel to smooth muscle cells; no evidence in IHBD	The first time demonstration of *Telocytes* in human EHBD. Possible role in biliary motility
Pasternak et al. (2012) [16]	Human	GB (with gallstones)	LM (H & E) (formalin fixed), double IHC (CD117 and mast cell tryptase)	Fusiform *telocytes* with sparse branches throughout the GB; more numerous in the body, exclusively in the muscularis propria	May be related to cholesterol gallstone disease
Rusu et al. (2011) [17]	Human cadaver	Hepatic and extrahepatic portal vein	H & E, Giemsa, IHC (CD117, CD34, α-SMA, GFAP, NF, desmin) on formalin-fixed tissues	*Telocytes* in the portal space, portal aspect of the limiting plate and periphery of the hepatic lobule	Function of portal *telocytes* unknown, possibly in the pathogenesis of idiopathic portal hypertension

(continued)

Table 14.4 (continued)

Author, year, reference	Study subjects (*n*)	Tissues studied	Methodology	Results	Remark/hypothesis
Huang et al. (2009) [55]	Guinea pig	EHBT, ampulla of Vater, duodenum (whole mount cryostat sections)	IHC (CD117), confocal laser scanning microscopy, IF	Pleomorphic to spindle-shaped *telocytes* with bipolar processes in EHBT (smooth muscle layer); maximum in distal CBD	Spontaneous rhythmic contraction in biliary system and maintenance of function of sphincter of Oddi
Pasternak et al. (2013) [56]	Human ($n=55$)	GB; 30 with stones and 25 controls	Formalin-fixed paraffin-embedded sections, LM (H & E, Toluidine blue), IHC (CD117, mast cell tryptase)	*Telocytes* present in muscularis propria; significantly reduced in those with stones compared to controls	*Telocytes* plays a role in GB motility; GB hypomotility is related to stone formation
Matyja et al. (2013) [57]	Human ($n=49$)	GBs; 24 with stones and 25 controls	Paraffin sections for IHC (CD117 and mast cell tryptase), measurement of cholesterol saturation index in GB bile	*Telocytes* (in muscularis propria) significantly reduced in study subjects than control; and correlated with increase in cholesterol saturation index	Bile composition is related to *telocyte* function and number
Xiao et al. (2013) [54]	Mice	Liver	TEM, IF double labelling (CD117/CD34, CD34/PDGFR-α, CD34/vimentin)	*Telocytes* in space of Disse, distinct from cells of Ito	Possible role in cell regeneration and liver diseases
Fu et al. (2015) [58]	Human	Liver biopsy with fibrosis	Masson trichrome stain, double labelling IF with CD117/CD34, or vimentin, PDGFR-α/β	Decrease in number (27–60 %) of *telocytes* in fibrotic areas	

GB gall bladder, *LM* light microscopy, *TEM* transmission electron microscopy, *EHBD* extrahepatic bile duct, *IHBD* intrahepatic bile duct, *IF* immunofluorescence, *H & E* haematoxylin and eosin stain, *IHC* immunohistochemistry, *SMA* smooth muscle actin, *GFAP* glial fibrillary acidic protein, *NF* neurofilament, *EHBT* extrahepatic biliary tree, *CBD* common bile duct, *PDGFR* platelet-derived growth factor receptor kinase inhibitors. On the other hand, the putative role of hepatic *telocytes* in stem cell regenerative medicine and the development of effective and targeted antifibrotic therapy in the human liver fibrosis is to be further re-explored in future studies.

Acknowledgement The authors would like to thank Professor Laurentiu M. Popescu for his inspiration and allowing us to use some of the figures related to *telocytes* from the *Journal of Cellular and Molecular Medicine*. The authors would also like to thank Professor Renu G'Boy Varghese, Department of Pathology at Pondicherry Institute of Medical Sciences, Pondicherry, India, for reviewing the cases and the tables in the manuscript.

Financial Support/Grant None declared.

Conflict of Interest The authors declare that there is no conflict of interest.

References

1. Tan CB, Zhi W, Shahzad G, Mustacchia P. Gastrointestinal stromal tumors: a review of case reports, diagnosis, treatment, and future directions. ISRN Gastroenterology. 2012. Article ID 595968.
2. Laurini JA, Carter JE. Gastrointestinal stromal tumors: a review of the literature. Arch Pathol Lab Med. 2010;134:134–41.
3. Kim KH, Nelson SD, Kim DH, Choi KU, Kim SJ, Min KW, et al. Diagnostic relevance of overexpression of PKC-θ and DOG-1 and KIT/PDGFRA gene mutations in extragastrointestinal stromal tumors: a Korean six center study of 28 cases. Anticancer Res. 2012;32:923–37.
4. Wang C, Jin MS, Zou YB, Gao JN, Li XB, Peng F, Wang HY, et al. Diagnostic significance of DOG-1 and PKC-θ expression and c-KIT/PDGFRA mutations in gastrointestinal stromal tumours. Scand J Gastroenterol. 2013;48:1055–65.
5. Kara T, Serinsoz E, Arpaci RB, Gubur O, Orekici G, Ata A, et al. Contribution of DOG-1 to the diagnosis of gastrointestinal stromal tumors. Pathol Res Pract. 2013;209:413–7.
6. Sun XW, Feng ZJ, Huang P, Hao W, Sui XL. Expression of DOG1, CD117 and PDGFRA in gastrointestinal stromal tumors and correlations with clinicopathology. Asian Pac J Cancer Prev. 2012;13:1389–93.
7. Miettinen M, Lasota J. Gastrointestinal stromal tumors: review on morphology, molecular pathology, prognosis, and differential diagnosis. Arch Pathol Lab Med. 2006;130:1466–78.
8. Weiss SW, Goldblum JR. Extra gastrointestinal stromal tumors. In: Enzinger and Weiss's soft tissue tumors. 5th ed. Mosby Elesevier; 2008; p. 565–79. ISBN 978-0-323-04628-2.
9. Rieth JD, Goldblum JR, Lyles RH, Weiss SW. Extragastrointestinal (soft tissue) stromal tumors: an analysis of 48 cases with emphasis on histologic predictors of outcome. Mod Pathol. 2000;13:577–85.
10. Popescu LM, Faussone-Pellegrini MS. TELOCYTES – a case of serendipity: the winding way from Interstitial Cells of Cajal (ICC), via Interstitial Cajal-Like Cells (ICLC) to TELOCYTES. J Cell Mol Med. 2010;14:729–40.
11. Nicolescu MI, Popescu LM. Telocytes in the interstitium of human exocrine pancreas: ultrastructural evidence. Pancreas. 2012;41:949–56.
12. Popescu LM, Hinescu ME, Ionescu N, Ciontea SM, Cretoiu D, Ardelean C. Interstitial cells of Cajal in pancreas. J Cell Mol Med. 2005;9:169–90.
13. Cretoiu SM, Cretoiu D, Marin A, Radu BM, Popescu LM. Telocytes: ultrastructural, immunohistochemical and electrophysiological characteristics in human myometrium. Reproduction. 2013. doi:10.1530/REP-12-0369.
14. Hinescu ME, Ardeleanu C, Gherghiceanu M, Popescu LM. Interstitial Cajal-like cells in human gall bladder. J Mol Hist. 2007;38:275–84.

15. Ahmadi O, Nicholson Mde L, Gould ML, Mitchell A, Stringer MD. Interstitial cells of Cajal are present in human extrahepatic bile ducts. J Gastroenterol Hepatol. 2010;25:277–85.
16. Pasternak A, Gajda M, Gil K, Matyja A, Tomaszewski KA, Walocha JA, Kulig J, Thor P. Evidence of interstitial Cajal-like cells in human gallbladder. Folia Histochem Cytobiol. 2012;50:581–5.
17. Rusu MC, Pop F, Hostiuc S, Curca GC, Streinu-Cercel. Extrahepatic and intrahepatic human portal interstitial Cajal cells. Anat Rec. 2011;294:1382–92.
18. Joseph P, Goyal R, Bansal P, Parmar R, Dutt S. Pancreatic extra-gastrointestinal stromal tumour with documentation of C-Kit mutation: a case report. J Clin Diagn Res. 2015;9:ED17–8.
19. Akbulut S, Yavuz R, Otan E, Hatipoglu S. Pancreatic extragastrointestinal stromal tumor: a case report and comprehensive literature review. World J Gastrointest Surg. 2014;6:175–82.
20. Padhi S, Sarangi R, Mallick S. Pancreatic extragastrointestinal stromal tumors, interstitial Cajal like cells, and telocytes. JOP. 2013;14:1–14.
21. Ortiz-Hidalgo C, De Leon Bojorge B, Albores-Saavedra J. Stromal tumor of the gallbladder with phenotype of interstitial cells of Cajal: a previously unrecognized neoplasm. Am J Surg Pathol. 2000;24:1420–3.
22. Mendoza-Marin M, Hoang MP, Albores-Saavedra J. Malignant stromal tumor of the gallbladder with interstitial cells of Cajal phenotype. Arch Pathol Lab Med. 2002;126:481–3.
23. Park JK, Choi SH, Lee S, Min KO, Yun SS, Jeon HM. Malignant gastrointestinal stromal tumor of the gallbladder. JKMS. 2004;19:763–7.
24. Peerlinck ID, Irvin TT, Sarsfield PT, Harington JM. GIST (gastro-intestinal stromal tumour) of the gallbladder: a case report. Acta Chir Belg. 2004;104:107–9.
25. Furihata M, Fujimori T, Imura J, Ono Y, Furihata T, Shimoda M, Kato M, Kita J, Ohkura Y, Kubota K. Malignant stromal tumor, so called "gastrointestinal stromal tumor", with rhabdomyomatous differentiation occurring in the gallbladder. Pathol Res Pract. 2005;201:609–13.
26. Al-Daraji WI, Prescott RJ, Al-Mahmoud RM, Husain EA, Haider SA. Cytological findings in a primary GIST of the gallbladder. Cytopathology. 2009;20:332–5.
27. Petrou A, Alexandrou P, Papalambros A, Saetta A, Fragkou P, Kontos M, et al. A malignant gastrointestinal stromal tumor of the gallbladder immunoreactive for PDGFRA and negative for CD 117 antigen (c-KIT). HPB Surg. 2011;2011:327192.
28. Bolanaki H, Delladetsima I, Argyropoulou P, Kapranou A, Kakolyris S, Simopoulos C, et al. Primary malignant gastrointestinal stromal tumor (GIST) of the gallbladder: report of a case. J Gastrointest Cancer. 2012;43:S151–5.
29. Kostov DV, Kobakov GL. Gastrointestinal stromal tumour of the gallbladder. HPB (Oxford). 2012;14:150.
30. Hu X, Forster J, Damjanov I. Primary malignant gastrointestinal stromal tumor of the liver. Arch Pathol Lab Med. 2003;127:1606–8.
31. Pai JJ, Khan N, Balart LA. Primary hepatic gastrointestinal stromal tumor. Am J Gastroenterol. 2003;98:S163.
32. De Chiara A, De Rosa V, Lastoria S, Franco R, Botti G, Iaffaioli VR, et al. Primary gastrointestinal stromal tumor of the liver with lung metastases successfully treated with STI-571 (imatinib mesylate). Front Biosci. 2006;11:498–501.
33. Ochiai T, Sonoyama T, Kikuchi S, Ikoma H, Kubota T, Nakanishi M, et al. Primary large gastrointestinal stromal tumor of the liver: report of a case. Surg Today. 2009;39:633–6.
34. Terada T. Gastrointestinal stromal tumor of the digestive organs: a histopathologic study of 31 cases in a single Japanese institute. Int J Clin Exp Pathol. 2010;3:162–8.
35. Luo XL, Liu D, Yang JJ, Zheng MW, Zhang J, Zhou XD. Primary gastrointestinal stromal tumor of the liver: a case report. World J Gastroenterol. 2009;15:3704–7.
36. Yamamoto H, Miyamoto Y, Nishihara Y, Kojima A, Imamura M, Kishikawa K, et al. Primary gastrointestinal stromal tumor of the liver with PDGFRA gene mutation. Hum Pathol. 2010;41:605–9.
37. Alecu L, Tulin A, Ursut B, Ursut B, Oproiu A, Obrocea F, et al. Gastrointestinal stromal tumor with primary hepatic unique location – clinical case. Chirurgia (Bucur). 2011;106:677–81.
38. Abedalthagafi M. Gastrointestinal stromal tumor originating from hepatic falciform ligament. BMJ Case Rep. 2012. doi:10.1136/bcr-03-2012-6136.

39. Li ZY, Liang QL, Chen GQ, Zhou Y, Liu QL. Extra-gastrointestinal stromal tumor of the liver diagnosed by ultrasound-guided fine needle aspiration cytology: a case report and review of the literature. Arch Med Sci. 2012;8:392–7.
40. Hayashi H, Iseki M, Kawabata M, Nonoshita M, Yamamoto H, Shimokwa I. Primary hepatic gastrointestinal stromal tumor showing remarkable myxoid change and hemangiopericytoma-like pattern. Acta Med Nagasaki. 2013;58:19–24.
41. Zhou B, Zhang M, Yan S, Zheng S. Primary gastrointestinal stromal tumor of the liver: report of a case. Surg Today. 2014;44:1142–6.
42. Louis AR, Singh S, Gupta SK, Sharma A. Primary GIST of the liver masquerading as primary intra-abdominal tumour: a rare extra-gastrointestinal stromal tumour (EGIST) of the liver. J Gastrointest Cancer. 2014;45:392–4.
43. Kim HO, Kim JE, Bae KS, Choi BH, Jeong CY, Lee JS. Imaging findings of primary malignant gastrointestinal stromal tumor of the liver. Jpn J Radiol. 2014;32:365–70.
44. Mao L, Chen J, Liu Z, Liu CJ, Tang M, Qiu YD. Extracorporeal hepatic resection and autotransplantation for primary gastrointestinal stromal tumor of the liver. Transplant Proc. 2015;47:174–8.
45. Lin XK, Zhang Q, Yang WL, Shou CH, Liu XS, Sun JY, et al. Primary gastrointestinal stromal tumor of the liver treated with sequential therapy. World J Gastroenterol. 2015;21:2573–6.
46. Long KB, Butrynski JE, Blank SD, Ibrahim KS, Dressel DM, Heinrich MC, et al. Primary extragastrointestinal stromal tumor of the pleura: report of a unique case with genetic confirmation. Am J Surg Pathol. 2010;34:907–12.
47. Papaspyros S, Papagiannopoulos K. Gastrointestinal stromal tumor masquerading as a lung neoplasm: a case presentation and literature review. J Cardiothorac Surg. 2008;3:31–4.
48. Peitsidis P, Zarganis P, Trichia H, Vorgias G, Smith JR, Akrivos T. Extragastrointestinal stromal tumor mimicking a uterine tumor. A rare clinical entity. Int J Gynecol Cancer. 2008;18:1115–8.
49. Lasota J, Jasinski M, Sarlomo-Rikala M, Miettinen M. Mutations in exon 11 of *c-kit* occur preferentially in malignant versus benign gastrointestinal stromal tumors and do not occur in leiomyomas or leiomyosarcomas. Am J Pathol. 1999;154:53–60.
50. Wardelmann E, Losen I, Hans V, Neidt I, Speidel N, Bierhoff E, et al. Deletion of Trp-557 and Lys-558 in the juxtamembrane domain of the *c-kit* protooncogene is associated with metastatic behavior of gastrointestinal stromal tumors. Int J Cancer. 2003;106:887–95.
51. Gramza AW, Corles CL, Heinrich MC. Resistance to tyrosine kinase inhibitors in gastrointestinal stromal tumors. Clin Cancer Res. 2009;15:7510–8.
52. Debiec-Rychter M, Lasota J, Sarlomo-Rikala M, Kordek R, Miettinen M. Chromosomal aberrations in malignant gastrointestinal stromal tumors: correlation with c-KIT gene mutation. Cancer Genet Cytogenet. 2001;128:24–30.
53. Fletcher CD, Berman JJ, Corless C, Gorstein F, Lasota J, Longley BJ, et al. Diagnosis of gastrointestinal stromal tumors: a consensus approach. Hum Pathol. 2002;33:459–65.
54. Xiao J, Wang F, Liu Z, Yang C. Telocytes in liver: electron microscopic and immunofluorescent evidence. J Cell Mol Med. 2013;17:1537–42.
55. Huang Y, Mei F, Yu B, Zhang HJ, Han J, Jiang ZY, et al. Distribution of the interstitial Cajal-like cells in the gallbladder and extrahepatic biliary duct of the guinea-pig. Acta Histochem. 2009;111:157–65.
56. Pasternak A, Gil K, Matyja A, Gajda M, Sztefko K, Walocha JA, et al. Loss of gallbladder interstitial Cajal-like cells in patients with cholelithiasis. Neurogastroenterol Motil. 2013;25:e17–24.
57. Matyja A, Gil K, Pasternak A, Sztefko K, Gajda M, Tomaszewski KA, et al. Telocytes: new insight into the pathogenesis of gallstone disease. J Cell Mol Med. 2013. doi:10.1111/jcmm.12057.
58. Fu S, Wang F, Cao Y, Huang Q, Xiao J, Yang C, et al. Telocytes in human liver fibrosis. J Cell Mol Med. 2015;19:676–83.
59. Franks I. Gallbladder: loss of interstitial Cajal-like cells in the gallbladder might contribute to gallstone formation. Nat Rev Gastroenterol Hepatol. 2012;9:689.
60. Wang F, Song Y, Bei Y, Zhao Y, Xiao J, Yang C. Telocytes in liver regeneration: possible roles. J Cell Ther. 2016;11:415–9.
61. Liu J, Cao Y, Song Y, Huang Q, Wang F, Yang W, Yang C. Telocytes in the liver. Curr Stem Cell Ther. 2016;11:415–9.

Somanath Padhi, M.D is presently working as an Associate Professor in the Department of Pathology at Pondicherry Institute of Medical Sciences, Pondicherry, India. His current fields of research include quality control and pathology of bone marrow, diagnostic pathobiology of multiple myeloma, gastrointestinal stromal tumours (GISTs), and extragastrointestinal stromal tumours (EGISTs). Inspired by Professor Popescu and his colleagues, he co-authored an editorial entitled 'Pancreatic extragastrointestinal stromal tumors, interstitial Cajal like cells, and telocytes' in January 2013 issue of *JOP Journal of Pancreas* (online), where 19 cases of pancreatic EGISTs till that time were systematically reviewed. He is the lifetime member of the Indian Association of Pathology and Microbiology, and since 2011, he has been the editorial board member and reviewer of several international journals, such as the *American Journal of Cancer Case Reports*, *Journal of Gastroenterology and Hepatology Research* (Hong Kong), *Urology Annals*, *Iranian Journal of Pathology*, *Frontiers of Medicine*, *BMJ Case Reports*, etc. He has authored more than 35 publications in indexed international and national journals.

SOMANATH PADHI

Hemanta Kumar Nayak, M.D., D.M is presently working as a DM resident in the Department of Gastroenterology at Sanjay Gandhi Postgraduate Institute of Medical Sciences, Lucknow, India. His current research areas include chronic pancreatitis, hepatobiliary malignancies, gastrointestinal bleed, and liver transplantation. He is currently involved in several research projects such as endoscopic management of PD stricture in chronic pancreatitis, endoscopic management of pancreatic pseudocyst, acute kidney injury in cirrhosis, etc. He is a member of the Indian Society of Gastroenterology and Indian National Association for the Study of Liver. He has also been a reviewer of several international journals such as ***BMJ Case Reports***, ***Journal of Postgraduate Medicine***, ***European Journal of Radiology***, etc. He has co-authored more than 25 publications in indexed national and international journals.

HEMANTA KUMAR NAYAK

Chapter 15
Cardiac Telocytes in Regeneration of Myocardium After Myocardial Infarction

Liao Zhaofu and Cai Dongqing

Abstract Recent research progress has revealed that a novel type of interstitial cells termed cardiac telocytes (CTs) is found in the interstitium of the heart. We demonstrated that CTs are distributed both longitudinally and within the cross network in the myocardium and that the density of CTs in the atrium-atria and base of the myocardium is higher than that in the middle of the myocardium, while the density of CTs in the epicardium is higher than that in the endocardium. In addition, we documented, for the first time, that the network of CTs in the infarct zone of the myocardium is destroyed during myocardial infarction (MI). This fact shows that, in addition to the death of cardiac myocytes, the previously unrecognized death of CTs is an important mechanism that contributes to the structural damage and poor healing and regeneration observed in the infarcted myocardium. Furthermore, we demonstrated, for the first time, that transplantation of CTs in cases of MI decreases the infarct size and improves myocardial function. The mechanisms behind the beneficial effects of CT transplantation are increased angiogenesis at the infarct site and the border zone, decreased fibrosis in the infarct and non-infarct zones, improved pathological reconstruction of the left ventricle, and increased regeneration of CTs in the infarct zone. Our findings reveal that CTs can be specifically identified by the following characteristics: very small cell bodies, extreme prolongation with some dilation, predisposition to cell death under ischemia, and expression of molecular markers such as c-Kit, CD34, vimentin, and PDGFR-β. CTs act as a structural and functional niche microenvironment in the myocardium and play an essential role in maintaining the integrity of the myocardium and in the regeneration of damaged myocardium.

L. Zhaofu • C. Dongqing (✉)
Key Laboratory for Regenerative Medicine, Ministry of Education, Jinan University, No. 601, Huang Pu Da Dao Zhong, Guangzhou 510632, China
e-mail: tdongbme@jnu.edu.cn

X. Wang, D. Cretoiu (eds.), *Telocytes*, Advances in Experimental Medicine and Biology, 913, DOI 10.1007/978-981-10-1061-3_15

15.1 Cardiac Telocytes Are Interstitial Cells that Play an Important Role in Regeneration of the Myocardium

The reconstruction of damaged myocardium remains a major challenge in regenerative medicine [1–4]. The use of cell therapy for MI has been reported to improve cardiac function and reduce the infarct size [5–12]. However, recent research that included more stringent placebo controls and randomized trials has raised questions as to whether cell therapy actually improves cardiac function, especially related to the mid- and long-term therapeutic effects of myocardial infarction (MI), in both animals and humans [13–23]. To date, the promise of complete cardiac regeneration has not been realized.

Recent studies have revealed that the interstitium informs responsiveness to physiopathological stimuli through continuous bidirectional cross talk between cardiomyocytes and noncardiac cells, such as cardiac fibroblasts, endothelial cells, and others, which act as a "cardiovascular unit" (CVU) and as building blocks of the heart to maintain the integrity of myocardial function [24–26]. During development and under pathological conditions, adult interstitial cells and microvessels control the proliferation, growth, and differentiation of cardiomyocytes in the myocardium [27–29]. Thus, we should consider a new perspective in which all cells of the myocardium are important for tissue homeostasis, development, disease, and regeneration [25, 26]. Therefore, understanding how cardiac myocytes and other cells of the myocardium work together to maintain structural and functional integrity in disease and regeneration is critical for elucidating the cellular and molecular mechanisms of regeneration following MI. Such knowledge may provide a starting point for the development of new therapies for regeneration following MI.

Recent studies have shown that the supporting niche cells within cardiac units of the myocardium might play an important role in myocardial regeneration [26]. One of the important discoveries is that of a novel type of interstitial cells named telocytes in the interstitium of the heart [30–37], intestine [38], uterus and fallopian tube [39], trachea and lung [40, 41], skeletal muscle [42], mammary gland [43], and placenta [44]. Within the cardiac stem cell niche, cardiac telocytes (CTs) play an essential role as niche-supporting cells that nurse the cardiac stem cells (CSCs) and angiogenic cells in the myocardium. Furthermore, they play an important role in regeneration following MI [45].

15.2 Morphology and Markers of CTs

Using transmission electron microscopy (TEM), we found that cardiac telocytes are located in the interstitial space of cardiac myocytes. CTs are characterized specifically as piriform, spindle, or triangular cells with a nucleus that occupies

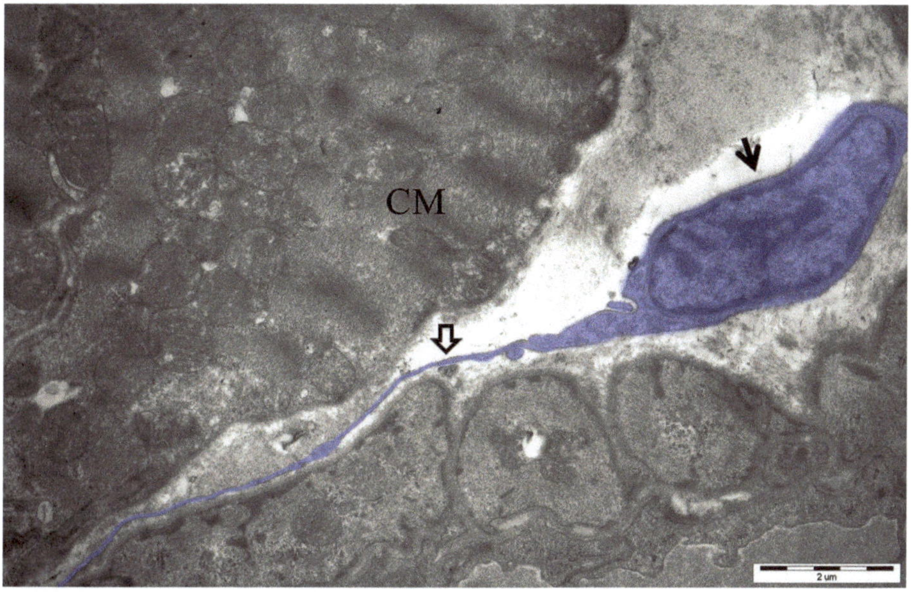

Fig. 15.1 Transmission electron microscopy revealed that cardiac telocytes are located in the interstitial space of cardiac myocytes (CMs). CTs are characterized specifically as piriform, spindle, or triangular cells with a nucleus (*arrow*), which occupies approximately 70 % of the cell volume and very long and thin, dichotomously branched prolongations called telopods (*open arrow*). Bar = 2 μm

approximately 25–30 % of the cell volume and very long and thin, dichotomously branched prolongations called telopods (Fig. 15.1), considered to be the "ultrastructural hallmark" [32, 36, 46].

Using double immunofluorescent staining for anti-c-kit and anti-CD34 (markers for CTs), we revealed that c-kit$^+$/CD34$^+$ CTs, with very small cell bodies (approximately 1:1 ratio for the cytoplasm and nucleus) and extremely thin prolongations, are present in the interstitial space of the cardiac myocytes in the myocardium. In addition, the prolongations of the CTs had a moniliform aspect (alternations and dilated segments). The 3D reconstruction of one represented CT also confirmed that the CT had a very small cell body with an approximately 1:1 ratio of the cytoplasm to the nucleus and one extremely long prolongation (diameter of approximately 1–2 μm) with some dilation. Our morphology studies using TEM and light microscopy evidence clearly indicate that the previously ignored CTs are important interstitial cells in the myocardium and play a physiopathological role in maintaining the integrity of the myocardium [47]. We also developed a magnetic bead technique for isolation of CTs from the myocardium [47]. Under phase-contrast microscopy, the primary culture of isolated c-kit$^+$ cardiac telocytes revealed that CTs are piriform/spindle/triangular cell bodies containing long and slender telopods, which present the alternation of thick segments, podoms, and thin segments, podomers. They are positive for expression of c-kit and CD34 [47].

Based on the morphological hallmarkers of CTs, which is piriform/spindle/triangular cell bodies containing long and slender telopods with the alternation of thick segments, podoms, and thin segments, podomers, and one of the membrane marker proteins – c-Kit – recently, we established a unique technique to characterize CT at the single-cell level. In this technique, c-Kit antibodies along with DiI Micro-label and flow cytometry were used to label and capture CTs. CTs were first isolated with c-Kit antibody-conjugated magnetic beads. Cells which exhibit the unique telocyte morphology were further labeled with DiI microinjection. The DiI^+ cells were sorted and recovered in a 96-well plate using single-cell flow cytometry. Our results revealed that CTs were c-Kit and CD34 positive as confirmed by immunofluorescence. In addition, reverse transcription-polymerase chain reaction (RT-PCR) results also showed that CTs were positive for c-Kit, CD34, vimentin, and PDGFR-β, which have been proposed as markers for CTs [48]. According to our and other studies [49, 50], c-Kit, CD34, vimentin, and PDGFR-β positivity along with the unique morphology described above should be used as hallmarks to identify CTs.

15.3 CTs Are Distributed Longitudinally and Within the Cross Network in Myocardium

Understanding the anatomic location and spatial distribution of CTs in the myocardium should help uncover their physiopathological functions. For this propose, using anti-c-kit immunofluorescent staining of the longitudinal and cross dimension of the whole heart, we demonstrated that many CTs were distributed in the longitudinal direction (Fig. 15.2). In addition, many CTs were found to be distributed in the cross direction (Fig. 15.3). This pattern indicated that CTs are distributed longitudinally and within the cross network in normal myocardium [47]. Furthermore, we also found that the density of CTs in the atrium-atria and base of the myocardium was higher than that in the middle, while the density of CTs in the epicardium was higher than that in the endocardium. These findings suggested that CTs are densely distributed around the epicardium and atrium-atria and base of the myocardium [51]. This unique pattern of distribution of the CTs in the myocardium might indicate that the structural niche and support from CTs are required for other cardiac cells, such as cardiac myocytes (CMs), endogenous cardiac stem cells (CSCs), and cardiac microvascular endothelial cells (CMECs) in this region, and provide morphological and structural evidence that CTs might play an important role in the structural and functional integrity of the myocardium and in the regeneration of injured myocardium.

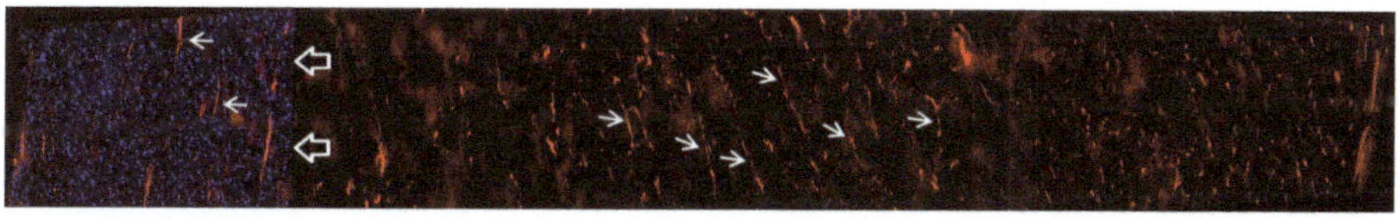

Fig. 15.2 The anti-c-kit immunofluorescent staining for the longitudinal dimension of the whole heart demonstrated that many CTs are distributed in the longitudinal direction (*arrow*). *Open arrow*: merged DAPI staining for nuclei

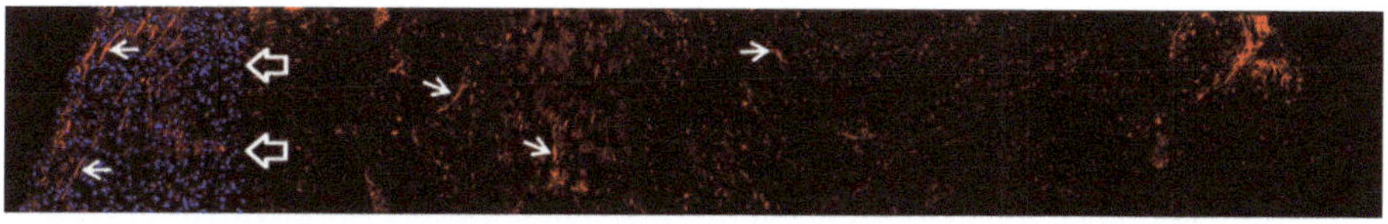

Fig. 15.3 The anti-c-kit immunofluorescent staining for the cross dimension of the whole heart documented that many CTs are distributed in the cross direction (*arrow*). *Open arrow*: merged DAPI staining for nuclei

15.4 The Network of CTs in the Myocardium Is Destroyed During Myocardial Infarction

Based on the hypothesis that CTs may function as niche-supporting cells, interact with other cardiac cells (such as CMs, CSCs, and CMECs) in the myocardium, might play an important role in maintaining the functional and structural integrity of the myocardium, and are involved in the regeneration of injured myocardium, we found that the network of CTs is destroyed during MI using an experimental left anterior descending (LAD) ligation myocardial infarction model. Under ischemic conditions of the medium and base sections of the heart in the infarcted zone, the density of CTs decreased progressively in the LAD ligated group compared with the same region in the non-LAD ligated control group. The density of CTs decreased significantly 1–3 days (1–4 days for base) after LAD ligation, and the cells were undetectable 4 days (5 days for base) after LAD ligation. In the border zone, the density of CTs decreased significantly 4–7 days (2–7 days for base) after LAD ligation when compared with the same region in the non-LAD ligated control group. In addition, the density of CTs reached its lowest level 7 days (6–7 days for the base) after LAD ligation. However, the CT density rebounded slightly 14 days after LAD ligation and was then maintained at a level similar to that observed 4 days after LAD ligation [47].

In the zone located opposite the infarcted zone (nonischemic area), the density of CTs decreased significantly 3–4 days (2–28 days for base) after LAD ligation when compared with the same region in the non-LAD ligated control group. The decrease in the density of CTs was progressive 2–4 days after LAD ligation, and the CT density reached its lowest point 3–4 days after LAD ligation. However, the density of

CTs increased 5 days after LAD ligation and was then maintained at a level similar to that measured 2 days after LAD ligation. In the atrium-atria part of the heart (above the ligated site; nonischemic area), the density of CTs progressively decreased 1–4 days after LAD ligation compared with the same region in the non-LAD-ligated control group. The density of CTs was lowest 3–4 days after LAD ligation and increased progressively 5–28 days after LAD ligation. At 14–28 days after LAD ligation, the density of CTs was maintained at a level similar to that measured 1 day after LAD ligation [47].

These results suggest that CTs undergo cell death in the permanently ischemic myocardium, and the level of damage is particularly high in the infarct zone as CTs were undetectable until 28 days after LAD ligation. This also suggests that the structural network of the CTs in the infarct zone of the myocardium was destroyed, and the regeneration of CTs in this area is rare. In view of the fact that the density of CTs in the border zone and nonischemic area decreases for approximately 1 week and then increases 2 weeks after LAD ligation, it is proposed that under permanent ischemic conditions (LAD permanent ligation), the regenerative capacity of CTs in the nonischemic zone is intact. These findings reveal that, in addition to the death of cardiac myocytes, the previously unrecognized death of CTs is an important mechanism that contributes to the structural damage of the myocardium in the infarcted zone and poor healing and regeneration in infarcted myocardium. This unique phenomenon led us to further hypothesize that regeneration and reconstruction of CTs and their network in the infarct zone might be a novel strategy to improve and promote better healing and regeneration after MI.

15.5 Transplantation of CTs After MI Decreases the Infarct Size and Improves Myocardial Function

According to this hypothesis, transplantation of CTs for MI has promising therapeutic effects for decreasing the infarct size and improving myocardial function. In the experimental LAD ligation rat MI model, we demonstrated, for the first time, that CT transplantation in the center and borders of the infarct site of ischemic myocardium decreased the size of the infarct significantly. In addition, the left ventricle function of CT-treated infarct hearts was significantly improved – which was characterized by a higher ejection fraction of the LV and lower diastolic and systolic diameters compared with the phosphate buffered saline (PBS)-treated controls. These therapeutic effects were confirmed both in the short term, 2 weeks after transplantation, and midterm, 14 weeks after transplantation. We found that the mechanisms behind the beneficial effects of CT transplantation are as follows: (1) CT transplantation significantly increased the vessel density at both the infarct site and the border zone; (2) CT transplantation was able to significantly decrease the fibrosis in the infarcted zone and non-infarcted zone and improve the reconstructed parameters such as the thickness of the LV border zone wall and infarcted myocardium wall as well as improve the pathological reconstruction of infarct myocardium and potentially decrease the risk of heart failure; and (3) CT transplantation could significantly increase the density of CTs in the infarct zone. These findings imply that the beneficial effects of CT transplantation are attributable to enhanced angiogenesis, decreased fibrosis in the ischemic myocardium, and improved pathological reconstruction, which support survival of CTs or regeneration of CTs in the ischemic myocardium [47, 52]. These results confirm the role of CTs as a structural and functional niche and microenvironment for endothelial cells. In fact, other studies have reported that CTs secrete vascular endothelial growth factor (VEGF), express angiogenic-associated microRNAs, and establish direct nano-contacts with newly derived endothelial cells at the border zone of MI [36, 37, 45]. In addition, our findings propose that CTs might provide a structural and functional niche microenvironment for cardiac fibroblasts and regulate the transition of fibroblasts to myofibroblasts and fibrogenesis in cardiac pathology. Therefore, the underlying molecular mechanisms would be an intriguing area for future investigation.

15.6 Cardiac Telocytes Are Unique Interstitial Cells in the Myocardium that Differ from Cardiac Fibroblasts

Even though our study and studies by others have demonstrated the unique ultrastructure by TEM and light microscopy to clearly reveal CTs as unique interstitial cells, specific markers for the identification of these cells are still lacking. In addition to demonstrating the hallmark morphology of CTs, we used c-Kit antibodies combined with DiI Micro-label and flow cytometry to reveal that CTs express c-Kit, CD34, vimentin, and PDGFR-β at the single-cell level [48]. Another unique hallmark of CTs is that they undergo cell death in the permanently ischemic myocardium [47, 52]. These characteristics imply that CTs are a unique type of interstitial cells in the myocardium that differ from cardiac fibroblasts. It is well established that under ischemic challenge, cardiac fibroblasts are able to survive and proliferate and then transform into myofibroblasts, whereas CTs undergo cell death in similar situations. Therefore, characteristics such as the unique morphology and predisposition to cell death under ischemia along with molecular markers such as c-Kit, CD34, vimentin, and PDGFR-β should enable the identification of CTs.

15.7 Summary

Ours and other studies clearly demonstrated that cardiac telocytes are interstitial cells that play a structural and functional niche microenvironmental role in the myocardium to regulate cardiac stem cells, cardiac angiogenesis, and fibrosis. According to our findings, CT transplantation after MI limits the size of the infarct and improves cardiac function significantly. Therefore, we propose the following intriguing issues for investigation in future studies: (1) the interaction of CTs with other cardiac cells, such as cardiac stem cells, endothelial cells, and cardiac fibroblasts; (2) the role of CTs in cardiac fibrosis; and (3) the possible synergic therapeutic effects with cardiac stem cells and other types of stem cells for MI. Information about these intriguing issues will assist in tailoring novel antiaging strategies to combat pathological phenotypes in the cardiovascular system and improve regeneration of the myocardium after MI.

Acknowledgments This work was supported by the Collaborated grant for HK-Macao-TW of Ministry of Science and Technology (2012DFH30060); the National Natural Science Foundation of China (30973158, 81170324, 81470433); the Research grant of the Department of Education of Guangdong (20122gjhz0003); the Guangdong Key grant for Natural Science Foundation (04105826; S2012020010895); the 863 grant (2007AA02Z105); the Guangdong grant for Science and Technology Development (2004B30601007); and the International collaborated grant of Guangdong (2009B050900007; 2013B051000062).

References

1. Bernstein HS, Srivastava D. Stem cell therapy for cardiac disease. Pediatr Res. 2012;71:491–9.
2. Ptaszek LM, Mansour M, Ruskin JN, Chien KR. Towards regenerative therapy for cardiac disease. Lancet. 2012;379:933–42.
3. Segers VF, Lee RT. Stem-cell therapy for cardiac disease. Nature. 2008;451:937–42.
4. Ieda M, Tsuchihashi T, Ivey KN, Ross RS, Hong TT, Shaw RM, Srivastava D. Cardiac fibroblasts regulate myocardial proliferation through beta1 integrin signaling. Dev Cell. 2009;16:233–44.
5. Assmus B, Schachinger V, Teupe C, Britten M, Lehmann R, Dobert N, Grunwald F, Aicher A, Urbich C, Martin H, Hoelzer D, Dimmeler S, Zeiher AM. Transplantation of progenitor cells and regeneration enhancement in acute myocardial infarction (TOPCARE-AMI). Circulation. 2002;106:3009–17.
6. Strauer BE, Brehm M, Zeus T, Kostering M, Hernandez A, Sorg RV, Kogler G, Wernet P. Repair of infarcted myocardium by autologous intracoronary mononuclear bone marrow cell transplantation in humans. Circulation. 2002;106:1913–8.
7. Britten MB, Abolmaali ND, Assmus B, Lehmann R, Honold J, Schmitt J, Vogl TJ, Martin H, Schachinger V, Dimmeler S, Zeiher AM. Infarct remodeling after intracoronary progenitor cell treatment in patients with acute myocardial infarction (TOPCARE-AMI): mechanistic insights from serial contrast-enhanced magnetic resonance imaging. Circulation. 2003;108:2212–8.
8. Fernandez-Aviles F, San Roman JA, Garcia-Frade J, Fernandez ME, Penarrubia MJ, de la Fuente L, Gomez-Bueno M, Cantalapiedra A, Fernandez J, Gutierrez O, Sanchez PL, Hernandez C, Sanz R, Garcia-Sancho J, Sanchez A. Experimental and clinical regenerative capability of human bone marrow cells after myocardial infarction. Circ Res. 2004;95:742–8.
9. Wollert KC, Meyer GP, Lotz J, Ringes-Lichtenberg S, Lippolt P, Breidenbach C, Fichtner S, Korte T, Hornig B, Messinger D, Arseniev L, Hertenstein B, Ganser A, Drexler H. Intracoronary autologous bone-marrow cell transfer after myocardial infarction: the BOOST randomised controlled clinical trial. Lancet. 2004;364:141–8.
10. Janssens S, Dubois C, Bogaert J, Theunissen K, Deroose C, Desmet W, Kalantzi M, Herbots L, Sinnaeve P, Dens J, Maertens J, Rademakers F, Dymarkowski S, Gheysens O, Van Cleemput J, Bormans G, Nuyts J, Belmans A, Mortelmans L, Boogaerts M, Van de Werf F. Autologous bone marrow-derived stem-cell transfer in patients with ST-segment elevation myocardial infarction: double-blind, randomised controlled trial. Lancet. 2006;367:113–21.
11. Schachinger V, Erbs S, Elsasser A, Haberbosch W, Hambrecht R, Holschermann H, Yu J, Corti R, Mathey DG, Hamm CW, Suselbeck T, Assmus B, Tonn T, Dimmeler S, Zeiher AM. Intracoronary bone marrow-derived progenitor cells in acute myocardial infarction. N Engl J Med. 2006;355:1210–21.

12. Plewka M, Krzeminska-Pakula M, Peruga JZ, Lipiec P, Kurpesa M, Wierzbowska-Drabik K, Korycka-Wolowiec A, Kasprzak JD. The effects of intracoronary delivery of mononuclear bone marrow cells in patients with myocardial infarction: a two year follow-up results. Kardiol Pol. 2011;69:1234–40.
13. Lunde K, Solheim S, Aakhus S, Arnesen H, Abdelnoor M, Egeland T, Endresen K, Ilebekk A, Mangschau A, Fjeld JG, Smith HJ, Taraldsrud E, Grogaard HK, Bjornerheim R, Brekke M, Muller C, Hopp E, Ragnarsson A, Brinchmann JE, Forfang K. Intracoronary injection of mononuclear bone marrow cells in acute myocardial infarction. N Engl J Med. 2006;355:1199–209.
14. Meyer GP, Wollert KC, Lotz J, Steffens J, Lippolt P, Fichtner S, Hecker H, Schaefer A, Arseniev L, Hertenstein B, Ganser A, Drexler H. Intracoronary bone marrow cell transfer after myocardial infarction: eighteen months' follow-up data from the randomized, controlled BOOST (BOne marrOw transfer to enhance ST-elevation infarct regeneration) trial. Circulation. 2006;113:1287–94.
15. de Silva R, Raval AN, Hadi M, Gildea KM, Bonifacino AC, Yu ZX, Yau YY, Leitman SF, Bacharach SL, Donahue RE, Read EJ, Lederman RJ. Intracoronary infusion of autologous mononuclear cells from bone marrow or granulocyte colony-stimulating factor-mobilized apheresis product may not improve remodelling, contractile function, perfusion, or infarct size in a swine model of large myocardial infarction. Eur Heart J. 2008;29:1772–82.
16. Hashemi SM, Ghods S, Kolodgie FD, Parcham-Azad K, Keane M, Hamamdzic D, Young R, Rippy MK, Virmani R, Litt H, Wilensky RL. A placebo controlled, dose-ranging, safety study of allogenic mesenchymal stem cells injected by endomyocardial delivery after an acute myocardial infarction. Eur Heart J. 2008;29:251–9.
17. Martin-Rendon E, Brunskill SJ, Hyde CJ, Stanworth SJ, Mathur A, Watt SM. Autologous bone marrow stem cells to treat acute myocardial infarction: a systematic review. Eur Heart J. 2008;29:1807–18.
18. Herbots L, D'Hooge J, Eroglu E, Thijs D, Ganame J, Claus P, Dubois C, Theunissen K, Bogaert J, Dens J, Kalantzi M, Dymarkowski S, Bijnens B, Belmans A, Boogaerts M, Sutherland G, Van de Werf F, Rademakers F, Janssens S. Improved regional function after autologous bone marrow-derived stem cell transfer in patients with acute myocardial infarction: a randomized, double-blind strain rate imaging study. Eur Heart J. 2009;30:662–70.
19. Tendera M, Wojakowski W, Ruzyllo W, Chojnowska L, Kepka C, Tracz W, Musialek P, Piwowarska W, Nessler J, Buszman P, Grajek S, Breborowicz P, Majka M, Ratajczak MZ. Intracoronary infusion of bone marrow-derived selected CD34+CXCR4+ cells and non-selected mononuclear cells in patients with acute STEMI and reduced left ventricular ejection fraction: results of randomized, multicentre Myocardial Regeneration by Intracoronary Infusion of Selected Population of Stem Cells in Acute Myocardial Infarction (REGENT) Trial. Eur Heart J. 2009;30:1313–21.
20. Maki MT, Koskenvuo JW, Ukkonen H, Saraste A, Tuunanen H, Pietila M, Nesterov SV, Aalto V, Airaksinen KE, Parkka JP, Lautamaki R, Kervinen K, Miettinen JA, Makikallio TH, Niemela M, Saily M, Koistinen P, Savolainen ER, Ylitalo K, Huikuri HV, Knuuti J. Cardiac function, perfusion, metabolism, and innervation following autologous stem cell therapy for acute ST-elevation myocardial infarction. A FINCELL-INSIGHT Sub-Study PET MRI Front Physiol. 2012;3:6.
21. Hirsch A, Nijveldt R, van der Vleuten PA, Tijssen JG, van der Giessen WJ, Tio RA, Waltenberger J, ten Berg JM, Doevendans PA, Aengevaeren WR, Zwaginga JJ, Biemond BJ, van Rossum AC, Piek JJ, Zijlstra F. Intracoronary infusion of mononuclear cells from bone marrow or peripheral blood compared with standard therapy in patients after acute myocardial infarction treated by primary percutaneous coronary intervention: results of the randomized controlled HEBE trial. Eur Heart J. 2011;32:1736–47.
22. Francis DP, Mielewczik M, Zargaran D, Cole GD. Autologous bone marrow-derived stem cell therapy in heart disease: discrepancies and contradictions. Int J Cardiol. 2013;168:3381–403.
23. Nowbar AN, Mielewczik M, Karavassilis M, Dehbi HM, Shun-Shin MJ, Jones S, Howard JP, Cole GD, Francis DP. Discrepancies in autologous bone marrow stem cell trials and enhancement of ejection fraction (DAMASCENE): weighted regression and meta-analysis. BMJ. 2014;348:g2688.

24. Goss RJ. Hypertrophy versus hyperplasia. Science. 1966;153:1615–20.
25. Anderson RH, Smerup M, Sanchez-Quintana D, Loukas M, Lunkenheimer PP. The three-dimensional arrangement of the myocytes in the ventricular walls. Clin Anat. 2009;22:64–76.
26. Ausoni S, Sartore S. The cardiovascular unit as a dynamic player in disease and regeneration. Trends Mol Med. 2009;15:543–52.
27. Thum T, Gross C, Fiedler J, Fischer T, Kissler S, Bussen M, Galuppo P, Just S, Rottbauer W, Frantz S, Castoldi M, Soutschek J, Koteliansky V, Rosenwald A, Basson MA, Licht JD, Pena JT, Rouhanifard SH, Muckenthaler MU, Tuschl T, Martin GR, Bauersachs J, Engelhardt S. MicroRNA-21 contributes to myocardial disease by stimulating MAP kinase signalling in fibroblasts. Nature. 2008;456:980–4.
28. Narmoneva DA, Vukmirovic R, Davis ME, Kamm RD, Lee RT. Endothelial cells promote cardiac myocyte survival and spatial reorganization: implications for cardiac regeneration. Circulation. 2004;110:962–8.
29. Zaglia T, Dedja A, Candiotto C, Cozzi E, Schiaffino S, Ausoni S. Cardiac interstitial cells express GATA4 and control dedifferentiation and cell cycle re-entry of adult cardiomyocytes. J Mol Cell Cardiol. 2009;46:653–62.
30. Popescu LM, Faussone-Pellegrini MS. TELOCYTES – a case of serendipity: the winding way from Interstitial Cells of Cajal (ICC), via Interstitial Cajal-Like Cells (ICLC) to TELOCYTES. J Cell Mol Med. 2010;14:729–40.
31. Popescu LM, Manole CG, Gherghiceanu M, Ardelean A, Nicolescu MI, Hinescu ME, Kostin S. Telocytes in human epicardium. J Cell Mol Med. 2010;14:2085–93.
32. Bani D, Formigli L, Gherghiceanu M, Faussone-Pellegrini MS. Telocytes as supporting cells for myocardial tissue organization in developing and adult heart. J Cell Mol Med. 2010;14:2531–8.
33. Kostin S. Myocardial telocytes: a specific new cellular entity. J Cell Mol Med. 2010;14:1917–21.
34. Gherghiceanu M, Manole CG, Popescu LM. Telocytes in endocardium: electron microscope evidence. J Cell Mol Med. 2010;14:2330–4.
35. Gherghiceanu M, Popescu LM. Heterocellular communication in the heart: electron tomography of telocyte-myocyte junctions. J Cell Mol Med. 2011;15:1005–11.
36. Manole CG, Cismasiu V, Gherghiceanu M, Popescu LM. Experimental acute myocardial infarction: telocytes involvement in neo-angiogenesis. J Cell Mol Med. 2011;15:2284–96.
37. Gherghiceanu M, Popescu LM. Cardiac telocytes – their junctions and functional implications. Cell Tissue Res. 2012;348:265–79.
38. Cantarero Carmona I, Luesma Bartolome MJ, Junquera Escribano C. Identification of telocytes in the lamina propria of rat duodenum: transmission electron microscopy. J Cell Mol Med. 2011;15:26–30.
39. Popescu LM, Ciontea SM, Cretoiu D. Interstitial Cajal-like cells in human uterus and fallopian tube. Ann N Y Acad Sci. 2007;1101:139–65.
40. Zheng Y, Li H, Manole CG, Sun A, Ge J, Wang X. Telocytes in trachea and lungs. J Cell Mol Med. 2011;15:2262–8.
41. Popescu LM, Gherghiceanu M, Suciu LC, Manole CG, Hinescu ME. Telocytes and putative stem cells in the lungs: electron microscopy, electron tomography and laser scanning microscopy. Cell Tissue Res. 2011;345:391–403.
42. Popescu LM, Manole E, Serboiu CS, Manole CG, Suciu LC, Gherghiceanu M, Popescu BO. Identification of telocytes in skeletal muscle interstitium: implication for muscle regeneration. J Cell Mol Med. 2011;15:1379–92.
43. Gherghiceanu M, Popescu LM. Interstitial Cajal-like cells (ICLC) in human resting mammary gland stroma. Transmission electron microscope (TEM) identification. J Cell Mol Med. 2005;9:893–910.
44. Suciu L, Popescu LM, Gherghiceanu M, Regalia T, Nicolescu MI, Hinescu ME, Faussone-Pellegrini MS. Telocytes in human term placenta: morphology and phenotype. Cells Tissues Organs. 2010;192:325–39.
45. Gherghiceanu M, Popescu LM. Cardiomyocyte precursors and telocytes in epicardial stem cell niche: electron microscope images. J Cell Mol Med. 2010;14:871–7.

46. Cretoiu SM, Cretoiu D, Marin A, Radu BM, Popescu LM. Telocytes: ultrastructural, immunohistochemical and electrophysiological characteristics in human myometrium. Reproduction. 2013;145:357–70.
47. Zhao B, Chen S, Liu J, Yuan Z, Qi X, Qin J, Zheng X, Shen X, Yu Y, Qnin TJ, Chan JY, Cai D. Cardiac telocytes were decreased during myocardial infarction and their therapeutic effects for ischaemic heart in rat. J Cell Mol Med. 2013;17:123–33.
48. Xiao H, Shen X, Zheng X, Li Y, Cai D. c-Kit + Combining with DiI Micro-label to Sort the Cardiac Telocytes. China Biotechnol. 2015;35:46–53.
49. Bei Y, Zhou Q, Fu S, Lv D, Chen P, Chen Y, Wang F, Xiao J. Cardiac telocytes and fibroblasts in primary culture: different morphologies and immunophenotypes. PLoS One. 2015;10:e0115991.
50. Chang Y, Li C, Lu Z, Li H, Guo Z. Multiple immunophenotypes of cardiac telocytes. Exp Cell Res. 2015;338(2):239–44.
51. Liu J, Shen X, Zheng X, Li Z, Wang J, Qi X, Cai D. Distribution of Telocytes in the rat heart. J Clin Rehabilitative Tissue Eng Res. 2011;15:3546–8.
52. Zhao B, Liao Z, Chen S, Yuan Z, Yilin C, Lee KK, Qi X, Shen X, Zheng X, Quinn T, Cai D. Intramyocardial transplantation of cardiac telocytes decreases myocardial infarction and improves post-infarcted cardiac function in rats. J Cell Mol Med. 2014;18:780–9.

Cai Dongqing *Education*: M.D. (Guangzhou Medical College; 1987), Ph.D. (The Chinese University of Hong Kong; 2000), Postdoctoral Associate (Weill Medical College of Cornell University [USA]; 2000–2003). *Employment*: Professor and Director, Key Laboratory for Regenerative Medicine, Ministry of Education, Ji Nan University; Director, Department of Developmental and Regenerative Biology, Ji Nan University. *Scientific interests*: (1) Cardiac aging and regeneration (focus on extracellular and intracellular microenvironments for cardiac microvascular endothelial cells, endogenous cardiac stem cells, cardiac telocytes, cardiac fibroblast, myocyte aging, and regeneration of myocardial infarction), (2) cardiac vascular-specific targeting angiogenesis and adult stem cell therapy for myocardial infarction, (3) adult cell therapy (endogenous cardiac stem cells, mesenchymal stem cells (MSCs), and cardiac interstitial cells) for myocardial infarction, (4) organ aging (focus on heart). *Grant*: 2003–present, total 23 government grants, included 863 International collaboration grant of Ministry of Science and Technology, five NSFC grants, two Key grants of GDZRKXJJ, six Guangdong and Guangzhou government grants, etc. *Publication*: 36 SCI papers have been published (including *JCMM*, *Proteomics*, *Am J Physiol Heart Circ Physiol*, *Physiological Genomics*, etc.). *Referee*: referee for six SCI journals (*JCMM*, *Physiological Genomics*, etc.). *Scientific society:* Vice Director, Society of cell therapy of Guangdong; Committee Member, Society of genetics of China.

Chapter 16
Myocardial Telocytes: A New Player in Electric Circuitry of the Heart

Winston Shim

Abstract The heart is an electrically conducting organ with networked bioelectric currents that transverse a large segment of interstitial space interspersed with the muscular parenchyma. Non-excitable connective cells in the interstitial space contributed importantly to many structural, biochemical, and physiological activities of cardiac homeostasis. However, contribution of interstitial cells in the cardiac niche has long been neglected. Telocyte is recently recognized as a distinct class of interstitial cell that resides in a wide array of tissues including in the epicardium, myocardium, and endocardium of the heart. They are increasingly described to conduct ionic currents that may have significant implications in bioelectric signaling. In this review, we highlight the significance of telocytes in such connectivity and conductivity within the interstitial bioelectric network in tissue homeostasis.

16.1 Background

All living cells and tissues are believed to be interlinked via bioelectric signaling mediated by ionic flow, electric fields, and voltage gradients to maintain interconnectivity [1]. The ability to maintain such bioelectric gradients across multidimensional networked cellular entities has important implications in health and disease state. Maintenance of bioelectric signaling, beyond electrical conduction per se, via differential resting membrane potentials among living cells, in particular, by non-excitable cells in the interstitial space has been highlighted to play important roles in developmental embryogenesis, tissue morphogenesis, and organ regeneration [2]. Indeed, gradients of membrane potentials within tissue niche are believed to direct

W. Shim
National Heart Research Institute Singapore,
5 Hospital Drive, National Heart Centre Singapore, Singapore 169609, Singapore

Cardiovascular and Metabolic Disorders Program, Duke-NUS Medical School,
8 College Road, Singapore 169857, Singapore
e-mail: winston.shim.s.n@nhcs.com.sg

X. Wang, D. Cretoiu (eds.), *Telocytes*, Advances in Experimental Medicine and Biology, 913, DOI 10.1007/978-981-10-1061-3_16

stem cell proliferation and differentiation [3] that could have far-reaching but often underappreciated influence on cellular physiology.

The heart is an electrically conducting organ whereby known players in the parenchyma and interstitial space bridge the bioelectric gradients by electrotonic conductivity and electrical circuitry that are interconnected through connexins and gap junction and via other hitherto poorly studied intercellular adaptors/connectors. Non-excitable connective cells in the interstitial space contributed importantly to many structural, biochemical, and physiological activities of cardiac homeostasis. Heterocellular electrotonic coupling between interstitial cells such as fibroblasts, endothelial cells, smooth muscle cells, stem cells, and cardiomyocytes is only beginning to be recognized. In fact, in situ membrane connectivity of fibroblasts that juxtaposed cardiomyocytes is clearly evidenced in studies involving dye transfer experiments in sinoatrial node that signals their direct participation in physiological and perhaps pathological processes of electrical conduction in the heart [4]. The ability of cardiac fibroblasts to act as a bridge between conducting cardiomyocytes has been widely demonstrated in various coupling experiments [5, 6].

Telocytes are recently described interstitial cells with exceptionally long cellular processes found within cardiac parenchyma of the epicardium, myocardium, and endocardium that are believed to mainly act as structural supporting and nursing cells in the heart [7–9]. Despite their implicated intermediary role in the electrical activities of cardiac rhythm [7] and atrial fibrillation [10], very limited is known about the electrophysiological property of cardiac telocytes. We recently reported that human atrial and ventricular telocytes responded to H_2S by attenuating TGF-β1-stimulated KCa1.1/Kv1.1 and Kir2.1 gene expression. However, the presence of competent K^+ channels in telocytes and their implications in myocardial physiopathology remain largely unexplored. Human myometrial telocytes have been reported to express calcium-dependent hyperpolarization-activated chloride inward current [11], and SK3 outward potassium rectifier channels were reported in uterine telocytes [12]. Furthermore, transient outward potassium current that exhibited pacemaker-like activity was reportedly present in gastrointestinal telocytes [13], and hyperpolarization-activated cyclic nucleotide-gated (HCN) channel was found in telocytes of murine gastric antrum [14]. In this review, we highlight the significance of telocytes in such connectivity and conductivity within the interstitial bioelectric network in cellular homeostasis.

16.2 Telocyte: A New Player Within Interstitial Network

The seminary work by Popescu and colleagues in the last decade has identified and established telocyte as a distinct cellular entity, separate from fibroblasts [15] in the interstitial space of many organs, including the heart [16], bladder [17], lungs [18], and skeletal muscle [19]. Such telocytes are intimately contacting parenchymal tissues within the organs and are implicated in electrical conduction beyond their traditional structural supporting role in the myocardial system due to their close proximity to nerve endings, cardiomyocytes [20], and pulmonary veins [10] in which atrial arrhythmias often originated in recurrent atrial fibrillation [21, 22].

Telocytes have previously been described as interstitial Cajal-like cells (ICLC) and were found in atrial and ventricular myocardium [23, 24] though very little was known of their function. Recently, telocytes have been increasingly recognized to participate beyond their passive supporting role but instead may play physiological diverse functions in the heart [25–27]. Telocytes have exceptionally long (10–1,000 μm), moniliform cellular processes named telopodes with intervening podoms and podomers that act as long-range cellular connector that transverses a vast interstitial network connecting different segments of the heart [28] including those of the epicardium [7, 29], myocardium [8, 26], and endocardium [9]. Their close proximity with cardiac progenitors and cardiomyocytes is believed to be important in the repair and regeneration of infarcted myocardium [7]. However, the mechanisms involving in such interactions are poorly understood.

The fact that telocytes were found widely in the developing and adult heart [16, 26] suggests that they are likely to have bioelectric modulatory function in their membrane potentials and capacitance due to their varying size and length in an adaptation towards microenvironment within cardiac niche. The extensive intercellular networks encompassed by telocytes with their long telopodes have been postulated to carry electrical signals and/or currents via intracellular cytoskeletal structures, perhaps in conjunction with transmitting electrical signals via gap junctions such as connexins that have been identified in a wide range of tissues [30, 31]. This unconventional concept is in congruent with the observed ionic species and specialized ion channels reported in telocytes found in a wide range of tissues. Indeed, telocytes were recently shown to exhibit voltage-gated ion channels whereby ion conductance characteristics of BKca and IKir currents were reported by our group [32] and ICaT by others [11, 33]. This supports that they may carry functional ionic currents that are likely contributing to important cellular cues and signaling within the interstitial network and in their electrical interaction with cardiomyocytes.

16.2.1 Telocyte: Physical Connector Bridging Electrotonic Conductivity

Human myocardial telocytes are clearly distinguishable from atrial or ventricular fibroblasts in culture whereby their characteristic long telopodes [27] with intersperse podomers and podoms that are often in contact with fibroblasts and other telocytes [34]. Consistent with distinctive telocyte identity [16, 26, 35], human telocytes express CD34, c-Kit, and PDGFR-β markers (Fig. 16.1)

We recently reported that the presence of inter-networking cardiac telocytes in the interstitial space improved myocardial strain post-myocardial transplantation of human-induced pluripotent stem cell (iPSC)-derived mesenchymal stem cells (iMSCs) into the infarcted myocardium [36]. Our results are consistent with the purported role of telocytes in mediating cardiac parenchymal and interstitial

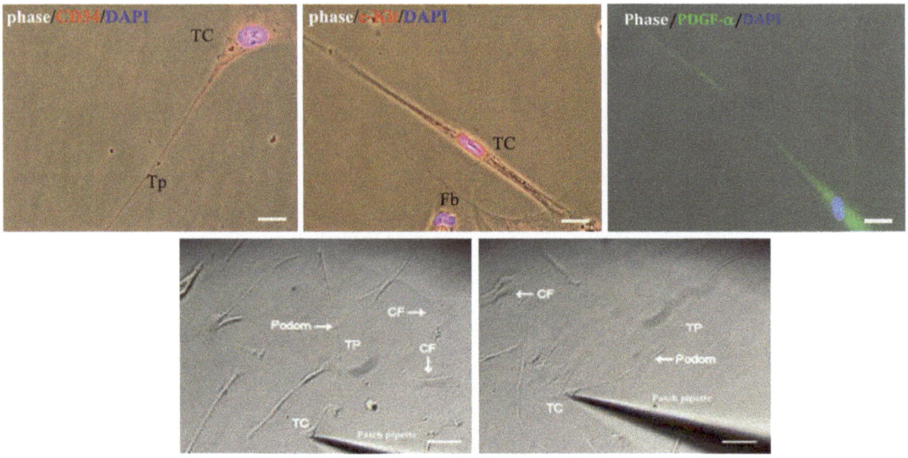

Fig. 16.1 Identification and characterization of myocardial telocytes in culture. Cells with extended telopodes that stained positive for CD34, c-kit, and PDGF-α that are characteristics of telocytes were patched with glass pipette in a whole-cell configuration for electrophysiology study (Images adapted from Sheng et al. [32])

interactions in promoting tissue repair [27]. Furthermore, we observed that myocardial telocytes intertwined with transplanted cardiac progenitors in the infarcted regions that coincided with increased angiogenesis [37] that is in agreement with beneficial effect of telocyte transplantation in recovering cardiac function [38]. The feasibility of long-range interaction between telocytes to maintain electrotonic conductivity within the cardiac niche cannot be discounted as the network of mechanical and biological interconnectivity from infarcted segment to other remote healthy segments of the myocardium was evidenced (Fig. 16.2) by close association of exogenously transplanted telocytes with resident telocytes [36, 37].

The decreased density of telocytes early post infarct (from 4 to 30 days) and their subsequent transplantation that decreased infarct size and improved cardiac function suggest physiological benefits of their present in the heart [38, 39]. Interestingly, such disappearance of telocytes coincides with the window of arrhythmogenic episodes often acutely experienced post-MI. In addition, heart failure patients experienced more than twofold decrease in the numbers of telocytes in the myocardium that coincided with remodeling of collagenous extracellular matrix in the cardiac niche [40] that is known to be a fertile substrate for pro-arrhythmogenic events [41]. Therefore, it may be tempting to surmise that loss of cardiac telocytes may compromise the three-dimensional spatial interconnectivity and perhaps conductivity of a continuous bioelectric gradient between the interstitial and parenchymal junctions within the myocardium that may further precipitate the electrical imbalance during post-MI period.

Heterocellular coupling of fibroblasts with cardiomyocytes in culture has been well documented to support wave propagation as far as 300 um through electrotonic interaction possibly via connexin43 (Cx43) and Cx45, though with major local

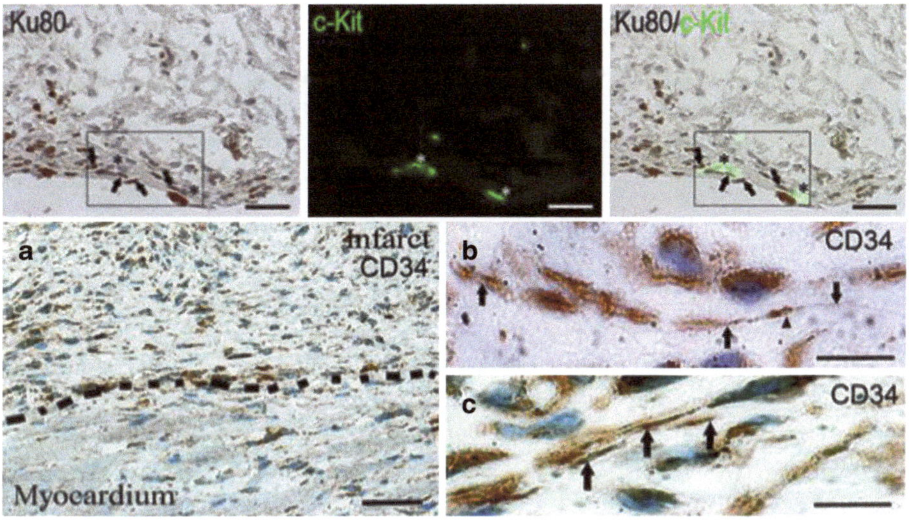

Fig. 16.2 Interconnected myocardial telocytes in infarcted murine myocardium showing connectivity of telocytes in the networked interstitial space in close proximity to cardiac progenitor cells and intact cardiac muscle fibers (Images adapted from Ja et al. [37] (*top panel*) and Miao et al. [36] (*bottom panel*))

conduction delays that may have implication in arrhythmias [5]. Similarly, telocytes are known to express C×43 [42] that may support electrotonic conduction and exert influence on cardiac arrhythmia. Therefore, more focused studies on coupling experiments between cardiomyocytes and telocytes for electrical interaction and integration within cardiac syncytium are warranted.

The elegant electron microscopy works provided by Popescu and colleagues attest to the possible existent of heterocellular coupling between telocytes and cardiomyocytes within myocardium. Indeed, beyond paracrine communication through telocytes released vesicles [43], direct heterocellular connection and coupling between telocytes and cardiomyocytes have been demonstrated previously such that atypical junctions formed by macromolecular complexes and nanopillars were observed between telopodes and sarcolemmal processes of cardiomyocytes [25]. This is in addition to the reported direct junctional connection between telocytes and cardiac progenitors [7, 9], between telocytes and cardiac stem cells via adherens junctions, puncta adherentia, and stromal synapses [44], and the existence of close apposition of telopodes against intercalated disks of cardiac muscle [22]. Collectively, these evidences support possible electrophysiological role of telocytes within the spatial architecture of cardiac syncytium.

Consistent with their intermediary function and role as bridging connector, a comprehensive proteomic analysis of cultured telocytes revealed that up to 4 % of cellular proteins were of cell junction components, thus affirming their key role in intercellular signaling and communication [15]. Tight adherens junctions were recently confirmed to exist between telopodes of telocytes in the myocardium [45], and specific junctional contacts were noted between telocytes and cardiomyocytes and between telocytes and other interstitial cells [28] in the myocardium supporting their extended connectivity that is structurally supportive of probable electrotonic conductivity. Whether such connectivity is present or changed throughout developmental phase and disease stage or differs between species and exists within other tissues that experience distinct mechanical stress than the myocardium is largely unknown.

Besides their interconnectivity through gap junctions and adherens junctions, it is believed that telocytes could connect with cardiomyocytes through recently described intercellular nanotubes that directly link these heterogeneous cells together. Such connectivity has been reported in coupling experiments between fibroblasts and cardiomyocytes [46]. In fact, intercellular tunneling nanotubes ranging from 50 to 200 nm in length, perhaps in conjunction with connexin junctional proteins, may impart electrical signals between fibroblasts and cardiomyocytes via activation of voltage-gated calcium channels [47] and exchange of mitochondria [48]. Such interactions may partake in arrhythmogenesis as density-dependent biphasic influence of fibroblasts on conduction velocity and upstroke velocity through partial depolarization of cardiomyocytes has been previously observed [49]. Interestingly, such heterocellular connections between fibroblasts and cardiomyocytes are reminiscent of telopodes of telocytes in contacting neighboring cells in a three-dimensional cardiac niche. Nevertheless, feasibility of interstitial telocytes acting as intervening electrotonic bridge between conducting cardiomyocytes has largely been neglected thus far. Intriguingly, the ratio of cardiomyocytes to telocytes in rat myocardium was reported to be 70:1 [42], which was similar to ratiometric density reported between cardiomyocytes and fibroblasts that were observed to exert influence on arrhythmogenic events in heterocellular coupling experiments [49].

Despite the mounting evidence suggesting heterocellular connectivity between telocytes and cardiomyocytes, it is unclear if telocytes are possibly acting in the overall electrical circuitry as passive insulating bystander or active conducting player. Transfection of gap junction proteins and voltage-gated ion channels into telocytes in coupling experiments with cardiomyocytes would be of crucial interest to further elucidate such connectivity in establishing bioelectric gradient and their importance in maintaining cardiac electrical homeostasis, at least spatially within a localized interstitial to myocardial niche.

16.2.2 Telocyte: Gated Ion Channels and Function

To date, voltage-gated ion channels are increasingly being recognized in telocytes isolated from a wide range of tissues. In the human myometrium, patch-clamp recordings revealed a calcium-dependent hyperpolarization-activated chloride inward current but absence of L-type calcium channels, which was postulated to modulate myometrial smooth muscle contractions [11]. Furthermore, uterine telocytes have been reported to express mibefradil-sensitive T-type calcium channels that is responsive to near-infrared low-level laser stimulation [50] that may be important in bioelectric signaling for modulating cellular behaviors during pregnancy and labor [45]. Furthermore, telocytes expressing calcium-activated chloride channels and inwardly rectifying chloride channels in gastrointestinal tract have been identified to mediate gut motility [51, 52]. Moreover, transient outward potassium current that mediated pacemaker-like activity was noted in gastrointestinal telocytes [13]. In addition, ionic currents were previously reported in telocytes whereby small conductance potassium SK3 channels in human myometrium [12] and in murine bladder [53] may be key to regulating muscular excitability and contractility. In addition, calcium-activated potassium channels in guinea pigs that regulate repolarization of stomach [54] and hyperpolarization-activated cyclic nucleotide-gated (HCN) channel of murine gastric antrum were identified [14].

We recently presented evidence of human ventricular and atrial telocytes that express large conductance Ca^{2+}-activated K^+ current (BK_{Ca}) and inwardly rectifying K^+ current (IK_{ir}) currents (Fig. 16.3) in a TGF-beta1-dependent manner indicating their importance in cardiac electrophysiology or electrotonic interactions in cardiac fibrosis [32]. We recorded and characterized ion currents in single telocytes with long telopodes using conventional whole-cell voltage clamp setup. Paxilline (a specific BK_{Ca} inhibitor)-sensitive and naringenin-responsive (a specific BK_{Ca} opener) currents were detected, confirming the presence of BK_{Ca} currents in atrial and ventricular telocytes. Furthermore, it was found that 4-aminopyridine-insensitive

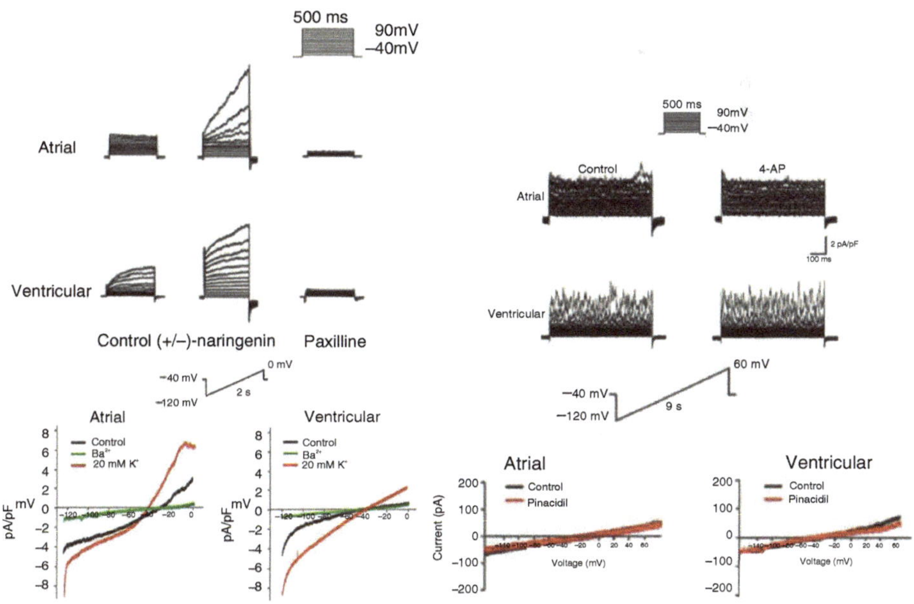

Fig. 16.3 Patch-clamp electrophysiology of atrial and ventricular telocytes. Both outward and inward potassium currents consistent with BKca and IKir were detected in the voltage-clamped cells (Images adapted from Sheng et al. [32])

and pinacidil (K_{ATP}-specific channel enhancer)-unresponsive currents were present, indicating the respective absence of transient outward currents (I_{to}) and ATP-sensitive K current (K_{ATP}) in myocardial telocytes. Nevertheless, inwardly rectifying K^+ currents that were attenuated by Ba^{2+} ion coupled with strongly spiking current amplitude in the presence of 20 mM K^+ bath solution were elicited in depolarizing telocytes, indicating the presence of IK_{ir} inward currents. The close proximity of telocytes to cardiac fibroblasts and cardiomyocytes suggests a probable role in facilitating mechano-electrical coupling of the heart. Consistently, calcium-releasing stores, such as caveolae, sarcoplasmic reticulum, and mitochondria that are typical of voltage-responsive cells, are present in telocytes [31]. Nevertheless, it is unclear if cellular resistance and capacitance of telocytes and resting membrane potentials would be sufficient to depolarize the heterocellularly coupled cardiomyocytes.

Human atrial fibroblasts are known to express a range of potassium channels that are important in proliferation and myofibroblast transformation Importantly, it is currently unclear the implications of greater potassium channel responsiveness observed in the atrial telocytes as compared to ventricular telocytes when stimulated with TGF-β1, a major mediator of myocardial fibrosis [32]. Although such dichotomy of cellular physiology coincided with the more robust fibrotic response of atrial fibroblasts as compared to ventricular fibroblasts in myocardial fibrosis [55, 56], it is unclear if the differential expression of such ion channels in atrial and ventricular telocytes has functional significance in cellular proliferation and overall interstitial conductivity or fibrotic response.

Telocytes are increasingly being recognized to carry ionic currents that may have important implications in behavioral response of local tissue niche. It is unclear if telocytes may exhibit regional-specific property in different organs. Experimental evidence supports the presence of functionally competent BK_{Ca} and IK_{ir} channels, but not I_{to} and K_{ATP} channels, in cardiac atrial and ventricular telocytes. It is unknown if telocytes could assume a different phenotype in different conditions such as those observed conversion of fibroblasts to myofibroblasts that may affect different cellular physiologies (e.g., smooth muscle actin expression [32]) that exerts distinct impact on the myocardial milieu.

16.3 Conclusion

There are mounting evidence that support functional present of a wide array of ion channels in telocytes found in various tissues. These channels are reported to contribute to telocyte electrophysiology but may also have other bioelectrical and mechanical significance in the tissue niche. Illustrating how these diverse ion channels contribute to function already well recognized in telocytes in mechanical supporting, topographical nursing, long-range sensing, microvesicle releasing, repair signaling, cytokine secreting, and immune system modulating would unveil a comprehensive road map of bioelectric traffic crisscrossing the intercellular highways of networked interstitial sphere in communicating with the parenchyma to sustain tissue homeostasis.

Acknowledgments This work was supported by the National Research Foundation Singapore (NRF-CRP-2008-02), Goh Foundation Gift (Singapore)/Duke-NUS Medical School (GCR/2013/008, GCR/2013/010, and GCR/2013/011), and Biomedical Research Council, Singapore (BMRC13/1/96/19/86).

References

1. Levin M. Endogenous bioelectrical networks store non-genetic patterning information during development and regeneration. J Physiol. 2014;592(Pt 11):2295–305.
2. Levin M. Molecular bioelectricity in developmental biology: new tools and recent discoveries: control of cell behavior and pattern formation by transmembrane potential gradients. Bioessays. 2012;34(3):205–17.
3. Edelstein L, Smythies J. The role of telocytes in morphogenetic bioelectrical signaling: once more unto the breach. Front Mol Neurosci. 2014;7:41.
4. Camelliti P, Green CR, LeGrice I, et al. Fibroblast network in rabbit sinoatrial node: structural and functional identification of homogeneous and heterogeneous cell coupling. Circ Res. 2004;94(6):828–35.
5. Gaudesius G, Miragoli M, Thomas SP, et al. Coupling of cardiac electrical activity over extended distances by fibroblasts of cardiac origin. Circ Res. 2003;93(5):421–8.
6. Sachse FB, Moreno AP, Abildskov JA. Electrophysiological modeling of fibroblasts and their interaction with myocytes. Ann Biomed Eng. 2008;36(1):41–56.
7. Popescu LM, Gherghiceanu M, Manole CG, et al. Cardiac renewing: interstitial Cajal-like cells nurse cardiomyocyte progenitors in epicardial stem cell niches. J Cell Mol Med. 2009;13(5):866–86.
8. Kostin S. Myocardial telocytes: a specific new cellular entity. J Cell Mol Med. 2010;14(7):1917–21.
9. Gherghiceanu M, Manole CG, Popescu LM. Telocytes in endocardium: electron microscope evidence. J Cell Mol Med. 2010;14(9):2330–4.
10. Kostin S, Popescu LM. A distinct type of cell in myocardium: interstitial Cajal-like cells (ICLCs). J Cell Mol Med. 2009;13(2):295–308.
11. Cretoiu SM, Cretoiu D, Marin A, et al. Telocytes: ultrastructural, immunohistochemical and electrophysiological characteristics in human myometrium. Reproduction. 2013;145(4):357–70.
12. Rosenbaum ST, Svalo J, Nielsen K, et al. Immunolocalization and expression of small-conductance calcium-activated potassium channels in human myometrium. J Cell Mol Med. 2012;16(12):3001–8.
13. Parsons SP, Huizinga JD. Transient outward potassium current in ICC. Am J Physiol Gastrointest Liver Physiol. 2010;298(3):G456–66.
14. Si X, Huang L, Gong Y, et al. Role of calcium in activation of hyperpolarization-activated cyclic nucleotide-gated channels caused by cholecystokinin octapeptide in interstitial cells of cajal. Digestion. 2012;85(4):266–75.
15. Zheng Y, Cretoiu D, Yan G, et al. Comparative proteomic analysis of human lung telocytes with fibroblasts. J Cell Mol Med. 2014;18(4):568–89.
16. Popescu LM, Manole CG, Gherghiceanu M, et al. Telocytes in human epicardium. J Cell Mol Med. 2010;14(8):2085–93.
17. Parajuli SP, Soder RP, Hristov KL, et al. Pharmacological activation of small conductance calcium-activated potassium channels with naphtho[1,2-d]thiazol-2-ylamine decreases guinea pig detrusor smooth muscle excitability and contractility. J Pharmacol Exp Ther. 2012;340(1):114–23.
18. Zheng Y, Li H, Manole CG, et al. Telocytes in trachea and lungs. J Cell Mol Med. 2011;15(10):2262–8.
19. Popescu LM, Manole E, Serboiu CS, et al. Identification of telocytes in skeletal muscle interstitium: implication for muscle regeneration. J Cell Mol Med. 2011;15(6):1379–92.

20. Zheng Y, Bai C, Wang X. Telocyte morphologies and potential roles in diseases. J Cell Physiol. 2012;227(6):2311–7.
21. Nguyen BL, Fishbein MC, Chen LS, et al. Histopathological substrate for chronic atrial fibrillation in humans. Heart Rhythm. 2009;6(4):454–60.
22. Mandache E, Gherghiceanu M, Macarie C, et al. Telocytes in human isolated atrial amyloidosis: ultrastructural remodelling. J Cell Mol Med. 2010;14(12):2739–47.
23. Hinescu ME, Popescu LM. Interstitial Cajal-like cells (ICLC) in human atrial myocardium. J Cell Mol Med. 2005;9(4):972–5.
24. Hinescu ME, Gherghiceanu M, Mandache E, et al. Interstitial Cajal-like cells (ICLC) in atrial myocardium: ultrastructural and immunohistochemical characterization. J Cell Mol Med. 2006;10(1):243–57.
25. Gherghiceanu M, Popescu LM. Heterocellular communication in the heart: electron tomography of telocyte-myocyte junctions. J Cell Mol Med. 2011;15(4):1005–11.
26. Bani D, Formigli L, Gherghiceanu M, et al. Telocytes as supporting cells for myocardial tissue organization in developing and adult heart. J Cell Mol Med. 2010;14(10):2531–8.
27. Popescu LM, Faussone-Pellegrini MS. TELOCYTES – a case of serendipity: the winding way from Interstitial Cells of Cajal (ICC), via Interstitial Cajal-Like Cells (ICLC) to TELOCYTES. J Cell Mol Med. 2010;14(4):729–40.
28. Gherghiceanu M, Popescu LM. Cardiac telocytes – their junctions and functional implications. Cell Tissue Res. 2012;348(2):265–79.
29. Gittenberger-de Groot AC, Winter EM, Poelmann RE. Epicardium-derived cells (EPDCs) in development, cardiac disease and repair of ischemia. J Cell Mol Med. 2010;14(5):1056–60.
30. Nielsen MS, Axelsen LN, Sorgen PL, et al. Gap junctions. Compr Physiol. 2012;2(3):1981–2035.
31. Smythies J, Edelstein L. Telocytes, exosomes, gap junctions and the cytoskeleton: the makings of a primitive nervous system? Front Cell Neurosci. 2014;7:278.
32. Sheng J, Shim W, Lu J, et al. Electrophysiology of human cardiac atrial and ventricular telocytes. J Cell Mol Med. 2014;18(2):355–62.
33. Cretoiu SM, Radu BM, Banciu A, et al. Isolated human uterine telocytes: immunocytochemistry and electrophysiology of T-type calcium channels. Histochem Cell Biol. 2015;143(1):83–94.
34. Mou Y, Wang Y, Li J, et al. Immunohistochemical characterization and functional identification of mammary gland telocytes in the self-assembly of reconstituted breast cancer tissue in vitro. J Cell Mol Med. 2013;17(1):65–75.
35. Xiao J, Wang F, Liu Z, et al. Telocytes in liver: electron microscopic and immunofluorescent evidence. J Cell Mol Med. 2013;17(12):1537–42.
36. Miao Q, Shim W, Tee N, et al. iPSC-derived human mesenchymal stem cells improve myocardial strain of infarcted myocardium. J Cell Mol Med. 2014;18(8):1644–54.
37. Ja KM, Miao Q, Zhen Tee NG, et al. iPSC-derived human cardiac progenitor cells improve ventricular remodelling via angiogenesis and interstitial networking of infarcted myocardium. J Cell Mol Med. 2015;20(2):323–32.
38. Zhao B, Liao Z, Chen S, et al. Intramyocardial transplantation of cardiac telocytes decreases myocardial infarction and improves post-infarcted cardiac function in rats. J Cell Mol Med. 2014;18(15):780–89.
39. Zhao B, Chen S, Liu J, et al. Cardiac telocytes were decreased during myocardial infarction and their therapeutic effects for ischaemic heart in rat. J Cell Mol Med. 2013;17(1):123–33.
40. Richter M, Kostin S. The failing human heart is characterized by decreased numbers of telocytes as result of apoptosis and altered extracellular matrix composition. J Cell Mol Med. 2015;19(11):2597–606.
41. Weber KT, Sun Y, Bhattacharya SK, et al. Myofibroblast-mediated mechanisms of pathological remodelling of the heart. Nat Rev Cardiol. 2013;10(1):15–26.
42. Popescu LM, Gherghiceanu M, Hinescu ME, et al. Insights into the interstitium of ventricular myocardium: interstitial Cajal-like cells (ICLC). J Cell Mol Med. 2006;10(2):429–58.
43. Fertig ET, Gherghiceanu M, Popescu LM. Extracellular vesicles release by cardiac telocytes: electron microscopy and electron tomography. J Cell Mol Med. 2014;18(10):1938–43.
44. Popescu LM, Fertig ET, Gherghiceanu M. Reaching out: junctions between cardiac telocytes and cardiac stem cells in culture. J Cell Mol Med. 2015;20(2):370–80.

45. Cretoiu D, Hummel E, Zimmermann H, et al. Human cardiac telocytes: 3D imaging by FIB-SEM tomography. J Cell Mol Med. 2014;18(11):2157–64.
46. Kohl P, Gourdie RG. Fibroblast-myocyte electrotonic coupling: does it occur in native cardiac tissue? J Mol Cell Cardiol. 2014;70:37–46.
47. Wang X, Veruki ML, Bukoreshtliev NV, et al. Animal cells connected by nanotubes can be electrically coupled through interposed gap-junction channels. Proc Natl Acad Sci U S A. 2010;107(40):17194–9.
48. He K, Shi X, Zhang X, et al. Long-distance intercellular connectivity between cardiomyocytes and cardiofibroblasts mediated by membrane nanotubes. Cardiovasc Res. 2011;92(1):39–47.
49. Miragoli M, Gaudesius G, Rohr S. Electrotonic modulation of cardiac impulse conduction by myofibroblasts. Circ Res. 2006;98(6):801–10.
50. Campeanu RA, Radu BM, Cretoiu SM, et al. Near-infrared low-level laser stimulation of telocytes from human myometrium. Lasers Med Sci. 2014;29(6):1867–74.
51. Zhu Y, Mucci A, Huizinga JD. Inwardly rectifying chloride channel activity in intestinal pacemaker cells. Am J Physiol Gastrointest Liver Physiol. 2005;288(4):G809–21.
52. Zhu MH, Sung IK, Zheng H, et al. Muscarinic activation of Ca2+-activated Cl- current in interstitial cells of Cajal. J Physiol. 2011;589(Pt 18):4565–82.
53. Lee H, Koh BH, Peri LE, et al. Functional expression of SK channels in murine detrusor PDGFR+ cells. J Physiol. 2013;591(Pt 2):503–13.
54. Kim YC, Suzuki H, Xu WX, et al. Ca2+ -activated K+ current in freshly isolated c-Kit positive cells in guinea-pig stomach. J Korean Med Sci. 2009;24(3):384–91.
55. Hanna N, Cardin S, Leung TK, et al. Differences in atrial versus ventricular remodeling in dogs with ventricular tachypacing-induced congestive heart failure. Cardiovasc Res. 2004;63(2):236–44.
56. Burstein B, Libby E, Calderone A, et al. Differential behaviors of atrial versus ventricular fibroblasts: a potential role for platelet-derived growth factor in atrial-ventricular remodeling differences. Circulation. 2008;117(13):1630–41.

Winston Shim is trained in molecular biology and has background in human adult mesenchymal stem cells and induced pluripotent stem cells. His group was among the first in the world to demonstrate ex vivo differentiation of human mesenchymal stem cells into cardiomyocytes. He has major interest in cardiac developmental biology and molecular biology of human cardiomyocytes. He has extensive experience in cardiac electrophysiology and regenerative medicine for heart failure. Dr Shim established the human-induced pluripotent stem cell program, cardiac tissue engineering program, and regenerative medicine program and launched platform technologies in cardiac electrophysiology, experimental animal surgery, tissue and plastic histology, cardiac ultrasound, and hemodynamic imaging core capabilities at the National Heart Centre Singapore for regenerative medicine in channelopathy, cardiomyopathy, and heart failure.

Chapter 17
Roles of Telocytes in the Development of Angiogenesis

Yonghua Zheng and Xiangdong Wang

17.1 Introduction

Telocytes were identified to be located on the extracellular matrix of blood vessels [1], e.g. arterioles, venues and capillaries in many organs and tissues, especially in the heart and lung. For example, the number of telocytes significantly increased around the neo-capillaries in the heart of mice with acute myocardial infarction, indicating that telocytes may contribute to the angiogenesis of capillaries [3, 5]. Telocytes have close relation with capillaries in the interstitial of lung tissues, indicating that telocytes might participate in the structure of air-blood barrier [6]. In this chapter, we try to elaborate the biological function of telocytes on the angiogenesis through studies from lung.

17.2 The Discovery of Lung Telocytes

Telocytes were firstly found, confirmed and isolated from mouse trachea and lung tissues through the methods of transmission/scanning electron microscope, immunohistochemistry and primary cell culture by Dr. Zheng et al. in 2011 [10]. Telocytes were found to be allocated in the interstitial space, between smooth muscle fibers, and/or between the cricoid cartilage and smooth muscles within the trachea (Fig. 17.1). Telocytes were observed in the interstitial space of terminal bronchioles,

Y. Zheng (✉)
Department of Respiratory Medicine, Xinhua Hospital, Jiaotong University, Shanghai, China
e-mail: zhengyonghua0118@hotmail.com

X. Wang, MD, PhD
Shanghai Institute of Clinical Bioinformatics, Fudan University Center for Clinical Bioinformatics, Shanghai, China

Shanghai Respiratory Research Institute, Biomedical Research Center, Fudan University Zhongshan Hospital, Shanghai, China
e-mail: Xiangdong.wang@clintransmed.org

X. Wang, D. Cretoiu (eds.), *Telocytes*, Advances in Experimental Medicine and Biology, 913, DOI 10.1007/978-981-10-1061-3_17

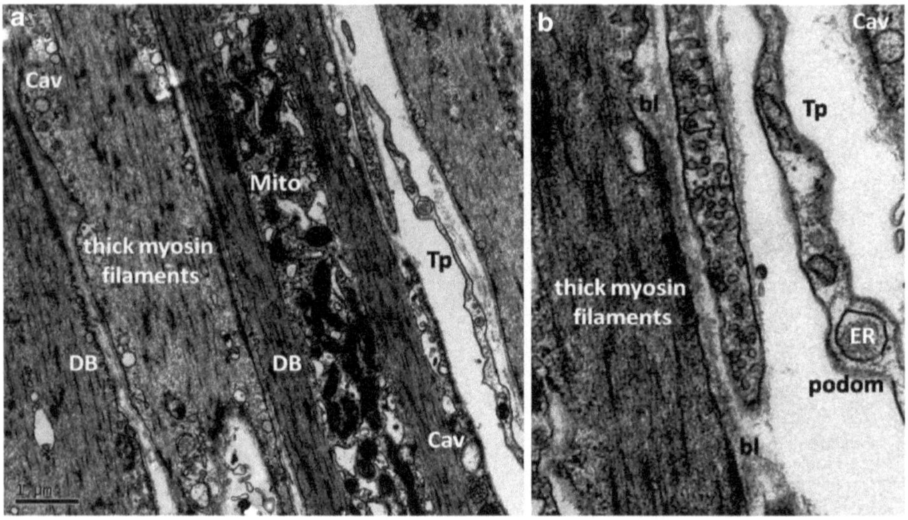

Fig. 17.1 (**a**) Mouse trachea; transmission electron microscopy of smooth muscle layer. Note, four muscle cells longitudinally cut and in between two fragments (one long, one short) of a Tp. It is to be mentioned the presence of thick myosin filaments in smooth muscle cells. *DB* dense bodies, *mito* mitochondria. (**b**) Higher magnification of a fragment from (**a**). Note, a smooth muscle cell with thick myosin filaments and basal lamina (*bl*); a fragment of a telopode, a podom containing element of ER. *cav* caveolae, *bl* basal lamina

of which Tps were connected with alveolar epithelial cells and other cells in the lung tissues, Telocytes were identified in the lung and were found to be related with stem-cell niches (Fig. 17.2), with positive staining of vimentin, c-kit and CD34 (Fig. 17.3).

17.3 Experience of Lung Telocyte Isolation

In order to study the biological function of telocytes, cell culture is the first step. Although several reports about the culture method have been published, there is still no detailed introduction. In this section, we try to elaborate the detailed primary culture method for telocytes according to published papers and our own study. Here we chose lung tissues of BABL/c mice as an example. After BABL/c mice were killed with an overdose of anesthetic, lung fragments were isolated under sterile conditions and collected into sterile tubes containing DMEM (Gibco, NY, USA), supplemented with 100 UI/ml penicillin, 0.1 mg/ml streptomycin (Sigma-Aldrich Shanghai Trading Co Ltd., Shanghai, China), placed on ice and transported to the cell culture laboratory. Samples were processed within 30 min. from surgery, and they were rinsed with sterile DMEM. The lung tissues were minced into fragments

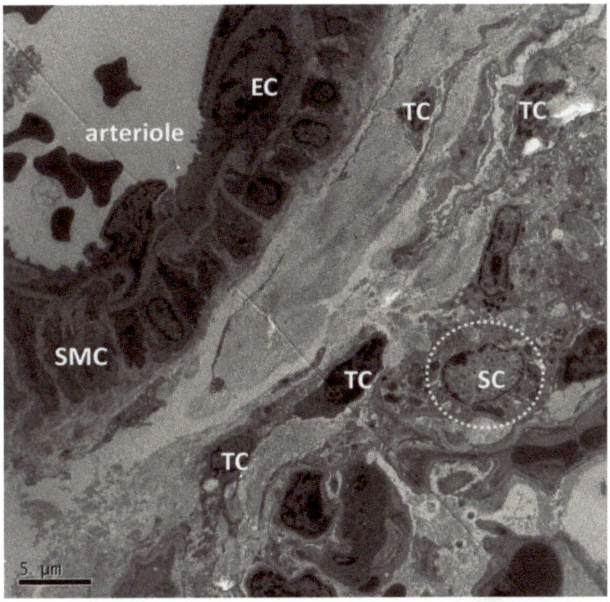

Fig. 17.2 Telocytes (*TCs*) in mice lung. About four TCs with telopodes (Tps) are visible in the interstitial of lung. A putative stem cell (SC, *yellow dotted circle*) is in close contacts with telocytes, establishing heterocellular junctions, visible under electronic microscope only. The tandem TCs-SC forms, presumably, a TC-SC niche. *EC* epithelial cells, *SMC* smooth muscle cells. Magnification is 2000×, scale bar is 5 μm

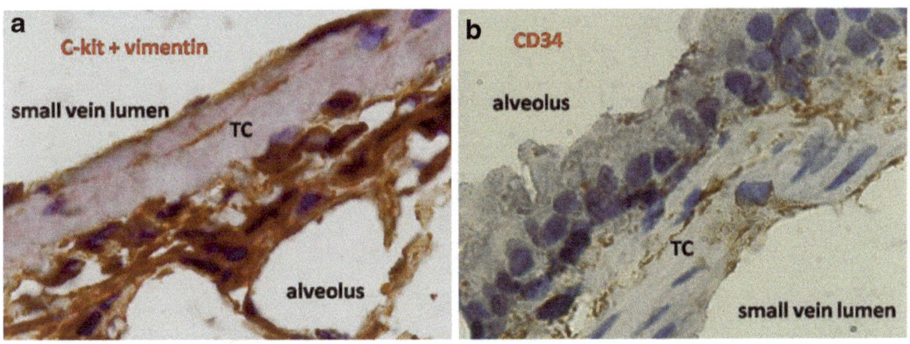

Fig. 17.3 (**a**) Immunohistochemical reaction for c-kit and vimentin in lung tissue. Long TC silhouette is seen positive near small vein lumen. (**b**) Positive reaction for CD34 of TC between airway epithelium and blood vessel endothelium. Magnification 1000×

of about 1 mm^3 and incubated on an orbital shaker for 4 h, at 37 °C, with 1 mg/ml collagenase type II (Gibco, NY, USA) in PBS (without Ca^{2+} and Mg^{2+}). Dispersed cells were separated from nondigested tissue by filtration through a 40 μm diameter cell strainer (BD Falcon, NJ, USA), collected by centrifugation at 2000 r.p.m., 5 min., re-suspended and cultured in a 75 cm^2 plastic culture flask containing DMEM, supplemented with 10 % foetal calf serum, 100 UI/ml penicillin and 0.1 mg/ml streptomycin (Sigma-Aldrich) for 2 h, and then the supernatant containing slowly adhering cells was collected and re-plated into a new 25 cm^2 plastic culture flask, at a density of 1×10^5 cells/cm^2, and maintained at 37 °C, in a humidified atmosphere (5 % CO_2 in air) until becoming semi-confluent (usually 4 days after plating), and the adherent cells were telocytes. Culture medium was changed every 48 h, and the mixed suspension cells were deleted through the changing of the medium. Cell viability was assessed by using the Trypan blue dye exclusion test and cells were examined by phase contrast microscope [2].

17.4 Morphology of Cultured Lung Telocytes

Primarily cultured telocytes could be identified clearly from 3rd to 4th days. Telocytes appear typical Tps with long and uneven caliber, podoms and/or podomers (Fig. 17.4a). Telocytes and Tps with very long and/or uneven caliber, or dilated portions resembling "beads on a string" were observed under the staining with methylene blue (Fig. 17.4b) and giemsa (Fig. 17.4c). Janus green B is a classic vital staining with high affinity for mitochondria and used to assess viability and localize mitochondria. Figure 17.4d illustrates the TC body and Tp with

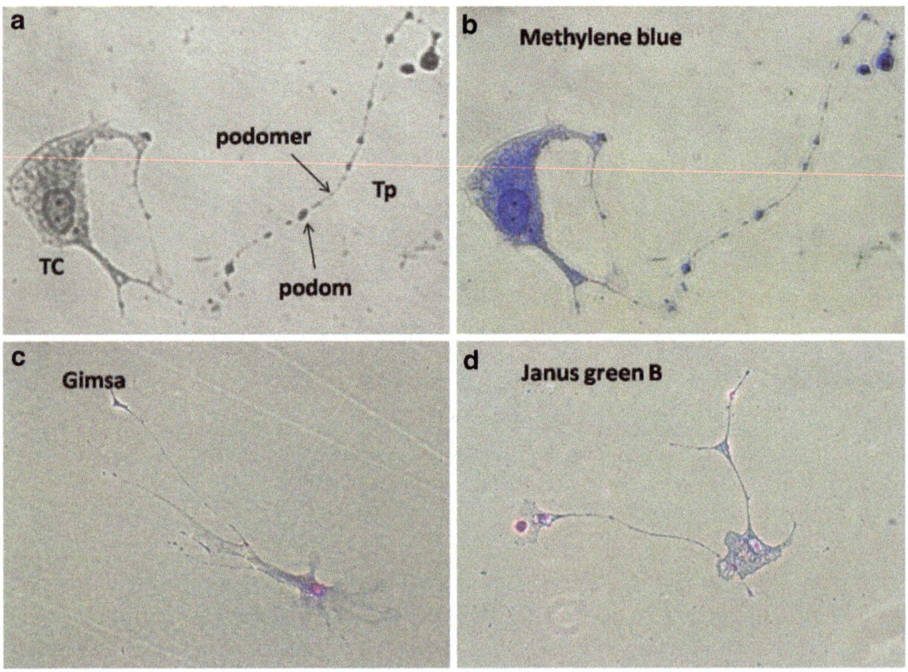

Fig. 17.4 Mouse lung telocytes (*TC*) in primary culture observed under phase-contrast microscopy. (**a**) TC is depicted with long and slender telopodes (*Tps*), which present the alternation of thick segments – podoms (*black arrows*) and thin segments – podomers (*black arrows*). Dichotomic branching of Tps is obvious. Magnification is 400×. (**b**) Methylene blue staining. Magnification is 400×. (**c**) Giemsa staining. Magnification is 100×. (**d**) Janus Green B staining Along Tp of another TC over-passed the cell body of the other Telocytes. Magnification is 100×

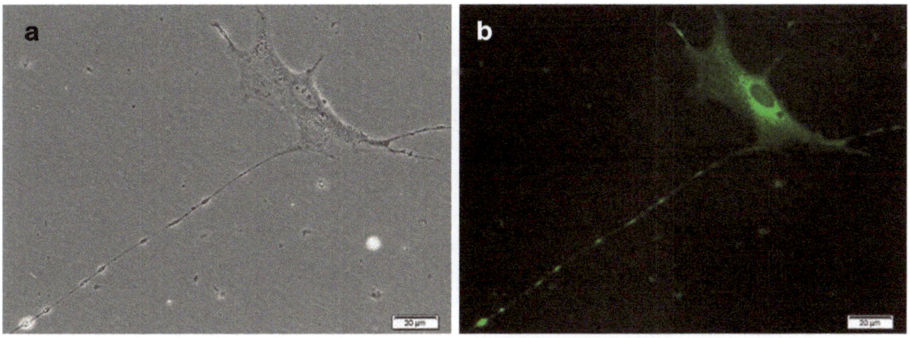

Fig. 17.5 Primary culture of human lung telocytes. (**a**) Telocytes are observed under phase-contrast microscopy. (**b**) Telocytes are observed under fluorescence microscopy. Telocytes were stained with Mito Tracker Green. The mitochondria are allocated mainly within Telocytes body and its podoms. The magnification was 400 ×, and the scale bar is 20 μm

podoms highly stained with Janus green B, indicating the rich presence of mitochondria in cell body and particularly in podoms. Mito Tracker Green FM is a molecular probe with high affinity for mitochondrial membranes and used to identify mitochondria in living cells. The strong staining of mitochondria in most Telocytes and Tps with Mito Tracker Green FM was observed (Fig. 17.5), demonstrating the accumulation of mitochondria in Telocytes and podoms. Those findings imply that the maintenance of Telocytes biological functions requires the large amount of energies, while Tps and podoms provide the energies for the extensions.

17.5 Proteomics Between Telocytes and Endothelial Cells

According to the study from Zheng et al, the proteomic analysis between telocytes and fibroblasts was studied by the method of iTRAQ and the function of protein expression profiles were analyzed with the aid of PANTHER Classification System.

By comparison between telocytes and microvascular endothelial cells, there are 38 proteins up-regulated in telocytes, especially Myosin-14, superoxide dismutase (SOD2), acid ceramidase (AC), envoplakin and epiplakin. Among these proteins, 18 of them are responsible for metabolic processes and 15 proteins in cellular processes, such as cell communication (4 proteins), cytokinesis (2 proteins), cellular component movement (2 proteins), and cell cycle (2 proteins). There are 60 proteins down-regulated in telocytes, especially cell surface glycoprotein MUC18, Ras-interacting protein 1, BTB/POZ domain-containing protein, peptidyl prolyl cis/trans isomerase and nestin and von Willebrand factor. The highly expressed proteins in telocytes are involved in important molecular functions such as: catalytic activity (17 proteins), structural molecule activity (13 proteins) compared to microvascular endothelial cells where significantly more proteins are involved in catalytic activity (30 proteins) and 29 proteins have molecular binding function. Ten up-regulated proteins in telocytes are involved in developmental processes: anatomical structure morphogenesis (10 proteins), mesoderm development (3 proteins), system development (2 proteins) and ectoderm development (1 protein). Moreover, 10 proteins in telocytes are related to localization processes such as vesicle mediated transport (4 proteins), protein transport (4 proteins) and ion transport (3 proteins). While in microvascular endothelial cells, two up regulated proteins are involved in nucleic acid-binding transcription and one has anti-oxidant activity.

The up-regulated telocytes proteins belong to the following pathways: nicotinic acetylcholine receptor (2 proteins), inflammation mediated by chemokines (2 proteins), de novo purine biosynthesis (2 proteins), cytoskeletal regulation by Rho GTPase (2 proteins), TCA cycle (1 protein), Parkinson disease (1 protein), integrin signaling (1 protein) and blood coagulation (1 protein) . In telocytes, the up-regulated proteins are related to the following cellular components: cell part (13 proteins), organelle (12 proteins), membrane (2 proteins), cell junction (2 proteins), extracellular region (1 protein) and extracellular matrix (1 protein).

The up-regulated proteins in microvascular endothelial cells are attributed to the following protein classes: enzyme modulator (11 proteins), cytoskeletal proteins (10 proteins), oxidoreductase (7 proteins), nucleic acid binding (6 proteins), transferase, isomerase and chaperone (5 proteins each), etc. The pathways map depicted the microvascular endothelial cells proteins are related to: integrin signalling pathway (3 proteins), Huntington disease (3 proteins), cytoskeletal regulation by Rho GTPase (3 proteins), pentose phosphate pathway (2 proteins), Parkinson disease (2 proteins), inflammation mediated by chemokines (2 proteins), glycolysis (2 proteins), etc. The cellular component of microvascular endothelial cells proteome demonstrated proteins related to: cell part (12 proteins), organelle (10 proteins), extracellular region (1 protein) and extracellular matrix (1 protein) [9]. Telocytes are completely different from microvascular endothelial cells. Protein expression profile showed that telocytes play specific roles in intercellular communication and intercellular signaling. Moreover, they might inhibit the oxidative stress and cellular ageing and may have pro-proliferative effects through the inhibition of apoptosis.

17.6 Telocytes and Angiogenesis

The angiogenesis is a common phenomenon both in physiological and pathological conditions, including proliferation of vascular endothelial cells, enzymatic degradation of basement membrane and interstitial matrices by endothelial cells, migration of vascular endothelial cells, or eventually formation of a blood vessel tube from sprouting vascular endothelial cells. Angiogenesis could be induced by activation of VEGF and EGF receptors by the binding of VEGF and EGF to induce the tube formation in the vascular endothelial cells, promote endothelial cell proliferation and vascular permeability, and maintain newly-formed blood vessels [4].

In the study of lung telocytes, Zheng et al. confirmed that lung telocytes were mainly distributed in the alveolar interstitial connected tightly with alveolar epithelia cells and participated in the structure of air-blood barrier, in the small vein and bronchioles and in the interstitial space of smooth muscle participated in the frame structure of the blood and bronchioles [7]. Telocytes are positive to CD34 and C-kit which expressed on the surface of hemopoietic stem cells, and are proposed to participate in the angiogenesis.

Distribution and structure characteristics of telocytes must fit to its biological functions. Zheng et al. also confirmed that production of VEGF and EGF from human lung telocytes increased significantly. Cultured medium of telocytes could promote the proliferation of HPMECs, and partially recover the ability of tube formation of HPMECs injured by LPS. Indicating that telocytes may participate in the angiogenesis of lung tissues, in both normal and pathological conditions and play roles as progenitor cells and nutrient cells during the regeneration and reparation of the injured tissues. Telocytes can be a new therapeutic target in lung diseases.

17.7 Perspectives

The angiogenesis is a common phenomenon both in physiological and pathological conditions, including proliferation of vascular endothelial cells, enzymatic degradation of basement membrane and interstitial matrices by endothelial cells, migration of vascular endothelial cells, or eventually formation of a blood vessel tube from sprouting vascular endothelial cells. Angiogenesis could be induced by activation of VEGF and EGF receptors by the binding of VEGF and EGF to induce the tube formation in the vascular endothelial cells, promote endothelial cell proliferation and vascular permeability, and maintain newly-formed blood vessels. Telocytes are a new type of interstitial cell with special biological functions, for the close relation with stem cells, they are also named as "stem cell helper cells". Telocyes play roles as progenitor cells and nutrient cells during the regeneration and reparation of the injured tissues, including the angiogenesis in both normal and pathological conditions through the secretion of VEGF and EGF [8]. Further studies are necessary for the application of cellular therapy with telocytes in clinic, such as lung injury, myocardial infarction, or tissue engineering.

References

1. Cretoiu SM, Popescu LM. Telocytes revisited. Biomol Concepts. 2014;5(5):353–69.
2. Hatta K, Huang ML, Weisel RD, Li RK. Culture of rat endometrial telocytes. J Cell Mol Med. 2012;16(7):1392–6.
3. Manole CG, Cismasiu V, Gherghiceanu M, Popescu LM. Experimental acute myocardial infarction: telocytes involvement in neo-angiogenesis. J Cell Mol Med. 2011;15(11):2284–96.
4. Reynolds LP, Borowicz PP, Caton JS, et al. Uteroplacental vascular development and placental function: an update. Int J Dev Biol. 2010;54(2-3):355–66.
5. Zhao B, Chen S, Liu J, et al. Cardiac telocytes were decreased during myocardial infarction and their therapeutic effects for ischaemic heart in rat. J Cell Mol Med. 2013;17(1):123–33.
6. Zheng Y, Bai C, Wang X. Telocyte morphologies and potential roles in diseases. J Cell Physiol. 2012;227(6):2311–7.
7. Zheng Y, Bai C, Wang X. Potential significance of telocytes in the pathogenesis of lung diseases. Expert Rev Respir Med. 2012;6(1):45–9.
8. Zheng Y, Chen X, Qian M, et al. Human lung telocytes could promote the proliferation and angiogenesis of human pulmonary microvascular endothelial cells in vitro. Mol Cell Ther. 2014;2:3.
9. Zheng Y, Cretoiu D, Yan G, et al. Protein profiling of human lung telocytes and microvascular endothelial cells using iTRAQ quantitative proteomics. J Cell Mol Med. 2014;18(6):1035–59.
10. Zheng Y, Li H, Manole CG, Sun A, Ge J, Wang X. Telocytes in trachea and lungs. J Cell Mol Med. 2011;15(10):2262–8.

Yonghua Zheng is a doctor in department of respiratory medicine, Xin Hua Hospital, Shanghai Jiao Tong University School of Medicine (Shanghai, China). He earned Ph.D. degree in Fudan University in 2013. Dr. Zheng's research interests are focused on pulmonary diseases such as acute lung injury, and biological function of lung telocytes such as its existence, morphology, proteomics, genomics, role in the tissue repair, and he has published many papers about telocytes in JCMM, JCP et al.

Xiangdong Wang is a Distinguished Professor of Medicine at Fudan University, Director of Shanghai Institute of Clinical Bioinformatics, Director of Fudan University Center of Clinical Bioinformatics, Deputy Director of Shanghai Respiratory Research Institute, Adjunct Professor of Clinical Bioinformatics at Lund University, and Visiting Professor of King's College of London. He serves as an Executive Dean of Institute of Clinical Science, Director of Biomedical Research Center, Fudan University Zhongshan Hospital. His main research is focused on clinical bioinformatics, disease-specific biomarkers, lung chronic diseases, cancer immunology, and molecular & cellular therapies. He is the author of more than 200 scientific publications with the impact factor about 600, citation number about 6642, h-index 49, i10-index 181, and cited journal impact factor about 5000.

Chapter 18
Telocytes in Inflammatory Gynaecologic Diseases and Infertility

Xiao-Jun Yang

18.1 Introduction

Infertility is a common disease in women of reproductive age. It can be divided into anatomical and functional disease. Anatomically, oviduct and uterine diseases, such as severe endometriosis, tubal ectopic pregnancy, para-tubal adhesion after pelvic inflammatory diseases, uterine septum, intrauterine adhesion, etc., frequently cause infertility. On the other hand, among the functional reasons, immune-mediated fertility problems and related diseases, such as autoimmune diseases, early stage of endometriosis, pelvic inflammatory disease and salpingitis, were the predominant diseases in the clinic. Generally, the female reproductive tract must maintain a unique immune micro-environment, in order to tolerate the semi-allogeneic sperm and foetus and protect against harmful pathogens [1, 2]. Further researches showed that immunocytes such as monocyte and macrophage, which were important multifunctional players in local peritoneal immune response in endometriosis or pelvic inflammatory disease, and their dysfunction or uncontrolled augmentations in quantity and/or activation might not only change smooth muscle motility, microcirculation and pelvic pain in endometriosis [3] but also lead to immune-mediated fertility problems, such as miscarriage, tubal infertility and tubal ectopic pregnancy [4, 5].

Endometriosis (EMs) is a kind of aseptic inflammatory, ischemic, oestrogen-dependent disease with many clinical manifestations [6, 7]. It was characterized by the presence of endometrium outside the uterine cavity and affects an estimated 8–10% of women of reproductive age in industrialized countries [6, 8]. Generally,

X.-J. Yang, PhD
Department of Obstetrics and Gynecology,
The First Affiliated Hospital of Soochow University,
188 Shizi Road, Suzhou City, Jiangsu Province 215006,
People's Republic of China
e-mail: yang.xiaojun@hotmail.com

X. Wang, D. Cretoiu (eds.), *Telocytes*, Advances in Experimental Medicine and Biology, 913, DOI 10.1007/978-981-10-1061-3_18

massive monocyte/macrophage and lymphocyte aggregation, which overproduce cytotoxic substance, such as inflammatory factors (iNOS, COX-2), oxidative stress (LPO) and estradiol, was the most frequent pelvic micro-environment changes in EMs [7–9]. Then, EMs causes many abnormalities either in local anatomy or reproductive endocrine and immunologic micro-environment disorders, subsequently leading to various clinical symptoms, such as chronic pelvic pain, hypermenorrhoea, dysmenorrhoea and sub- or infertility [8].

Current knowledge regarding EMs-related fertility disorders remains insufficient. Generally, in women with obvious macroscopic anatomical changes of the pelvis, mechanical interference will be adversely affected by oocyte pickup, transport and tubal peristalsis. However, women with absence of macroscopic pelvic alterations also complained of unexplained functional sub- or infertility [8], of which, many steps of the fertilization process were believed to be disturbed by inflammatory peritoneal fluid which were thought to exert direct cytotoxic effects or release various cytokines/enzymes into the pelvic milieu, such as ovulatory dysfunction, altered folliculogenesis, sperm phagocytosis, luteal-phase defects, impaired fertilization, defective implantation and early embryo growth [10]. Interestingly, the phenomenon of impaired utero-tubal sperm transport in early-stage EMs and adenomyosis, strongly suggested that "tubal motility disorder" might be a potential cause for EMs-associated tubal factor sub- or infertility [11]. We suggested that tubal dysperistalsis might develop from a subset of clinically affected known or unknown oviduct cells that were functionally dysregulated, probably due to aforementioned toxic pelvic micro-environment.

Acute pelvic inflammatory disease (PID) is a serious, infectious disease of the upper female genital tract, which frequently causes severe pelvic damage and leads to fertility problems among women of reproductive age, such as tubal ectopic pregnancy and tubal factor infertility [12]. Generally, PID was caused by ascending polymicrobial infection, particularly sexually transmitted organisms, such as *Chlamydia trachomatis*, mycoplasma and *Neisseria gonorrhoeae* or vaginal dysbacteriosis. Clinically, acute PID can be divided into tubo-ovarian abscess, pelvic peritonitis, acute endometritis and acute salpingitis (AS) which was the most important component of PID spectrum. AS can cause severe anatomical damage of a subset of oviduct cells, manifested by oviduct acute edema, pyosalpinx, obstruction of the lumen, peritubal adhesion, interstitial fibrosis and rigidity of the wall [12]; frequently induced oviduct dysfunction, such as chronic tubal spasm, immunologically mediated mechanism in *Chlamydia trachomatis* infection. All of these changes were suggested as the underlying mechanisms for AS-induced female fertility problems. In our opinion, the clinical sequelae of fertility problems were developed from a subset of AS-affected oviduct cells that were structurally and functionally dysregulated, including the widely known ciliated cells and various types of interstitial cells such as smooth muscle cells, fibroblasts, etc. All of these cell components play definite roles and were indispensable in a successful reproductive process. Nevertheless, besides the well-known impair of classically described oviduct cells, other damages that simultaneously occurred within the oviduct wall, which also structurally and functionally affect oviduct fertility capacity, are worthy of further investigation.

With recent progress in electron microscope, telocytes (TCs), previously known as interstitial Cajal-like cells (ICLC), were identified as a distinct interstitial cell component by Popescu et al., in interstitial space of a wide variety of cavitary and non-cavitary human and mammalian organs, including non-pregnant and pregnant myometrium, endometrium, oviduct, placenta, etc. (www.telocytes.com). TCs have particular ultrastructure which can be clearly distinguished from classical interstitial cell of Cajal (ICC), fibroblasts, etc. Morphologically, TCs have a small piriform-/spindle-/triangular-shaped cell body, with two to five extremely long and slim telopodes (Tps), which contained thin segments (podomers) and dilated segments (podoms). By their Tps, TCs provide visible "short-distance" direct structural support for homocellular or heterocellular junctions between themselves and other type of cells. And based on their distribution, identification, immunophenotype and ultrastructure features in different normal organs/tissues, TCs were supposed to play an essential role in the maintenance of structural and functional integrity, by intercellular signalling, spreading slow waves generated by the pacemaker ICC, involved in local tissue homeostasis/angiogenesis, nurse stem cells (SCs) and mediated tissue repair/remodelling, immunoregulation or immunosurveillance, etc. On the other hand, increasing number of reports described "long-distance" indirect intercellular contacts for TCs, such as through chemical [13–15], paracrine/juxtacrine signalling [13, 14, 16–19], extracellular vesicles (EVs) [14, 16–20] [15, 21–30], sex hormone [14, 16, 31–33] and/or microRNAs [23, 34–37]. These paracrine effects were believed to play important roles in function-specific intercellular communication and regulate activity of "long-distance" neighbouring cells.

Recently, disease-induced TC damage was reported in fibrotic lesions of skin, cardiac, ulcerative colitis, Crohn's disease and gallstone disease, and multiple potential pathophysiological roles have been speculated for TCs [37–42]. Nevertheless, as a new type of interstitial cell, the exact alterations of TC population in disease-affected oviduct tissue, such as in aseptic inflammatory disease of EMs, and infectious AS; and potential involvement in structural and functional abnormalities of oviduct, further engagement in oviduct fertility capacity or adverse reproductive outcome (sub- or infertility), need more detailed evidence.

On the other hand, increasing studies showed that both in normal and disease-affected tissues, TCs developed heterocellular junctions with various immunocytes and potentially modulated their activities, such as macrophages, mast cells, lymphocytes, etc. [1, 26–28, 38–41, 43–53]. In particular, in inflammatory-affected rat oviduct tissue [45, 46], TCs connected to the activated immunocytes, including mononuclear cells, mast cell, eosinophils and neutrophils. All of these studies supposed that TCs might act as active players or "data suppliers" and involved in local immunoregulation/immunosurveillance [26, 28, 50, 51]. Progressive damage and loss of TCs might impair intercellular communication or immune homeostasis in immunoinflammatory process, such as multiple autoimmune, chronic inflammatory and fibrotic disorders [26, 38–41, 44–46, 51]. We speculated that through direct junctional complexes or possibly indirect paracrine messages, TCs might potentially involve in immunological signal presenting and/or transduction, influence and contribute to subsequent immunoresponse and then change normal physiological process and lead to immune-mediated gynaecologic diseases or reproductive abnormalities. Nevertheless, currently, no reliable cytological evidences are available for this hypothesis.

Regarding the aforementioned questions, we have designed researches and provided related evidence in the following three aspects. We believed that such knowledge will be helpful to elucidate pathophysiological role of TCs in female genital tract disease and fertility problems, with the aim of providing a potential target for genetic, pharmaceutical and clinical interventions.

18.2 In Vivo TC Damage, Molecular Mechanisms and Potential Impact in Fertility in EMs-Affected Rat Oviduct Tissue

Women with EMs frequently complained with unexplained fertility problems. The recently identified TCs were found to participate in the maintenance of structural and functional integrity of female oviduct, but so far whether TCs were involved in EMs-affected oviduct tissue and potentially influence female fertility capacity remains to be elucidated.

TCs were studied in an oviduct EMs model and in sham control Sprague-Dawley rat, respectively, together with determination of inducible nitric oxide synthase (iNOS), COX-2, lipid peroxide (LPO) and estradiol. Briefly, the autotransplantation rat model of oviduct EMs was surgically constructed by removing uterine horn and transplanted towards both surfaces of contralateral mesosalpinx, with endometrial side adjacent to the arteries that irrigate the oviduct [54, 55]. Rats received the same surgery with removal of the uterine horn and blank sutures, without any tissue masses, serving as the sham control. Then, EMs-affected oviduct segment with grade III ectopic endometriotic vesicles (larger than 4 mm) was harvested at 2 months [56]. Proceed for an integrated technique of haematoxylin and eosin staining, in situ immunohistochemistry (IHC), double-labelled immunofluorescence IHC staining and transmission electron microscopy (TEM) observation.

After confirming by routine pathologic observation for disease-affected and disease-unaffected oviduct tissues (Fig.18.1), in situ IHC was performed on consecutive sections from sham oviduct, and presumably TCs with special morphology and immunophenotype were observed: perivascular stellate-shaped cells with prolonged cell body and double-positive expression for CD34/vimentin (Fig. 18.2a, b) and negative for c-kit (Fig.18.2c).

Double-labelled fluorescent IHC confirmed the existence of typical TCs in sham group (Fig.18.3a), with characteristic appearance: one or more extremely long/thin cellular prolongations located around perivascular space, with specific immunophenotype of CD34 positive/vimentin positive/c-kit negative, consistent with in situ IHC (Fig.18.2), while in EMs-affected oviduct tissue, cell populations with typical TC morphology and immunophenotype significantly decreased and were sparse or even completely undetectable ($P=0.000$; Fig. 18.3b, c).

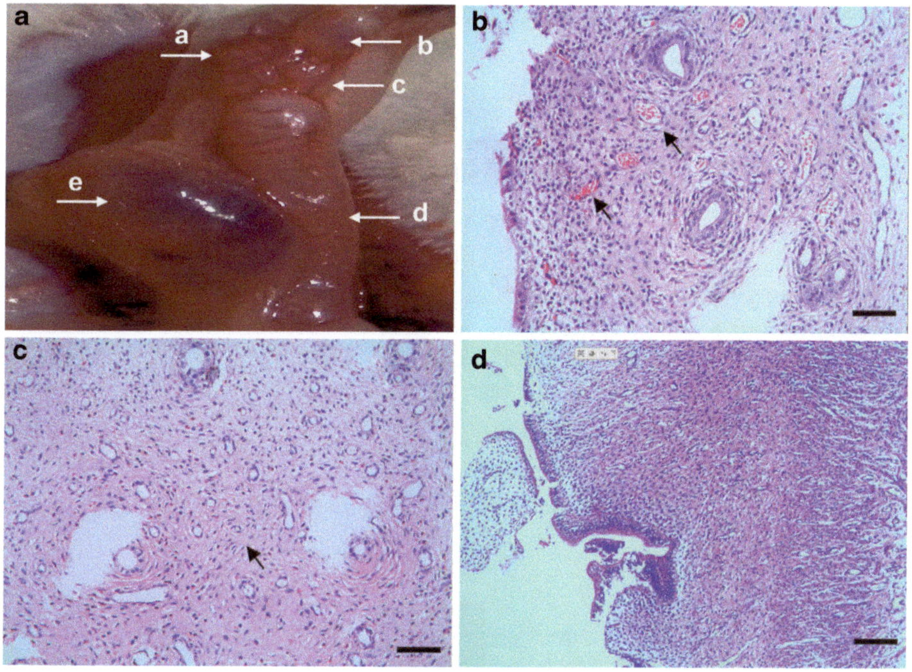

Fig. 18.1 Macroscopical and microscopical view of EMs-affected rat oviduct, scale bar = 100 μm. (**a**) Ectopic endometrial vesicles (>4 mm in diameter), located in mesosalpinx under naked eyes in study group. (**a**) ovary; (**b**) ectopic vesicle; (c) oviduct; (**d**) uterine horn; (**e**) oviduct swelling and edema after removal of the opposite uterine horn. (**b**) Inflammatory tissue reaction in EMs-affected oviduct, manifested by hyperplasia and disturbance of capillaries (*black arrows*). (**c**) Chronic inflammation and interstitial fibrosis, manifested by hyperplasia capillaries with excessive interstitial lymphocyte infiltration and fibre contents (*black arrow*) in EMs-affected oviduct wall. (**d**) Nearly normal oviduct tissue from the sham control (Yang et al. [45] adapted with permission)

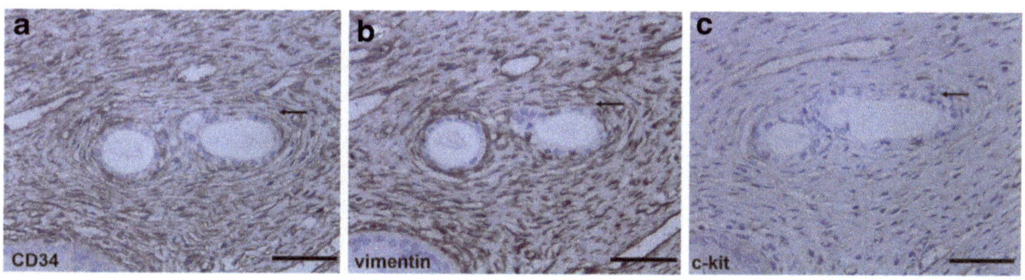

Fig. 18.2 Single-labelled in situ IHC on serial sections from the sham control (*black arrow*); scale bar = 200 μm. (**a**) Perivascular TC-like CD34 (+) cells with slender cell body and prolongations. (**b**) TC-like CD34 (+) vimentin (+) cells. (**c**) c-Kit (−) in corresponding site of serial slides (Yang et al. [45] adapted with permission)

Meanwhile, quantitative analysis by using single-labelled fluorescent IHC suggested that in EMs-affected oviduct tissue, the contents of iNOS, COX-2, LPO and estradiol were elevated significantly than in the sham group, respectively (all $P<0.01$; Fig. 18.4a–d), thus suggestive of intra-tubal inflammation and ischaemia state.

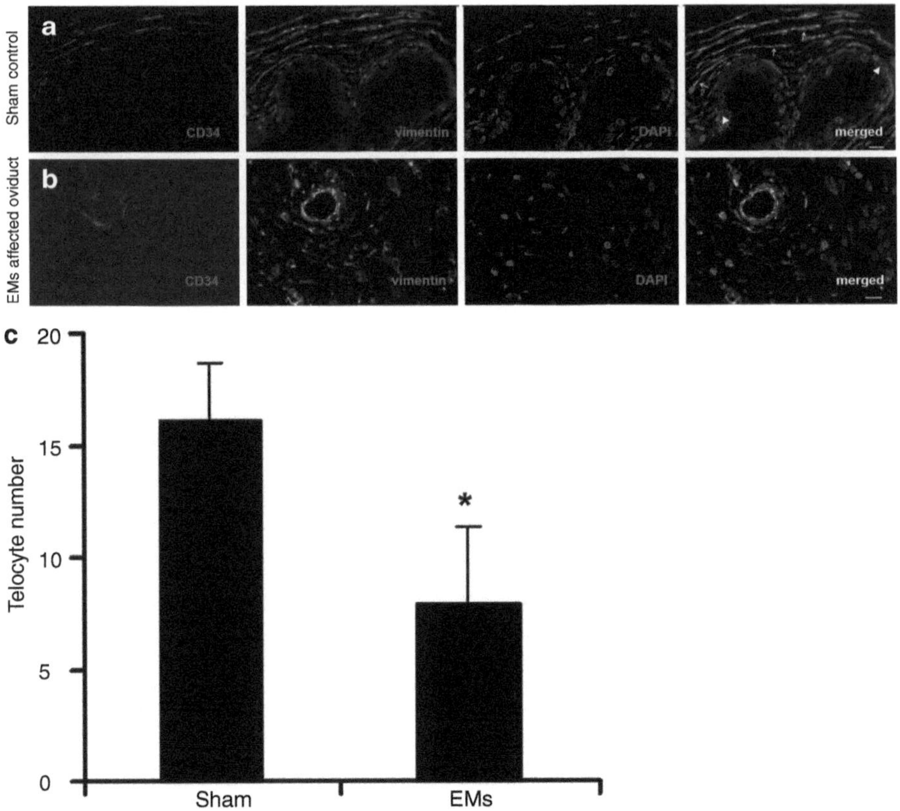

Fig. 18.3 Double-labelled immunofluorescence IHC confirmed TCs with typical morphology and immunophenotype of CD34 (+)/vimentin (+); c-kit-negative images not shown; scale bar = 20 μm. (**a**) In sham control, CD34 (*red*) overlying vimentin (*green*) both in cellular body and telopodes (Tps) (solid *arrows*) around perivascular space (CD34-positive capillary cells, indicated by *dotted arrows*). (**b, c**) In EMs-affected oviduct, TCs with intact structure were scarce or even undetectable, decreased significantly in number. $*P<0.05$. Error bars = SD (Yang et al. [45] adapted with permission)

Ultrastructure observation in sham oviduct confirmed that normal TCs display typical ultrastructure features: a slender piriform/spindle/triangular cell body, with one or more extremely long, thin, very sinuous telopodes (Tps), alteration with thin (podomers) and thick segments (podoms), stretched from cell bodies to different directions. In addition, a rich amount of organelles, such as mitochondria, rough endoplasmic reticulum, cytoskeletal elements, caveolae and microvesicles, is located within podoms (Fig. 18.5). TCs are distributed around the perivascular space; two or three layers of TCs formed almost a complete sheath with Tps and surrounded the vascular endothelial cells, with homocellular or heterocellular junctions among Tps or fibrocytes and pericytes (Fig. 18.5a). Moreover, TCs also scattered among smooth muscle cells (SMC), with heterocellular junctions between them, and microvesicles are contained in Tps and synaptic cleft (Fig. 18.5b).

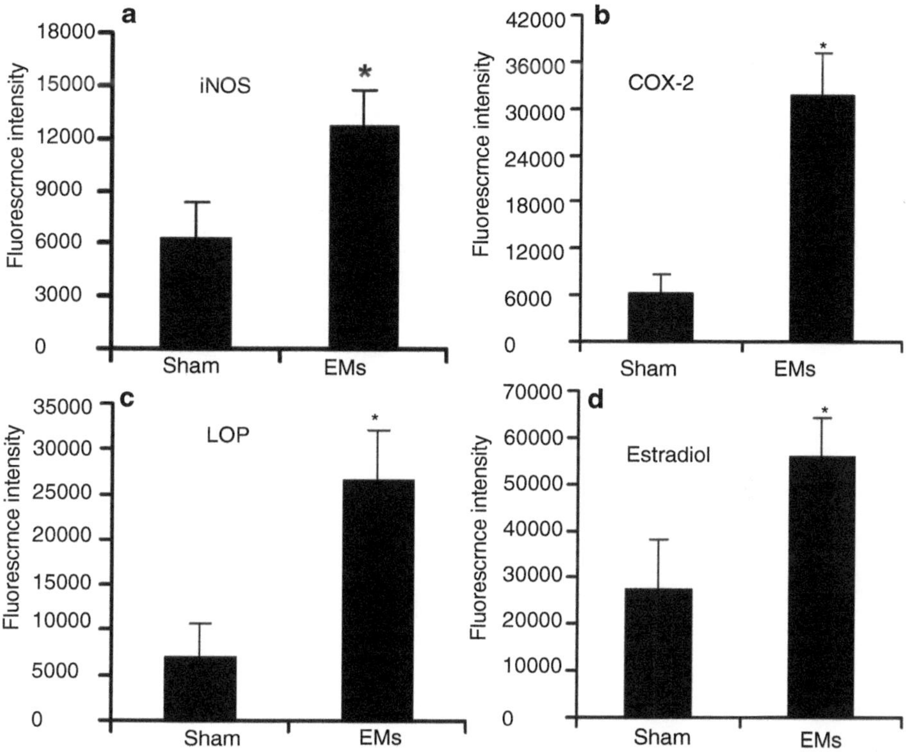

Fig. 18.4 iNOS (**a**), COX-2 (**b**), LPO (**c**) and estradiol (**d**) increased significantly in EMs-affected oviduct tissue, as compared to sham control. $*P<0.05$. Error bars = SD (Yang et al. [45] adapted with permission)

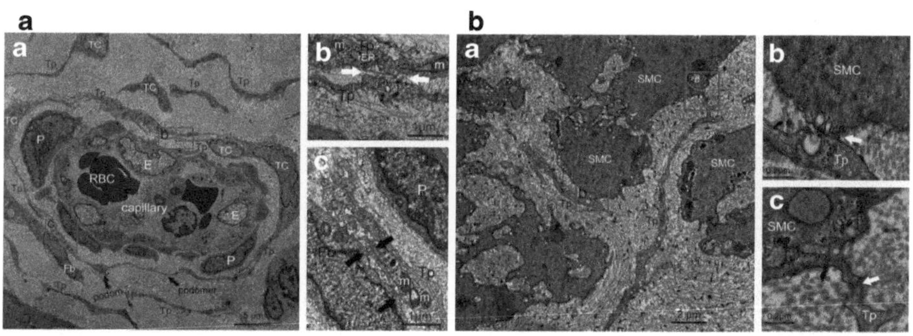

Fig. 18.5 Normal TCs. (**a**) Perivascular TCs. (**a**) By Tps (podom and podomer, *black arrows*), TCs completely enwrap vascular endothelial cells (E) and pericyte (P). Mitochondria (M), rough endoplasmic reticulum (rER). (**b** and **c**) Higher magnifications of the *blue boxed* areas within “a”: homocellular and heterocellular junctions between TCs (*white arrows* in “b”) and fibrocyte (Fb; *black arrows* in “c”). (**b**) TCs among smooth muscle cells (SMCs). (**a**) Heterocellular junctions between Tp and SMC. (**b** and **c**) Higher magnifications of the blue boxed areas within “a”: extracellular vesicles in synaptic cleft (*white arrows*) (Yang et al. [45] adapted with permission)

However, in EMs-affected oviduct tissue, which is confirmed by typical ectopic endometriotic glands (Fig. 18.6a), extensive ultrastructural damage or complete loss of TCs was observed (Fig. 18.6b–e), such as loss of organelles, swollen cell nucleus and mitochondria, cytoplasmic vacuolization, endoplasmic reticulum dilatation and swollen

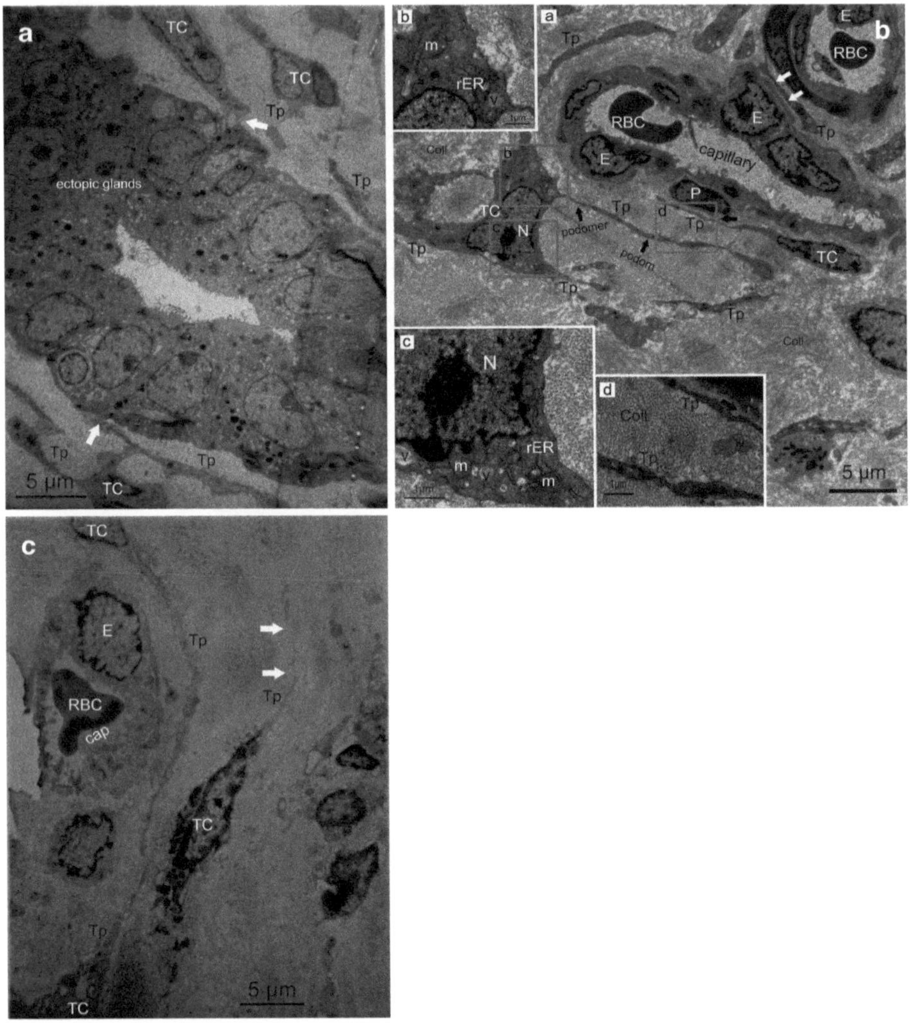

Fig. 18.6 TC damage in EMs-affected oviduct tissue. (**a**) Ectopic endometriotic glands with dense secretory granules, with close contact to Tps (*white arrows*). (**b**) Perivascular TC damage and tissue fibrosis, with nearly normal endothelial cells (*E*). (**a**) Disintegration of TC network, with swollen cell junctions (*white arrows*). (**b–d**) Higher magnifications of the *blue boxed* areas; (**b** and **c**) cellular organelles damage, cell nucleus (*N*) and mitochondria (*m*) swollen, dilatation of rough endoplasmic reticulum (rER), cytoplasmic vacuolization (*v*). (**d**) Excessive amount of collagen fibres (Coll) embedded by damaged Tps. (**c**) Perivascular TC damage (*white arrows*), with endothelium (*E*) damage. (**d**) Heterocellular junctions between slightly damaged TCs and mast cells (MC; *white arrows*), together with tissue fibrosis (collagen fibril, Coll). (**e**) Damaged TCs nursed a group of putative normal stem cells (SC) to develop a possible SC niche, with heterocellular junctions between them (*white arrows*) (Yang et al. [45] adapted with permission)

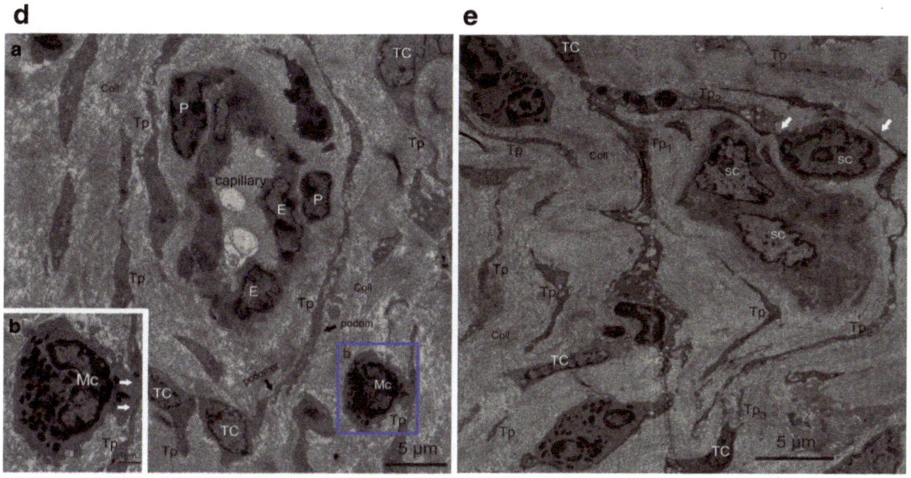

Fig. 18.6 (continued)

cell junctions. Interestingly, TC damage with normal endothelial cells and SC, nearly normal TCs with severely damaged microvessels, abundant collagen fibres and tissue fibrosis can also be observed (Fig. 18.6b, d, e). In addition, TCs developed heterocellular synapse to mast cells (MCs) (Fig. 18.6d) and stem cell (SC) (Fig. 18.6e), suggesting potential participation in local immunoregulation and stem cell-mediated tissue repair.

This successfully constructed rat model displays ectopic endometriotic vesicles macroscopically and inflammation microscopically, with overproduced iNOS, COX-2, LPO and estradiol in oviduct tissue. This kind of rat model resembled clinic physiopathology of EMs and thus suitable for TCs and infertility research.

Classically, oviduct stromal cells are mainly composed of myocyte and immunocytes such as dendritic cells, macrophages, mast cells, plasma cells, eosinophils and lymphocytes, plus interstitial cells of Cajal (ICC) as pacemaker for smooth muscles [57]and fibroblasts as effector cells for tissue fibrosis [40]. Each of them was essential in reproductive process. However, we reported for the first time that the newly discovered TCs participate in the maintenance of structural and functional integrity of normal oviduct tissue. And in EMs-affected oviduct, we found inflammation and reactive oxygen species-induced extensive TC damage and demonstrated a broad involvement of TCs with tissue fibrosis in EMs-affected oviduct wall and parallel trends of TC damage with the severity of EMs.

Although the exact roles of TCs are supposed merely based on their distribution and intercellular connections, oviduct TC damage might cause (i) dysregulation of intercellular signalling, such as impaired immunoregulation/immunosurveillance, attenuated intercellular signalling and oviduct contractility; (ii) impaired stem cell-mediated tissue repair/regeneration and remodelling of interstitial fibrosis; and (iii) derangement of 3-D interstitial architecture, which were structural support for intercellular signalling, tissue repair/remodelling and homeostasis. All in all, the newly identified TCs provide a new explanation for structural and functional oviduct disease.

In the future, animal models with TC network depletion will be more helpful to clarify what exactly happened on the pathway, by which TCs mediate cell interactions with other structural components of oviduct, and unveil the real functional consequences of TC damage on reproductive process. In addition, comparative high-throughput analysis for TCs between disease-affected and disease-unaffected tissues would provide more insights into potential roles of TCs in oviduct pathophysiology. On the other hand, whether uterine TCs simultaneously underwent cell damage and cause uterine dysfunction may be another critical topic. Interestingly, as accompanied by normal endothelial and SC, the mechanism of why TCs were seemingly less tolerant in the disturbed pelvic milieu needs further elucidation. Finally, whether TC transportation (together with or without stem cells), rebuilding TC network, can promote regeneration and reparation of disease-affected tubal damage, oviduct fibrosis and fertility disorders is worthy of future research.

18.3 In Vivo Ultrastructure Damage of TCs in Acute Salpingitis-Affected Rat Oviduct Tissues

Acute salpingitis (AS) is a kind of acute pelvic inflammatory diseases; it causes severe damage to a subset of classically described cells lining in the oviduct wall and finally leads to interstitial fibrosis and tubal factor sub- or infertility. However, with recent increasing reports regarding TC damage in various disease-affected tissues, potential involvement and pathophysiologic role of TCs in AS-induced tubal factor fertility problems remain unknown.

In order to get AS-affected rat oviduct tissues, the rat model of AS was established, the injection of bacterial liquid (0.1 ml, 2×10^7 E. coli) into both sides of oviduct lumens was performed only after entering the abdominal cavity via lower midline trans-abdominal incision under aseptic conditions. The open surgery was done before injection [58]. The same amount of sterile saline instead of bacterial liquid served as the sham control. Then, 7 days later, both sides of oviducts were obtained for HE, IHC, immunofluorescence IHC and TEM observation.

AS was confirmed by acute inflammation within oviduct tissues under HE staining, by massive neutrophil infiltration under CD177 immunostaining (Fig. 18.7a–d) and simultaneously by single-labelled immunofluorescence IHC and quantitative analysis, which showed significant overproduced inflammatory factors (iNOS, COX-2) (both $P=0.000$; Fig. 18.7e–f).

TC immunodiagnosis was performed by double-labelled immunofluorescence IHC and confirmed the existence of rich amount of CD34/vimentin double-positive TCs around perivascular space of the sham group. TCs showed typical cell body and two or more extremely long/thin prolongations, as well as double immunofluorescence in its full length (Fig. 18.8a). While there were no co-expression of vimentin and c-kit in cell bodies or prolongations of TCs (images not shown). Nevertheless, in sections from AS-affected oviduct tissues, CD34/vimentin double-positive TCs with intact structure significantly decreased and were sparse or completely absent ($P=0.000$; Fig. 18.8b, c).

Ultrastructure observation under TEM identified normal TCs by their characteristic ultrastructure features (Fig. 18.9a, b). TCs enwrap the whole capillaries with

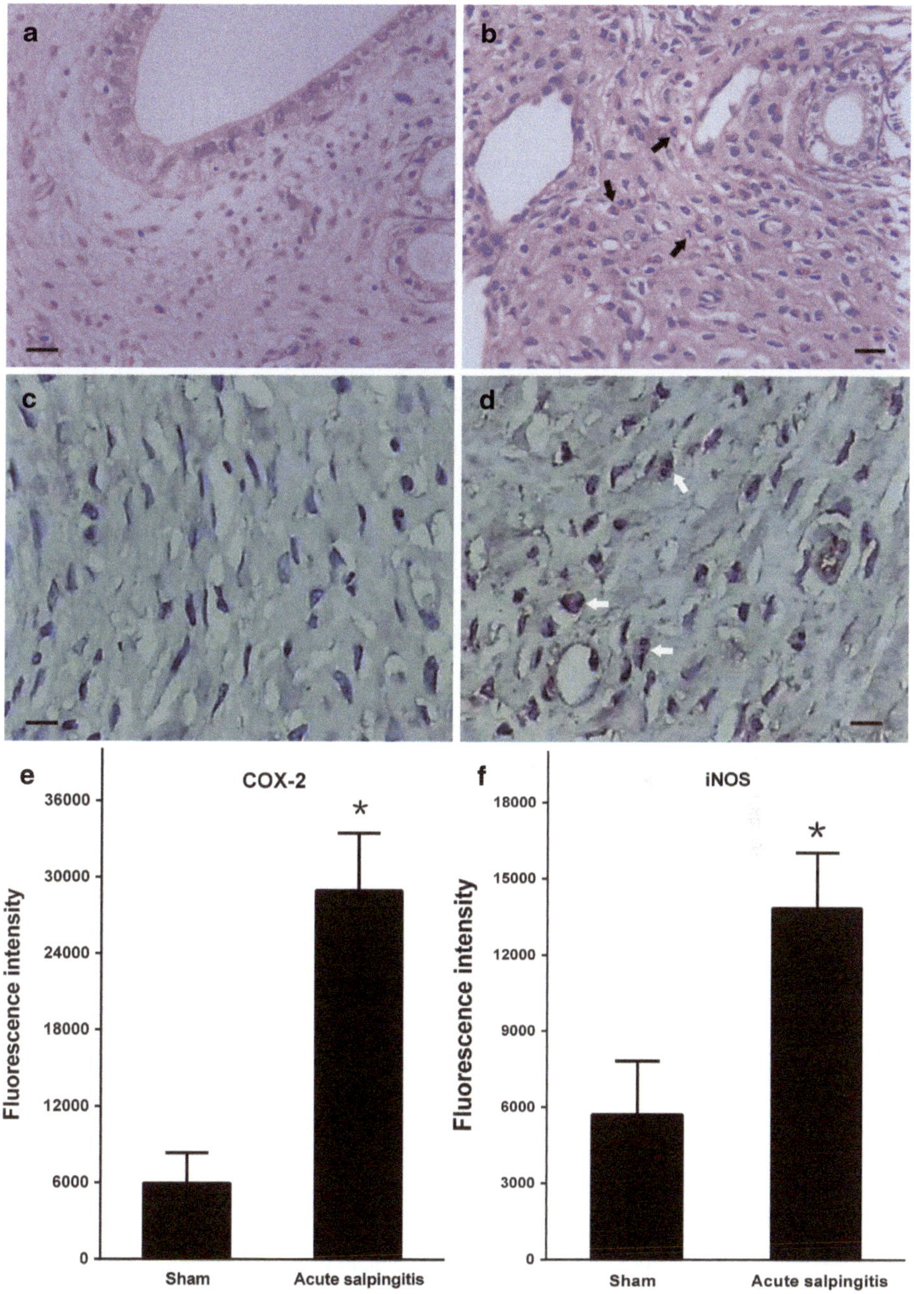

Fig. 18.7 Oviduct inflammation confirmed by HE staining (**a**, **b**, scale bar = 40 μm), IHC (**c**, **d**, scale bar=20 μm) and single-labelled immunofluorescence IHC and quantitative analysis (e, f). (**a**, **c**) Normal oviduct tissues, with no signs of inflammatory cells infiltration indicated by negative CD177 IHC. (**b**, **d**) Acute inflammation in AS-affected oviduct tissues showed signs of SMC and capillary swelling, interstitial congestion, exudation, oedema and interstitial fibrosis (**b**, *black arrows*) and massive neutrophil infiltrations, suggested by strong positive CD177 IHC (**d**, *white arrows*). (**e**, **f**) COX-2 (e) and iNOS (f) increased significantly in AS-affected oviduct tissue, as compared to sham control. *$P<0.05$. Error bars=SD (Yang et al. [46] adapted with permission)

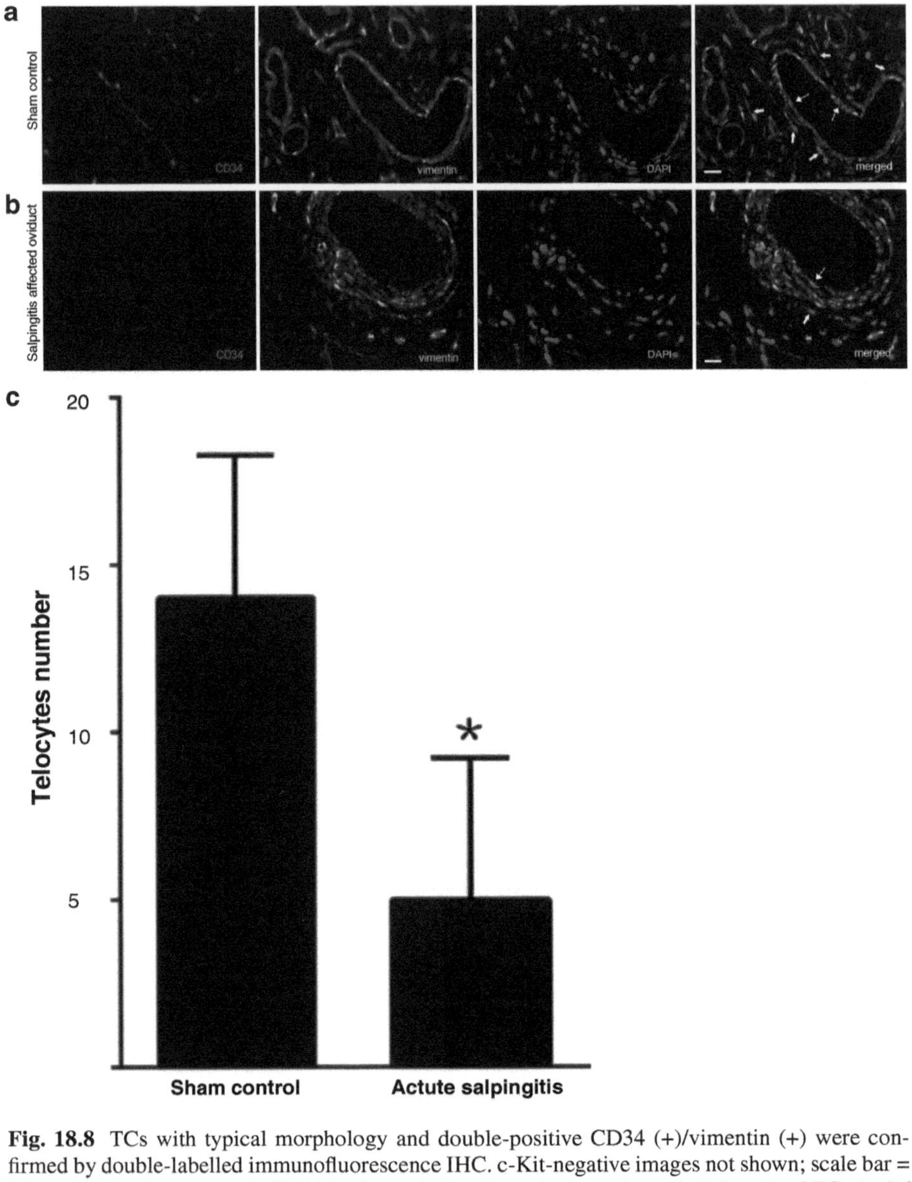

Fig. 18.8 TCs with typical morphology and double-positive CD34 (+)/vimentin (+) were confirmed by double-labelled immunofluorescence IHC. c-Kit-negative images not shown; scale bar = 20 μm. (**a**) In sham control, CD34 (*red*) overlying vimentin (*green*) in whole length of TCs (*solid arrows*). (**b** and **c**) In AS-affected oviduct tissues, CD34/vimentin double-positive TCs with well-defined nuclei were less densely stained (solid arrows), significantly decreased in cell number. $^*P<0.05$, error bars = SD (Yang et al. [46] adapted with permission)

long Tps or scattered in SMC bundles in mucosa and muscular layer of the sham oviduct, with complex homocellular and heterocellular junctions with adjacent cell components, thus organizing a unique 3-D network, providing mechanical support

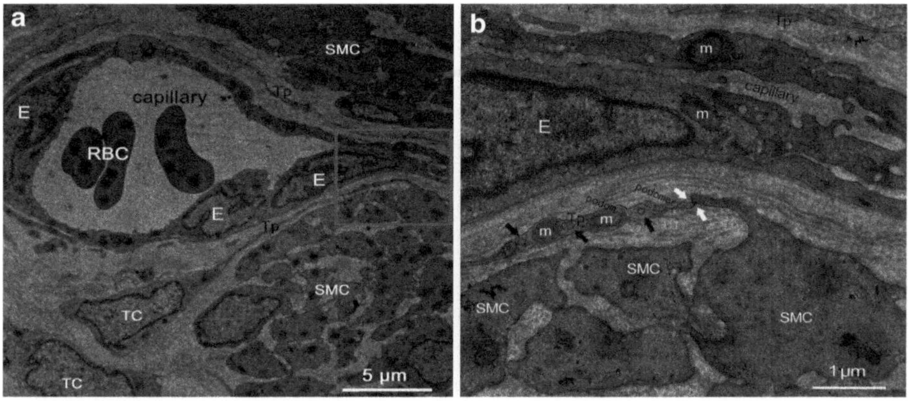

Fig. 18.9 Normal TCs in perivascular space or among smooth muscle bundles. *RBC* red blood cells, *E* vascular endothelial cells, *m* mitochondria, *rER* rough endoplasmic reticulum. (**A**, **a**) By Tps, which was composed of podoms and podomers, TCs surrounded and developed heterocellular contact with SMCs (*white arrows*) in perivascular space. (**b**) Higher magnification of the *blue boxed* area; homocellular contacts (*black arrow*) between TCs and heterocellular connections with *E* (*white arrows*). (**B**, **a**) Perivascular TCs scattered among SMCs. (**b**) Higher magnification of the *blue boxed* area: abundant mitochondria (*m*) and microvesicles (*black arrows*) within podom and heterocellular junctions between Tps and SMCs (*white arrows*) (Yang et al. [45] adapted with permission)

or structural basis for interstitial compartment and communicating and potentially affecting any of nearby different cell types.

Nevertheless, in AS-affected oviduct tissues, TCs displayed multiple ultrastructural damage both in cellular body and Tps, together with obvious TC loss, disrupted TC-SC niches (TC-SCNs) and excessive collagen content (Fig. 18.10). Furthermore, impaired homocellular or heterocellular junctions connected by TCs and adjacent cells also cause dearrangement of interstitial 3-D network. And especially, TCs can also be observed to connect with the activated immunocytes with dense secretory granules (mononuclear cells, MO; eosinophils, Eo; Fig. 18.10a, c) and thus might be involved in local immunoregulation and functionally affect (repression or activation) local immune state.

These results provided the first evidence that oviduct TCs displayed inflammatory-induced extensive ultrastructure damage, with obvious cell loss, decrease or almost complete absence in the successfully constructed AS-affected oviduct tissues, which were confirmed by acute inflammation, interstitial fibrosis, numerous neutrophils infiltration and overproduced iNOS and COX-2.

Similarly, in the first part of this chapter, the aseptic-based inflammation also causes TC damage in EMs-affected oviduct tissues [45], herein strengthening the broad involvement of TC damage in infection-based inflammatory AS-affected oviduct tissues. Nevertheless, although none of the supposed pathophysiologic functions have yet been proven definitive for oviduct TCs, based on their distinct ultrastructure, distribution and immunophenotype, we suggested that TC damage

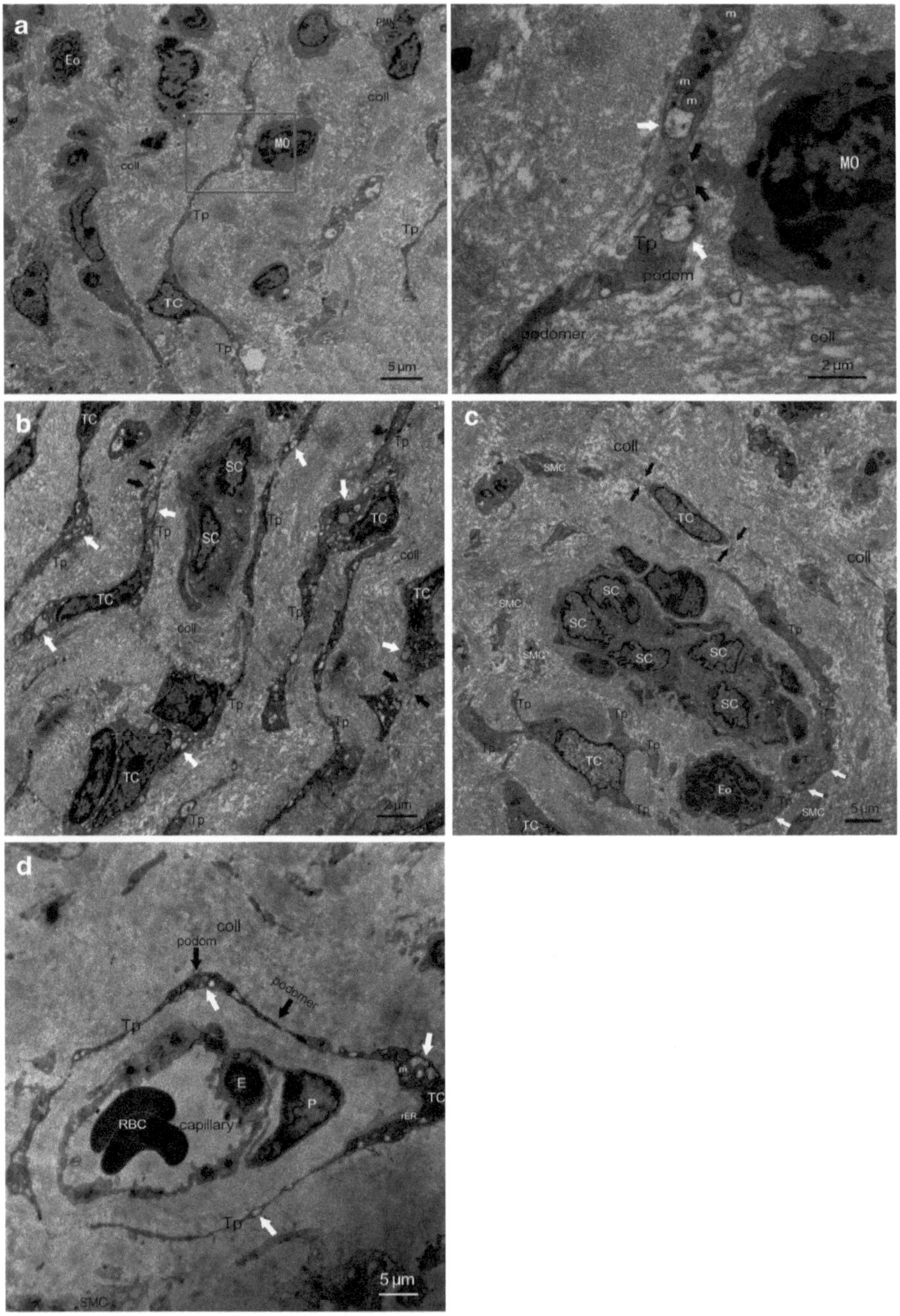
a
Eo
PMN
coll
MO
Tp
TC
5 μm
m
MO
Tp
podom
podomer
coll
2 μm
b
TC
SC
Tp
coll
2 μm
c
coll
SMC
TC
SC
Tp
Eo
5 μm
d
coll
podom
podomer
Tp
TC
E
P
RBC
capillary
rER
5 μm
SMC

might consequently lead to structural and reproductive functional abnormalities of oviduct, probably via the following mechanisms:

(i) Ultrastructure damage, such as loss of multiple subcellular organelles within cellular body and Tps, will subsequently cause decreased synthesis of energy and substances, presumably, e.g. TC-specific genome, microRNAs and proteomic profiles, ion channels, neurotransmitter and cytoskeletal elements, which were all key elements for complex functions and activities of TCs, such as electrophysiology, dynamics of Tps in terms of adherence, spreading/extension and ramification, etc.

(ii) Swollen or loss of heterocellular (planar or point) junctions between Tps and nearby cells, e.g. capillaries, SMCs, activated immunocytes and SCs, will inevitably decrease TC-mediated intercellular signalling and proposed functions, such as tissue repair/regeneration, homeostasis/angiogenesis, immunoregulation, neurotransmission/muscular contraction, etc.

(iii) Loss of strategic position and disintegrating of extracellular 3-D architecture which resulted from TC damage, will change correct organization or mechanical support for different structural components of interstitium; disturb cell migration or chemotaxis (SCs, activated immunocytes, etc.); impair intercellular signal transduction, coordination and integration between TCs and adjacent cells; and finally cause interstitial fibrosis. Interestingly, although previous studies suggested that TC loss might occur before the onset of fibrotic reconstruction of the intestinal wall [41], now it still lacks molecular evidence to tell whether TC loss happened before the onset of oviduct fibrosis or it is just being merely a consequence of the fibrotic process.

(iv) Finally, a special intra-tubal immune state was essential for a successful fertilization, early embryonic development and transportation. However, heterocellular junctions between impaired TCs and the activated immunocytes (MO, Eo) might affect local immune state or inflammatory response by either repression or activation of secretion of various cytotoxic cytokines/enzymes. Subsequently they will likely to cause immune-mediated early pregnancy failure, such as intra-tubal sperm phagocytosis, impaired fertilization, defective early embryo growth, abnormal transportation and implantation.

Fig. 18.10 TC damage in AS-affected oviduct tissues, with tissue fibrosis. (**A, a**) Degenerated Tps developed heterocellular contact with activated mononuclear cells (MO), which contained dense secretory granules, together with eosinophil (Eo) and neutrophil (PMN) infiltration. (**b**) Higher magnification of the blue boxed area: degenerated Tps connected to activated MO (*black arrows*), with swollen mitochondria (m) and vacuoles (*white arrows*) in Tps. (**B, C**) Intercellular contacts became wider or disappeared (*black asterisks*) between a group of putative stem cells (SCs) and damaged TCs and Tps (*black arrows*), between Tps and activated eosinophils (Eo) containing dense secretory granules (*white asterisk*) in disrupted TC-SC niches. Swollen mitochondria (m), cytoplasmic vacuolization (*white arrows*) in Tps and dissolution of SMCs can also be observed. (**D**) Perivascular TCs and Tps, endothelial cell (*E*) and pericyte (*P*) damage: swollen mitochondria (m), rough endoplasmic reticulum (rER) dilatation and cytoplasmic vacuolization (*white arrows*) (Yang et al. [45] adapted with permission)

Nevertheless, future work should take the following into consideration, such as the underlying molecular and cellular mechanisms for TC damage in the setting of different oviduct diseases and downstream pathway of TC involvement in differentiation/activation of fibroblasts/fibrocytes and accompanied fibrosis, in local inflammatory process/immunoregulation and possibly in immune-mediated early pregnancy failure.

18.4 In Vitro Immunomodulation of Mouse Peritoneal Macrophages by Uterine TCs

The first two parts of this chapter describe in vivo heterocellular junctions between TCs and various adjacent immunocytes, including mononuclear cells, mast cells, eosinophils and neutrophils, both in normal and inflammatory-affected oviduct tissue in SD rat model, suggesting potential involvement of TCs in local immunoregulation or immunosurveillance. Therefore, we hypothesized that through direct heterocellular junctions or indirect paracrine effects, TCs might have the ability to influence the activity of immunocytes and then be involved in local inflammatory process or in immune-mediated various reproductive abnormalities.

Nevertheless, so far, no reliable in vitro cytological evidence is available to support the proposed immunoregulation/immunosurveillance role for TCs. The aim of this part of study is to confirm the aforementioned in vivo findings, by evaluating in vitro paracrine effect of uterine TCs on mouse peritoneal macrophages (pMACs). Therefore, mouse pMACs were designed to coculture with uterine TC-conditioned media (TCM) for 48 h, followed with study of in vitro morphology, viability and cytokines/enzyme production by pMACs [59]. Meanwhile, applying the same amount of DMEM/F12 or lipopolysaccharide (LPS) instead of TCM served as negative and positive controls, respectively.

To harvest uterine TCs, mice uterine tissues were used for primary culture, and in vitro TC identification was carried out by methylene blue staining, mitochondrial labelling and double immunofluorescence cytochemistry. And then uterine TCs were clearly identified based on its typical morphology: small bipolar or multipolar cell body with extremely long Tps, which composed of podomers and podoms; and by Tps, TCs formed homocellular contacts (Fig. 18.11a–c). In addition, active energy metabolism was indicated by mitochondria labelling (Fig. 18.11d). Furthermore, double-positive CD34/vimentin was obvious both in the cell body and its alternating thick and thin segments of Tps (Fig. 18.11e).

Morphological study of pMACs demonstrated that after 48 h of coculture with TCM, DMEM/F12 or LPS (0.5 μg m L-1), obvious activation/immunoresponse of pMACs was elicited by TCM, in contrast to overstimulation or cell death by LPS exposure and no sign of activation by DMEM/F12 (Fig. 18.12a–c). Further quantitative cell viability assay of pMACs by a cell counting kit 8 (CCK-8) indicated significant activation of pMACs by TCM and LPS, as compared to DMEM/F12 (Fig. 18.12d) (both $P<0.05$), with no significant difference between TCM and LPS ($P>0.05$), thus verifying obvious cell morphological differences among three groups.

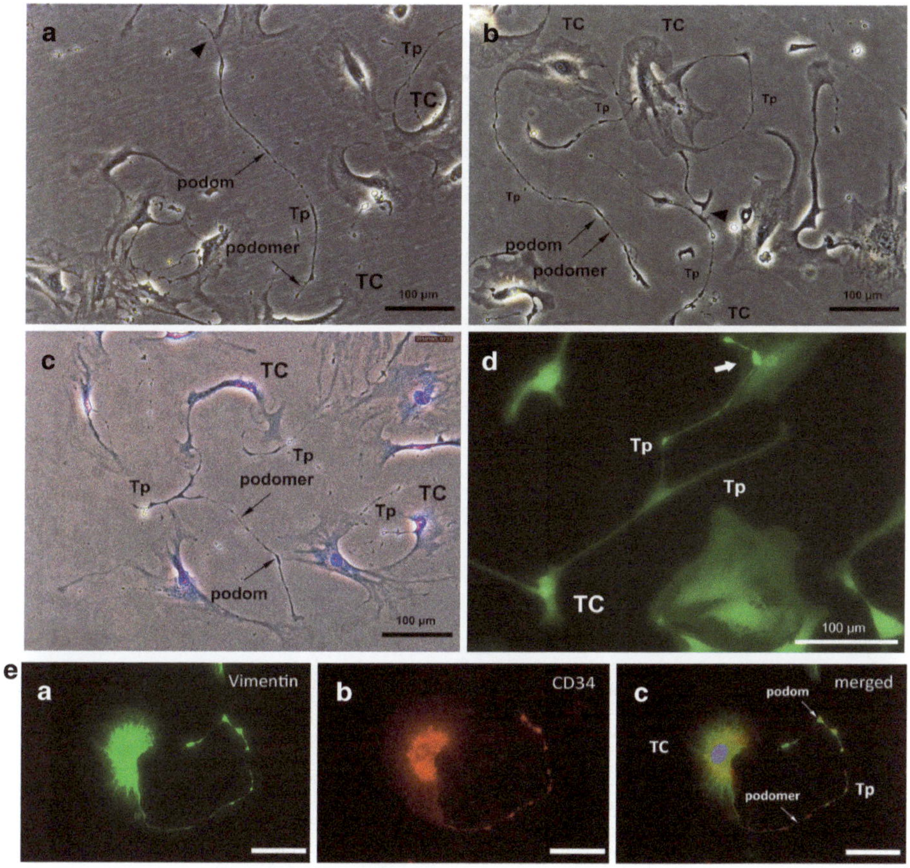

Fig. 18.11 Typical TCs in primary culture from the mouse uterus demonstrated characteristic cell body and Tps, alternating with podoms and podomers that developed homocellular network (*black arrowhead*). (**a** and **b**) Phase-contrast microscopy. (**c**) Methylene blue staining under phase-contrast microscopy. (**d**) MitoTracker Green staining indicated extensive fluorescence and cell metabolism around intercellular connections between Tps and other cells (*white arrow*). (**e**) Typical TC morphology with immunophenotype of CD34 (+)/vimentin (+)/c-kit (−) indicated by double-labelled immunofluorescence IHC. Scale bar = 50 µm (Chi et al. [59] adapted with permission)

Quantitative analysis of a panel of pMAC-derived cytokines/enzyme showed that interleukin-6 (IL-6) and iNOS were significantly elevated in TCM-treated pMACs; tumour necrosis factor a, IL1-R1 and IL-10 were also significantly, but slightly, up-regulated (all $P<0.05$; Fig. 18.13a, b). Meanwhile, no changes were observed for transforming growth factor-b1, IL-1b, IL-23a and IL-18 (all $P>0.05$; Fig. 18.13b).

These data provided preliminary in vitro evidence and support our hypothesis that TCs are not merely innocent bystanders in the interstitial compartment but are instead potential functional active players in local immunoregulation or immunosurveillance. TCs display the ability to trigger the activation of pMACs and potentially induce a subsequent in vitro immune response, manifested by morphology and viability alterations, and secretion of multiple cytokines/enzyme, likely

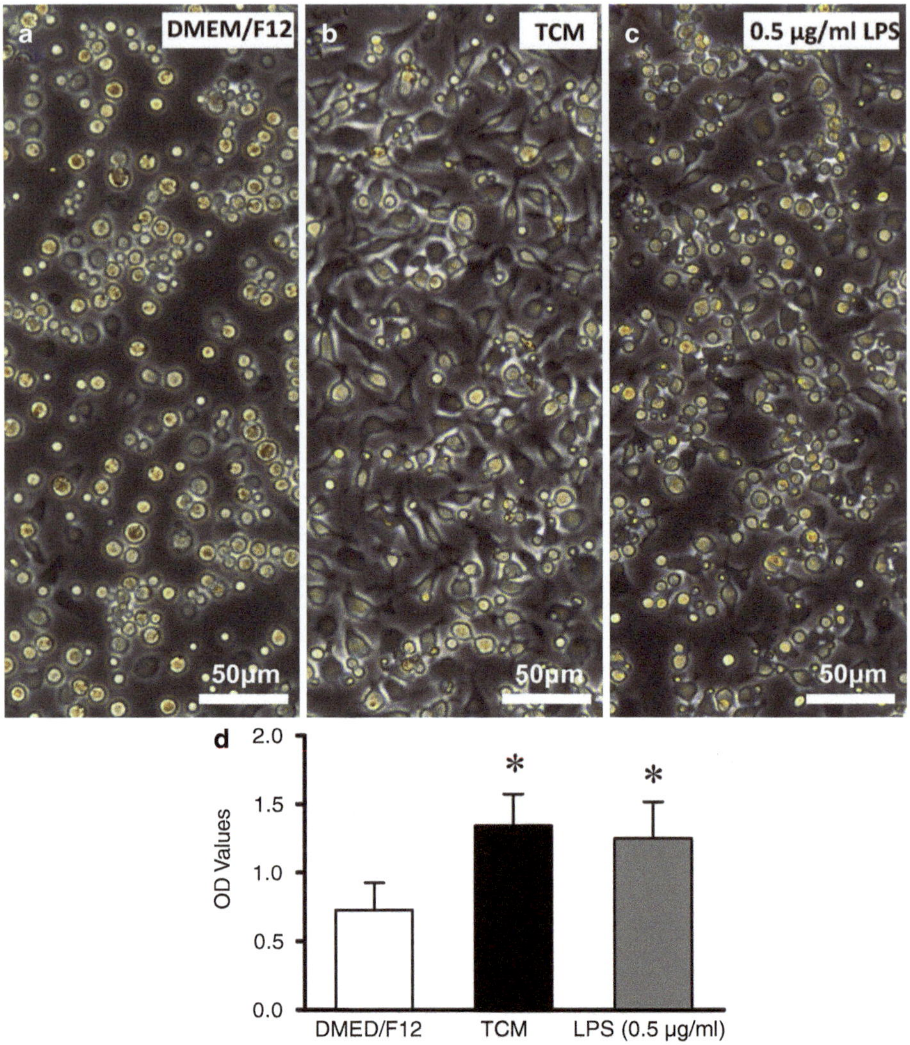

Fig. 18.12 Morphology and viability changes of mouse pMACs after exposure to TCM, DMEM/F12 or LPS (0.5 μg mL-1) for 48 h. (**a**) No signs of cell activation in DMEM/F12 group, manifested by normal morphology with regular round shape, abundant clear cytoplasm and wider intercellular spaces. (**b**) Moderate activation/immune response in TCM group, manifested by obvious morphological changes, including polyhedron shape, large and sufficient pseudopodia, abundant granules within the cytoplasm and narrow intercellular spaces. (**c**) Excessive activation in LPS group, manifested by irregular, doublet or multiple shapes and cell death, indicated by cell membrane blebbing, cell body atrophy and nuclear condensation or fragmentation. (**d**) TCM and LPS both significantly activated pMACs (*$P<0.05$ versus DMEM/F12), with no significant difference between TCM and LPS ($P>0.05$), error bars = SD (Chi et al. [59] adapted with permission)

through indirect paracrine effects. Such ability was similar to but slightly weaker than the classical stimulus, LPS.

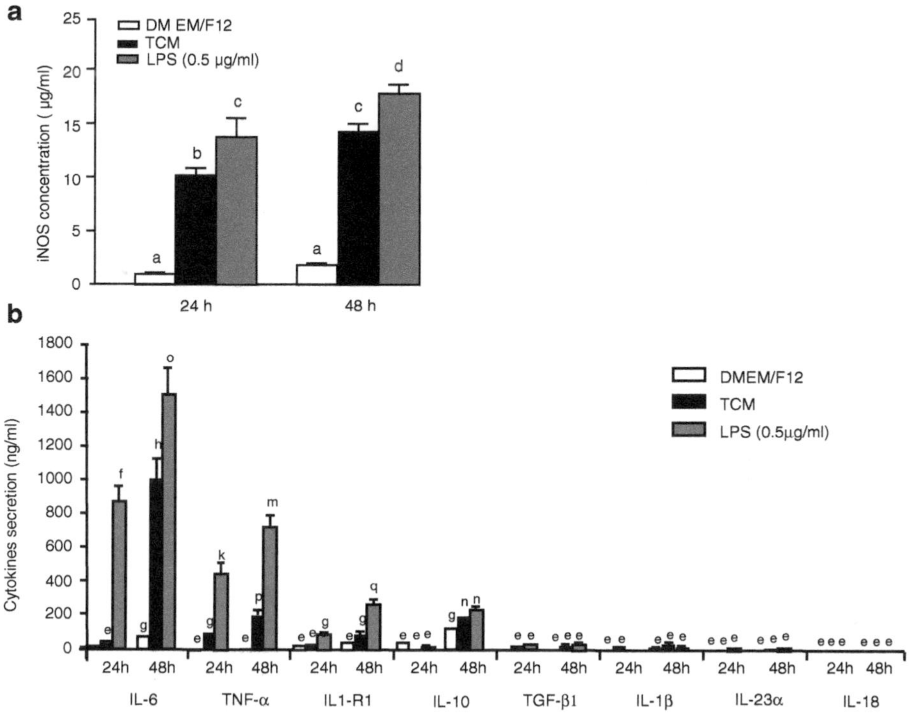

Fig. 18.13 ELISA analysis of nine pMACs-related cytokines/enzymes after exposure to TCM, DMEM/F12 or LPS (0.5 μg mL-1) for 24 and 48 h, respectively. The bars that share a common letter represent either non-significant difference ($P>0.05$), or their values were too low for any biological behaviour. (**a** and **b**) iNOS and IL-6 were overproduced significantly, together with TNF-a, IL1-R1 and IL-10 slightly, but significantly increased in TCM and LPS groups, as compared to DMEM/F12 (all $P<0.05$), either at 24 or 48 h, by one-way ANOVA, followed by significant differences between DMEM/F12 and test values by Dunnett's test (all $P<0.05$) and significant differences within TCM and LPS groups at the 24 and 48 h time points, respectively, by *t*-test (all $P<0.05$). (**b**) No significant changes in TGF-b1, IL-1b, IL-23a and IL-18 were observed (all $P>0.05$) (Chi et al. [59] adapted with permission)

We believe that TCM inducing the panel of elevated cytokines/enzymes might play important roles in reproduction. iNOS is an enzyme catalysing the production of nitric oxide (NO) from L-arginine. iNOS produces large quantities of NO upon stimulation, such as by pro-inflammatory cytokines (e.g. interleukin-1, tumour necrosis factor alpha and interferon gamma). iNOS has been suggested to participate in host immunity, antimicrobial and antitumour activities as part of the oxidative burst of macrophages. Furthermore, it has been proposed that excessive NO generated through increased iNOS production may decrease tubal ciliary beats and smooth muscle contractions and thus affect embryo transport, which may consequently result in tubal ectopic pregnancy and tubal factor infertility and probably also in uterine contractility disorders [60].

Moreover, high IL-6 levels might cause improper endometrial environments and implantation failure. And overexpression of TNF-a not only stimulates the pathological production of NO but also causes toxicity in multiple reproductive processes, such as implantation failure and immune-mediated abortion. Also, up-regulation of IL-1R1 by TCM might cause abnormal peri-implantation state and finally lead to failure of blastocyst implantation. Moreover, overproduction of IL-10, which was essential for a successful pregnancy, might lead to an adverse pregnancy outcome.

Generally, TCs release three types of extracellular vesicles (EVs): exosomes, ectosomes and multivesicular bodies. EVs contain many secretomes and transfer complex multimolecular biological messages from Tps [14–30, 61]. Paracrine secretion mediated cell proliferation, differentiation and tissue repair has been proposed for TCs in many studies. Specifically, uterine TCs can shed or released EVs and/or exosomes from Tps [16, 20, 25]. TCs from other organs/tissues also release soluble mediators, containing IL-6, VEGF and nitric oxide [34, 50, 61]; and IL-6, VEGF, macrophage inflammatory protein 1a (MIP-1a), MIP-2 and monocyte chemoattractant protein 1 (MCP-1), IL-2, IL-10, IL-13 and growth-related oncogene/keratinocyte-derived chemokine (GRO-KC) were significantly expressed in the secretome of cultured rodent cardiac TCs [61]. These multiple paracrine mediators might be essential for intercellular communication (long distance) and regulate the activity of neighbouring immunocytes, thus achieving the immunoregulation/immunosurveillance roles of TCM. However, behind the general concepts of paracrine effects, we still wondered what were the underlying cellular and molecular mechanisms. The exact pathway (nuclear factor kappa B, NF-kB, or else) and/or complex networks, which are responsible for the crosstalk between pMACs and TCM, maintenance of pMAC activation and subsequent overproduction of the panel of cytokines/enzymes, remain to be fully elucidated.

More questions still need to be answered, whether this immunomodulation role is only specific for uterine TCs or exists widely in other organs/tissues. And whether the observed near-infrared low-level laser stimulation of human myometrium TC growth [62] might also provide a potential choice for TC-associated abnormalities related to local reproductive immunity remains to be determined. Last but not least, to confirm this in vitro result, animal models will be more helpful to better clarify or strengthen in vivo functional consequences of the proposed immunoregulation/immunosurveillance roles of TCs. Then we can forward clinical applications of TCs in immune-mediated fertility problems and other related diseases.

References

1. Popescu LM, Ciontea SM, Cretoiu D, Hinescu ME, Radu E, Ionescu N, et al. Novel type of interstitial cell (Cajal-like) in human fallopian tube. J Cell Mol Med. 2005;9(2):479–523.
2. Popescu LM, Ciontea SM, Cretoiu D. Interstitial Cajal-like cells in human uterus and fallopian tube. Ann N Y Acad Sci. 2007;1101(1):139–65.
3. Anaf V, Chapron C, El Nakadi I, De Moor V, Simonart T, Noël J-C. Pain, mast cells, and nerves in peritoneal, ovarian, and deep infiltrating endometriosis. Fertil Steril. 2006;86(5):1336–43.
4. Woidacki K, Zenclussen AC, Siebenhaar F. Mast cell-mediated and associated disorders in pregnancy: a risky game with an uncertain outcome? Front Immunol. 2014;5:231. doi:10.3389/fimmu.2014.00231.

5. Sandvei R, Anker C, Wollen AL, Flood PR. Mast cells in the tubal wall in women using an intrauterine contraceptive device. BJOG Int J Obstet Gynaecol. 1986;93(7):758–64.
6. Burney RO, Giudice LC. Pathogenesis and pathophysiology of endometriosis. Fertil Steril. 2012;98(3):511–9.
7. Ren Q-Z, Qian Z-H, Jia S-H, Xu Z-Z. Vascular endothelial growth factor expression up-regulated by endometrial ischemia in secretory phase plays an important role in endometriosis. Fertil Steril. 2011;95(8):2687–9.
8. Macer ML, Taylor HS. Endometriosis and infertility: a review of the pathogenesis and treatment of endometriosis-associated infertility. Obstet Gynecol Clin North Am. 2012;39(4): 535–49.
9. Mier-Cabrera J, Jiménez-Zamudio L, García-Latorre E, Cruz-Orozcoc O, Hernández-Guerrero C. Quantitative and qualitative peritoneal immune profiles, T-cell apoptosis and oxidative stress-associated characteristics in women with minimal and mild endometriosis. BJOG Int J Obstet Gynaecol. 2011;118(1):6–16.
10. Giudice LC, Evers JL, Healy DL. Endometriosis: science and practice. Singapore: Wiley; 2012.
11. Kissler S, Zangos S, Wiegratz I, Kohl J, Rody A, Gaetje R, et al. Utero-tubal sperm transport and its impairment in endometriosis and adenomyosis. Ann N Y Acad Sci. 2007;1101(1):38–48.
12. Nowicki S, Ram P, Pham T, Goluszko P, Morse S, Anderson GD, et al. Pelvic inflammatory disease isolates of Neisseria gonorrhoeae are distinguished by C1q-dependent virulence for newborn rats and by the sac-4 region. Infect Immun. 1997;65(6):2094–9.
13. Ceafalan L, Gherghiceanu M, Popescuo L, Simionescu O. Telocytes in human skin–are they involved in skin regeneration? J Cell Mol Med. 2012;16(7):1405–20.
14. Smythies J, Edelstein L. Telocytes, exosomes, gap junctions and the cytoskeleton: the makings of a primitive nervous system? Front Cell Neurosci. 2013;7:278. doi:10.3389/fncel.2013.00278.
15. Zheng Y, Cretoiu D, Yan G, Cretoiu SM, Popescu LM, Fang H, et al. Protein profiling of human lung telocytes and microvascular endothelial cells using iTRAQ quantitative proteomics. J Cell Mol Med. 2014;18(6):1035–59.
16. Cretoiu SM, Cretoiu D, Marin A, Radu BM, Popescu LM. Telocytes: ultrastructural, immunohistochemical and electrophysiological characteristics in human myometrium. Reproduction. 2013;145(4):357–70.
17. Nicolescu MI, Popescu LM. Telocytes in the interstitium of human exocrine pancreas: ultrastructural evidence. Pancreas. 2012;41(6):949–56.
18. Fertig ET, Gherghiceanu M, Popescu LM. Extracellular vesicles release by cardiac telocytes: electron microscopy and electron tomography. J Cell Mol Med. 2014;18(10):1938–43.
19. Mandache E, Popescum L, Gherghiceanu M. Myocardial interstitial Cajal-like cells (ICLC) and their nanostructural relationships with intercalated discs: shed vesicles as intermediates. J Cell Mol Med. 2007;11(5):1175–84.
20. Creţoiu SM, Creţoiu D, Popescu LM. Human myometrium–the ultrastructural 3D network of telocytes. J Cell Mol Med. 2012;16(11):2844–9.
21. Popescu LM, Gherghiceanu M, Suciu LC, Manole CG, Hinescu ME. Telocytes and putative stem cells in the lungs: electron microscopy, electron tomography and laser scanning microscopy. Cell Tissue Res. 2011;345(3):391–403.
22. Rusu M, Mirancea N, Mănoiu V, Valcu M, Nicolescu M, Păduraru D. Skin telocytes. Ann Anat Anat Anz. 2012;194(4):359–67.
23. Cismasiu VB, Popescu LM. Telocytes transfer extracellular vesicles loaded with microRNAs to stem cells. J Cell Mol Med. 2015;19(2):351–8.
24. Bei Y, Wang F, Yang C, Xiao J. Telocytes in regenerative medicine. J Cell Mol Med. 2015;19(7):1441–54.
25. Roatesi I, Radu BM, Cretoiu D, Cretoiu SM. Uterine telocytes: a review of current knowledge. Biol Reprod. 2015;93(5):1–13.
26. Cretoiu SM, Popescu LM. Telocytes revisited. Biomolecular Concepts. 2014;5(5):353–69.
27. Nicolescu MI, Bucur A, Dinca O, Rusu MC, Popescu LM. Telocytes in parotid glands. Anat Rec. 2012;295(3):378–85.
28. Popescu LM, Gherghiceanu M, Cretoiu D, Radu E. The connective connection: interstitial cells of Cajal (ICC) and ICC-like cells establish synapses with immunoreactive cells. Electron microscope study in situ. J Cell Mol Med. 2005;9(3):714–30.

29. Popescu L, Manole E, Şerboiu CS, Manole C, Suciu LC, Gherghiceanu M, et al. Identification of telocytes in skeletal muscle interstitium: implication for muscle regeneration. J Cell Mol Med. 2011;15(6):1379–92.
30. Pellegrini M-SF, Popescu LM. Telocytes. BioMolecular Concepts. 2011;2(6):481–9.
31. Cretoiu SM, Cretoiu D, Suciul L, Popescu L. Interstitial Cajal-like cells of human fallopian tube express estrogen and progesterone receptors. J Mol Histol. 2009;40(5):387–94.
32. Cretoiu D, Ciontea SM, Popescu L, Ceafalanc L, Ardeleanu C. Interstitial Cajal-like cells (ICLC) as steroid hormone sensors in human myometrium: immunocytochemical approach. J Cell Mol Med. 2006;10(3):789–95.
33. Gevaert T, Vos R, Aa F, Joniau S, Oord J, Roskams T, et al. Identification of telocytes in the upper lamina propria of the human urinary tract. J Cell Mol Med. 2012;16(9):2085–93.
34. Manole C, Cismaşiu V, Gherghiceanu M, Popescu L. Experimental acute myocardial infarction: telocytes involvement in neo-angiogenesis. J Cell Mol Med. 2011;15(11):2284–96.
35. Gherghiceanu M, Popescu LM. Cardiac telocytes—their junctions and functional implications. Cell Tissue Res. 2012;348(2):265–79.
36. Cismasiu V, Radu E, Popescu L. miR-193 expression differentiates telocytes from other stromal cells. J Cell Mol Med. 2011;15(5):1071–4.
37. Zhao B, Chen S, Liu J, Yuan Z, Qi X, Qin J, et al. Cardiac telocytes were decreased during myocardial infarction and their therapeutic effects for ischaemic heart in rat. J Cell Mol Med. 2013;17(1):123–33.
38. Manetti M, Guiducci S, Ruffo M, Rosa I, Faussone-Pellegrini MS, Matucci-Cerinic M, et al. Evidence for progressive reduction and loss of telocytes in the dermal cellular network of systemic sclerosis. J Cell Mol Med. 2013;17(4):482–96.
39. Milia AF, Ruffo M, Manetti M, Rosa I, Conte D, Fazi M, et al. Telocytes in Crohn's disease. J Cell Mol Med. 2013;17(12):1525–36.
40. Manetti M, Rosa I, Messerini L, Guiducci S, Matucci-Cerinic M, Ibba-Manneschi L. A loss of telocytes accompanies fibrosis of multiple organs in systemic sclerosis. J Cell Mol Med. 2014;18(2):253–62.
41. Manetti M, Rosa I, Messerini L, Ibba-Manneschi L. Telocytes are reduced during fibrotic remodelling of the colonic wall in ulcerative colitis. J Cell Mol Med. 2015;19(1):62–73.
42. Matyja A, Gil K, Pasternak A, Sztefko K, Gajda M, Tomaszewski KA, et al. Telocytes: new insight into the pathogenesis of gallstone disease. J Cell Mol Med. 2013;17(6):734–42.
43. Ullah S, Yang P, Zhang L, Zhang Q, Liu Y, Chen W, et al. Identification and characterization of telocytes in the uterus of the oviduct in the Chinese soft-shelled turtle, Pelodiscus sinensis: TEM evidence. J Cell Mol Med. 2014;18(12):2385–92.
44. Bosco C, Díaz E, Gutiérrez R, González J, Parra-Cordero M, Rodrigo R, et al. A putative role for telocytes in placental barrier impairment during preeclampsia. Med Hypotheses. 2015;84(1):72–7.
45. Yang XJ, Yang J, Liu Z, Yang G, Shen ZJ. Telocytes damage in endometriosis-affected rat oviduct and potential impact on fertility. J Cell Mol Med. 2015;19:452–62.
46. Yang J, Chi C, Liu Z, Yang G, Shen Z-J, Yang X-J. Ultrastructure damage of oviduct telocytes in rat model of acute salpingitis. J Cell Mol Med. 2015;19(7):1720–8.
47. Chen X, Zheng Y, Manole CG, Wang X, Wang Q. Telocytes in human oesophagus. J Cell Mol Med. 2013;17(11):1506–12.
48. Luesma MJ, Gherghiceanu M, Popescu LM. Telocytes and stem cells in limbus and uvea of mouse eye. J Cell Mol Med. 2013;17(8):1016–24.
49. Popescu B, Gherghiceanu M, Kostin S, Ceafalan L, Popescu L. Telocytes in meninges and choroid plexus. Neurosci Lett. 2012;516(2):265–9.
50. Cretoiu D, Cretoiu SM, Simionescu AA, Popescu L. Telocytes, a distinct type of cell among the stromal cells present in the lamina propria of jejunum. Histol Histopathol. 2012;27(8):1067–78.
51. Alunno A, Ibba-Manneschi L, Bistoni O, Rosa I, Caterbi S, Gerli R, et al. Telocytes in minor salivary glands of primary Sjögren's syndrome: association with the extent of inflammation and ectopic lymphoid neogenesis. J Cell Mol Med. 2015;19(7):1689–96.
52. Carmona IC, Bartolomé M, Escribano CJ. Identification of telocytes in the lamina propria of rat duodenum: transmission electron microscopy. J Cell Mol Med. 2011;15(1):26–30.

53. Manole C, Gherghiceanu M, Simionescu O. Telocyte dynamics in psoriasis. J Cell Mol Med. 2015;19(7):1504–19.
54. Vernon MW, Wilson E. Studies on the surgical induction of endometriosis in the rat. Fertil Steril. 1985;44(5):684–94.
55. Appleyard CB, Cruz ML, Rivera E, Hernandez GA, Flores I. Experimental endometriosis in the rat is correlated with colonic motor function alterations but not with bacterial load. Reprod Sci. 2007;14(8):815–24.
56. Quereda F, Barroso J, Acien P. Individual and combined effects of triptoreline and gestrinone on experimental endometriosis in rats. Eur J Obstet Gynecol Reprod Biol. 1996;67(1):35–40.
57. Dixon RE, Hwang SJ, Hennig GW, Ramsey KH, Schripsema JH, Sanders KM, et al. Chlamydia infection causes loss of pacemaker cells and inhibits oocyte transport in the mouse oviduct. Biol Reprod. 2009;80(4):665–73.
58. Pearlman MD, Mcneeley SG, Frank TS, Hoeft-Loyer C. Antiendotoxin antibody is protective against tubal damage in an Escherichia coli rabbit salpingitis model. Am J Obstet Gynecol. 1994;171(6):1588–93.
59. Chi C, Jiang XJ, Su L, Shen ZJ, Yang XJ. In vitro morphology, viability and cytokine secretion of uterine telocyte-activated mouse peritoneal macrophages. J Cell Mol Med. 2015. doi:10.1111/jcmm.12711.
60. Perez Martinez S, Viggiano M, Franchi A, Herrero M, Ortiz M, Gimeno M, et al. Effect of nitric oxide synthase inhibitors on ovum transport and oviductal smooth muscle activity in the rat oviduct. J Reprod Fertil. 2000;118(1):111–7.
61. Albulescu R, Tanase C, Codrici E, Popescu DI, Cretoiu SM, Popescu LM. The secretome of myocardial telocytes modulates the activity of cardiac stem cells. J Cell Mol Med. 2015;19(8): 1783–94.
62. Campeanu R-A, Radu BM, Cretoiu SM, Banciu DD, Banciu A, Cretoiu D, et al. Near-infrared low-level laser stimulation of telocytes from human myometrium. Lasers Med Sci. 2014;29(6):1867–74.

Xiao-Jun Yang Xiao-Jun Yang is an assistant professor in the Department of Obstetrics and Gynecology, the First Affiliated Hospital of Soochow University. He is interested in "oviduct diseases and infertility". Currently, he is focusing on "telocytes in inflammatory gynaecological diseases and relation to immune-mediated reproductive abnormalities". His study is important not only for clarifying involvement of TCs in multiple gynaecologic diseases but also unveiling molecular mechanisms for TC-mediated immunoregulation/immunosurveillance of local immunocytes, immune response and inflammatory process. Ultimately, he hopes to provide potential target for immune-mediated reproductive abnormalities. Dr. Xiao-Jun Yang has published nine scientific papers within internationally well-recognized journals.

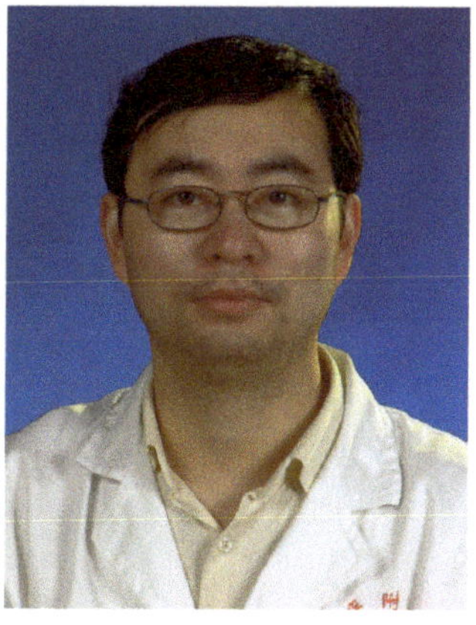

Chapter 19
Electrophysiological Features of Telocytes

Daniel Dumitru Banciu, Adela Banciu, and Beatrice Mihaela Radu

Abstract Telocytes (TCs) are interstitial cells described in multiple structures, including the gastrointestinal tract, respiratory tract, urinary tract, uterus, and heart. Several studies have indicated the possibility that TCs are involved in the pacemaker potential in these organs. It is supposed that TCs are interacting with the neighboring muscular cells and their network contributes to the initiation and propagation of the electrical potentials. In order to understand the contribution of TCs to various excitability mechanisms, it is necessary to analyze the plasma membrane proteins (e.g., ion channels) functionally expressed in these cells. So far, potassium, calcium, and chloride currents, but not sodium currents, have been described in TCs in primary cell culture from different tissues. Moreover, TCs have been described as sensors for mechanical stimuli (e.g., contraction, extension, etc.). In conclusion, TCs might play an essential role in gastrointestinal peristalsis, in respiration, in pregnant uterus contraction, or in miction, but further highlighting studies are necessary to understand the molecular mechanisms and the cell-cell interactions by which TCs contribute to the tissue excitability and pacemaker potentials initiation/propagation.

19.1 Introduction

Telocytes (TCs) have been described in a variety of tissues/organs, including the gastrointestinal tract (colon, small intestine, etc.), urinary tract (urethra, prostate, etc.), respiratory tract (lungs, trachea, esophagus, etc.), skeletal muscle, heart, liver,

D.D. Banciu • A. Banciu
Department of Anatomy, Animal Physiology and Biophysics, Faculty of Biology, University of Bucharest, Splaiul Independentei 91-95, Bucharest 050095, Romania

B.M. Radu (✉)
Department of Anatomy, Animal Physiology and Biophysics, Faculty of Biology, University of Bucharest, Splaiul Independentei 91-95, Bucharest 050095, Romania

Department of Neurological and Movement Sciences, University of Verona, Strada Le Grazie 8, Verona 37134, Italy
e-mail: beatricemihaela.radu@univr.it; beatrice.radu@bio.unibuc.ro

X. Wang, D. Cretoiu (eds.), *Telocytes*, Advances in Experimental Medicine and Biology, 913, DOI 10.1007/978-981-10-1061-3_19

kidney, skin, eye, etc. [1–16]. Extensive immunohistochemical and electron microscopy studies have been conducted on TCs, but their unique features and dynamic changes in cell culture limited the in vitro electrophysiological analysis [17, 18]. In the last years, several patch-clamp studies have been done on TCs in order to identify the ion channels functionally expressed in these cells and to understand their possible contribution to pacemaker potentials. In this chapter will be presented the most recent in vitro electrophysiological studies conducted on TCs from different tissues and the current conclusions based on these studies. At the end of the chapter are discussed some open questions concerning the role played by TCs in the various organs.

19.2 Which Ion Channels Are Functionally Expressed in Telocytes?

Interstitial cells (IC), interstitial cells of Cajal (ICC), interstitial Cajal-like cells (ICLC), and telocytes (TC) were studied in terms of ion channels in a manner which highlights the functions of these cells (Table 19.1). Their distribution in different tissues influenced the evaluation of potential roles played by these cells. It can be assumed that the existence of long extensions, the cell motility, and the ability to develop gap junctions with neighboring cells influence the type of ion channels expressed and their degree of activation and local functions.

19.2.1 Potassium Currents

Voltage-sensitive potassium channels $K_v7.5$ are involved in the excitation of the interstitial cells of Cajal (ICC) associated with the myenteric plexus but not with those associated with the submuscular plexus of the colon [19]. Immunohistochemical and qRT-PCR characterization revealed the presence of $K_v7.5$ channels in the colonic ICC. Carbachol, a muscarinic acetylcholine receptor agonist, inhibited the potassium currents, indicating a cholinergic-dependent activation of the voltage-sensitive potassium channels. Moreover, XE991, a specific Kv7 channel blocker, was able to abolish completely the potassium currents [19]. These currents are very similar to the inwardly rectifying maxi-chloride currents that were previously described in the ICC associated with the mouse myenteric plexus from the small intestine [29]. Generally, the M-current is carried through heteromeric Kv7.2 and Kv7.3, Kv7.3 and Kv7.5, or Kv7.4 and Kv7.5, but in the case of colonic ICC, the current seems to be carried exclusively through homomeric Kv7.5 channels [19, 30–32]. By contrast to the myocytes, the ICC in the prostate are characterized by the absence of sensitive outward potassium currents [20]. Hyperpolarization-activated cyclic nucleotide currents (HCN) have been also described in mouse colonic ICC, but only HCN1 and HCN3 channel transcripts were detected [24].

Table 19.1 Ion channels functionally expressed described in TCs or ICC with a role in cellular excitability

Ion channels	Isoform	Patch-clamp evidence	Pharmacology	IHC/IF evidence	qRT-PCR evidence	Cellular description	Anatomical localization	Species	Reference
Outward rectifying K^+ channels	Kv7.5	Yes	XE991 (20 μM) Carbachol (1 μM)	Yes	Yes	ICC	Colonic intramuscular	Balb/c mice	[19]
	–	Yes	CsCl (130 mM or 15 mM) pipette	No	No	ICC	Prostate	Guinea pig	[20]
	Kv4.3	No	4-aminopyridine (4-AP; 5 mM)	No	No	TC	Atrial and ventricular	Human	[21]
	–	Yes	No	No	No	Vimentin(+) c-KIT(–) ICC-like	Myometrium	Human	[22]
Inwardly rectifying K^+ channels	Kir2.1	Yes	Ba^{2+} (0.5 mM)	No	No	TC	Atrial and ventricular	Human	[21]
ATP-sensitive K^+ channels (K_{ATP})	–	No	Pinacidil (30 μM)	No	No	TC	Atrial and ventricular	Human	[21]
Ca^{2+}-activated K^+ channels	Kv1.1	Yes	Paxilline (100 μM) Na-ringenin (10 μM)	No	Yes	TC	Atrial and ventricular	Human	[21]
	SK3	No	No	Yes	Yes	CD34(+) ICC-like	Myometrium	Human	[23]

(continued)

Table 19.1 (continued)

Ion channels	Isoform	Patch-clamp evidence	Pharmacology	IHC/IF evidence	qRT-PCR evidence	Cellular description	Anatomical localization	Species	Reference
Hyperpolarization-activated cyclic nucleotide (HCN) channels	HCN1 HCN3	Yes	CsCl (5 mM) ZD7288 (10 μM) Zatebradine (10 μM) Clonidine (100 μM) Genistein (10 μM)	No	Yes	ICC	Mid colon	CD1 mice	[24]
	–	Yes	CsCl (5 μM) ZD7288 (5 μM)	No	No	ICC	Gastric antrum	Balb/c mice	[25]
L-type Ca^{2+} channels	–	Yes	Nifedipine (1 μM) Bay K 8644 (1 μM) Ba^{2+} (1.8 mM) ext	No	No	IC	Urinary bladder	Guinea pig	[26]
	–	Yes	Nifedipine (1 μM) Ba^{2+} (5 mM) ext	No	No	IC	Prostate	Guinea pig	[20]
	–	Yes	No	No	No	TC	Myometrium	Human	[27]
T-type Ca^{2+} channels	–	No	Ni^{2+} (100 μM)	No	No	ICC	Urinary bladder	Guinea pig	[26]
	–	Yes	Ni^{2+} (10 μM)	No	No	IC	Prostate	Guinea pig	[20]
	Cav3.1 Cav3.2	Yes	Mibefradil (1 μM)	Yes	No	TC	Myometrium	Human	[27]
	–	Yes	Mibefradil (1 μM)	No	No	TC	Myometrium	Human	[28]
Ca^{2+}-activated Cl^- channels		Yes	Niflumic acid (10 μM)	No	No	IC	Prostate	Guinea pig	[20]
	–	Yes	$CdCl_2$ (50μM) CsCl (85 mM ext; insensitive)	No	No	TC	Myometrium	Human	[9]

ICC's presence was previously described in the atrial and ventricular myocardium [33, 34]. In order to understand the role played by TCs in heart contractility, a detailed analysis of the existing potassium currents was done. In atrial and ventricular TCs have been evidenced large conductance Ca^{2+}-activated potassium currents (BK_{Ca}) and inwardly rectifying potassium currents (Kir), by applying paxilline and naringenin, or Ba^{2+}, respectively. However, in these cells neither potassium outward currents nor the ATP-sensitive potassium currents have been identified by applying 4-aminopyridine, or pinacidil, respectively [21]. The presence of small-conductance calcium-activated potassium channels (SK_{Ca}) was also confirmed in human uterine TCs; their mRNA levels were significantly lower in pregnant myometrium compared to nonpregnant myometrium; and SK activators were suggested to reduce contractility in human myometrium [23].

19.2.2 Calcium Currents

Voltage-gated L-type calcium channels were evidenced in the urinary bladder based on nifedipine and Bay K 8644 response [26], prostate based on nifedipine response [20], and myometrium [27] based on TCs. These channels have different roles in other cell types. Voltage-gated L-type calcium channels are involved in heart automaticity [35–38]. TCs were identified in the heart [39], and it is possible that these cells possess such properties. Changes in urinary bladder function are made by streptozotocin-induced diabetes [40] and hypercholesterolemia [41] via voltage-gated L-type calcium channels. This may correlate with the involvement of these channels in cellular electrical automatism, in a manner relatively independent of tissue. TCs have the ability to achieve gap junctions with surrounding cells, and linking these issues may explain the existence of this automatism on the tissue. Androgens induce intracellular calcium increase via voltage-gated L-type calcium channels in prostate cancer cells [42]. This could explain the involvement of these channels in cell growth and multiplication. By extension, it is important to verify if these channels are involved in the cellular growth and multiplication in physiological conditions.

Voltage-gated L-type calcium channels are involved in augmentation of spontaneous uterine contractility in pregnant rat modulated by protease-activated receptor 2 [43]. This modulation of rhythmic contractions can be extrapolated that is important at TC level and not just at tissue and organ levels, due to long cell extensions that are influenced by mechanical forces and due to gap junctions of TCs. Voltage-gated L-type calcium channels are modulated by alpha5beta1 integrin-fibronectin interactions, with a role in myogenic tone and vascular wall remodeling [44]. Modulation of L-type Ca^{2+} channels by hypoxia [45] can create a logical link in the pathophysiological mechanism, explaining the importance of these channels at TC in the muscle tissue under the influence promoting automatism, but also on the tissue by integrating interactions between TCs and muscle cells, and achieving feedback loops that include various other factors such as the vascular factors and humoral factors consecutively.

Voltage-gated T-type calcium channels (Cav3.1 and Cav3.2 subunits) were evidenced in the urinary bladder [26], prostate [20], and myometrium [27] on cultured TC and tissue strips. These channels play various roles in the smooth muscle wall of the blood vessels. Cav3.1 has a role on the blood vessel relaxation in an NO-dependent manner [46] and can induce myogenic constriction in the mesenteric vessels [47] and hypoxia-dependent pulmonary vasoconstriction [48]. These dual behaviors and the dependence on hypoxia may be extrapolated to myometrium, emphasizing the role of TC in uterus development and growth under hypoxic conditions, or even in birth control. Cav3.2 has a role in the relaxation of coronary vascular smooth muscles [49] and augmented contractility during oxidative stress [50]. Cav3.2 functions of myometrium TCs may be involved in fetal growth vascular adaptation.

Intracellular Ca^{2+} concentration is an important excitability regulator in ICC, and besides the L-type and T-type calcium channels, the sodium-calcium exchanger (NCX3) contributes to the calcium homeostasis in the rat bladder [51].

19.2.3 Chloride Currents

Human myometrial TCs have been described to present calcium-dependent hyperpolarization-activated chloride inward currents [9]. Ca^{2+}-activated Cl^- channels on ICC were highlighted indirectly by chloride concentration modulation [52] and subsequently by response to $CdCl_2$ and CsCl [9] and response to niflumic acid [20].

19.2.4 Sodium Currents

Electrophysiological studies on the human myometrium failed to prove the presence of Na^+ currents in TCs [9, 22].

19.3 Are Telocytes Involved in the Pacemaker Activity?

Different types of cells, including ICC/TCs, can induce these changes through intracellular mediators, dynamic changes of ionic concentrations, or other local stimulating factors and can contribute to the existence of spontaneous electrical activity in various areas associated with initiation/propagation of pacemaker activities (Table 19.2).

Table 19.2 Spontaneous electrical activity[a] (pacemaker potentials) in ICC and TC

Anatomical localization	Cellular type	Recording technique	Pharmacology	Molecular mechanisms involved	Species	Reference
Urinary bladder	ICC	Patch clamp	Carbachol (10 nM) Atropine (1 μM) Nifedipine (1 μM) ATP (5 μM)	Muscarinic receptors L-type Ca^{2+} channels	Balb/c mice	[53]
	ICC	Patch clamp	Carbamylcholine (1 μM; 10 μM) Atropine (1 μM) 4-DAMP (1 μM)	M2 and M3 cholinergic receptors	Sprague-Dawley rats	[54]
Intestine (embryonic)	c-KIT(+) ICC	Calcium imaging	Nifedipine (1 μM)	L-type Ca^{2+} channels	ICR mice	[55]
Small intestine	ICC	Patch clamp Calcium imaging	Y25130 (10 μM) YRS39604 (10 μM) SB269970 (10 μM) DIDS (10 μM) Thapsigargin (5 μM) PD98059 (10 μM) SB203580 (10 μM) JNK II inhibitor (10 μM)	Serotonin (5-HT_3, 5-HT_4, 5-HT_7) receptors Cl^- channel Ca^{2+} ATPase (endoplasmic reticulum) Mitogen-activated protein kinases (p42/44; p38; c-jun NH_2-terminal kinase)	Balb/c mice	[56]
	ICC	Patch clamp	Histamine (10, 50 and 100 μM) 2-Pyridylethylamine (2-PEA; 50 μM) Cetirizine (100 μM) Dimaprit (50 μM) R-alpha-methylhistamine (R-alpha-MeHa; 50 μM) 4-methylhistamine (4-MH; 50 μM) U-73122 (5 μM) 5-fluoro-2-indolyl des-chlorohalopemide (FIPI; 1 μM) KT5720 (1 μM) Pyr3 (2 μM)	Histamine (H_1, H_2, H_3, H_4) receptors Ca^{2+} ATPase (endoplasmic reticulum) Phospholipase A Phospholipase C Phospholipase D TRPC3	Balb/c mice	[57]

(continued)

Table 19.2 (continued)

Anatomical localization	Cellular type	Recording technique	Pharmacology	Molecular mechanisms involved	Species	Reference
Urethra	ICC	Calcium imaging	H-89 (10 μM) Forskolin (10 μM) 8-Bromo-cAMP (1 mM)	Protein kinase A (PKA)	New Zealand white rabbits	[58]
	ICC	Calcium imaging	KB-R 7943 (3 μM)	NCX	New Zealand white rabbits	[59]
			SEA-0400 (1 μM) Caffeine (1 mM)	NCX		
			Tetracaine (100 μM)	RyR		
			2-APB (100 μM) U73122 (10 μM)	IP3R		
Fallopian tube	ICC-like/TC	Field potentials	–	–	Human	[60]
Myometrium	ICC-like/TC	Field potentials	–	–	Human	[61]

[a]Only the papers describing the electrical activity recorded on cell cultures were considered and not those dedicated to electrical recordings on strips of tissue

19.3.1 Role of Telocytes in the Pacemaker Activity of Internal Cavitary Organs

ICC-like cells isolated from the urethra [58, 59], bladder [62], prostate [63], *corpus cavernosum* [64], small intestine [56], embryonic intestine [55], Fallopian tube [60], and myometrium [61] present rhythmic firing activities acting also as neuromodulators [65]. These cells which act as pacemakers can modulate the activities of muscular, nervous, or secretory systems. In ICC from the guinea pig prostate, the spontaneous transient depolarizations are initiated with the opening of a small number of Ca^{2+}-activated Cl^- channels followed by a small membrane depolarization which triggers the calcium influx through T-type calcium channels, and furthermore the summation of calcium transients would manifest in the pacemaker potential that opens L-type calcium channels in ICC and their neighboring smooth muscle cells [20].

Patch-clamp experiments using cultured ICC from Balb/C mice urinary bladder revealed that these cells act as pacemaker, presenting individual spikes and bursting potentials similar to those observed in intact bladder tissues [53]. These spontaneous electrical potentials were inhibited by nifedipine (L-type voltage-gated calcium channel blocker) suggesting the involvement of these types of calcium channels [53]. Another study conducted on the rat bladder shows that spontaneous calcium activity is not influenced by L-type Ca^{2+} channels but rather by the T-type calcium channels [66]. The carbachol-induced calcium oscillations were blocked by atropine (a muscarinic receptor antagonist) [53, 54]. These findings suggested the possible role of voltage-gated calcium channels and muscarinic receptors in generating the pacemaker behavior in bladder ICC.

Although, electrophysiological recordings on TCs from human myometrium failed to prove the presence of Na^+ currents [9, 22], studies on transgenic mice have proved that longitudinal contractions of the uterus depend on a KIT signaling pathway of ICC-like cells [67]. On the other hand, the spontaneous electrical activity recorded on the ICC urinary bladder, small intestine, or urethra is based on intracellular calcium changes [53–59].

Calcium imaging on urethra ICC suggests that PKA, RyR, IP3R, and NCX are involved in ICC pacemaker activity [58, 59, 65]. The pacemaker activity in small intestine ICC is modulated through 5-HT_3 and 5-HT_4 receptors, chloride channels, Ca^{2+}-ATPase from the endoplasmic reticulum, phospholipase A, phospholipase C, phospholipase D, and TRPC3 [56, 57]. Calcium imaging studies on embryonic mouse intestinal ICC showed the role of L-type voltage-gated calcium channels in rhythmic electric activities [55]. These results revealed that the pacemaker mechanism is more complex and needs to be studied in an integrated manner. Calcium imaging studies on embryonic mouse intestinal ICC showed the role of L-type voltage-gated calcium channels in rhythmic electric activities [55]. In human myometrium and Fallopian tubes, Cajal-like cells/TC present spontaneous electric activity without a rhythmical pattern [60, 61].

ICC have been proposed to be the pacemaker cells in the gastrointestinal tract. In the small intestine, ICC associated with the myenteric plexus are generating slow waves that contribute to the rhythmic muscle contractions in the proximal intestine

[68]. In the colon, ICC associated with the submuscular plexus contribute to slow waves in canine, rat, mouse, and human [69–73], while ICC associated with the myenteric plexus do not contribute to slow waves in rat and mouse but generate rhythmic transient depolarizations of low and variable frequency as a result of L-type calcium channels activation [72, 73]. In the colon, ICC associated with myenteric plexus and intramuscular plexus are cooperating for the generation of pacemaker activity, and their excitability is regulated by the cholinergic inhibition of $K_v7.5$ channels [19]. Hyperpolarization-activated cyclic nucleotide currents from mouse colonic ICC are tonically activated by basal cAMP production and participate in the regulation of the pacemaker activity [24].

19.3.2 Role of Telocytes in the Pacemaker Activity of the Heart

Cav3.1 plays a role in sinoatrial node automaticity [74] and atrioventricular node automaticity [75]. Extrapolation of Cav3.1 involvement in TC automatism or in tissues containing TC automatism requires further investigations. Cardiac TCs have been suggested to be implicated in cardiac rhythm and atrial fibrillation [76, 77].

Sodium-calcium exchanger is involved in the pacemaker activity of the sinoatrial node [78] and in the overactive bladder in transgenic mice overexpressing NCX1.3 [79]. It is assumed that the activity of NCX on TC is modulated by ionic concentrations, by gap junctions, and telepods length that can develop the pacemaker functionality of a tissue rather than of individual and independent cells.

Ca^{2+}-activated Cl^- channels are found in many cell types and have different roles, among them being cardiac rhythmic depolarization, modulation of smooth muscle contraction, and taste receptor modulation. Their role is unknown in TCs, but the ability of these cells to have long extensions, gap junctions, and an increased dynamic of telepods apparently without stimulus creates opportunities for studying TC behavior under the influence of Ca^{2+}-activated Cl^- channels. TC feature to have a rhythmic electrical activity could be attributed to these channels. In vivo gap junctions between TC and adjacent cells could explain TC calcium dynamics changes in the presence of neighboring muscle cells and also the lack of in vivo electric automatism in the absence of gap junctions. Further studies are needed to assess the role of Ca^{2+}-activated Cl^- channels in TCs and their possible modulatory effect on chemoreceptors.

19.4 What's Next?

19.4.1 TCs Are Sensitive to Stimuli That Modulate Membrane Fluidity

It was evidenced that TC uterine motility can be modulated by mechanical stimuli via optical tweezer [28]. Blocking the voltage-gated T-type calcium channels, which have a degree of mechanosensitivity, decreases this optical modulation by means of mechanical forces. Channels that have a degree of mechanosensitivity are sensitive to external mechanical stimulation but also to changes in membrane fluidity. Imatinib, a tyrosine kinase inhibitor that can be used in treating patients with chronic myeloid leukemia, may alter membrane fluidity through alteration of lipid metabolism [80]. Imatinib can decrease spontaneous contractile activity in guinea pig models and in human nonpregnant myometrium [81, 82]. Moreover, any change in the concentration of steroid hormones can cause changes in membrane fluidity and influence the functionality of mechanosensitive-like ion channels.

Future studies should focus on cell-cell communication and to explore the influence exerted on TC function by the surrounding myocytes through various factors that might affect membrane fluidity. Our hypothesis is that TCs could be involved in the stimulation of muscle development where mechanical stress is elevated (e.g., uterine musculature). Besides the uterus, TCs may represent a mechanical sensor that contributes to the pacemaker activity in the heart, gastrointestinal tract, or urinary tract.

19.4.2 Hormonal Regulation of TCs: Role in Birth Delivery

We hypothesize the TCs involvement in a feedback loop control of uterus that triggers the contraction initiation in birth delivery. The plasticity of the uterine musculature in pregnancy [83] might be under the influence of TCs that can modulate the activity of the surrounding myocytes. Rhythmic muscular activities are associated to significant vascular changes in the uterus involving endocrine and humoral factors release that could be detected by TCs.

The increase in placental volume and the level of secreted steroid hormones can modulate the cellular membrane fluidity [84, 85] and subsequently the activity of TCs, the function of myocytes, the frequency and force of contractions, and finally the endocrine feedback loop leading to fetal expulsion. These hypotheses should be tested in future studies on the interactions of TCs with surrounding cells in a such manner that can integrate mechanical and hormonal modulation with the therapeutic purpose of preventing the premature birth.

19.4.3 Integrating Information About TCs

There is a lack of comprehensive knowledge on the ion channels functionally expressed in TCs due to the variety of experimental approaches, including species, age differences, and methodology of analysis (in vitro and in vivo studies, staining on living and fixed tissue, presence or absence of neighboring cells, etc.). An important limitation in clinical studies is represented by the reduced number of samples from patients.

The studies conducted so far led to the hypothesis that TCs play a role in intercellular communication. However, the role of TCs in tissue excitability and pacemaker activity is still unclear. Therefore, it is imperative to connect all the information available on TCs and to understand their physiological role and their involvement in a variety of pathological conditions (e.g., psoriasis, myocardial infarction, focal lymphocytic sialadenitis, preeclampsia, ulcerative colitis, etc.) [86–90]. In conclusion, it is a great challenge to explain how TCs with distinct protein expression profiles (e.g., ion channels) from different tissues are correlated with similar functions, morphology, and dynamics of these cells.

Acknowledgments We are grateful for the financial support received through the program Partnerships in priority areas – PN II, UEFISCDI, Project No. 82/2012. B M Radu has a PhD fellowship from the Italian Ministry of Research (MIUR). DD Banciu is financed by the Sectoral Operational Programme Human Resources Development (SOPHRD), the European Social Fund, and the Romanian Government under the contract number POSDRU/159/1.5/S/141531.

References

1. Ceafalan L, Gherghiceanu M, Popescu LM, Simionescu O. Telocytes in human skin – are they involved in skin regeneration? J Cell Mol Med. 2012;16(7):1405–20.
2. Gevaert T, De Vos R, Van Der Aa F, Joniau S, van den Oord J, Roskams T, De Ridder D. Identification of telocytes in the upper lamina propria of the human urinary tract. J Cell Mol Med. 2012;16(9):2085–93.
3. Koh BH, Roy R, Hollywood MA, Thornbury KD, McHale NG, Sergeant GP, Hatton WJ, Ward SM, Sanders KM, Koh SD. Platelet-derived growth factor receptor-α cells in mouse urinary bladder: a new class of interstitial cells. J Cell Mol Med. 2012;16(4):691–700.
4. Rusu MC, Nicolescu MI, Jianu AM, Lighezan R, Mănoiu VS, Păduraru D. Esophageal telocytes and hybrid morphologies. Cell Biol Int. 2012;36(12):1079–88.
5. Rusu MC, Jianu AM, Mirancea N, Didilescu AC, Mănoiu VS, Păduraru D. Tracheal telocytes. J Cell Mol Med. 2012;16(2):401–5.
6. Suciu LC, Popescu BO, Kostin S, Popescu LM. Platelet-derived growth factor receptor-β-positive telocytes in skeletal muscle interstitium. J Cell Mol Med. 2012;16(4):701–7.
7. Qi G, Lin M, Xu M, Manole CG, Wang X, Zhu T. Telocytes in the human kidney cortex. J Cell Mol Med. 2012;16(12):3116–22.
8. Corradi LS, Jesus MM, Fochi RA, Vilamaior PS, Justulin Jr LA, Góes RM, Felisbino SL, Taboga SR. Structural and ultrastructural evidence for telocytes in prostate stroma. J Cell Mol Med. 2013;17(3):398–406.
9. Cretoiu SM, Cretoiu D, Marin A, Radu BM, Popescu LM. Telocytes: ultrastructural, immunohistochemical and electrophysiological characteristics in human myometrium. Reproduction. 2013;145(4):357–70.
10. Díaz-Flores L, Gutiérrez R, Sáez FJ, Díaz-Flores Jr L, Madrid JF. Telocytes in neuromuscular spindles. J Cell Mol Med. 2013;17(4):457–65.
11. Luesma MJ, Gherghiceanu M, Popescu LM. Telocytes and stem cells in limbus and uvea of mouse eye. J Cell Mol Med. 2013;17(8):1016–24.
12. Matyja A, Gil K, Pasternak A, Sztefko K, Gajda M, Tomaszewski KA, Matyja M, Walocha JA, Kulig J, Thor P. Telocytes: new insight into the pathogenesis of gallstone disease. J Cell Mol Med. 2013;17(6):734–42.
13. Vannucchi MG, Traini C, Manetti M, Ibba-Manneschi L, Faussone-Pellegrini MS. Telocytes express PDGFRα in the human gastrointestinal tract. J Cell Mol Med. 2013;17(9):1099–108.

14. Xiao J, Wang F, Liu Z, Yang C. Telocytes in liver: electron microscopic and immunofluorescent evidence. J Cell Mol Med. 2013;17(12):1537–42.
15. Zhao B, Chen S, Liu J, Yuan Z, Qi X, Qin J, Zheng X, Shen X, Yu Y, Qnin TJ, Chan JY, Cai D. Cardiac telocytes were decreased during myocardial infarction and their therapeutic effects for ischaemic heart in rat. J Cell Mol Med. 2013;17(1):123–33.
16. Li L, Lin M, Li L, Wang R, Zhang C, Qi G, Xu M, Rong R, Zhu T. Renal telocytes contribute to the repair of ischemically injured renal tubules. J Cell Mol Med. 2014;18(6):1144–56.
17. Niculite CM, Regalia TM, Gherghiceanu M, Huica R, Surcel M, Ursaciuc C, Leabu M, Popescu LM. Dynamics of telopodes (telocyte prolongations) in cell culture depends on extracellular matrix protein. Mol Cell Biochem. 2015;398(1–2):157–64.
18. Roatesi I, Radu BM, Cretoiu D, Cretoiu SM. Uterine telocytes: a review of current knowledge. Biol Reprod. 2015;93(1):10.
19. Wright GW, Parsons SP, Loera-Valencia R, Wang XY, Barajas-López C, Huizinga JD. Cholinergic signalling-regulated KV7.5 currents are expressed in colonic ICC-IM but not ICC-MP. Pflugers Arch. 2014;466(9):1805–18.
20. Lang RJ, Tonta MA, Takano H, Hashitani H. Voltage-operated Ca2+ currents and Ca2+ -activated Cl- currents in single interstitial cells of the guinea pig prostate. BJU Int. 2014;114(3): 436–46.
21. Sheng J, Shim W, Lu J, Lim SY, Ong BH, Lim TS, Liew R, Chua YL, Wong P. Electrophysiology of human cardiac atrial and ventricular telocytes. J Cell Mol Med. 2014;18(2):355–62.
22. Duquette RA, Shmygol A, Vaillant C, Mobasheri A, Pope M, Burdyga T, Wray S. Vimentin-positive, c-kit-negative interstitial cells in human and rat uterus: a role in pacemaking? Biol Reprod. 2005;72(2):276–83.
23. Rosenbaum ST, Svalø J, Nielsen K, Larsen T, Jørgensen JC, Bouchelouche P. Immunolocalization and expression of small-conductance calcium-activated potassium channels in human myometrium. J Cell Mol Med. 2012;16(12):3001–8.
24. Shahi PK, Choi S, Zuo DC, Kim MY, Park CG, Kim YD, Lee J, Park KJ, So I, Jun JY. The possible roles of hyperpolarization-activated cyclic nucleotide channels in regulating pacemaker activity in colonic interstitial cells of Cajal. J Gastroenterol. 2013;49(6):1001–10.
25. Si X, Huang L, Gong Y, Lu J, Lin L. Role of calcium in activation of hyperpolarization-activated cyclic nucleotide-gated channels caused by cholecystokinin octapeptide in interstitial cells of cajal. Digestion. 2012;85(4):266–75.
26. McCloskey KD. Calcium currents in interstitial cells from the guinea-pig bladder. BJU Int. 2006;97(6):1338–43.
27. Cretoiu SM, Radu BM, Banciu A, Banciu DD, Cretoiu D, Ceafalan LC, Popescu LM. Isolated human uterine telocytes: immunocytochemistry and electrophysiology of T-type calcium channels. Histochem Cell Biol. 2015;143(1):83–94.
28. Campeanu RA, Radu BM, Cretoiu SM, Banciu DD, Banciu A, Cretoiu D, Popescu LM. Near-infrared low-level laser stimulation of telocytes from human myometrium. Lasers Med Sci. 2014;29(6):1867–74.
29. Parsons SP, Kunze WA, Huizinga JD. Maxi-channels recorded in situ from ICC and pericytes associated with the mouse myenteric plexus. Am J Physiol Cell Physiol. 2012;302:C1055–69.
30. Lerche C, Scherer CR, Seebohm G, Derst C, Wei AD, Busch AE, Steinmeyer K. Molecular cloning and functional expression of KCNQ5, a potassium channel subunit that may contribute to neuronal M-current diversity. J Biol Chem. 2000;275:22395–400.
31. Schroeder BC, Hechenberger M, Weinreich F, Kubisch C, Jentsch TJ. KCNQ5, a novel potassium channel broadly expressed in brain, mediates M-type currents. J Biol Chem. 2000;275:24089–95.
32. Bal M, Zhang J, Zaika O, Hernandez CC, Shapiro MS. Homomeric and heteromeric assembly of KCNQ(Kv7) K+ channels assayed by total internal reflection fluorescence/fluorescence resonance energy transfer and patch clamp analysis. J Biol Chem. 2008;283:30668–76.
33. Hinescu ME, Popescu LM. Interstitial Cajal-like cells (ICLC) in human atrial myocardium. J Cell Mol Med. 2005;9(4):972–5.

34. Hinescu ME, Gherghiceanu M, Mandache E, Ciontea SM, Popescu LM. Interstitial Cajal-like cells (ICLC) in atrial myocardium: ultrastructural and immunohistochemical characterization. J Cell Mol Med. 2006;10(1):243–57.
35. Platzer J, Engel J, Schrott-Fischer A, Stephan K, Bova S, Chen H, Zheng H, Striessnig J. Congenital deafness and sinoatrial node dysfunction in mice lacking class D L-type Ca2+ channels. Cell. 2000;102:89–97.
36. Zhang Z, Xu Y, Song H, Rodriguez J, Tuteja D, Namkung Y, Shin HS, Chiamvimonvat N. Functional roles of Ca(v)1.3 (alpha(1D)) calcium channel in sinoatrial nodes: insight gained using gene-targeted null mutant mice. Circ Res. 2002;90:981–7.
37. Mangoni ME, Couette B, Bourinet E, Platzer J, Reimer D, Striessnig J, Nargeot J. Functional role of L-type Cav1.3 Ca2+ channels in cardiac pacemaker activity. Proc Natl Acad Sci U S A. 2003;100:5543–8.
38. Mesirca P, Torrente AG, Mangoni ME. Functional role of voltage gated Ca(2+) channels in heart automaticity. Front Physiol. 2015;6:19.
39. Popescu LM, Manole CG, Gherghiceanu M, Ardelean A, Nicolescu MI, Hinescu ME, Kostin S. Telocytes in human epicardium. J Cell Mol Med. 2010;14(8):2085–93.
40. Leiria LO, Mónica FZ, Carvalho FD, Claudino MA, Franco-Penteado CF, Schenka A, Grant AD, De Nucci G, Antunes E. Functional, morphological and molecular characterization of bladder dysfunction in streptozotocin-induced diabetic mice: evidence of a role for L-type voltage-operated Ca2+ channels. Br J Pharmacol. 2011;163(6):1276–88.
41. Balkanci ZD, Pehlivanoğlu B, Bayrak S, Karabulut I, Karaismailoğlu S, Erdem A. The effect of hypercholesterolemia on carbachol-induced contractions of the detrusor smooth muscle in rats: increased role of L-type Ca2+ channels. Naunyn Schmiedeberg's Arch Pharmacol. 2012; 385(11):1141–8.
42. Sun YH, Gao X, Tang YJ, Xu CL, Wang LH. Androgens induce increases in intracellular calcium via a G protein-coupled receptor in LNCaP prostate cancer cells. J Androl. 2006; 27(5):671–8.
43. Kim YH, Chung S, Lee YH, Kim EC, Ahn DS. Increase of L-type Ca2+ current by protease-activated receptor 2 activation contributes to augmentation of spontaneous uterine contractility in pregnant rats. Biochem Biophys Res Commun. 2012;418(1):167–72.
44. Gui P, Chao JT, Wu X, Yang Y, Davis GE, Davis MJ. Coordinated regulation of vascular Ca2+ and K+ channels by integrin signaling. Adv Exp Med Biol. 2010;674:69–79.
45. Peers C, Scragg JL, Boyle JP, Fearon IM, Taylor SC, Green KN, Webster NJ, Ramsden M, Pearson HA. A central role for ROS in the functional remodelling of L-type Ca2+ channels by hypoxia. Philos Trans R Soc Lond B Biol Sci. 2005;360(1464):2247–54.
46. Svenningsen P, Andersen K, Thuesen AD, Shin HS, Vanhoutte PM, Skott O, Jensen BL, Hill C, Hansen PB. T-type Ca channels facilitate NO-formation, vasodilatation and NO-mediated modulation of blood pressure. Pflügers Arch. 2014;466:2205–14.
47. Ball CJ, Wilson DP, Turner SP, Saint DA, Beltrame JF. Heterogeneity of L- and T-channels in the vasculature: rationale for the efficacy of combined L- and T-blockade. Hypertension. 2009;53:654–60.
48. Thuesen A, Andersen H, Cardel M, Toft A, Walter S, Marcussen Jensen B, Bie P, Hansen P. Differential effect of T-type voltage-gated calcium channel disruption on renal plasma flow and glomerular filtration rate in vivo. Am J Physiol Ren Physiol. 2014;307:F445–52.
49. Chen CC, Fan YP, Shin HS, Su CK. Basal sympathetic activity generated in neonatal mouse brainstem-spinal cord preparation requires T-type calcium channel subunit 1H. Exp Physiol. 2011;96:486–94.
50. Hofmann F, Lacinova L, Klugbauer N. Voltage-dependent calcium channels: from structure to function. Rev Physiol Biochem Pharmacol. 1999;139:33–87.
51. Zhong X, Deng J, He P, You N, Wang Q, Song B, Li L. Reverse mode of the sodium/calcium exchanger subtype 3 in interstitial cells of Cajal from rat bladder. Urology. 2013;82(1):254. e7–12.
52. Zhu Y, Mucci A, Huizinga JD. Inwardly rectifying chloride channel activity in intestinal pacemaker cells. Am J Physiol Gastrointest Liver Physiol. 2005;288(4):G809–21.

53. Kim SO, Jeong HS, Jang S, Wu MJ, Park JK, Jiao HY, Jun JY, Park JS. Spontaneous electrical activity of cultured interstitial cells of cajal from mouse urinary bladder. Korean J Physiol Pharmacol. 2013;17(6):531–6.
54. Wu Y, Shi C, Deng J, Zhang X, Song B, Li L. Expression and function of muscarinic subtype receptors in bladder interstitial cells of cajal in rats. Urol J. 2014;11(3):1642–7.
55. Ito-Dufros Y, Funakoshi Y, Uehara A, Oishi K. In vitro development of gut-like tissue demonstrating rhythmic contractions from embryonic mouse intestinal cells. Neurogastroenterol Motil. 2007;19(4):288–300.
56. Kim BJ, Kim HW, Lee GS, Choi S, Jun JY, So I, Kim SJ. Poncirus trifoliate fruit modulates pacemaker activity in interstitial cells of Cajal from the murine small intestine. J Ethnopharmacol. 2013;149(3):668–75.
57. Kim BJ, Kwon YK, Kim E, So I. Effects of histamine on cultured interstitial cells of cajal in murine small intestine. Korean J Physiol Pharmacol. 2013;17(2):149–56.
58. Drumm BT, Sergeant GP, Hollywood MA, Thornbury KD, McHale NG, Harvey BJ. The role of cAMP dependent protein kinase in modulating spontaneous intracellular Ca2+ waves in interstitial cells of Cajal from the rabbit urethra. Cell Calcium. 2014;56(3):181–7.
59. Drumm BT, Large RJ, Hollywood MA, Thornbury KD, Baker SA, Harvey BJ, McHale NG, Sergeant GP. The role of Ca(2+) influx in spontaneous Ca(2+) wave propagation in interstitial cells of Cajal from the rabbit urethra. J Physiol. 2015;593(15):3333–50.
60. Popescu LM, Ciontea SM, Cretoiu D, Hinescu ME, Radu E, Ionescu N, Ceausu M, Gherghiceanu M, Braga RI, Vasilescu F, Zagrean L, Ardeleanu C. Novel type of interstitial cell (Cajal-like) in human fallopian tube. J Cell Mol Med. 2005;9(2):479–523.
61. Ciontea SM, Radu E, Regalia T, Ceafalan L, Cretoiu D, Gherghiceanu M, Braga RI, Malincenco M, Zagrean L, Hinescu ME, Popescu LM. C-kit immunopositive interstitial cells (Cajal-type) in human myometrium. J Cell Mol Med. 2005;9(2):407–20.
62. McCloskey KD. Interstitial cells in the urinary bladder—localisation and function. Neurourol Urodyn. 2010;29:82–7.
63. Nguyen DT, Dey A, Lang RJ, Ventura S, Exintaris B. Contractility and pacemaker cells in the prostate gland. J Urol. 2011;185:347–51.
64. Hashitani H, Suzuki H. Identification of interstitial cells of Cajal in corporal tissues of the guinea-pig penis. Br J Pharmacol. 2004;141:199–204.
65. Drumm BT, Koh SD, Andersson KE, Ward SM. Calcium signalling in Cajal-like interstitial cells of the lower urinary tract. Nat Rev Urol. 2014;11(10):555–64.
66. Deng J, He P, Zhong X, Wang Q, Li L, Song B. Identification of T-type calcium channels in the interstitial cells of Cajal in rat bladder. Urology. 2012;80(6):1389.e1–e7.
67. Allix S, Reyes-Gomez E, Aubin-Houzelstein G, Noël D, Tiret L, Panthier JJ, Bernex F. Uterine contractions depend on KIT-positive interstitial cells in the mouse: genetic and pharmacological evidence. Biol Reprod. 2008;79(3):510–7.
68. Thomsen L, Robinson TL, Lee JC, Farraway LA, Hughes MJ, Andrews DW, Huizinga JD. Interstitial cells of Cajal generate a rhythmic pacemaker current. Nat Med. 1998;4:848–51.
69. Smith TK, Reed JB, Sanders KM. Interactions of two electrical pacemakers in muscularis of canine proximal colon. Am J Physiol. 1987;252:C290–9.
70. Liu LW, Huizinga JD. Electrical coupling of circular muscle to longitudinal muscle and interstitial cells of Cajal in canine colon. J Physiol. 1993;470:445–61.
71. Rae MG, Fleming N, McGregor DB, Sanders KM, Keef KD. Control of motility patterns in the human colonic circular muscle layer by pacemaker activity. J Physiol. 1998;510(Pt1):309–20.
72. Pluja L, Alberti E, Fernandez E, Mikkelsen HB, Thuneberg L, Jiminez M. Evidence supporting presence of two pacemakers in rat colon. Am J Physiol Gastrointest Liver Physiol. 2001;281:G255–66.
73. Yoneda S, Fukui H, Takaki M. Pacemaker activity from submucosal interstitial cells of Cajal drives high-frequency and low-amplitude circular muscle contractions in the mouse proximal colon. Neurogastroenterol Motil. 2004;16:621–7.
74. Mangoni ME, Traboulsie A, Leoni AL, Couette B, Marger L, Le Quang K, Kupfer E, Cohen-Solal A, Vilar J, Shin HS, Escande D, Charpentier F, Nargeot J, Lory P. Bradycardia and slow-

ing of the atrioventricular conduction in mice lacking CaV3.1/alpha1G T-type calcium channels. Circ Res. 2006;98(11):1422–30.
75. Marger L, Mesirca P, Alig J, Torrente A, Dubel S, Engeland B, Kanani S, Fontanaud P, Striessnig J, Shin HS, Isbrandt D, Ehmke H, Nargeot J, Mangoni ME. Functional roles of Ca(v)1.3, Ca(v)3.1 and HCN channels in automaticity of mouse atrioventricular cells: insights into the atrioventricular pacemaker mechanism. Channels (Austin). 2011;5:251–61.
76. Popescu LM, Gherghiceanu M, Manole CG, Faussone-Pellegrini MS. Cardiac renewing: interstitial Cajal-like cells nurse cardiomyocyte progenitors in epicardial stem cell niches. J Cell Mol Med. 2009;13(5):866–86.
77. Kostin S, Popescu LM. A distinct type of cell in myocardium: interstitial Cajal-like cells (ICLCs). J Cell Mol Med. 2009;13(2):295–308.
78. Torrente AG, Zhang R, Zaini A, Giani JF, Kang J, Lamp ST, Philipson KD, Goldhaber JI. Burst pacemaker activity of the sinoatrial node in sodium-calcium exchanger knockout mice. Proc Natl Acad Sci U S A. 2015;112(31):9769–74.
79. Yamamura H, Cole WC, Kita S, Hotta S, Murata H, Suzuki Y, Ohya S, Iwamoto T, Imaizumi Y. Overactive bladder mediated by accelerated Ca2+ influx mode of Na+/Ca2+ exchanger in smooth muscle. Am J Physiol Cell Physiol. 2013;305(3):C299–308.
80. Iurlo A, Orsi E, Cattaneo D, Resi V, Bucelli C, Orofino N, Sciumè M, Elena C, Grancini V, Consonni D, Orlandi EM, Cortelezzi A. Effects of first- and second-generation tyrosine kinase inhibitor therapy on glucose and lipid metabolism in chronic myeloid leukemia patients: a real clinical problem? Oncotarget. 2015;6(32):33944–51.
81. Lam M, Dey A, Lang RJ, Exintaris B. Effects of imatinib mesylate on the spontaneous activity generated by the guinea-pig prostate. BJU Int. 2013;112(4):E398–405.
82. Cretoiu SM, Simionescu AA, Caravia L, Curici A, Cretoiu D, Popescu LM. Complex effects of imatinib on spontaneous and oxytocin-induced contractions in human non-pregnant myometrium. Acta Physiol Hung. 2011;98(3):329–38.
83. Wu X, Morgan KG, Jones CJ, Tribe RM, Taggart MJ. Myometrial mechanoadaptation during pregnancy: implications for smooth muscle plasticity and remodelling. J Cell Mol Med. 2008;12(4):1360–73.
84. Jansson T, Powell TL, Illsley NP. Gestational development of water and non-electrolyte permeability of human syncytiotrophoblast plasma membranes. Placenta. 1999;20(2–3):155–60.
85. Whiting KP, Restall CJ, Brain PF. Steroid hormone-induced effects on membrane fluidity and their potential roles in non-genomic mechanisms. Life Sci. 2000;67(7):743–57.
86. Manole CG, Gherghiceanu M, Simionescu O. Telocyte dynamics in psoriasis. J Cell Mol Med. 2015;19(7):1504–19.
87. Manole CG, Cismaşiu V, Gherghiceanu M, Popescu LM. Experimental acute myocardial infarction: telocytes involvement in neo-angiogenesis. J Cell Mol Med. 2011;15(11):2284–96.
88. Alunno A, Ibba-Manneschi L, Bistoni O, Rosa I, Caterbi S, Gerli R, Manetti M. Telocytes in minor salivary glands of primary Sjögren's syndrome: association with the extent of inflammation and ectopic lymphoid neogenesis. J Cell Mol Med. 2015;19(7):1689–96.
89. Bosco C, Díaz E, Gutiérrez R, González J, Parra-Cordero M, Rodrigo R, Barja P. A putative role for telocytes in placental barrier impairment during preeclampsia. Med Hypotheses. 2015;84(1):72–7.
90. Manetti M, Rosa I, Messerini L, Ibba-Manneschi L. Telocytes are reduced during fibrotic remodelling of the colonic wall in ulcerative colitis. J Cell Mol Med. 2015;19(1):62–73.

Chapter 20
The Cutaneous Telocytes

Catalin G. Manole and Olga Simionescu

20.1 Introduction

The milestone paper from Professors L.M. Popescu and M.S. Faussone-Pellegrini entitled "*TELOCYTES – a case of serendipity: the winding way from Interstitial Cells of Cajal (ICC)*, via *Interstitial Cajal-Like Cells (ICLC) to TELOCYTES*" described and characterized a new population of interstitial cells – telocytes (TCs). Two years later, Popescu's team extended their research to the presence of TCs in the skin. Human skin is the largest organ, having a surface of 1.75 m^2. Therefore, the description and documentation of the presence of a new cell type in the dermal stroma is very important for skin metabolism and homeostasis. In the past 6 years (2010–2016), the interest in TCs has increased progressively (see Fig. 20.1). According to the ISI Web of Knowledge, at the beginning of 2016 there were 175 published papers focused on TCs, with more than 3,205 citations, leading to a Hirsch index of 34. In addition, the dedicated website www.telocytes.com was accessed 3,950 times (according to Google Analytics) during the last 2 months.

C.G. Manole, MD, PhD (✉)
1st Clinic of Dermatology, 'Carol Davila' University of Medicine and Pharmacy, Colentina Clinical Hospital, Bucharest, Romania

Department of Cell Biology and Histology, 'Carol Davila' University of Medicine and Pharmacy, Bucharest, Romania

Laboratory of Molecular Medicine, 'Victor Babeş' National Institute of Pathology, Bucharest, Romania
e-mail: catalin.manole@jcmm.org

O. Simionescu, MD, PhD
1st Clinic of Dermatology, 'Carol Davila' University of Medicine and Pharmacy, Colentina Clinical Hospital, Bucharest, Romania
e-mail: contact@topdermuni.ro

X. Wang, D. Cretoiu (eds.), *Telocytes*, Advances in Experimental Medicine and Biology, 913, DOI 10.1007/978-981-10-1061-3_20

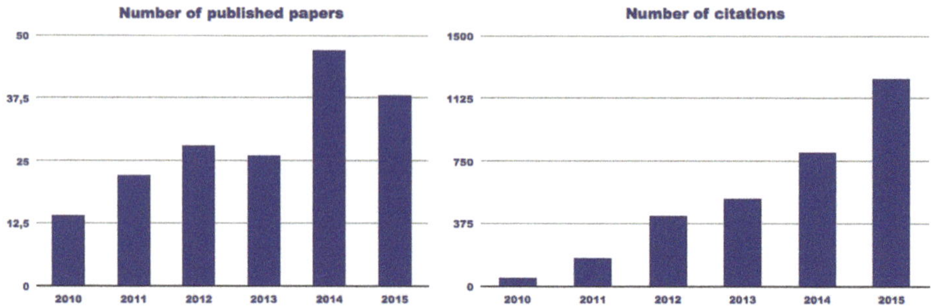

Fig. 20.1 Current publications and citations regarding telocytes in the ISI Web of Knowledge

20.1.1 Where Are TCs Located?

The TCs are found within the stroma of many human and mammalian cavitary and non-cavitary organs (see Table 20.1). There is not an accurate single immunohistochemical marker for TCs; therefore, they are best identified using a combination of markers (CD34 and PDGFRα) along with transmission electron microscopy data. The cardiovascular system, respiratory system, musculoskeletal system, digestive system, nervous system, skin, eyes, and genitourinary system are influenced by the presence of TCs within their stroma [19].

20.1.2 Why TCs Were Previously Overlooked?

A natural question was raised from the very beginning by Prof. L.M. Popescu, the "Father of Telocytes": "Why telocytes were previously overlooked?". In the era of scientific and technologic progress, histology is a morphological science classically considered "closed". Perhaps the physical dimensions of TCs, especially of their (ultra)structural hallmarks – telopodes, prevented observation due to the physical limits of light (or even electron) microscopy. It is still difficult to view TC differentiation from the rest of stromal cells by light microscopy. In addition, TCs share immunophenotypical features with other interstitial cells and research is still ongoing to identify specific skin TC markers to individualize their immunophenotype [7, 37].

It is well-known that the resolving power of light microscopy is 0.2 μm. Incidentally, this value represents the mean diameter of podomers. Moreover, the best histological sections obtained from a microtome could be about 2–4 μm, and would not offer the best resolution. However, the three-dimensionality of TCs and distribution of telopodes might prevent capturing an entire cell in the same plane of section. Low magnifications could not discriminate the details of TCs; while, higher magnifications (immersion objectives) significantly decrease the area of observation in favor of more details. Hence, in light microscopy, capturing an entire TC with its telopodes is a lucky incident.

Table 20.1 The cavitary and non-cavitary organs where interstitium telocytes are located

	Organ	Reference
Cardiovascular system	Heart (epicardium, myocardium, endocardium)	[9, 23, 30, 51, 53, 62]
	Heart valves	[69]
	Pulmonary veins	[49]
	Aorta, blood vessels	[6, 32, 70]
Respiratory system	Trachea and bronchi	[56, 72, 74]
	Lungs	[21, 50, 59, 71, 72, 74]
	Pleura	[27]
Nervous system	Meninges and choroid plexus	[48]
Musculoskeletal system	Striated muscle	[3, 17, 61]
	Fascia lata	[16]
Skin	Papillary dermis and reticular dermis	[7, 13, 29, 57]
Eye	Limbus and uvea	[35]
Digestive system	Esophagus	[10]
	Duodenum	[4]
	Jejunum	[12]
	Colon	[65]
	Salivary glands	[1, 43]
	Liver	[66, 67]
	Gallbladder	[38]
	Pancreas	[44, 45]
Hematopoietic organs	Spleen	[8]
	Bone marrow	[33]
Genitourinary system	Mammary gland	[39, 41, 47]
	Fallopian tube	[49, 68]
	Myometrium	[14, 15, 54]
	Placenta	[60]
	Kidney	[34, 52, 73]
	Ureter and urinary bladder	[22, 55, 64]
	Prostate	[11]

When using light microscopy in usual histology, another major obstacle in discriminating TCs is their similarity with the surrounding fibroblast-like cells, and also their close apposition with these cells. Hence, the best combination of immunohistochemistry markers must be used for TCs. Based on their positive marker expression for CD34, c-kit/CD117, and vimentin, two TC subtypes were documented in the human dermis [7, 37]. These data are supported by the most recent results of Vannucchi et al. regarding the co-existence of multiple subpopulations of TCs (perhaps with organ function adapted morphologies) in the urinary bladder [64], and the existence of multiple immunophenotypes of cardiac TCs [9].

The resolution of electron microscopy is limited by the size of the observation area, since there is a higher range of magnifications. However, a low transmission electron microscopy magnification does not permit a clear discrimination of TCs from the other surrounding cells. Also, a higher transmission electron microscopy magnification offers ultrastructural details, but drastically decreases the area of observation, not allowing the identification of a TC. To obtain ultrastructural details of a complete TC, collages were made from many increased resolution images. However, the most reliable method for identification of TCs remains transmission electron microscopy [5].

20.2 Telocytes in Human Normal Skin

20.2.1 The Ultrastructure of Skin TCs

20.2.1.1 General Aspect

Ultrastructurally, dermal TCs recapitulate the features of TCs described in other organs [7]. As observed in Fig. 20.2a, in transmission electron microscopy, dermal TCs have:

(a) A small ovoid cell body;
(b) A large nucleus (~25 % of cell volume) with a considerable amount of euchromatin, but also heterochromatin (appears as clusters attached to the inner face of the nuclear envelope) [49];
(c) Scarcely perinuclear cytoplasm with a small number of mitochondria (~5 % of cell volume), small amount of Golgi cisterns, rough and smooth endoplasmic reticulum, as well as elements of the cytoskeleton [49];
(d) The existence of an intermittent thin basal lamina;
(e) Considerable amount of caveolae;
(f) Several cellular prolongations. The number of cellular prolongations give different shapes to the cellular body: piriform (1 prolongation), spindle shape (2 prolongations), triangular (3 prolongations), and stellate (4 or more prolongations).

The cellular prolongations of dermal TCs represent their (ultra)structural hallmarks, and were termed *telopodes*. In fact, considering the proximate genus and specific difference, we may state that the shortest definition for a TC is "a cell with telopodes" (Popescu 2010).

20.2.1.2 Telopodes

The cellular prolongations of TCs, telopodes, have the following distinctive ultrastructural attributes (Fig. 20.2b):

1. The *number* of visible telopodes is usually two to three. However, depending on the angle and how the plane of section is intercepting the cell, there are two to five visible telopodes within transmission electron microscopy images.

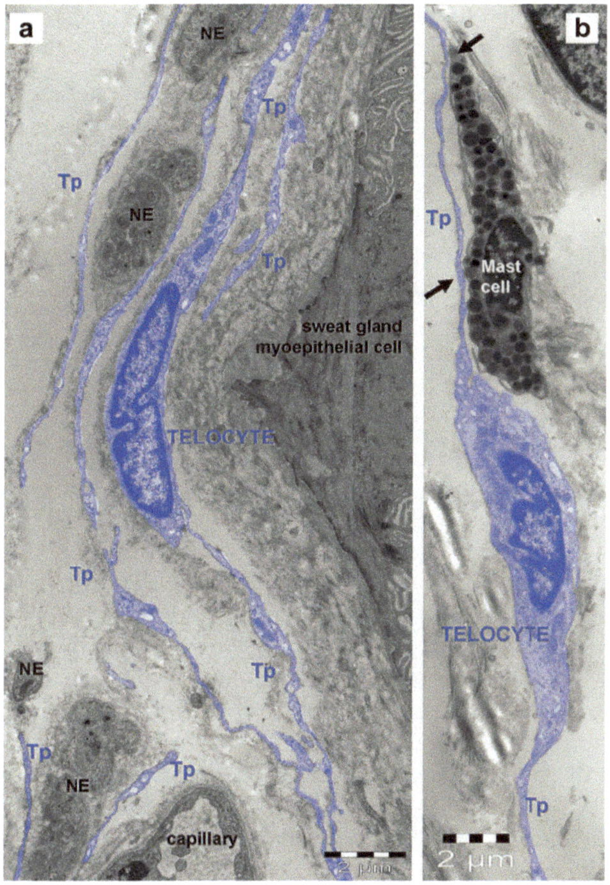

Fig. 20.2 Telocyte layers in human skin appendages. (**a**) Telocytes (*blue*) form layers around an eccrine sweat gland mixed with nerve endings (NE). (**b**) Telocytes are typically associated with mast cells. *Tp* telopodes (Reproduced with permission from Ceafalan et al. [7])

2. The *length* of telopodes, measured on electron microscopy images, ranged from 29.88 to 74.52 μm (mean: 46.37 ± 12.63 μm). Moreover, telopodes are slender cellular prolongations, with a width ranging between 47.40 and 129.51 nm (mean: 90.86 ± 30.79 nm).
3. Telopodes *suddenly emerge* from the cell body and following a sinuous trajectory within tissue.
4. In terms of caliber, telopodes appear to be *uneven caliber structures* (mostly below the 0.2 μm resolving power of light microscopy, but entirely visible under transmission electron microscopy).
5. Telopodes are made up of alternating thin segments (*podomers*; 40–100 nm diameters) and dilated segments (*podoms*; 200–500 nm diameters). This alternation creates the "*beads-on-string*" *appearance* of telopodes.
6. Podoms typically contain elements of endoplasmic reticulum, mitochondria, and caveolae. The association of these ultrastructural elements is specific to "*Ca^{2+} uptake/releasing units*".

7. Through telopodes, TCs are involved in a *3D network*, either physically (by junctions) or chemically (by paracrine secretion).

This ultrastructural profile shows that TCs are completely different from fibroblasts (Fig. 20.3). In the dermis, by homocellular and heterocellular junctions (e.g. *puncta adhaerentia minima*, *recessus adhaerente*, or gap junctions; details in subchapter "Telocytes: connecting cells"), TCs are organized in a 3D network. They form a spatially distributed labyrinthine network by convolutions and overlapping [7]. On the other hand, the connection between TCs (or between TCs and other cells) could be mediated by paracrine secreted signaling molecules or microRNA (via shed vesicles) [13].

20.2.2 FIB-SEM Tomography of Ultrastructural TCs

A recent study [13] took advantage of a very new and powerful technique, focus ion beam scanning electron microscopy (FIB-SEM), and showed the 3D spatial conformation of TCs in the human dermis. The computer-assisted reconstruction of serial sections revealed subcellular volumetric details of TCs and provided additional valuable findings regarding the spatial conformation of telopodes. Thus, under FIB-SEM, telopodes appear either as uneven ribbon-like segments (flattened veils with knobs as equivalents for podoms), or uneven tubular structures with irregular dilations (homologue for podoms). Moreover, the volumetric 3D reconstruction after FIB-SEM tomography showed that dermal TCs are responsible for shedding microvesicles (dimensions >100 nm) rather than exosomes (dimensions <100 nm). In normal skin, TCs are predominantly distributed in the reticular dermis. The density of TCs is reduced in the papillary dermis. In the reticular dermis, TCs were found around blood vessels, within the perifollicular sheath, outside the glassy membrane, and surrounding sebaceous glands, arrector pili muscles, and eccrine sweat glands (secretory and excretory parts). Around skin adnexa, TCs create two to three incomplete concentric sheaths [7].

20.2.3 The Immunohistochemical Profile of TCs

The TCs could have different immunophenotypes, depending on their subtype and distribution in skin. Immunohistochemistry was used to distinguish and characterize TCs from other interstitial cells of the dermis. Using immunofluorescence, it was shown that skin TCs were positive for CD34 or c-kit/CD117 expression, mainly in the reticular (but also papillary) dermis [7]. These cells were also positive for vimentin. These data indicate there are two subtypes of skin TCs that share the same morphology: CD34/vimentin positive TCs (situated mainly in the reticular dermis,

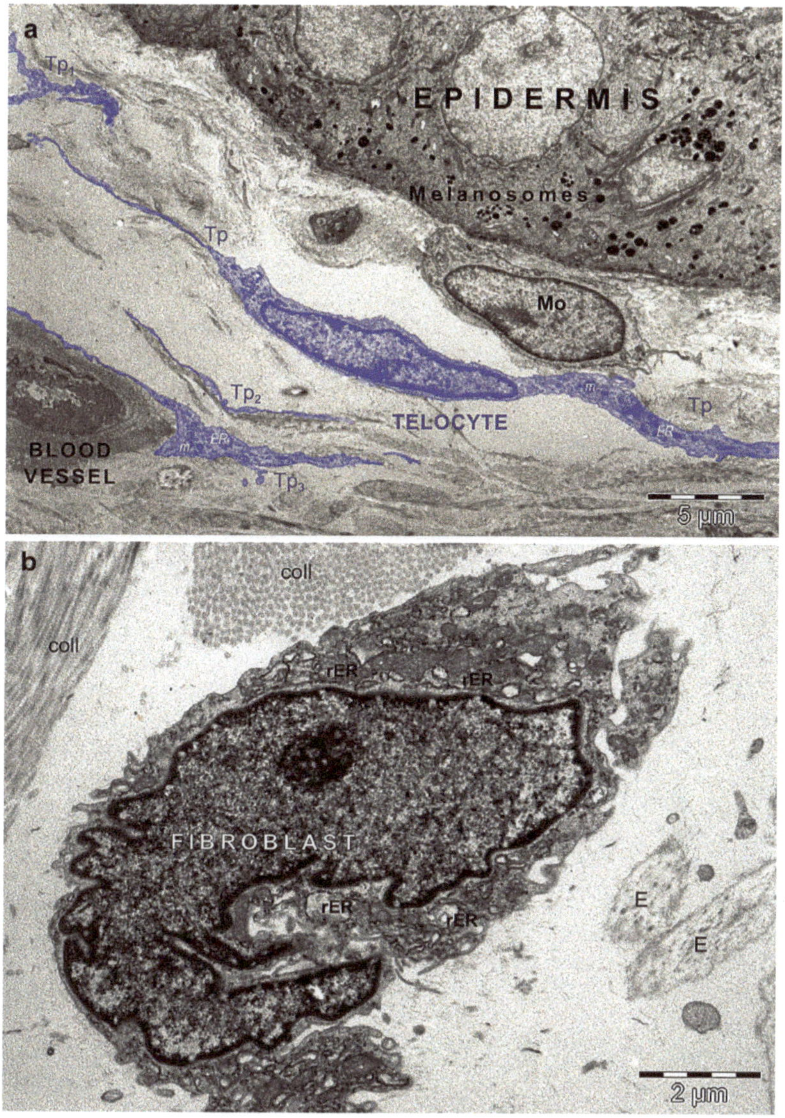

Fig. 20.3 Transmission electron microscopy of the normal human dermis. (**a**) A telocyte (and its telopodes – Tp) is visible in the papillary dermis, parallel with the basal membrane of the epidermis. Podoms – the dilated segments of Tp – show mitochondria (m) and endoplasmic reticulum (ER) within. A close relation of Tp with a mononuclear cell is visible. However, other TPs – Tp1, Tp2, Tp3 (most probably belonging to other different TCs) run parallel to each other. Moreover, Tp3 is establishing close contacts with the ab luminal face of endothelial cells of a capillary. (**b**) A typical fibroblast is shown in the human papillary dermis. The prominent euchromatic nucleus with visible nucleolus is characteristic. Also, the presence of abundant rough endoplasmic reticulum (rER) is seen. The fibroblast is surrounded by collagen fibres (coll) and elastic fibres (E) (Reproduced with permission from Kang et al. [29])

around deep segments of hair follicles, sebaceous glands, arrector pili muscles, and secretory or ductal parts of the sweat glands) and CD117/vimentin positive TCs (surrounding hair follicles and sweat glands) (Fig. 20.4). Complementary data also indicated positivity for PDGFRα for cells with morphologies suggestive of TCs, situated within the papillary dermis. Currently, a single marker cannot unequivocally identify the TCs. The most appropriate combination of markers for TCs could be CD34/PDGFRα [37] (Fig. 20.5).

20.2.4 The Differential Diagnosis of TCs in Normal Skin

It is well-known that fibroblasts are the most common cells of connective tissue; thus, the other cells are sometimes classified as fibroblast-like cells. However, TCs represent a distinct cellular population of interstitial cells. In the attempt to discriminate TCs from other skin cells, several microscopy techniques were performed (histology, immunohistochemistry, immunofluorescence, transmission electron microscopy, FIB-SEM tomography), as well as cytokine profile analysis. To date, the best immune identification of TCs by light microscopy is using double immunostaining (for CD34 and PDGFRα). The immune expression of TCs differs from the immune expression of any other dermal dendritic cell. The TCs do not express CD1a (like Langerhans cells), mast cell tryptase (like mast cells), αSMA (like

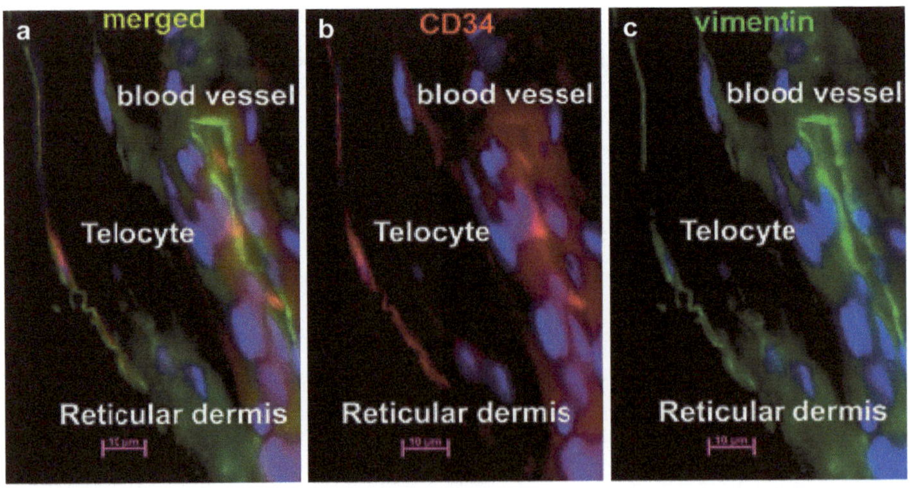

Fig. 20.4 Telocytes in human reticular dermis vasculature. Epifluorescence microscopy: double immunofluorescence labelling shows a telocyte located in the perivascular space of a blood vessel that (**a**) coexpresses CD34 and vimentin (*yellow* indicates merged colours) (**b**) CD34 label (*red*); (**c**) vimentin label (*green*). Nuclei are counterstained with DAPI (*blue*). Original magnification 400× (Reproduced with permission from Ceafalan et al. [7])

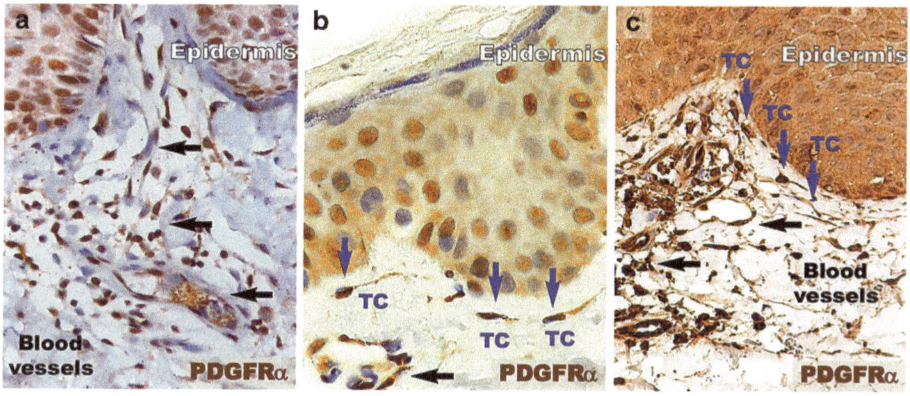

Fig. 20.5 Higher magnifications of immunohistochemistry for PDGFRα in the papillary dermis show the absence of telocytes but the presence of inflammatory cells (mainly lymphocytes) in a psoriatic plaque (**a**). However, in the distant uninvolved skin (**b**), telocytes (TC, *blue arrows*) with very long telopodes are situated in the papillary dermis and run parallel to the line of the dermal-epidermal junction. (**c**) Telocytes (TC; *blue arrows*) were also observed in the papillary dermis of treated skin, with their telopodes running parallel to the basement membrane. *Black arrows* indicate blood vessels. Magnification 400× (Reproduced with permission from Manole et al. [37])

myofibroblasts), etc. (Table 20.2). Moreover, TCs do not express procollagen or CD90, making them distinct from fibroblasts [7, 37].

The differences between TCs and other interstitial cells were documented using various techniques. The cytokine antibody array analyses indicated several points of difference between skin TCs and fibroblasts. The expression of ENA-78 (Epithelial Neutrophil Activated peptide 78) and GCP-2 (Granulocyte Chemotactic protein 2) were higher in TCs than fibroblasts. These results are supportive for the presumed roles of TCs in vascular homeostasis and neoangiogenesis, intercellular signaling, immune modulation response, and skin cancer progression. In contrast, the cytokine profile of fibroblasts revealed 37 highly expressed cytokines with documented structural roles in interstitium homeostasis and initiation of wound healing [29].

20.2.5 Skin TCs: Connecting Cells

Skin TCs are interconnected by homocellular junctions (*puncta adhaerentia minima*, *recessus adhaerente*, and also gap junctions) and heterocellular junctions in a 3D network [7]. *Puncta adhaerentia minima* are very small junctions, with role(s) in adhesion, and are especially found on cells with mesenchymal origin. Usually, the diameters of such junctions do not exceed 50 nm. Some proteins were identified at the level of these junctions, among them N-cadherin, cadherin-11, catenin α- and β-, etc. [20]. *Recessus adhaerens* represent the tight-fitting insertion of a plasma membrane expansion of one cell into the deep plasma membrane invaginations of

Table 20.2 Immune expression of different skin cells

Cell type	Immunohistochemical marker	References
Endothelial cells	CD31 (PECAM1)	[28]
Fibroblasts	Procollagen-1	[31]
Myofibroblasts	αSMA	[18]
Pericytes	αSMA, PDGFRβ, NG2, CD146, nestin	[2, 42, 63]
Stem cells	SCA-1, CD29, CD90, CD34, CD44	[46]
Melanocytes	S-100, HMB45, Melan-A	[58]
Macrophages	CD11c, CD14, CD40, CD11b, CD64, CD68	[40]
Mast cells	c-kit/CD117	[26]
Telocytes	CD34+PDGFRα	[37]
Langerhans cells	CD1a, Langerin/CD207	[37]
Dermal dendritic cells	CD11c, CD83, CD208	[37]
Inflammatory dendritic cells	CD11c, CD14, CD209, NOS	[37]
Plasmacytoid dendritic cells	CD11c, CD123, CD205, TNFα	[37]

CD31 (*PECAM-1*) –usually used for determining the presence of endothelial cells in histological tissue sections
Procollagen type 1 – represents a useful marker for activated fibroblasts
αSMA (*alpha Smooth Muscle Actin*) – role in cell motility, structure, and integrity
PDGFRβ (*Platelet Derived Growth Factor Receptor beta*) – role in blood vessel formation. Stimulates the recruitment, migration, and proliferation of pericytes and smooth muscle cells to endothelial cells
PDGFRα (*Platelet Derived Growth Factor Receptor alpha*) – roles in the differentiation of bone marrow-derived mesenchymal stem cells
CD146 (*Melanoma Cell Adhesion Molecule – MCAM*) – indicates the endothelial cell lineage
Nestin –marker for neural stem cells, for newly formed/proliferating endothelial cells
SCA-1 (*Stem Cells Antigen – 1*) – used for identifying hematopoietic stem cells (together with other markers)
CD9 – protein found on the surface of exosomes; regulates and modulates cell migration and adhesion
CD90 (*Thy-1*) – marker for different stem cells
CD34 (*Hematopoietic progenitor cell antigen CD34*) – mediates the attachment of stem cells to bone marrow extracellular matrix/stromal cells
CD44 – roles in lymphocyte functions, hematopoiesis, and tumor metastasis
S100 – protein that indicates the presence of melanocytes, myoepithelial cells, macrophages, Langerhans cells, dendritic cells, and keratinocytes. Involved in cell growth and differentiation in histological sections. Also indicates the inflammatory process. *S100A7* (also known as *psoriasin*) acts like an inflammatory cytokine, particularly in psoriasis
CD11b (*Integrin alpha M*) – identifies macrophages, involved in immune processes such as phagocytosis
CD11c (*Integrin*) –induces cellular activation. Also, identifies human dendritic cells, but also on monocytes and macrophages
CD14 – identifies macrophages, neutrophils, and dendritic cells
CD40 – identifies endothelial cells, smooth muscle cells, and fibroblasts
CD64 – identifies macrophages and monocytes
CD68 – identifies macrophages and monocytes

(continued)

Table 20.2 (continued)

CD117 – identifies a few types of hematopoietic progenitor cells; is also expressed by mast cells, melanocytes, and ICC
CD207 (*Langerin*) – expressed by skin or mucosal Langerhans cells
CD209 – identifies macrophages and dendritic cells; it mediates dendritic interactions with vascular endothelium
CD83 – plays a role in antigen presentation
CD208 – is involved in dendritic cell function and adaptive immunity
NOS – involved in cell signaling, angiogenesis, insulin secretion, and vascular tone modulation
CD123 (*interleukin-3 receptor*) – roles in proliferation and differentiation of hematopoietic cell lines
TNFα – protein involved in inflammation. It is produced by macrophages, neutrophils, mast cells, neurons, and lymphocytes

another cell [24]. Ultrastructurally, the gap junctions are aggregates of intercellular channels responsible for the intercellular transfer of ions and small molecules between neighboring cells. The junctions between TCs support the idea that skin TCs do integrate and modulate the various signaling inputs, thus contributing to tissue homeostasis and pathology (Fig. 20.6).

Dermal TCs are also connected with other types of interstitial cells (e.g. immune reactive cells as mononuclear cells or mast cells) by heterocellular junctions (point contacts and planar contacts) [7]. It is tempting to believe that TCs act as mast cell modulators in allergic conditions. Also, close spatial relations were documented with macrophages. Within the human dermis, telopodes have a special position relative to blood vessels (having direct contact with the *abluminal* surfaces of the endothelial cells) and nerve endings. Immunohistochemistry (expression for nestin and c-kit/CD117) and transmission electron microscopy demonstrated that TCs cooperate with stem cells in the human dermis. The TCs were found near the hair follicle bulge, enwrapping clusters of stem cells with their telopodes. Stem cells showed the ultrastructural characteristics of undifferentiated cells: a remarkable large euchromatic nucleus, abundant cytoplasm with few mitochondria, endoplasmic reticulum, and free ribosomes. The direct interaction between TCs and stem cells appeared as point contacts and planar contacts. These data indicate that, as in other organs, TCs and stem cells are involved in tandem structures, mostly found in stem cell niches [25, 48, 50, 61]. This is highly suggestive that TCs are involved in regeneration/repair processes at the skin level.

20.2.6 Presumptive Roles of TCs in Normal Skin

The specific roles of TCs in skin are unclear (resulting in a very exciting research field). Their localization in the interstitium (near structures), their involvement in a 3D network, and their junctions with immunoreactive cells suggest TCs have important roles. Some possible roles are:

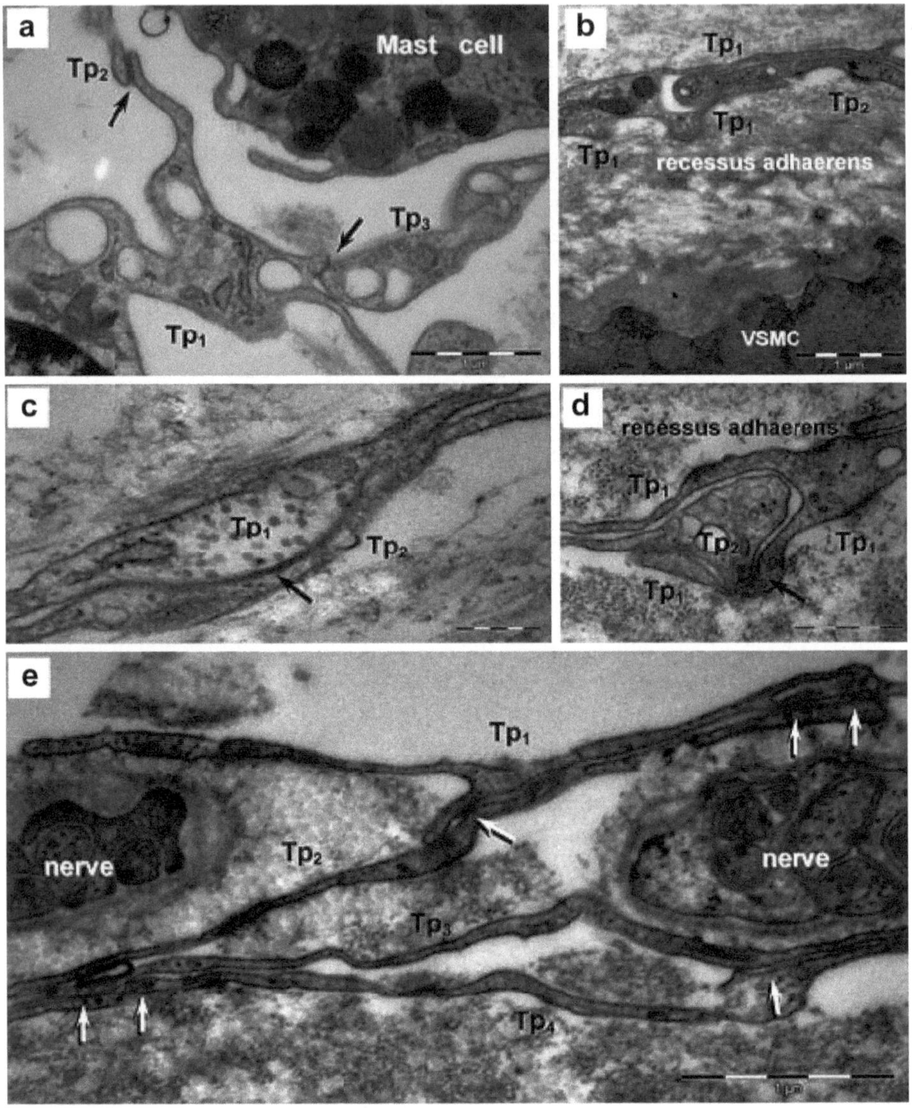

Fig. 20.6 Homocellular junctions between telocytes in human skin (TEM). (**a**) Three telopodes (Tp1–Tp3) are connected by small adhaerens junctions (*arrows*). (**b**) A gap junction is visible between two telopodes (Tp1, Tp2). (**c, d**) The telopode (Tp1) of one telocyte embraces the cytoplasmic extension (Tp2) of another telocyte, forming a *recessus adhaerens*. A *punctum adhaerentia minimum* is visible in D (*arrow*). *VSMC* vascular smooth muscle cell. (**e**) Overlapping telopodes (Tp1–Tp4) are connected by *puncta adhaerentia minima* (*arrows*) (Reproduced with permission from Ceafalan et al. [7])

1. *The (ultra)structure and 3D configuration of TCS within the interstitium may involve them in the architectural network between collagen and elastic fibers.* Failure of TCs, by dystrophy or disintegration in systemic sclerosis, is sustained by disorganized fibers [36].
2. *The tandem between TCs and stem cells indicates that TCs are acting as nurse cells for mesenchymal and epithelial stem cells* [7]. Failure in epidermal turnover (as in psoriasis vulgaris) is associated with severe dystrophic changes of TCs in terms of nuclear degeneration features (including nuclear extrusion) [37].
3. *TCs may represent immunologic targets and might subsequently be involved in cutaneous autoimmune disorders.*
4. *By their paracrine secretion, TCs may be involved in controlling and modulating the functions of fibroblasts* [36]. The secretion of paracrine signaling molecules (including microRNAs) and their shuttle by shedding vesicles represents the indirect way TCs interact with surrounding cells, and also the way they modify the microenvironment [7].
5. *The close vicinity between TCs and endothelial cells indicates the role of TCs in vascular physiology, and moreover in skin regeneration/repair processes.* The loss of TCs could represent the background mechanism of vascular pathology in psoriasis. This may explain the most characteristic sign of psoriasis: the Auspitz's sign (or the "*bloody dew*" sign).
6. *TCs may modulate mast cell activity during allergic reactions.* Since TCs are located in close vicinity with mast cells, they share some immunophenotypical features [37].

20.3 Telocytes in Skin Pathology

To date, TCs have been studied in systemic sclerosis and psoriasis vulgaris. These two conditions lead to severe TC dystrophy and loss of TCs in areas of affected skin.

20.3.1 Systemic Sclerosis

Systemic sclerosis is a multi-systemic disease, featuring excessive fibrosis (extensive activation of fibroblasts), inflammation, and vasculopathy (through endothelial cell injury). The most frequently involved organs are skin, esophagus, lung, heart, and kidneys. The TCs were described in both limited and diffuse systemic sclerosis [36]. The intensive fibrotic process disarranges the normal structures of skin creating severe ultrastructural damage, which leads to progressive loss of TCs. Histopathologically, systemic sclerosis is characterized by extensive accumulation of collagen I and III in the lower dermis and subcutaneous fibrous trabeculae. The early stages of systemic sclerosis are illustrated by pathologic accumulations of collagen bundles in the reticular dermis, parallel to the epidermis. Panniculitis, mucoid edema, and lymphocytic infiltrate may also be present at this stage, along with epidermal atrophy, eccrine gland entrapment, capillary dilations with endothelial proliferation, and complete obstruction. As systemic sclerosis progresses to the later stages, the dermis becomes more avascular and inflammation decreases. Thus, the later stages of systemic sclerosis feature more densely packed collagen bundles, the disappearance of the pilosebaceous units, and sweat glands.

In systemic sclerosis TCs show severe ultrastructural changes including swollen mitochondria, cytoplasmic vacuolization, and lipofuscin bodies. These changes are highly suggestive of ischemia changes, since decreasing tissue oxygen is a result of endothelial damage and represents the *primum movens* of systemic sclerosis. The loss of TCs was documented in the reticular dermis, but also in the papillary dermis, and was proportionally related to the evolving early stages of systemic sclerosis [36]. In early stages of systemic sclerosis, TCs have a patchy distribution, being absent in the papillary dermis and numerically reduced in the reticular dermis. Mostly, TCs are situated around arteriole, nerves, hair follicles, and sebaceous glands. Conversely, in later stages of systemic sclerosis, TCs were absent in the reticular dermis, in the connective tissue around arteriole, and in most of the adnexal structures, but TCs were preserved around the excretory parts of the eccrine glands. Thus, Manetti et al. raised the question "Why telocytes are more sensitive to ischemia in comparison with other stromal cells like fibroblasts, myofibroblasts, or mast cells?" Altering the 3D network of TCs also affects the extracellular matrix, since TCs in the dermis enwrap collagen and elastic fibers with their telopodes. It is noteworthy that the investigated patients did not receive immunosuppressive treatment or disease-modifying drugs.

20.3.2 Psoriasis Vulgaris

Psoriasis is a keratinization disorder, a chronic immunologic disease with a genetic background that affects normal epidermal turnover. Four populations of dendritic cells were previously documented to be involved in the immune response of psoriasis: Langerhans cells, dermal dendritic cells, inflammatory dendritic epidermal cells, and plasmacytoid dendritic cells. In vulgaris psoriasis, there is acanthosis of the epidermis ridges, which are elongated and club-shaped at their bases. This false acanthosis is due to the lack of a granular layer. The papillae are thin and elongated and contain dilated capillaries. Parakeratosis may alternate with orthokeratosis and hyperkeratosis. The blood vessels are surrounded by inflammatory infiltrate, mainly consisting of lymphocytes, mast cells, macrophages, and dendritic cells. Lymphocytes could be also observed in the epidermis in mature lesions. Munro's microabcesses represent the accumulation of neutrophils in the stratum corneum, and this is a histopathological hallmark of psoriasis.

A recent comparative study on psoriasis employed biopsies (a) from the mature psoriatic lesion; (b) from non-lesional skin (40 cm distance from any lesion); and (c) after the clearance induced by local treatment plan with keratolytic (urea and salicylic acid) and cytoreductive (based on Anthraline) therapies, followed by calcipotriol and steroid ointments. This study demonstrated that in psoriasis vulgaris, TCs displayed severe dystrophy features (other than those reported in systemic sclerosis by Professor Faussone-Pellegrini's team), or were lost at the lesion level. On the other hand, TCs were ultrastructurally and numerically normal in perilesional skin. Unlikely systemic sclerosis, the psoriatic TCs had different dystrophy features, most likely due to a shorter turnover of keratinocytes. Using transmission electron microscopy, we demonstrated that TCs had apoptotic nuclei, or nuclear extrusions and cytoplasmic disintegration (Fig. 20.7). The extruded nuclei were frequently observed in

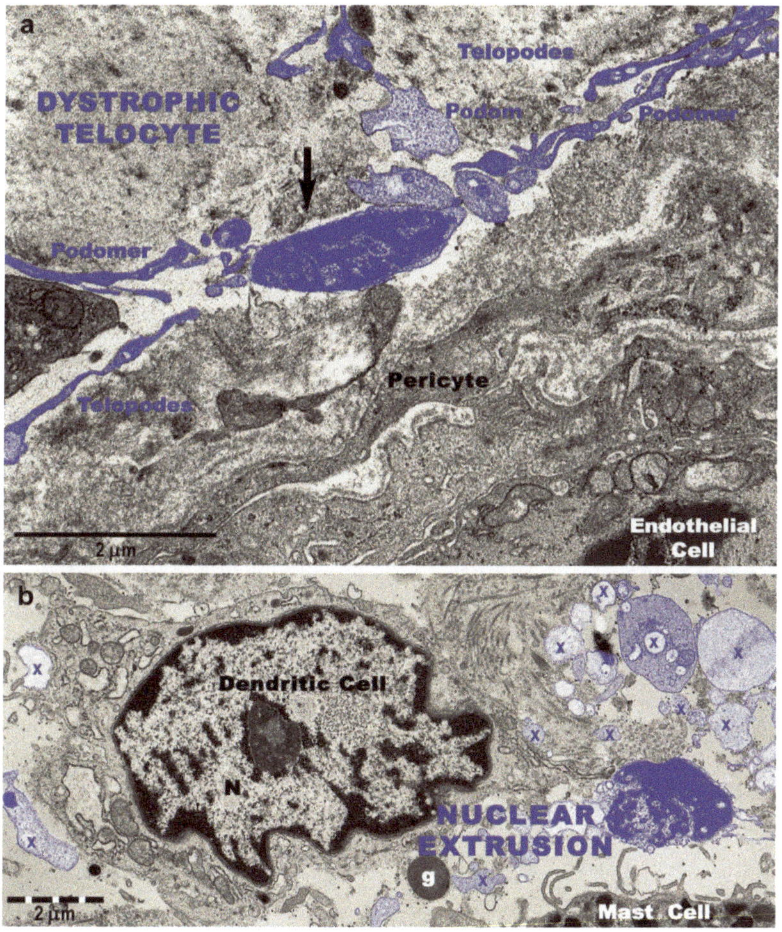

Fig. 20.7 Transmission electron microscope images show degenerative changes in telocytes (digitally coloured in *blue*) from a psoriatic plaque. (**a**) A telocyte with shrivelled nucleus and detached telopodes. The *arrow* indicates dissolution of the cellular membrane and the cytoplasmic content surrounding the nucleus. (**b**) An extruded nucleus and cytoplasmic fragments (X) of a telocyte are visible in the vicinity of a dendritic cell. *g* granule (of a mast cell) (Reproduced with permission from Manole et al. [37])

close contact with dendritic cells. Also, dystrophic TCs with fragmented telopodes were found in psoriatic lesions. Additionally, no homocellular junctions between TCs were found in psoriatic skin, but there were heterocellular junctions with mast cells [37]. Moreover, frequent interruptions were found in the basement membrane, allowing Langerhans cells to migrate from the epidermis to dermis. Previous evidence suggested that a T-cell-mediated immune reaction is pivotal to the pathogenesis of psoriasis [75]. In our study we found no proof of any relation between TCs and T-cells.

Another important change found in psoriasis is the modification of the vascular smooth muscle cell phenotype that is translated into a loss of the contractile phenotype with consequent dilation of the arteries. Also, the close vicinity of TCs with blood vessels could have implications in vascular pathology. Thus, the loss of TCs surrounding blood vessels could lead to the characteristic psoriatic vascular pathology [37]. This could be a sustainable argument for the Auspitz's sign. An ongoing study is dedicated to the involvement of TCs in pustular psoriasis, since these patients do not typically present the Auspitz's sign. The impairment of the TC network (either homocellular or heterocellular communication of TCs) could significantly impact the lesion initiation as well as disease progression.

20.4 Conclusions

Telocytes represent a distinct stromal cell population in the skin. In the papillary dermis, TCs are scarce, but their density in increased in the reticular dermis. Currently, the best method for identifying TCs is staining with CD34 and PDGFRα. In the skin, TCs are involved in a 3D stromal network, either by homocellular junction (Telocyte-Telocyte junction) or by heterocellular junctions (Telocyte – other interstitial cell). This network is important for maintaining the normal architecture of the dermis. However, by their close contacts either with immunoreactive cells and stem cells or with endothelial cells of the vascular structures, TCs seem to be involved in skin homeostasis and/or regeneration/repair processes. In our opinion, research should be developed in two directions in the near future: a specific immunohistochemical marker for TCs and further documentation of the role(s) of TCs by studying them in different dermatoses and skin neoplasms.

References

1. Alunno A, Ibba-Manneschi L, Bistoni O, Rosa I, Caterbi S, Gerli R, Manetti M. Telocytes in minor salivary glands of primary Sjögren's syndrome: association with the extent of inflammation and ectopic lymphoid neogenesis. J Cell Mol Med. 2015;19(7):1689–96.
2. Bergers G, Song S. The role of pericytes in blood-vessel formation and maintenance. Neuro Oncol. 2005;7(4):452–64.

3. Bojin FM, Gavriliuc OI, Cristea MI, Tanasie G, Tatu CS, Panaitescu C, Paunescu V. Telocytes within human skeletal muscle stem cell niche. J Cell Mol Med. 2011;15(10):2269–72.
4. Cantarero Carmona I, Luesma Bartolomé MJ, Junquera Escribano C. Identification of telocytes in the lamina propria of rat duodenum: transmission electron microscopy. J Cell Mol Med. 2011;15(1):26–30.
5. Cantarero I, Luesma MJ, Alvarez-Dotu JM, Muñoz E, Junquera C. Transmission electron microscopy as key technique for the characterization of telocytes. Curr Stem Cell Res Ther. 2016;11(5):410–4.
6. Cantarero I, Luesma MJ, Junquera C. The primary cilium of telocytes in the vasculature: electron microscope imaging. J Cell Mol Med. 2011;15(12):2594–600.
7. Ceafalan L, Gherghiceanu M, Popescu LM, Simionescu O. Telocytes in human skin – are they involved in skin regeneration? J Cell Mol Med. 2012;16(7):1405–20.
8. Chang Y, Li C, Gan L, Li H, Guo Z. Telocytes in the Spleen. PLoS One. 2015;10(9):e0138851.
9. Chang Y, Li C, Lu Z, Li H, Guo Z. Multiple immunophenotypes of cardiac telocytes. Exp Cell Res. 2015;338(2):239–44.
10. Chen X, Zheng Y, Manole CG, Wang X, Wang Q. Telocytes in human oesophagus. J Cell Mol Med. 2013;17(11):1506–12.
11. Corradi LS, Jesus MM, Fochi RA, Vilamaior PS, Justulin Jr LA, Góes RM, Felisbino SL, Taboga SR. Structural and ultrastructural evidence for telocytes in prostate stroma. J Cell Mol Med. 2013;17(3):398–406.
12. Cretoiu D, Cretoiu SM, Simionescu AA, Popescu LM. Telocytes, a distinct type of cell among the stromal cells present in the lamina propria of jejunum. Histol Histopathol. 2012;27(8):1067–78.
13. Cretoiu D, Gherghiceanu M, Hummel E, Zimmermann H, Simionescu O, Popescu LM. FIB-SEM tomography of human skin telocytes and their extracellular vesicles. J Cell Mol Med. 2015;19(4):714–22.
14. Cretoiu SM, Cretoiu D, Marin A, Radu BM, Popescu LM. Telocytes: ultrastructural, immunohistochemical and electrophysiological characteristics in human myometrium. Reproduction. 2013;145(4):357–70.
15. Creţoiu SM, Creţoiu D, Popescu LM. Human myometrium – the ultrastructural 3D network of telocytes. J Cell Mol Med. 2012;16(11):2844–9.
16. Dawidowicz J, Szotek S, Matysiak N, Mielańczyk Ł, Maksymowicz K. Electron microscopy of human fascia lata: focus on telocytes. J Cell Mol Med. 2015;19(10):2500–6.
17. Díaz-Flores L, Gutiérrez R, Sáez FJ, Díaz-Flores Jr L, Madrid JF. Telocytes in neuromuscular spindles. J Cell Mol Med. 2013;17(4):457–65.
18. Ehrlich HP, Allison GM, Leggett M. The myofibroblast, cadherin, alpha smooth muscle actin and the collagen effect. Cell Biochem Funct. 2006;24(1):63–70.
19. Faussone Pellegrini MS, Popescu LM. Telocytes. Biomol Concepts. 2011;2(6):481–9.
20. Franke WW, Rickelt S, Barth M, Pieperhoff S. The junctions that don't fit the scheme: special symmetrical cell-cell junctions of their own kind. Cell Tissue Res. 2009;338(1):1–17.
21. Galiger C, Kostin S, Golec A, Ahlbrecht K, Becker S, Gherghiceanu M, Popescu LM, Morty RE, Seeger W, Voswinckel R. Phenotypical and ultrastructural features of Oct4-positive cells in the adult mouse lung. J Cell Mol Med. 2014;18(7):1321–33.
22. Gevaert T, De Vos R, Van Der Aa F, Joniau S, van den Oord J, Roskams T, De Ridder D. Identification of telocytes in the upper lamina propria of the human urinary tract. J Cell Mol Med. 2012;16(9):2085–93.
23. Gherghiceanu M, Manole CG, Popescu LM. Telocytes in endocardium: electron microscope evidence. J Cell Mol Med. 2010;14(9):2330–4.
24. Gherghiceanu M, Popescu LM. Cardiac telocytes – their junctions and functional implications. Cell Tissue Res. 2012;348(2):265–79.
25. Gherghiceanu M, Popescu LM. Cardiomyocyte precursors and telocytes in epicardial stem cell niche: electron microscope images. J Cell Mol Med. 2010;14(4):871–7.
26. Hauswirth AW, Florian S, Schernthaner GH, Krauth MT, Sonneck K, Sperr WR, Valent P. Expression of cell surface antigens on mast cells: mast cell phenotyping. Methods Mol Biol. 2006;315:77–90.

27. Hinescu ME, Gherghiceanu M, Suciu L, Popescu LM. Telocytes in pleura: two- and three-dimensional imaging by transmission electron microscopy. Cell Tissue Res. 2011;343(2):389–97.
28. Jackson DE. The unfolding tale of PECAM-1. FEBS Lett. 2003;540(1–3):7–14.
29. Kang Y, Zhu Z, Zheng Y, Wan W, Manole CG, Zhang Q. Skin telocytes versus fibroblasts: two distinct dermal cell populations. J Cell Mol Med. 2015;19(11):2530–9.
30. Kostin S. Myocardial telocytes: a specific new cellular entity. J Cell Mol Med. 2010;14(7):1917–21.
31. Krustrup D, Rossen K, Thomsen HK. Procollagen 1 – a marker of fibroblastic and fibrohistiocytic skin tumors. J Cutan Pathol. 2006;33(9):614–8.
32. Li H, Lu S, Liu H, Ge J, Zhang H. Scanning electron microscope evidence of telocytes in vasculature. J Cell Mol Med. 2014;18(7):1486–9.
33. Li H, Zhang H, Yang L, Lu S, Ge J. Telocytes in mice bone marrow: electron microscope evidence. J Cell Mol Med. 2014;18(6):975–8.
34. Li L, Lin M, Li L, Wang R, Zhang C, Qi G, Xu M, Rong R, Zhu T. Renal telocytes contribute to the repair of ischemically injured renal tubules. J Cell Mol Med. 2014;18(6):1144–56.
35. Luesma MJ, Gherghiceanu M, Popescu LM. Telocytes and stem cells in limbus and uvea of mouse eye. J Cell Mol Med. 2013;17(8):1016–24.
36. Manetti M, Guiducci S, Ruffo M, Rosa I, Faussone-Pellegrini MS, Matucci-Cerinic M, Ibba-Manneschi L. Evidence for progressive reduction and loss of telocytes in the dermal cellular network of systemic sclerosis. J Cell Mol Med. 2013;17(4):482–96.
37. Manole CG, Gherghiceanu M, Simionescu O. Telocyte dynamics in psoriasis. J Cell Mol Med. 2015;19(7):1504–19.
38. Matyja A, Gil K, Pasternak A, Sztefko K, Gajda M, Tomaszewski KA, Matyja M, Walocha JA, Kulig J, Thor P. Telocytes: new insight into the pathogenesis of gallstone disease. J Cell Mol Med. 2013;17(6):734–42.
39. Mihalcea CE, Moroşanu AM, Murăraşu D, Puiu L, Cinca S, Voinea SC, Mirancea N. Particular molecular and ultrastructural aspects in invasive mammary carcinoma. Rom J Morphol Embryol. 2015;56(4):1371–81.
40. Misharin AV, Morales-Nebreda L, Mutlu GM, Budinger GR, Perlman H. Flow cytometric analysis of macrophages and dendritic cell subsets in the mouse lung. Am J Respir Cell Mol Biol. 2013;49(4):503–10.
41. Mou Y, Wang Y, Li J, Lü S, Duan C, Du Z, Yang G, Chen W, Zhao S, Zhou J, Wang C. Immunohistochemical characterization and functional identification of mammary gland telocytes in the self-assembly of reconstituted breast cancer tissue in vitro. J Cell Mol Med. 2013;17(1):65–75.
42. Nakagawa S, Miki Y, Miyashita M, Hata S, Takahashi Y, Rai Y, Sagara Y, Ohi Y, Hirakawa H, Tamaki K, Ishida T, Watanabe M, Suzuki T, Ohuchi N, Sasano H. Tumor microenvironment in invasive lobular carcinoma: possible therapeutic targets. Breast Cancer Res Treat. 2016;155(1):65–75.
43. Nicolescu MI, Bucur A, Dinca O, Rusu MC, Popescu LM. Telocytes in parotid glands. Anat Rec (Hoboken). 2012;295(3):378–85.
44. Nicolescu MI, Popescu LM. Telocytes in the interstitium of human exocrine pancreas: ultrastructural evidence. Pancreas. 2012;41(6):949–56.
45. Padhi S, Sarangi R, Mallick S. Pancreatic extragastrointestinal stromal tumors, interstitial Cajal like cells, and telocytes. JOP. 2013;14(1):1–14.
46. Lui PPY. Markers for the identification of tendon-derived stem cells in vitro and tendon stem cells in situ – update and future development. Stem Cell Res Ther. 2015;6:106.
47. Petre N, Rusu MC, Pop F, Jianu AM. Telocytes of the mammary gland stroma. Folia Morphol (Warsz). 2015. doi:10.5603/FM.a2015.0123.
48. Popescu BO, Gherghiceanu M, Kostin S, Ceafalan L, Popescu LM. Telocytes in meninges and choroid plexus. Neurosci Lett. 2012;516(2):265–9.
49. Popescu LM, Faussone-Pellegrini MS. TELOCYTES – a case of serendipity: the winding way from Interstitial Cells of Cajal (ICC), via Interstitial Cajal-Like Cells (ICLC) to TELOCYTES. J Cell Mol Med. 2010;14(4):729–40.

50. Popescu LM, Gherghiceanu M, Suciu LC, Manole CG, Hinescu ME. Telocytes and putative stem cells in the lungs: electron microscopy, electron tomography and laser scanning microscopy. Cell Tissue Res. 2011;345(3):391–403.
51. Popescu LM, Manole CG, Gherghiceanu M, Ardelean A, Nicolescu MI, Hinescu ME, Kostin S. Telocytes in human epicardium. J Cell Mol Med. 2010;14(8):2085–93.
52. Qi G, Lin M, Xu M, Manole CG, Wang X, Zhu T. Telocytes in the human kidney cortex. J Cell Mol Med. 2012;16(12):3116–22.
53. Richter M, Kostin S. The failing human heart is characterized by decreased numbers of telocytes as result of apoptosis and altered extracellular matrix composition. J Cell Mol Med. 2015;19(11):2597–606.
54. Roatesi I, Radu BM, Cretoiu D, Cretoiu SM. Uterine telocytes: a review of current knowledge. Biol Reprod. 2015;93(1):10.
55. Rusu MC, Folescu R, Mănoiu VS, Didilescu AC. Suburothelial interstitial cells. Cells Tissues Organs. 2014;199(1):59–72.
56. Rusu MC, Jianu AM, Mirancea N, Didilescu AC, Mănoiu VS, Păduraru D. Tracheal telocytes. J Cell Mol Med. 2012;16(2):401–5.
57. Rusu MC, Mirancea N, Mănoiu VS, Vâlcu M, Nicolescu MI, Păduraru D. Skin telocytes. Ann Anat. 2012;194(4):359–67.
58. Shinohara MM, Deubner H, Argenyi ZB. S100, HMB-45, and Melan-A negative primary melanoma. Dermatol Online J. 2009;15(9):7.
59. Song D, Cretoiu D, Zheng M, Qian M, Zhang M, Cretoiu SM, Chen L, Fang H, Popescu LM, Wang X. Comparison of Chromosome 4 gene expression profile between lung telocytes and other local cell types. J Cell Mol Med. 2015. doi:10.1111/jcmm.12746.
60. Suciu L, Popescu LM, Gherghiceanu M, Regalia T, Nicolescu MI, Hinescu ME, Faussone-Pellegrini MS. Telocytes in human term placenta: morphology and phenotype. Cells Tissues Organs. 2010;192(5):325–39.
61. Suciu LC, Popescu BO, Kostin S, Popescu LM. Platelet-derived growth factor receptor-β-positive telocytes in skeletal muscle interstitium. J Cell Mol Med. 2012;16(4):701–7.
62. Tao L, Wang H, Wang X, Kong X, Li X. Cardiac telocytes. Curr Stem Cell Res Ther. 2016; 11(5):404–9.
63. Tripodo C, Di Bernardo A, Ternullo MP, Guarnotta C, Porcasi R, Ingrao S, Gianelli U, Boveri E, Iannitto E, Franco G, Florena AM. CD146(+) bone marrow osteoprogenitors increase in the advanced stages of primary myelofibrosis. Haematologica. 2009;94(1):127–30.
64. Vannucchi MG, Traini C, Guasti D, Del Popolo G, Faussone-Pellegrini MS. Telocytes subtypes in human urinary bladder. J Cell Mol Med. 2014;18(10):2000–8.
65. Vannucchi MG, Traini C, Manetti M, Ibba-Manneschi L, Faussone-Pellegrini MS. Telocytes express PDGFRα in the human gastrointestinal tract. J Cell Mol Med. 2013;17(9): 1099–108.
66. Wang F, Song Y, Bei Y, Zhao Y, Xiao J, Yang C. Telocytes in liver regeneration: possible roles. J Cell Mol Med. 2014;18(9):1720–6.
67. Xiao J, Wang F, Liu Z, Yang C. Telocytes in liver: electron microscopic and immunofluorescent evidence. J Cell Mol Med. 2013;17(12):1537–42.
68. Yang J, Chi C, Liu Z, Yang G, Shen ZJ, Yang XJ. Ultrastructure damage of oviduct telocytes in rat model of acute salpingitis. J Cell Mol Med. 2015;19(7):1720–8.
69. Yang Y, Sun W, Wu SM, Xiao J, Kong X. Telocytes in human heart valves. J Cell Mol Med. 2014;18(5):759–65. doi:10.1111/jcmm.12285. PubMed PMID: 24674389, PubMed Central PMCID: PMC4119382, Epub 2014 Mar 26.
70. Zhang HQ, Lu SS, Xu T, Feng YL, Li H, Ge JB. Morphological evidence of telocytes in mice aorta. Chin Med J (Engl). 2015;128(3):348–52.
71. Zheng Y, Chen X, Qian M, Zhang M, Zhang D, Bai C, Wang Q, Wang X. Human lung telocytes could promote the proliferation and angiogenesis of human pulmonary microvascular endothelial cells in vitro. Mol Cell Ther. 2014;2:3.
72. Zheng Y, Li H, Manole CG, Sun A, Ge J, Wang X. Telocytes in trachea and lungs. J Cell Mol Med. 2011;15(10):2262–8.

73. Zheng Y, Zhu T, Lin M, Wu D, Wang X. Telocytes in the urinary system. J Transl Med. 2012;10:188.
74. Zheng YH, Li H, Ge JB, Gao HJ, Wang XD. Location of telocytes in mouse bronchial and pulmonary tissues. Zhonghua Bing Li Xue Za Zhi. 2012;41(3):172–5.
75. Calonje JE, Brenn T, Lazar AJ, McKee PH. Spongiotic, psoriasiform and pustular dermatoses. In: Calonje JE, Brenn T, Lazar AJ, McKee PH, editors. McKee's pathology of the skin. 4th ed. Philadelphia: Elsevier; 2011. p. 180–218.

Catalin G. Manole is currently Research Assistent at 'Victor Babeş' National Institute of Pathology, Bucharest in the Department of Molecular Medicine. He is also Assistant Professor at the Department of Cell Biology and Histology, 'Carol Davila' University of Medicine and Pharmacy, Bucharest. The topic of his PhD thesis was about the presence of telocytes in heart interstitium and their involvement in orchestration of angiogenesis after 30 days of experimental acute myocardial infarction. Between 2009–2015, Dr. Catalin G. Manole co-authored 18 ISI-indexed papers and Conference Abstracts. However, according to ISI Web of Knowledge, the total number of citations of his papers is currently heading 680 and his Hirsch Index is 11. Six of the published papers are dealing with cardiac ultrastructure, cardiac regeneration/repair as well as neo-angiogenesis in border zone of myocardial infarction. Moreover, two of the papers are regarding the presence and distinctiveness of telocytes in human skin. Notably, the paper entitled "Telocyte dynamics in psoriasis" is indicating possible crucial role(s) of telocytes in psoriasis pathogenesis.

He participated in many national and European research projects. In between 2010–2016 Dr. Manole lectured in China, Hungary and Romania each time for a notably receptive audience. In June 2014, in Switzerland, at 4th edition of the Lugano Stem Cell Meeting Dr. Catalin G. Manole won the First Prize of the Jury for the scientific work entitled "Telocytes and stem cells in experimental myocardial infarction in rats".

Currently, Dr. Manole is Project Manager at 'Victor Babeş' National Institute of Pathology, Bucharest. The topic of his project is dealing with the most anticipated *in situ* collaboration between transplanted telocytes and cardiac stem cells for prevention ultrastructural remodeling of the ischemic heart failure.

Olga Simionescu is the Head of the first Clinic of Dermatology in Colentina Hospital, Bucharest and Clinical Professor of Dermatology at Carol Davila University of Medicine and Pharmacy, Bucharest, Romania.

Professor Simionescu has been working over the last 21 years within the Department of Dermatology, Colentina Hospital-Bucharest and published along the years important papers on "epiluminiscent microscopy(dermoscopy)" and skin cancer, as well a variety of research works in the field of cutaneous melanoma.

As a clinical practitioner in Dermatology and Venereology, Professor Simionescu started the basic research in 1996 in the field of cutaneous apoptosis and signalling molecules in sun damaged human skin-what later became the main subject of her Doctoral thesis. It follows the description of stem cells (p63+) in keratinocytes cultures from human adult skin. Another highlight of Professor Simionescu's clinical research activity was a successful publishing of a dermoscopical scoring scale in the clinical diagnosis of seborrheic keratosis, based on the correlation algorithm between tumoral apoptosis and dermoscopy.

On February 2010, Professor Laurentiu M. Popescu (Prof. Olga Simionescu's late husband) together with Professor Maria Simonetta Faussone-Pellegrini published a significant milestone paper on telocytes under the title "Telocytes- a case of serendipity". This research became a cornerstone of Professor Simionescu's common work with her husband, resulting eventually by publishing together the accurate description of telocytes into the dermis of human normal skin(2012). Three years after publishing the above description, the couple displayed (Dragos Cretoiu, Mihaela Gherghiceanu, Eric Hummel, Hans Zimmermann, Olga Simionescu, Laurentiu M.Popescu) a 3D reconstruction of human dermal TCs by FIB-SEM tomography.

In 2015, with the "Telocyte dynamics in psoriasis", Professor Olga Simionescu succeeded to prove for the first time the loss and distrophy of telocytes in psoriasis vulgaris, emphasizing the involvement of these cells in skin homeostasis, skin remodelling, skin regeneration and repair.

Chapter 21
The Third Dimension of Telocytes Revealed by FIB-SEM Tomography

Dragos Cretoiu

Abstract Lately, spatial three-dimensional (3D) identity of cells and their interrelations with the environment that surrounds it represent a challenging trend with the purpose to achieve a holistic view over the functions. Combining data from different imaging of cells in the third dimension can offer insight into behavior modalities making a world of difference. This chapter outlines a breakthrough in telocyte research by volume electron microscopy with the aid of focused ion beam scanning electron microscopy (FIB-SEM). Reconstructing 3D (three-dimensional) appearance of telocytes from a set of two-dimensional (2D) images by FIB-SEM tomography allowed to extract valuable data about their volume in nanoscale dimensions such as the three-dimensional morphology of telopodes and extracellular vesicles.

21.1 Introduction to Telocytes

Telocytes (TCs) are seen nowadays as a new cell population residing in the stromal space of many organs of vertebrates from reptiles to humans (for additional details, visit www.telocytes.com). Shortly defined as cells with telopodes (Tps), TCs can be identified with certainty only by electron microscopy – the technique by which they were initially discovered [1, 2]. TCs are ultrastructurally characterized by a small cellular body with different shapes according to the number of telopodes: piriform (1), spindle (2), triangular (3), stellate (4,5) [1, 3]. Telopodes (Tps), which are the main characteristic for these special cells, are extremely long reaching sometimes about 100 μm, tortuous, and described as having alternative thin and thick regions called podomers and podoms, respectively [4]. Cells need external environment and interactions with their neighbors to function properly. In this regard, it was demonstrated that TCs are able to exchange information at short- and long-distance by

D. Cretoiu
Victor Babes National Institute of Pathology, Bucharest 050096, Romania
Division of Cellular and Molecular Biology and Histology,
Carol Davila University of Medicine and Pharmacy, Bucharest 050474, Romania
e-mail: dragos@cretoiu.ro

X. Wang, D. Cretoiu (eds.), *Telocytes*, Advances in Experimental Medicine and Biology, 913, DOI 10.1007/978-981-10-1061-3_21

homo- and heterocellular junctions and by extracellular vesicle release, respectively [5–8]. Recent advance in TC research, such as determining their genetic profile [9–13], miRNA imprint [14], proteome, and secretome [15–17], as well as revealing some of their electrophysiological properties [18, 19], brings us closer to discovering their functions. Although many of them are just hypothetical, few seem to be closer to the truth, based on numerous evidences pointing in the same direction. Therefore, we could list, not necessarily in order of importance, TCs' involvement in modulating local homeostasis, in morphogenesis, in regeneration, and in repair, probably nursing the stem cells, as well as being capable of intervening in the inflammatory or carcinogenetic processes (see extensive up-to-date reviews [20–22]).

21.2 The Three-Dimensional (3D) Appearance of TCs

Usually, we are accustomed to see cells in images depicting static two-dimensional (2D) structures. However, cells are three-dimensional structures having complex interactions with surrounding structure in their close proximity. Over time, reconstruction techniques have evolved progressively from those involving passionate electron microscopists willing to dedicate a lot of time on manual or semiautomated reconstruction of serial ultrathin sections to the most promising technology at the moment FIB-SEM (focused ion beam scanning electron microscopy) tomography.

The lack of a 3D viewpoint hampered the discovery of TCs, scientists, and electron microscopists in particular being used to describe only entirely seen structures. In this context, TCs represent a special case because of the unusually long telopodes that have contacted with different structures (blood capillaries, nerve bundles) and with different cellular types (stem cells, smooth muscle cells, cardiomyocytes, mast cells, basophils, lymphocytes, eosinophils, plasma cells, or macrophages) found in the immediate vicinity [23–25]. Discovered by serendipity, TCs were previously overlooked mainly because of their thin (average 0.10 ± 0.05 μm, with a min = 0.03 μm and a max = 0.24 μm) and tortuous labyrinthine telopodes, which were impossible to be observed in light microscopy (0.2 μm resolving power) [1]. Moreover, not even with the help of electron microscopy, TCs could not be seen in the past or better to say they were overlooked, because often images captured only telopode fragments and not the entire cell. Electron microscopy uses ultrathin sections of about 60–80 nm and higher magnifications. At low magnification (×2,000), a TC is too small to be observed as a different peculiar cell, and afterward, an increase of the magnification to gain access to the ultrastructural details decreases the diameter of the observed field in such a way that at a higher magnification, only a part of the telopode can be surprised in the image. Many of the details definitory for a TC can only be observed after two-dimensional (2D) reconstruction of successive electron microscopy fields (usually 8–15 photomicrographs are concatenated), depicting the cellular network in the tissue [26] (Fig. 21.1).

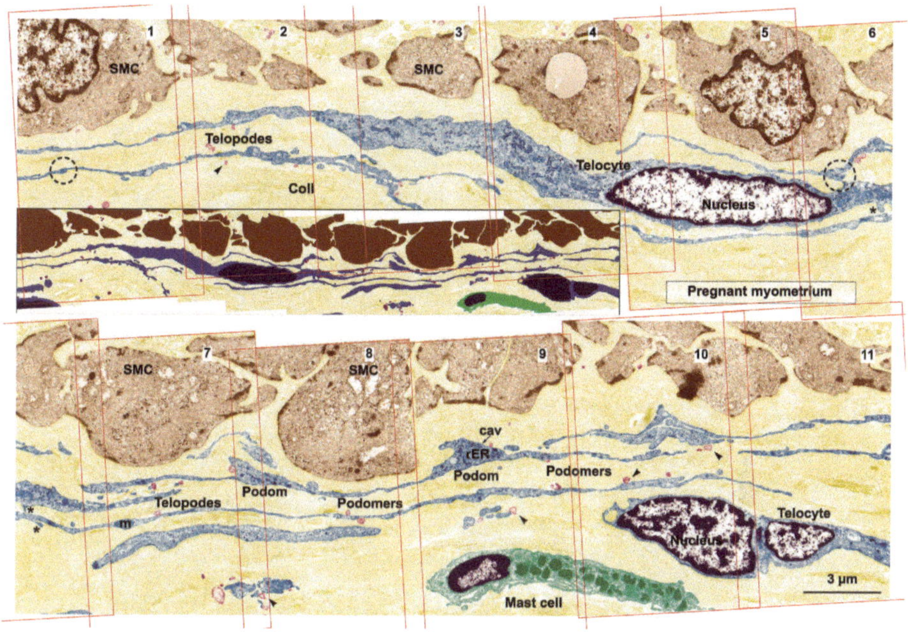

Fig. 21.1 Representative ultrathin section of human pregnant myometrium. Two-dimensional sequenced concatenation from 11 serial electron micrographs (*red rectangles*) depicting the 3D network of TCs (*blue*) interconnected by homocellular junctions (*dotted circles*). Smooth myocytes (SMC) are depicted in cross section digitally colored in *brown*. In their vicinity, numerous Tps (*blue*) establish a network and release extracellular organelles (exosomes and shedding vesicles) (*arrowheads*) digitally colored in purple. One mast cell (*green*) is in the vicinity of this network. Some vesicles are captured at the moment of being shed from Tps (marked with *). *Cav* caveolae, *coll* collagen, *m* mitochondria, *RER* rough endoplasmic reticulum, *N* nucleus. Scale bar = 3 μm (Reproduced with permission from Cretoiu et al. [26])

Since TC discovery, I was concerned about their spatial dimension, and we started to rebuild these cells by means that were somehow rudimentary and ended with the use of the latest technologies.

Typical image processing algorithm required to obtain three-dimensional reconstruction can be described as follows: image acquisition, identifying areas of interest, identifying objects of interest, debugging alignment, gauging dimensions, 3D reconstruction, and final processing of the images obtained. At the beginning of TC research, serial electron photomicrographs (film negatives) were digitized at 1,200 dpi with the aid of a scanner, and images were processed using Adobe Photoshop© to manually outline cell contours. With the aid of the Reconstruct software (Reconstruct 1.0.4.0., 1996–2005 John C. Fiala; http://synapses.bu.edu), images were aligned with respect to specific cells of interest [23]. Three-dimensional reconstruction and rendering were performed using the cell contours arranged into objects (Figs. 21.2a, b and 21.3a, b).

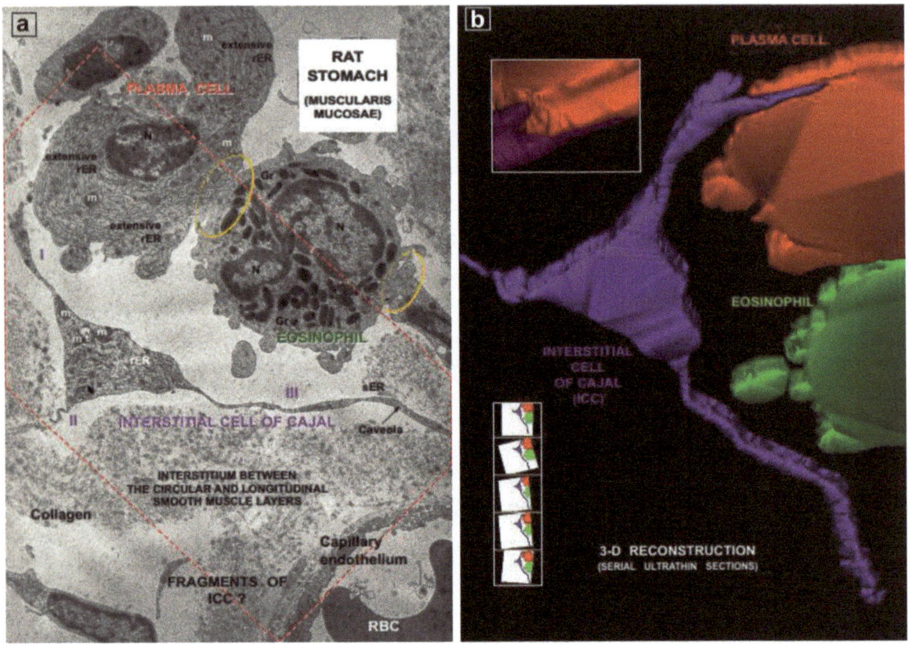

Fig. 21.2 (**a**, **b**) Rat stomach muscularis mucosae: TEM (**a**, original magnification 1,150×) and 3D image reconstruction from serial ultrathin sections (**b**). An ICC (with triangular cell body and three processes I–III) appears located near a plasma cell and an eosinophil (**a**). *N* nucleus, *SER* smooth endoplasmic reticulum, *RER* rough endoplasmic reticulum, *m* mitochondria, *RBC* red blood cell. *Yellow* ovals in (**a**) highlight two contact areas between the eosinophil, the plasma cell, and another cell. A series of five ultrathin sections revealed that the ICC process (I) approaches the plasma cell and establishes a contact point with it. 3D image reconstruction (**b**) from five serial sections (the area marked with *red dashed line* in 2A) shows that ICC processes are rather flattened than cylindrical and that they branch in a 3D pattern. A short branch establishes a synapse with the plasma cell. In the inset, the same synapse is viewed from a different angle (90°rotation). Apparently, different-angle stereoscopic views are necessary to establish if such a synapse belongs to PS or MS type. Ab initio, an intermediate type of synapse (IS) cannot be excluded (Reproduced with permission from Popescu et al. [23])

Lately, TCs were examined using ion beam milling (FIB-SEM) technique to collect data sets of many microns in the z-axis, considered as a true revolution for ultrastructural volume 3D visualization and reconstruction [27–29]. In one of our studies, acquired images were sorted into stacks according to section alignments and processed using Adobe Photoshop CS6 (Adobe Systems Incorporated, San Jose, CA, USA) for realignment, noise detection and removal, and luminance level adjustment. Images were cropped by regions of interest and loaded by batches into 3D Slicer 4.3.1 software package (http://www.slicer.org) and reconstructed using volume rendering module [30]. To better observe the third dimension of TCs in the heart, we used a stack of 500 serial images at 50 nm z-interval to reconstruct a volume of 10,908.57 μm^3 corresponding to the following cardiac tissue dimensions: x – 20.77 μm; y – 21.01 μm; z – 25 μm, and covering a 436.38 μm^2 area. The

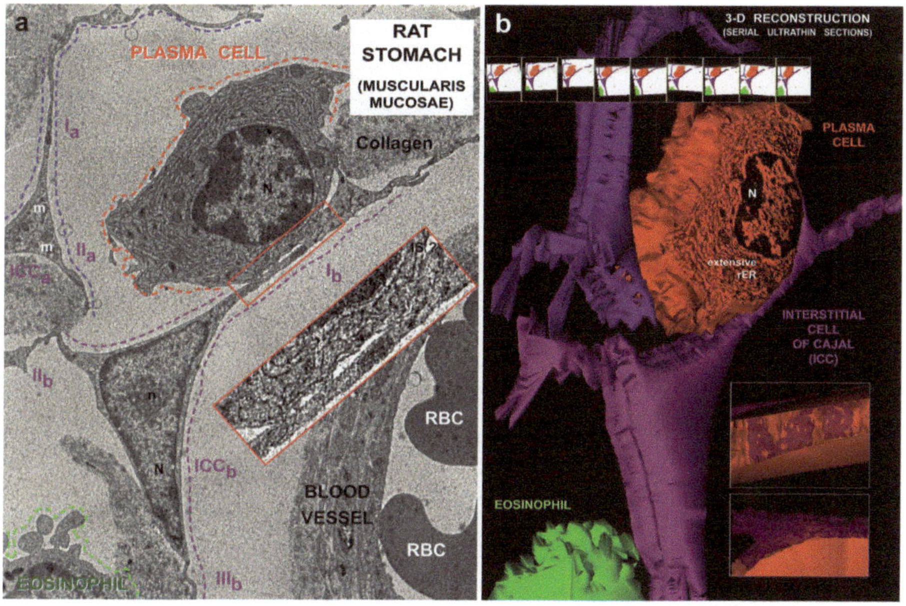

Fig. 21.3 (**a**, **b**) Rat stomach muscularis mucosae: 3D image reconstruction from nine serial ultrathin sections (**a**) and TEM (**b**), original magnification 1,500×. (**a**) Computer-aided volume rendering shows flattened ICC processes emerging from the cell body (*violet*) and branching around the plasma cell (*red*). Other ICC prolongations (*blue*), probably belonging to another cell, show the same 3D morphology. The ICC b-plasma cell synapse is uneven, with "key-and-lock" areas. Upper inset in (**a**) shows contact points where the distance between the two cell membranes is 15 nm or less (in *violet*), seen from the plasma cell cytoplasm. In the lower inset, the ICC process has been rendered transparent in order to depict the same synapse. (**b**) The ICC b (*violet dotted outline*) seems to establish an intermediate synapse (*black rectangle*) with a plasma cell (*red dotted outline*). The inset presents a magnification of this synapse, showing that it comprises, besides contact points, also a longer plasmalemmal apposition. Another ICC process (*blue dotted outline*) and an eosinophil (*green dotted outline*) are also present in close proximity (Reproduced with permission from Popescu et al. [23])

reconstructed volume showed TCs with long and thin telopodes, present in the perivascular space of atrial tissue (Fig. 21.4). With the aid of FIB-SEM tomography, we were able to observe a third dimension of telopodes, which in fact appear as narrow but flattened (ribbonlike) structures with swellings generated by the podoms. Moreover, we once more confirm the presence of adherens junctions between telopodes participating in the formation of TC network in the cardiac interstitium.

We used also the FIB-SEM technology to investigate the human dermis [31]. This time, the reconstructed volume was of 2,270 μm^3, and the images were loaded into Amira 5.0.1 (Visage Imaging, Berlin, Germany) software package. A stack of 275 serial images were assembled to obtain a 3D reconstruction (Fig. 21.5), 360° orthogonal rotation and a 3D digitally colored volume rendering. 3D reconstruction of dermal TCs by FIB-SEM tomography revealed the same conformation for the telopodes which appeared as long, flattened irregular veils (ribbonlike segments)

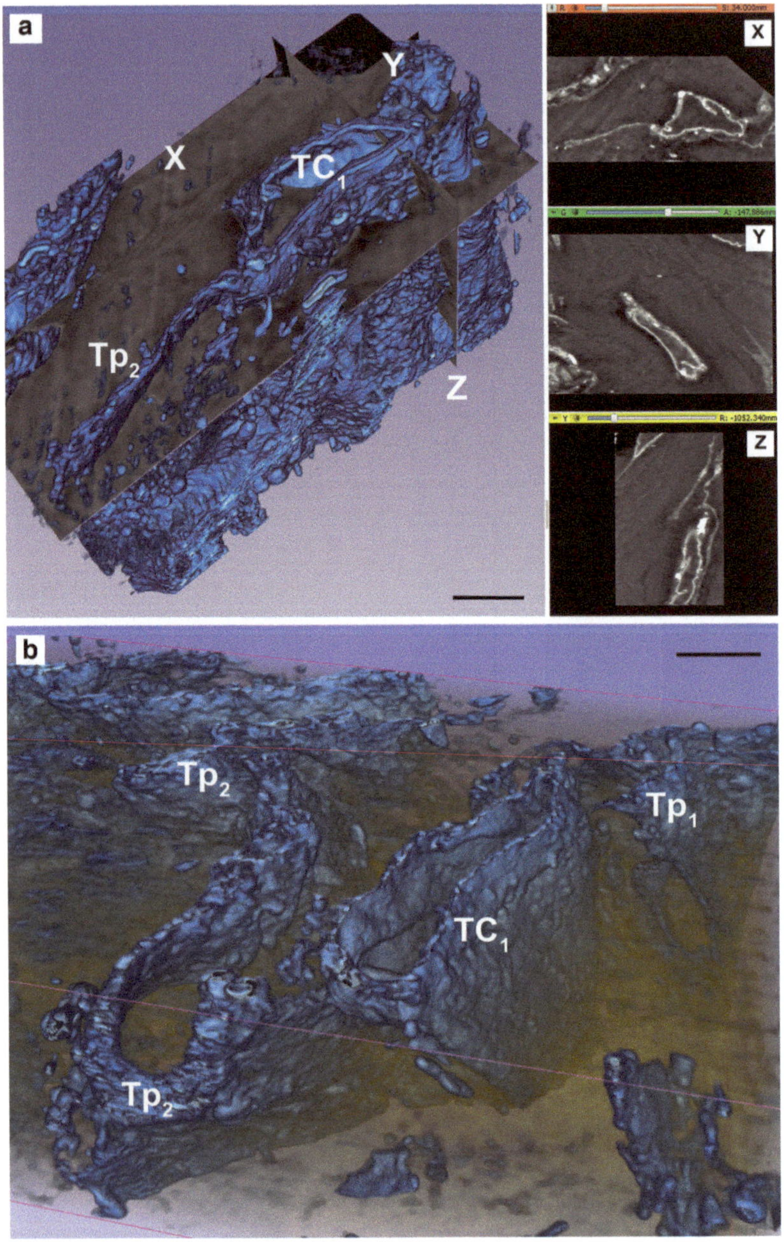

Fig. 21.4 (**a**, **b**) Automated segmentation of the stack containing a telocyte (TC1) shows that the telopode (Tp2) is long (20 μm), narrow (0.2–1 μm), and flat, given a ribbon appearance of the cell. X-Y-Z slice projections from volume rendering could be seen in the right side of A. Scale bars: 2 μm (Reproduced with permission from Cretoiu et al. [30])

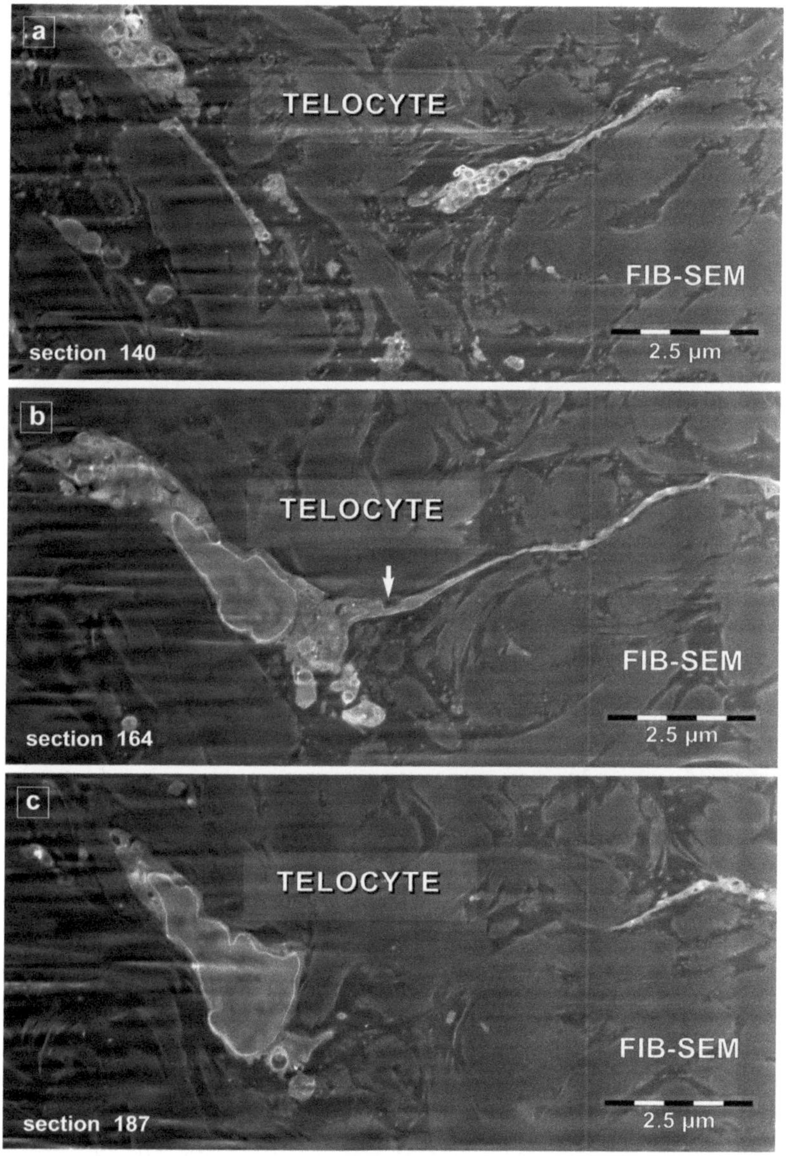

Fig. 21.5 (**a**–**c**) FIB-SEM backscattered electron images. Three nonconsecutive serial images obtained at ~1.2 μm z-interval show the narrow emergence (*arrow*) of a telopode from the cellular body of a telocyte (Reproduced with permission from Cretoiu et al. [31])

with knobs, corresponding to podoms, but also displayed tubular structures (podomers) with uneven caliber because of irregular dilations (knobs) – the podoms (Fig. 21.6). In addition, we were able to capture in 3D pictures the extracellular vesicles released by a human dermal TCs [31].

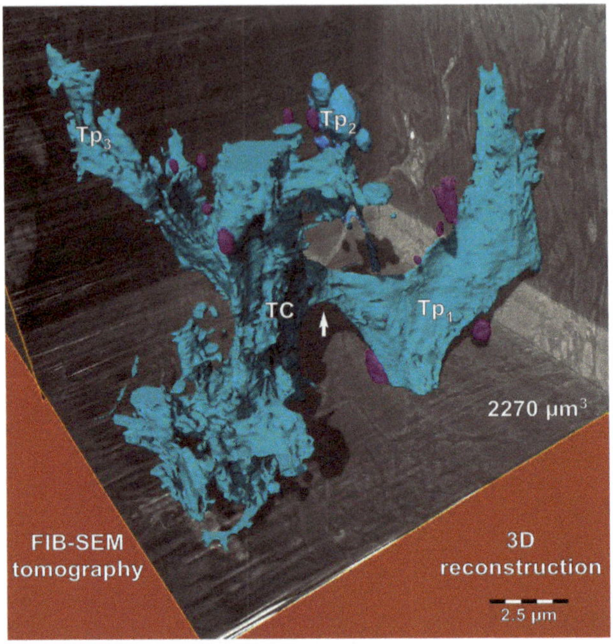

Fig. 21.6 FIB-SEM tomography of a 2,270 μm^3 volume from human papillary dermis encompasses a segment from a telocyte reconstructed in *blue*. Three-dimensional reconstruction of the stack containing the telocyte shows a "wing-like" telopode (Tp1), a telopode with typical appearance (Tp2), and a telopode with anfractuous contour (Tp3). The *arrow* indicates the narrow emergence of Tp1 suggested by serial imaging. A portion of the cell body (TC) is located in the center. At least ten extracellular vesicles appear reconstructed in *purple* (Reproduced with permission from Cretoiu et al. [31])

21.3 Concluding Remarks

Nowadays, it became more and more obvious that cell biological research is oriented to the dynamically three-dimensional investigations. These new concepts already started to modify scientist perception about inner architecture of the cell and the outer dimensional complexity of its interactions. In the case of TCs, the first steps were made in deciphering their third dimension with FIB-SEM technology capable to offer the highest voxel resolution of automated serial imaging modalities. Since we are at the beginning of perceiving TC elaborate 3D structure in the future, our aim will be to use more advanced technologies, where both sample preparation and imaging have been automated, to obtain unprecedented details about TC organelles and spatial relationships.

Acknowledgments This project was supported by a grant of the Romanian National Authority for Scientific Research, CNCS–UEFISCDI, project number 82/2012 and 194/2014.

References

1. Popescu LM, Faussone-Pellegrini MS. TELOCYTES – a case of serendipity: the winding way from Interstitial Cells of Cajal (ICC), via Interstitial Cajal-Like Cells (ICLC) to TELOCYTES. J Cell Mol Med. 2010;14:729–40.
2. Cantarero I, Luesma MJ, Alvarez-Dotu JM, Munoz E, Junquera C. Transmission electron microscopy as key technique for the characterization of telocytes. Curr Stem Cell Res Ther. 2015;11:410–4.
3. Faussone Pellegrini MS, Popescu LM. Telocytes. Biomol Concepts. 2011;2:481–9.
4. Roatesi I, Radu BM, Cretoiu D, Cretoiu SM. Uterine telocytes: a review of current knowledge. Biol Reprod. 2015;93:10.
5. Gherghiceanu M, Popescu LM. Cardiac telocytes – their junctions and functional implications. Cell Tissue Res. 2012;348:265–79.
6. Cretoiu SM, Cretoiu D, Marin A, Radu BM, Popescu LM. Telocytes: ultrastructural, immunohistochemical and electrophysiological characteristics in human myometrium. Reproduction. 2013;145:357–70.
7. Fertig ET, Gherghiceanu M, Popescu LM. Extracellular vesicles release by cardiac telocytes: electron microscopy and electron tomography. J Cell Mol Med. 2014;18:1938–43.
8. Popescu LM, Fertig ET, Gherghiceanu M. Reaching out: junctions between cardiac telocytes and cardiac stem cells in culture. J Cell Mol Med. 2016;20:370–80.
9. Sun X, Zheng M, Zhang M, Qian M, Zheng Y, Li M, et al. Differences in the expression of chromosome 1 genes between lung telocytes and other cells: mesenchymal stem cells, fibroblasts, alveolar type II cells, airway epithelial cells and lymphocytes. J Cell Mol Med. 2014;18:801–10.
10. Zheng M, Sun X, Zhang M, Qian M, Zheng Y, Li M, et al. Variations of chromosomes 2 and 3 gene expression profiles among pulmonary telocytes, pneumocytes, airway cells, mesenchymal stem cells and lymphocytes. J Cell Mol Med. 2014;18:2044–60.
11. Song D, Cretoiu D, Zheng M, Qian M, Zhang M, Cretoiu SM, et al. Comparison of Chromosome 4 gene expression profile between lung telocytes and other local cell types. J Cell Mol Med. 2016;20:71–80.
12. Liu J, Cao Y, Song Y, Huang Q, Wang F, Yang W, et al. Telocytes in liver. Curr Stem Cell Res Ther. 2016;11:415–9.
13. Zhu Y, Zheng M, Song D, Ye L, Wang X. Global comparison of chromosome X genes of pulmonary telocytes with mesenchymal stem cells, fibroblasts, alveolar type II cells, airway epithelial cells, and lymphocytes. J Transl Med. 2015;13:318.
14. Cismasiu VB, Radu E, Popescu LM. miR-193 expression differentiates telocytes from other stromal cells. J Cell Mol Med. 2011;15:1071–4.
15. Zheng Y, Cretoiu D, Yan G, Cretoiu SM, Popescu LM, Wang X. Comparative proteomic analysis of human lung telocytes with fibroblasts. J Cell Mol Med. 2014;18:568–89.
16. Zheng Y, Cretoiu D, Yan G, Cretoiu SM, Popescu LM, Fang H, et al. Protein profiling of human lung telocytes and microvascular endothelial cells using iTRAQ quantitative proteomics. J Cell Mol Med. 2014;18:1035–59.
17. Albulescu R, Tanase C, Codrici E, Popescu DI, Cretoiu SM, Popescu LM. The secretome of myocardial telocytes modulates the activity of cardiac stem cells. J Cell Mol Med. 2015;19:1783–94.
18. Cretoiu SM, Radu BM, Banciu A, Banciu DD, Cretoiu D, Ceafalan LC, et al. Isolated human uterine telocytes: immunocytochemistry and electrophysiology of T-type calcium channels. Histochem Cell Biol. 2015;143:83–94.
19. Campeanu RA, Radu BM, Cretoiu SM, Banciu DD, Banciu A, Cretoiu D, et al. Near-infrared low-level laser stimulation of telocytes from human myometrium. Lasers Med Sci. 2014; 29:1867–74.
20. Cretoiu SM, Popescu LM. Telocytes revisited. Biomol Concepts. 2014;5:353–69.

21. Bei Y, Wang F, Yang C, Xiao J. Telocytes in regenerative medicine. J Cell Mol Med. 2015;19:1441–54.
22. Vannucchi MG, Bani D, Faussone-Pellegrini MS. Telocytes contribute as cell progenitors and differentiation inductors in tissue regeneration. Curr Stem Cell Res Ther. 2016;11:383–9.
23. Popescu LM, Gherghiceanu M, Cretoiu D, Radu E. The connective connection: interstitial cells of Cajal (ICC) and ICC-like cells establish synapses with immunoreactive cells. Electron microscope study in situ. J Cell Mol Med. 2005;9:714–30.
24. Cretoiu D, Cretoiu SM, Simionescu AA, Popescu LM. Telocytes, a distinct type of cell among the stromal cells present in the lamina propria of jejunum. Histol Histopathol. 2012;27:1067–78.
25. Dawidowicz J, Szotek S, Matysiak N, Mielanczyk L, Maksymowicz K. Electron microscopy of human fascia lata: focus on telocytes. J Cell Mol Med. 2015;19:2500–6.
26. Cretoiu SM, Cretoiu D, Popescu LM. Human myometrium – the ultrastructural 3D network of telocytes. J Cell Mol Med. 2012;16:2844–9.
27. Peddie CJ, Collinson LM. Exploring the third dimension: volume electron microscopy comes of age. Micron. 2014;61:9–19.
28. Kizilyaprak C, Daraspe J, Humbel BM. Focused ion beam scanning electron microscopy in biology. J Microsc. 2014;254:109–14.
29. Narayan K, Danielson CM, Lagarec K, Lowekamp BC, Coffman P, Laquerre A, et al. Multi-resolution correlative focused ion beam scanning electron microscopy: applications to cell biology. J Struct Biol. 2014;185:278–84.
30. Cretoiu D, Hummel E, Zimmermann H, Gherghiceanu M, Popescu LM. Human cardiac telocytes: 3D imaging by FIB-SEM tomography. J Cell Mol Med. 2014;18:2157–64.
31. Cretoiu D, Gherghiceanu M, Hummel E, Zimmermann H, Simionescu O, Popescu LM. FIB-SEM tomography of human skin telocytes and their extracellular vesicles. J Cell Mol Med. 2015;19:714–22.

Dragos Cretoiu, MD, PhD Current position: Division of Cellular and Molecular Biology and Histology, Department of Morphological Sciences, Carol Davila University of Medicine and Pharmacy, Bucharest, Romania; and Department of Molecular Medicine, Victor Babes National Institute of Pathology, Bucharest, Romania.

Education: M.D. from Carol Davila University, Bucharest (2003); Ph.D. from Carol Davila University, Bucharest (2012); Postdoctoral Fellowship at Victor Babes National Institute of Pathology, Bucharest, Romania (2014-2016).

Research Interests: I am particularly interested in telocytes, a newly discovered type of cell, and I have spent the past years working on computer 3D reconstructions, and medical image processing and genomics/proteomics data analysis. My primary interest is in understanding telocytes' 3D configuration and the interrelations between these cells and the surrounding structures. Moreover, I have a great interest in studying telocytes in cell culture by various microscopy techniques, such as fluorescence and scanning confocal microscopy, time-lapse videomicroscopy, and electron microscopy.

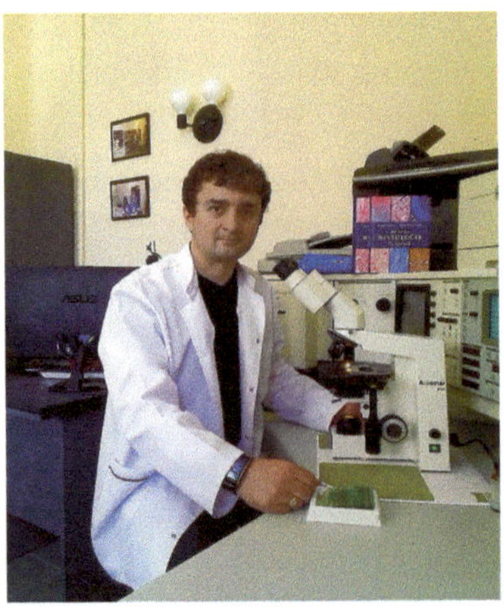

Chapter 22
Immunohistochemistry of Telocytes in the Uterus and Fallopian Tubes

Sanda Maria Cretoiu

Abstract The seminal work of Popescu and colleagues first demonstrated the existence of a new cell type – the telocytes. We were among the first who reported the presence of such cells in the female genital tract and performed TEM examinations, as well as immunohistochemical staining in the attempt to find a specific marker. Telocytes from rat and from the human uterus and from human fallopian tube were extensively investigated initially by comparison with interstitial cells of Cajal. Progress in telocyte research led to the identification of different subtypes suggestive for a heterogeneous telocyte population which can even coexist in the same location. As a consequence, the functions of TCs are still elusive and can be considered a versatile phenomenon that depends on a variety of conditions, including signal reception and transmission of information via extracellular vesicles or by direct intercellular contact.

22.1 Introduction to the History of This Discovery: From Interstitial Cajal-Like Cells to Telocytes

The last decade was marked by the discovery of a new type of cell – the telocyte – which is a starting point in rethinking the importance of the interstitial space. This serendipity event belongs to Professor Popescu and his team in Bucharest, Romania. All started in 2005 when Popescu focused on interstitial cells of Cajal (ICC) research [1]. Popescu and co-workers described, for the first time, in the rat and human exocrine pancreas, cells resembling with ICC, which suggested to be pancreatic interstitial Cajal-like cells (Fig. 22.1) [1]. In the next 5 years, he and his team described and characterized these cells in several organs [2–10]. In 2010, with the support of the world-known expert in ICC, Professor Faussone-Pellegrini, he decided that

S.M. Cretoiu
Victor Babeş National Institute of Pathology, Bucharest 050096, Romania

Division of Cell Biology and Histology, Carol Davila University of Medicine and Pharmacy, 8 Eroii Sanitari Blvd., Bucharest 050474, Romania
e-mail: sanda@cretoiu.ro

X. Wang, D. Cretoiu (eds.), *Telocytes*, Advances in Experimental Medicine and Biology, 913, DOI 10.1007/978-981-10-1061-3_22

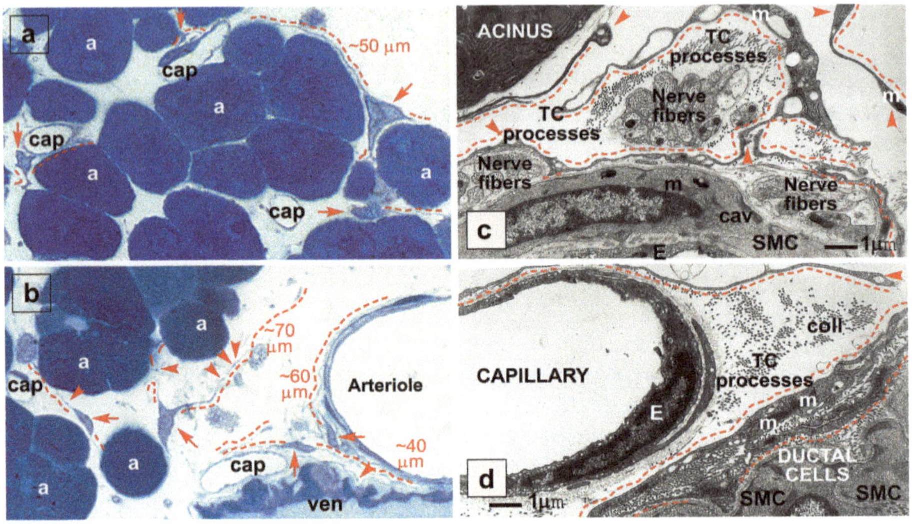

Fig. 22.1 Rat exocrine pancreas. (**a, b**) Nonconventional light microscopy – semi-thin sections of Epon-embedded material were cut on ultramicrotome and stained with toluidine blue, tissue fixed by perfusion with glutaraldehyde and postfixed in OsO4, objective 100×. Note that only the TC body (*arrows*) and the emerging cytoplasmic processes (*dashed lines*) are evident using the above-mentioned methodology. The length of the cytoplasmic processes is very impressive: 20–50–100 µm (!). In addition, the cytoplasmic processes are very thin (less than 0.5 µm) and have a characteristic "moniliform" aspect (*arrowheads*). (**c, d**) TEM images showing the vicinity relationships with acini, unmyelinated terminal nerve fibers, arteriolar smooth muscle cells (*C*), capillaries, or ductal cells (*D*). The cytoplasmic processes of TC (*dashed lines*) can be easily recognized: very long and thin and with dilatations ("moniliform" aspect), containing mitochondria and other organelles (*arrowheads*). *a* acini, *cap* capillary, *art* arteriole, *ven* venule, *m* mitochondria, *cav* caveolae, *SMC* smooth muscle cells, *E* endothelium, *coll* collagen (Reproduced and modified with permission from Popescu et al. [1])

ICLC was, in fact, a new type of cell and coined the name of telocyte [11]. The new term aims to prevent further confusion with ICC, myofibroblasts, or any other interstitial cell space residents. Nowadays, telocytes are described in the interstitial space of multiple organs and tissues in mammals, birds, fish, and reptiles (Table 22.1).

Telocytes are characterized by unusual-length tubular and/or veillike cytoplasmic extensions called telopodes, which might be present in a variable number, usually between 2 and 5 [80]. Therefore, the shortest appropriate definition for telocytes is cells with telopodes. Another distinctive feature is that these telopodes build 3D scaffolds by homocellular contacts [81]. Furthermore, telocytes establish important connections with distinct structures, e.g., blood vessels, nerve bundles, and covering and glandular epithelia, as well as with several different cell populations such as stem cells, immune cells, and muscle fibers by means of heterocellular junctions [82]. Telocytes' main definitive characteristics are seen in Table 22.2 by comparison with interstitial cells of Cajal and fibroblasts.

Recent papers underline the capacity of telocytes to communicate at long distance by releasing extracellular vesicles. There are three types of vesicles released

Table 22.1 List of telocytes' localization and species in which telocytes' presence was demonstrated

Organ	Location	Species	References
Heart	Epicardium	Hu, ro	[12, 13]
	Myocardium	Hu, ro, fi, re	[14–20]
	Endocardium	Ro	[21]
	Myocardial sleeves	Hu	[22]
	Heart valves	Hu, ro	[23, 24]
Digestive tract	Esophagus	Hu, ro	[25, 26]
	Duodenum	Ro	[27]
	Jejunum	Ro	[28]
	Ileum	Bi	[29]
	Colon	Hu	[30]
	Liver	Hu, ro	[31, 32]
Annex glands	Gallbladder	Hu	[33, 34]
	Salivary gland	Hu, ro	[35, 36]
	Exocrine pancreas	Hu, ro	[1, 37–39]
Respiratory system	Trachea	Ro	[40, 41]
	Lungs	Ro	[40, 42]
Urinary system	Kidney	Hu, ro	[43, 44]
	Renal pelvis	Hu	[45]
	Ureters	Hu, ro	[44, 45]
	Bladder	Hu, ro	[44, 46]
	Urethra	Hu	[45]
Female reproductive system	Fallopian tube	Hu, ro, re	[7, 47, 48]
	Uterus	Hu, ro	[49–53]
	Placenta	Hu	[10, 54–56]
	Mammary gland	Hu	[9, 57, 58]
Vasculature	Blood vessels	Ro, pig	[59–61]
	Thoracic duct	Hu	[62]
Serous membranes	Mesentery	Ro	[63]
	Pleura	Hu, ro	[64]
Muscle	Skeletal muscle	Hu, ro	[65, 66]
	Neuromuscular spindles	Hu	[67]
Sensory organs	Skin	Hu	[68–71]
	Eye	Ro	[72]
Other organs	Prostate	Ro	[73]
	Bone marrow	Ro	[74]
	Meninges and choroid plexus	Ro	[75]
	Dura mater	Dog	[76]
	Fascia lata	Hu	[77]
	Testis	Re	[78]
	Spleen	Ro	[79]

hu human, *ro* rodent, *fi* fish, *re* reptiles, *bi* birds

Table 22.2 Morphological aspects, semiquantitative data concerning the ultrastructural elements (transmission electron microscopy) and specific markers of telocytes compared to archetypal enteric interstitial cells of Cajal (ICC) and fibroblasts

		Interstitial cells of Cajal (ICC)	Myometrial telocytes	Fallopian tube telocytes	Fibroblasts
Cell shape		Oval- or spindle-shaped body	Spindle or stellate body		Polymorphic body
Nucleus		Oval, mostly euchromatic	Oval, heterochromatic under nuclear membrane		Oval, euchromatic with 1–2 visible nucleoli
Cytoplasm	Smooth ER	+ +	+	+	+–
	Rough ER	+	+	+	+ + +
	Golgi apparatus	+	+		+ + +
	Mitochondria	+ + +	+ +	+	+
	Intermediate filaments	+ +	+	+	+
	Microtubules	+	+	+	+
	Thin filaments	+	+	+	+
	Calcium-releasing units	n.a.	Present	n.a.	n.a.
Other structures	Caveolae	+	+	++	–
	Basal lamina	0	+–	+–	–
Immunohistochemical markers		c-Kit	Co-localization of c-kit, CD34, and connexin 43, lack of prolyl 4-hydroxylase		Prolyl 4-hydroxylase
Intercellular contacts	Nerve endings	+ +	+	+	–
	Blood vessels	n.a.	+		–
	Immune cells	n.a.	+ + +		–
	Smooth muscle cells	+	+	+	–
	Other interstitial cells	+	+	+	–
	Gap junctions	+	+		–

Adapted from Hutchings et al. [52]

by telocytes: exosomes (released from endosomes, 45 ± 8 nm), ectosomes (budding directly from the plasma membrane, 128 ± 28 nm), and multivesicular cargos (MVCs, 1 ± 0.4 μm), both in situ and in vitro [83]. These vesicle-shedding properties could be responsible for controlling the extracellular environment and in turn for rapid phenotype adjustment, explaining why some immunohistochemical markers are inconstantly expressed.

Recent advances allowed the characterization of telocytes concerning their microRNA imprint, gene profile, and proteomic analysis [84–88]. In addition, their secretome was also analyzed, and it was demonstrated to influence the stem cells [89] and macrophages [90]. Genes from several chromosomes (1, 2, 3, 4, 17, 18, X) were globally analyzed by comparison between pulmonary telocytes, pneumocytes, airway cells, mesenchymal stem cells, and lymphocytes [91–95].

22.2 Telocytes in the Female Genital Tract

Reproductive functions are essential for the beginning of a new life. One of the key contributors is the integrity of female genital tract. In this view, the structure, microanatomy, and physiology of the fallopian tubes and uterus were closely analyzed, and remarkable progress has been made. However, their state of contractility still represents an unsolved equation for scientists all over the world. This is also reflected by the lack of advances in clinical treatments of many disorders such as infertility by implantation failure, spontaneous miscarriage, or preterm birth.

In 2005, our attention was drawn by the hypothetic existence of ICCs outside the gastrointestinal tract, in hollow organs with smooth muscle coats, which theoretically need a pacemaker to trigger the contraction. We described, then, the existence of some peculiar cells in human and rat myometrium and in human fallopian tubes [2, 3]. These cells proved to be of a distinct type than any other cell already described and were later, in 2010, designated under the name of telocytes [11].

22.3 Telocytes' Immunohistochemical Characterization

Immunohistochemistry (IHC) is an imaging technique that combines the histology techniques with biochemical and immunological ones in order to identify cell and tissue components by a particular antigen-antibody reaction. At cellular level, IHC is usually used to identify and classify specific cells within a cell population whose morphology is heterogeneous or apparently homogenous. The antigen-antibody complex is routinely visualized under the microscope through the addition of either a fluorochrome conjugate or an enzyme to the antibody. It is well known that the antibodies are not entirely specific and can cross-react with more than one antigen or the reactions can be false positive and false negative. Sometimes one can observe cells, which express antigens typically associated with a completely different class of cells, and therefore, we need to use appropriate sections of control tissues.

Although electron microscopy is essential for telocyte identification, it was thought that IHC would bring valuable support for this. Nevertheless, for the time being, none of the cell-surface markers which were identified proved to be specific for telocyte demonstration. Until now, there are several cell-surface markers found on telocytes in different locations.

As concerns the female genital tract, the phenotype of telocytes has been progressively investigated. The immunophenotype was determined both on tissue samples from nonpregnant and pregnant uteri and from human fallopian tubes, as well as in cell cultures usually obtained from the muscle coat of the fallopian tube and uterus and analyzed by IHC and/or immunofluorescence (IF).

22.3.1 Uterus

Several surface markers were identified on myometrial telocytes. The first report about myometrial telocytes tested these cells by immunofluorescence assays for c-kit tyrosine receptor kinase, CD34, vimentin, and the specific α-smooth muscle actin (α-SMA) expressions [2]. Telocytes were found to express CD117/c-kit (Fig. 22.2a, b) in distinct membrane areas, sometimes in association with CD34 (Fig. 22.2c), while telopodes were intensely positive for vimentin (Fig. 22.2d, e) and tested negative for α-SMA (Fig. 22.3) [2]. In a further study, immunocytochemistry performed on telocytes in culture depicted telocytes with intense immunostaining for estrogen receptor (ER) and progesterone receptor (PR), strong on the nuclei and weak at the cytoplasmic level (Fig. 22.4a, b) [4]. These findings were strengthened by double immunolabeling for CD117/c-kit and ER and CD117/c-kit and PR (Fig. 22.4c, d) and confirmed the same distribution of PR and ER on c-kit-positive cells (Fig. 22.5) [4].

Salama (2013) performed a c-kit immunohistochemical characterization of telocytes in immature, adult nonpregnant/pregnant, and postpartum rats [96]. She detected c-kit-positive cells interpreted as telocytes and showed that the mean counts were significantly different in the endometrium and myometrium of distinct groups (Table 22.3).

Hatta et al. reported in human myometrium telocytes positive for both vimentin and connexin 43 and also demonstrated the presence of rat endometrial telocytes in cell culture labeled for CD34, vimentin, and connexin 43 [50]. Rosenbaum et al. studied the SK3 expression in the myometrium from pregnant and nonpregnant women using immunohistochemistry mRNA and demonstrate that SK3 channels are localized in CD34-positive telocytes and that SK3 channels were detected at higher levels in nonpregnant compared to pregnant myometrium [97].

Nowadays, we consider that telocytes' existence can be demonstrated by double immunolabeling for CD34 and PDGFR-α [98]. Moreover, CD34 expression in cells with very long telopodes is considered to be suggestive for telocytes [99]. Additionally, uterine TCs were found to be double positive for CD34 and PDGFR-α and also express T-type Ca^{2+}channels (Ca_v3.1 (α1G), Cav3.2 (α1H) members) (Figs. 22.6 and 22.7) [100].

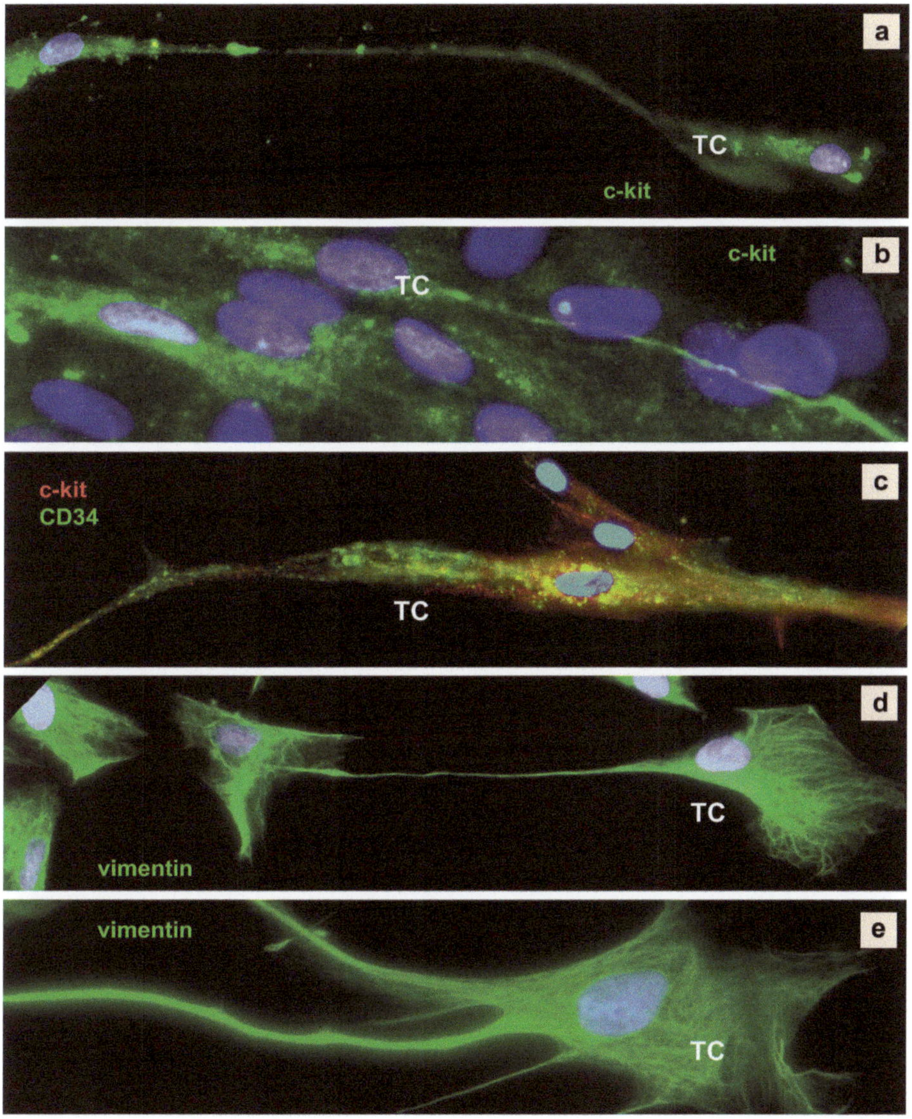

Fig. 22.2 (**a–e**) Human myometrial cells in culture (the second passage): IF for c-kit (*green* in **a**, **b**), c-kit and CD34 (*red* and *green*, respectively, in **c**), and vimentin (*green* in **d**, **e**). Cells that display the TC morphologic feature (long, moniliform processes) express c-kit and contact adjacent cells (**a–c**). Some cells suggestive for TC co-express c-kit and CD34 (**c**). The characteristic cell processes are immunoreactive for vimentin and establish connections with nearby cells (**d**, **e**). Original magnification 60×, nuclear counterstaining with Hoechst 33342 (*blue*) (Reproduced and modified with permission from Ciontea et al. [2])

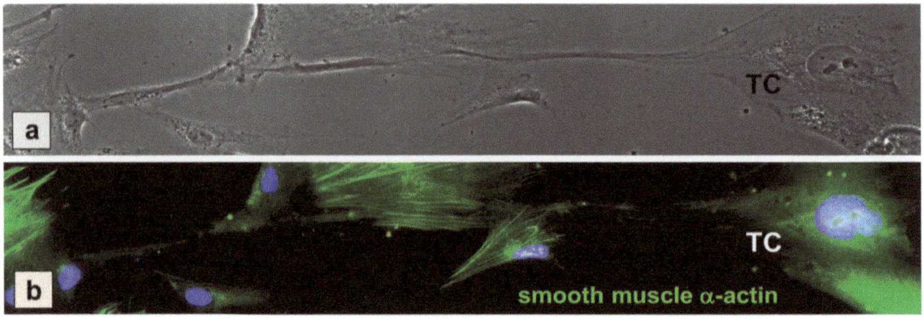

Fig. 22.3 Human myometrial cell culture (the second passage). (**a**) Control phase-contrast microscopy and (**b**) immunofluorescence for α-smooth muscle actin (*green*) of the same microscopic fields; typical TC has one long SMA-negative process that contacts several SMA-positive myocytes. Although the TC cell body is reactive for SMA, no filamentous pattern is seen. Image reconstruction, original magnification 60×, nuclear counterstaining with Hoechst 33342 (*blue*) (Reproduced and modified with permission from Ciontea et al. [2])

22.3.1.1 Fallopian Tube

In 2005, our laboratory discovered and described a novel type of interstitial cells in the fallopian tube tissue [3] which are now identified to be telocytes. We then analyzed the telocytes' percentage in the fallopian tube wall and found the following results: in the mucosa, telocytes represent 18 ± 2 % in the lamina propria underneath the basement membrane and ~8 % in the entire lamina propria thickness, and in the muscularis, telocytes represent 7.8 ± 1.2 %.

Human tubal telocytes were initially tested in tissue samples using 15 different antibodies for single or double immunolabeling (see Table 22.4). Telocytes positive for c-kit were revealed at high density in the fallopian tube mucosa (in the lamina propria) (Fig. 22.8a–c). In the same location, CD34-positive cells were also reported (Fig. 22.8d–f). c-Kit/CD34 double IHC reaction was positive for both antigens (Fig. 22.9a) and showed that c-kit was expressed mainly on the cell body, while CD34 stained preferentially the telopodes (Fig. 22.9b). In addition, some of the tubal telocytes from the lamina propria were found to express S-100, while the majority of telocytes from the lamina propria and muscularis were nonreactive for SMA. Double IHC for S-100 and SMA revealed two different types of telocytes: expressing both S-100 and SMA in the lamina propria or expressing either S-100 or SMA in the muscularis. NK-1 and nestin were inconstantly and weakly positive on tubal telocytes, while desmin showed a lack of expression on telocytes. Vimentin was found to be inconstantly positive, sometimes co-expressed with c-kit. IHC with antibodies against some additional antigens showed negative immunostaining for CD1a, CD62P, CD68, GFAP, NSE, chromogranin A, and PGP 9.5.

The immunophenotype of tubal telocytes was assessed in vitro, by IF, and the same markers were found to be (co-)expressed: c-kit and CD34 (Fig. 22.10a, b, c, e). Occasionally in the same cell, desmin is co-expressed with c-kit (Fig. 22.10d). In

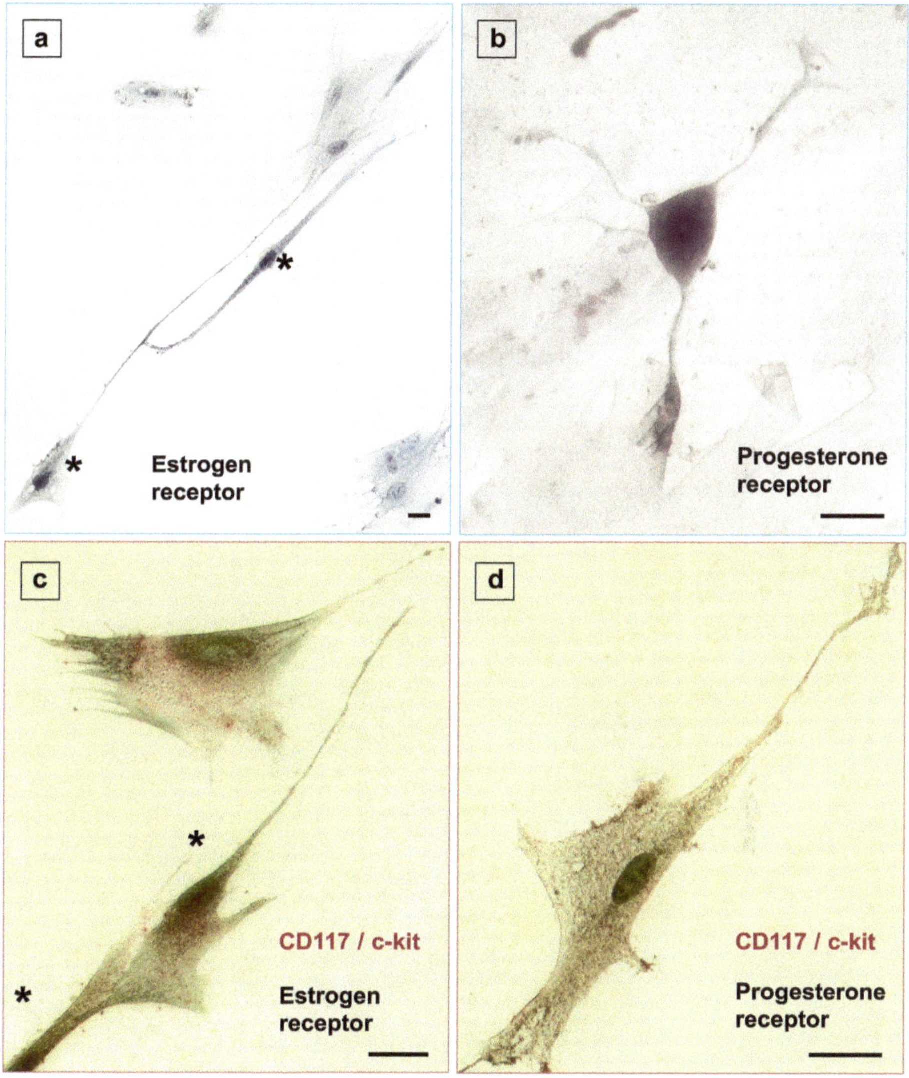

Fig. 22.4 Human myometrial cell culture, fourth passage. Immunocytochemical staining for the estrogen and progesterone receptors. (**a**) Immunocytochemical detection of estrogen receptor – dark-stained nuclei (*), counterstaining with methyl green for negative nuclei. (**b**) ICLC stained positive for progesterone receptor. (**c**) Double staining (*) for CD117/c-kit (*red*) and estrogen receptor (*black*). (**d**) Double staining for CD117/c-kit (*red*) and progesterone receptor (*black*). Scale bar = 10 μm (Reproduced and modified with permission from Cretoiu et al. [4])

addition, caveolin-1 and caveolin-2 cellular distribution was also detected, with greater intensity in telopodes (Fig. 22.11a, b). SMA was detected in cell cultures only in cells with myofibroblastic morphology (Fig. 22.11c). Moreover, tubal telocytes also express ER and PR (Figs. 22.12 and 22.13) [101].

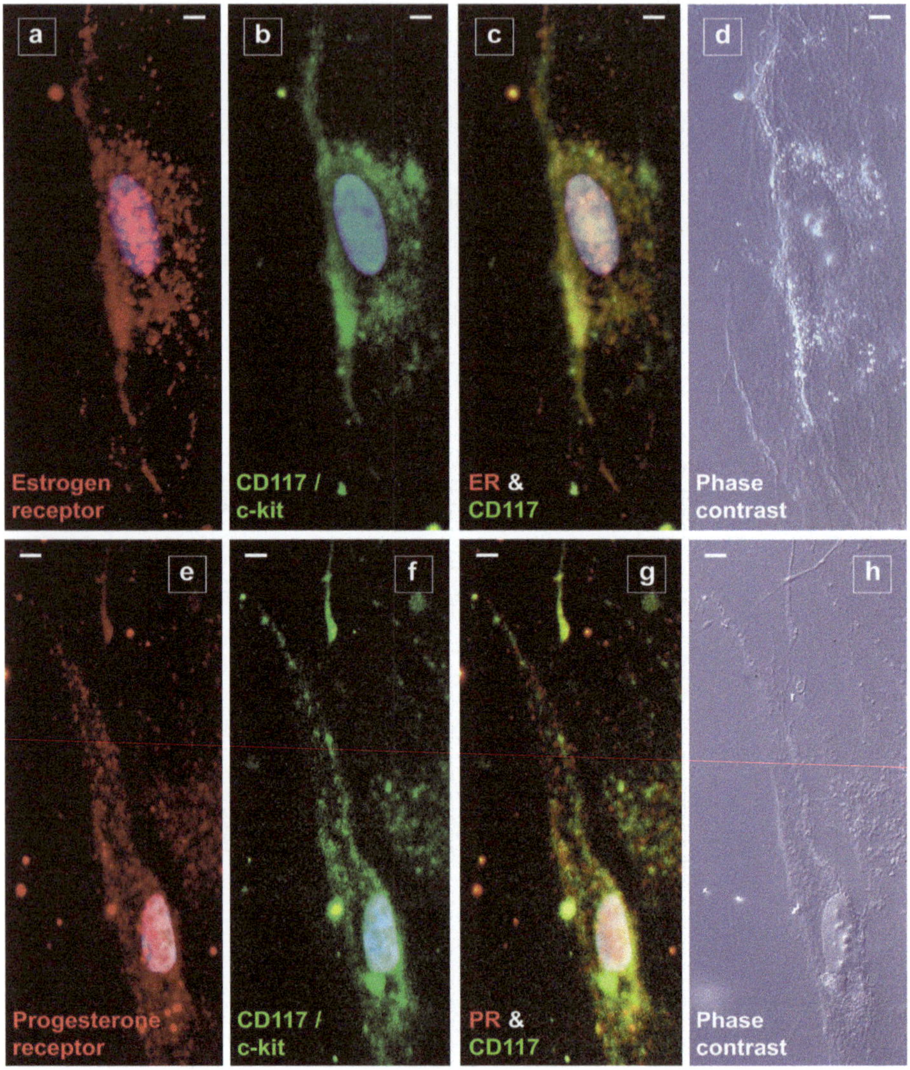

Fig. 22.5 Human myometrial cell culture, fourth passage. Immunofluorescent labeling for estrogen (**a**) and progesterone (**e**) receptor (*red*) which appears both inside the nucleus and in the cytoplasm. c-Kit/CD117 (*green*) only in the cytoplasm (**b**, **f**) and double labeling for both markers (**c**, **g**), where the co-expression appears as yellow areas. Hoechst 33342 (*blue*) for nuclear counterstaining. Phase-contrast microscopy focused on the same cells, typical ICLCs with long, moniliform prolongations (**d**, **h**). Scale bar = 2 μm (Reproduced and modified with permission from Cretoiu et al. [4])

The above determined markers were inconstantly demonstrated by other studies. Yang et al. demonstrated on rabbit isthmic segment of the oviduct the presence of typical telocytes with strong c-kit immunoreactivity forming “a complex, dense, and highly anastomosing network between the adjacent muscle layers” [102]. The

Table 22.3 Mean counts of telocytes (c-kit-positive cells) in the rat endometrium and myometrium

Rat group	Endometrium	Myometrium
Immature	3.9±0.74	3.3±0.95
Adult nonpregnant	10.4±1.17	11.90±1.37
Pregnant	13.9±1.29	6.8±1.48
Postpartum	10.2±1.03	15.4±2.41

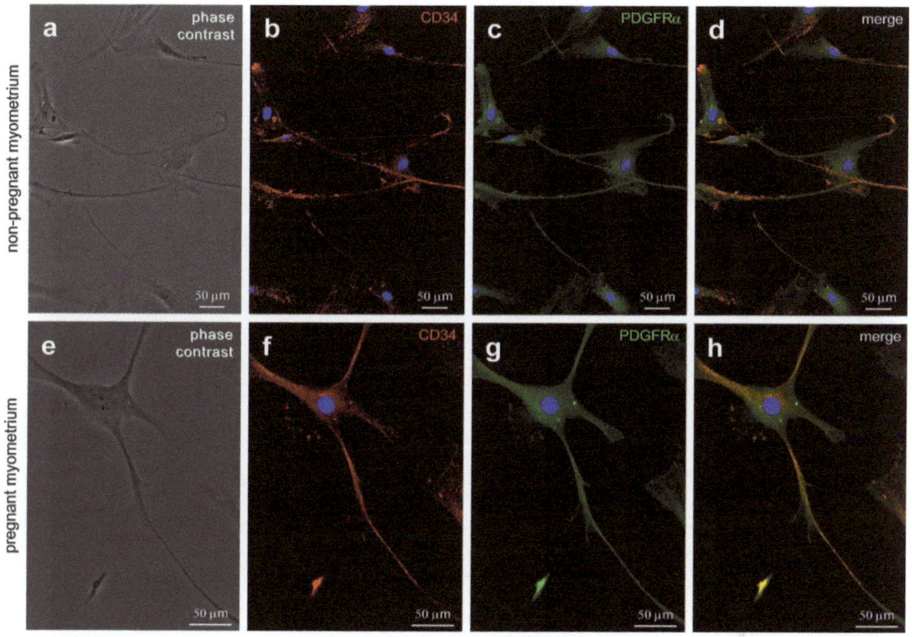

Fig. 22.6 TCs in cell cultures from human nonpregnant/pregnant myometrium after 72 h in culture at the first passage. (**a**, **e**) Phase-contrast microscopy depicts cell morphologies very evocative for TCs: small cell body with very long Tps characterized by a moniliform silhouette (alternation of podoms and podomers). Fluorescence microscopy shows CD34 (**b**, **f**) and PDGFR-α (**c**, **g**)-positive TCs. (**d**, **h**) Double labeling for both markers showing a different expression pattern for PDGFR-α and CD34 in the same TC. Nuclei were stained with DAPI. Bar = 50 μm (Reproduced with permission from Faussone-Pellegrini and Popescu [99])

same author demonstrated by c-kit immunohistochemistry that telocytes' network is affected in human fallopian tubes from endometriosis and tubal ectopic pregnancy [103]. A recent study performed on rat oviducts demonstrated that in normal oviducts, telocytes are characterized by a positive immunophenotype for CD34 and vimentin, while some are negative for c-kit possibly because of immunophenotype variation among organ and/or animal species [48], thus confirming the existence of different telocyte subpopulations [46]. Therefore, for an illustrative and efficient immunodiagnosis of the telocytes, it is advisable to use a combination of markers such as CD34, c-kit, vimentin, and PDGFR-α or PDGFR-β [80] regardless of location.

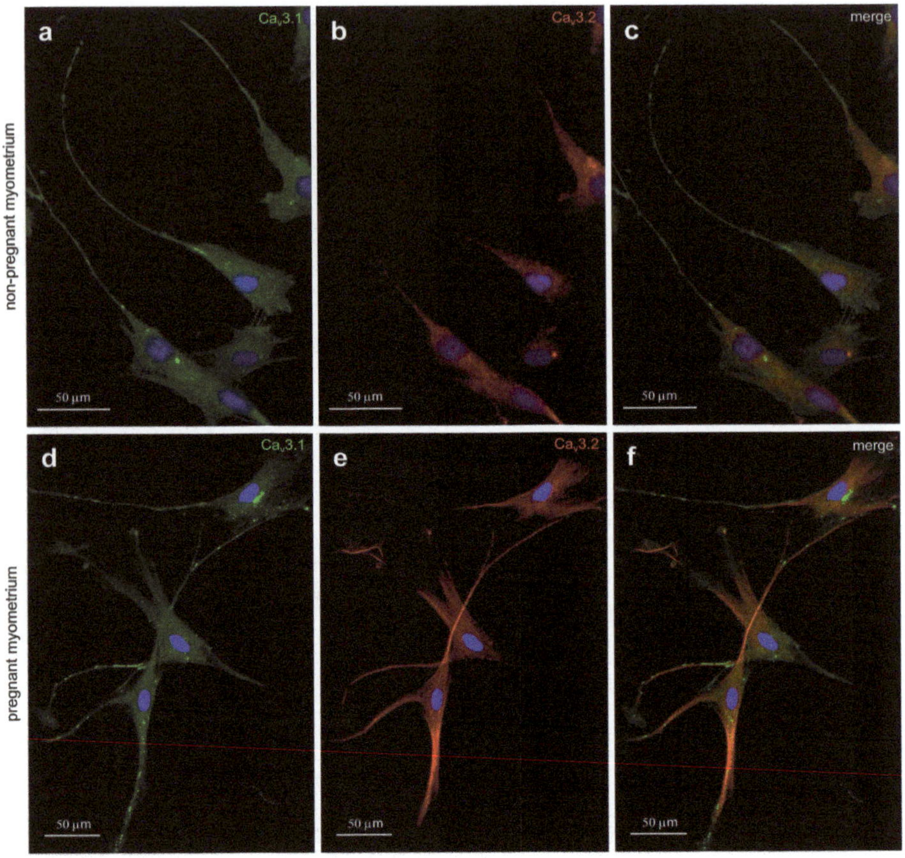

Fig. 22.7 Double immunolabeling for T-type calcium channels in cell cultures from nonpregnant/pregnant myometrium. (**a**) CaV3.1 immunolabeling (*green*) was detected on TC cell body, but it was stronger at Tps level. (**b**) CaV3.2 expression was detected only at the cell body level. (**c**) Co-expression of CaV3.1 positive (*green*) and CaV3.2 (*red*) is presented on merged images. (**d**) Immunolabeling for T-type calcium channels in cell cultures from pregnant myometrium. Strong staining for CaV3.1 was found in the cytoplasm, adjacent to the nucleus and in the Tps of the TCs. (**e**) CaV3.2 immunolabeling was found throughout the cytoplasm of TCs and predominantly within Tps. (**f**) Merged images show a co-expression pattern for CaV3.1 (*green*) and CaV3.2 (*red*). Note that both TCs and SMCs (cells with a widened cell body without extensions) express CaV3.1 and CaV3.2. Nuclei were stained with DAPI. Bar = 50 μm (Reproduced with permission from Faussone-Pellegrini and Popescu [99])

22.3.1.2 Functional Correlations

In the uterus, telocytes were reported to be located in the endometrium and myometrium [50, 81]. In a rat model, telocytes were shown to suffer numerical variations, which might explain, for example, the quiescent myometrium during pregnancy (decreased number of telocytes) and the postpartum involution of the uterus at first

Table 22.4 Summary of immunohistochemical results for telocytes from human fallopian tube

Antibody	Clone	Dilution	Source	IHC positivity
CD117/c-kit	Polyclonal	1:100	DAKO	+ + + +
CD34	QBEnd10	1:100	BioGenex	+ + +
S-100	Polyclonal	1:500	DAKO	+ +
α-SMA	1A4	1:1,500	Sigma	+
CD57	NK1	1:50	DAKO	+
Nestin	5,326	1:100	Santa Cruz	+
Desmin	D33	1:50	DAKO	–
Vimentin	V9	1:50	DAKO	+
NSE	BBS/NC/VI-H14	1:50	DAKO	+
GFAP	6F2	1:50	DAKO	–
CD68	PG-M1	1:50	DAKO	±
CD62P	1E3	1:25	DAKO	–
CD1a	CD1a-235	1:30	Novocastra	–
Chromo A	LK2H10	1:50	Novocastra	–
PGP9.5	10A1	1:40	Novocastra	–

The intensity of telocytes' reactivity was assessed semiquantitatively using an adaptation of quick score method [76]
Intensity, negative (no staining of any cellular part at high magnification): –
Occasionally weak positive: ±
Low (only visible at high magnification): +
Medium (readily visible at low magnification): + +
High (strikingly positive at high magnification): + + +
Strong (strikingly positive even at low magnification): + + + +

by controlling the bleeding and next by reverting back to its initial state (increased number of telocytes) [96]. Moreover, Horn et al. similarly proposed a role for telocytes in preventing premature uterine contractility [104].

Telocytes also had been reported to be the starting point for the gastrointestinal stromal tumor of the uterus [105], a fact reinforced by the presence of the common markers with this type of tumors and with perivascular epithelioid cell tumors [106]. Many hypothetical functions were emitted for uterine telocytes (for details, see review [107]). However none of them were demonstrated. Besides the most obvious one, that of intercellular signaling, other possible functions might be considered, e.g., initiators and coordinators of myometrial contraction, participants in immune surveillance or in morphogenetic bioelectrical signaling, etc.

Based on the observation that methotrexate treatment for tubal pregnancy was followed either by recurrent cornual pregnancy or by infertility, Yang et al. demonstrated a reduced number of telocytes after methotrexate exposure, in a dose-dependent manner, and a disruption of the 3D network, followed by a decrease in spontaneous tubal motility [102].

In endometriosis-affected rat oviducts, TCs are reduced or absent and might be involved in deregulation of intercellular signaling and immune regulation [69].

Moreover, in acute salpingitis-affected oviduct tissues, telocytes lose out their interstitial 3D network in response to inflammatory conditions, suggesting that telocytes are important in tissue homeostasis, their integrity preventing interstitial fibrosis [47].

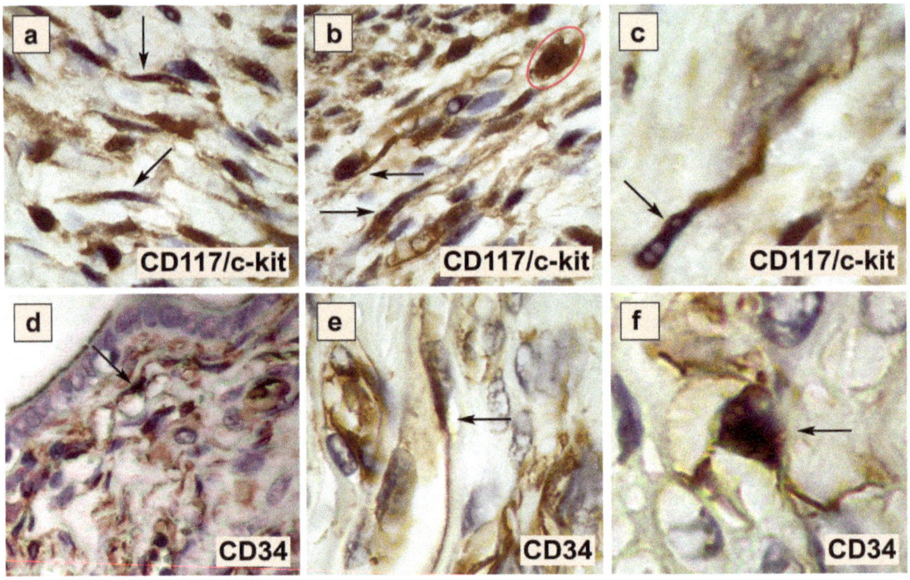

Fig. 22.8 Human fallopian tube. Note the strong CD117/c-kit immunoreactivity of the spindle-shaped telocytes (*arrows*) in the lamina propria (**a–c**) and the isolated rounded c-kit-positive mast cells (magenta encircled areas in (**b**). The immunohistochemical staining for CD34 also shows immunoreactive telocytes in the lamina propria (*arrows*). Counterstaining with Mayer's hematoxylin. Original magnification 40× in (**a, b, d**) and 100× in (**c, e, f**) (Reproduced with permission from Popescu et al. [3])

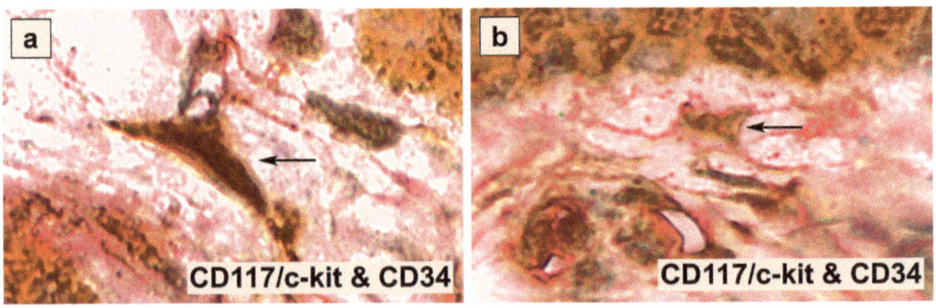

Fig. 22.9 (**a, b**) Human fallopian tube and lamina propria of mucosal layer. Double IHC staining. Brown staining (DAB) indicates CD117/c-kit immunoreactivity and red staining (*fast red*) shows CD34-positive telocytes (*arrows*). Moniliform cell processes have a stronger reactivity for CD34 than for CD117/c-kit, the second marker being preferentially distributed on cell bodies. Original magnification 100× (Reproduced with permission from Popescu et al. [3])

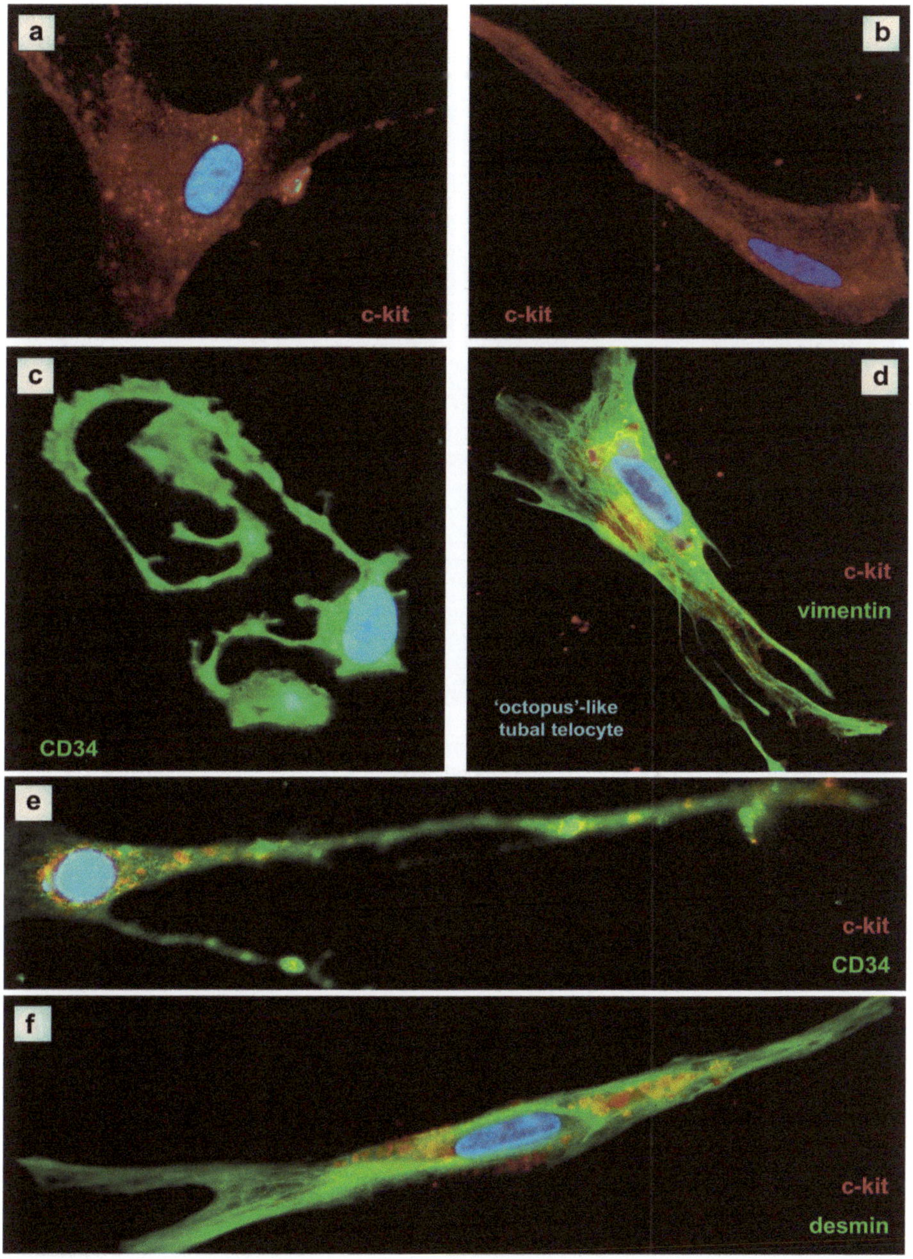

Fig. 22.10 Human fallopian tube primary cultures. (**a, b**) Telocytes assessed by IF for CD117/c-kit (*red*) and (**c**) for CD34 (*green*); (**d**) double immunofluorescent labeling of an "octopus"-like telocyte: vimentin (*green*) and CD117/c-kit (*red*). (**e**) Double immunofluorescent labeling for CD117/c-kit (*red*) and CD34 (*green*). This cell is positive for both markers, and co-localization appears as *yellow* areas. (**f**) Immunostaining for desmin and CD117/c-kit. The cell is positive for both desmin (*green*) and CD117/c-kit (*red*). Nuclei counterstained with Hoechst 33342 (*blue*). Original magnification, 60× (Reproduced with permission from Popescu et al. [3])

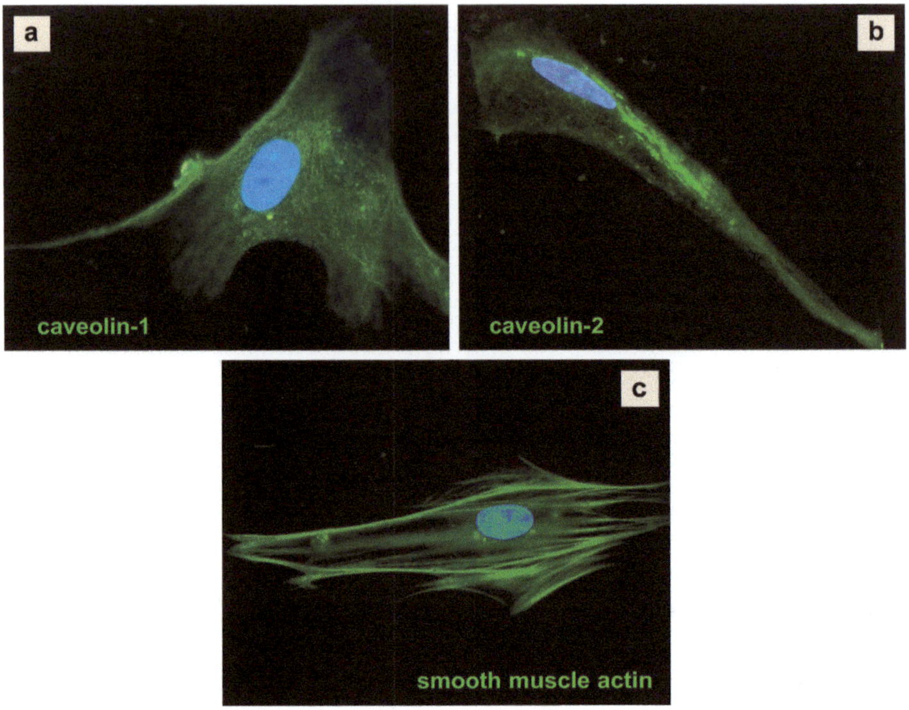

Fig. 22.11 Human fallopian tube. Telocytes in primary culture (day 4). Immunofluorescent labeling for caveolin-1 (**a**), caveolin-2 (**b**), and smooth muscle actin (**c**). FITC-conjugated secondary antibodies (*green*) were used to visualize the reactions; Hoechst 33342 (*blue*) for nuclear counterstaining. Antibodies against caveolin-1 (**a**) and caveolin-2 (**b**) stain preferentially telocyte processes. The cell depicted in (**c**) does not display characteristic telocyte morphology and stains positive for SMA (reveals stress fibers characteristic for myofibroblasts). Original magnification 60× (Reproduced with permission from Popescu et al. [3])

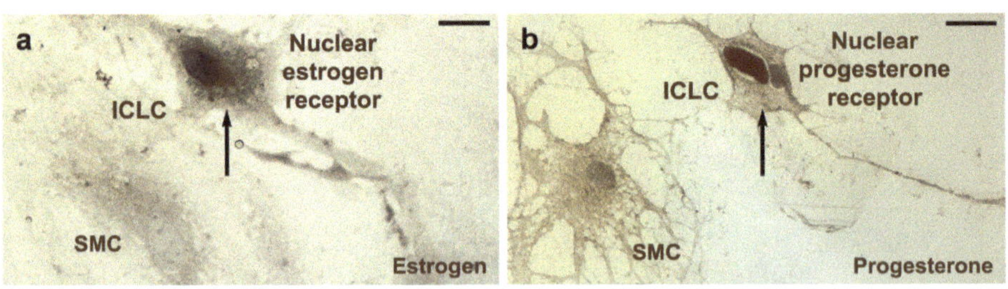

Fig. 22.12 (**a, b**) Human fallopian tube cell culture, fourth passage. The expression of ER-α (**a**) and PR-A (**b**) demonstrated by immunocytochemical staining. Telocytes (*arrows*) stained positive for ER and PR (brown nuclei). Scale bar = 5 μm (Reproduced with permission from Cretoiu et al. [100])

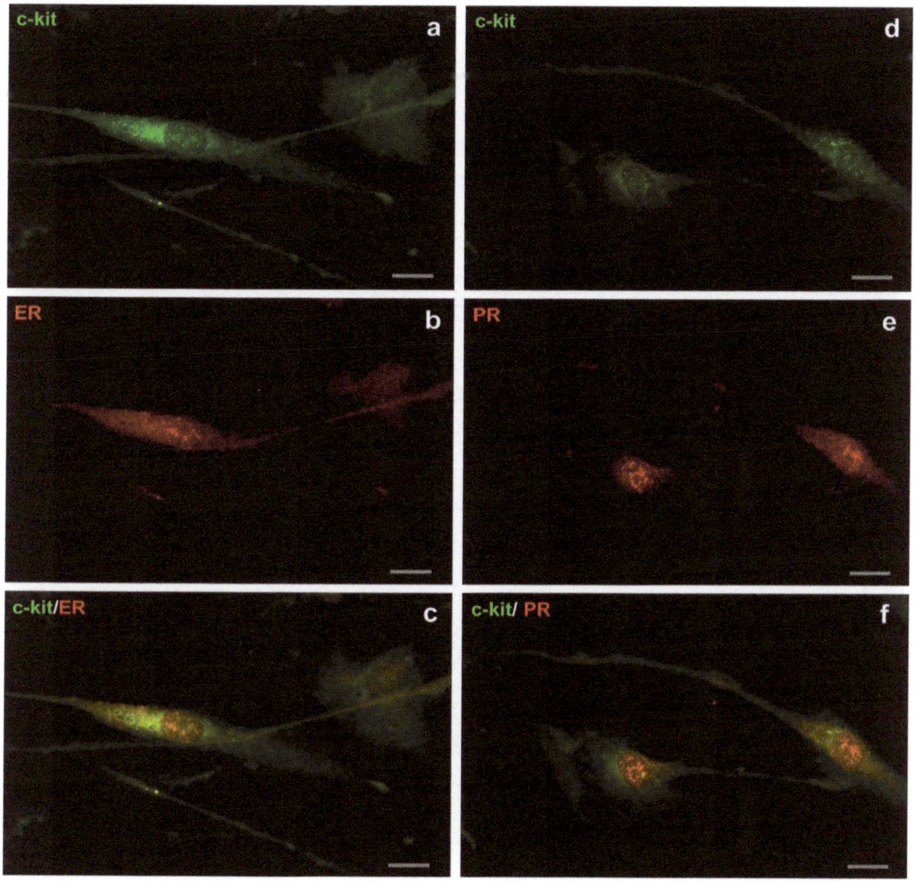

Fig. 22.13 Human fallopian tube cell culture, sixth passage. Immunofluorescent labeling for c-kit/FITC (**a**) and ER-α/Alexa Fluor 546 (**b**) and superimposed images to show co-localization (**c**). c-Kit fluorescence of telocyte (**d**) and PR-A fluorescence (**e**). Superimposed labeling for both markers (**f**), where c-kit (*green*) is localized only in the cytoplasm and PR-A is expressed in the telocyte nuclei (*red*). Scale bar = 5 μm (Reproduced with permission from Cretoiu et al. [100])

22.4 Conclusion

This chapter has highlighted several described markers for telocytes identification, focusing on the immunohistochemistry of telocytes in the uterus and fallopian tubes. Some of the detected markers are characteristic for stem cells (c-kit, CD34, PDGFR-α, and PDGFR-β) and might help to explain the telocytes' phenotype shift along with their differential localization, a concept that lately gains more and more credibility. The tissue-specific phenotypes of telocytes might best be described by the influence of the local combination of signaling molecules that may determine their differentiation into various cell types and in addition their ability to generate local provision of growth and survival factors necessary for angiogenesis. Different phenotypes of tissue-specific telocytes might also explain tissue-specific functions, including tissue regeneration, signal transduction, or a role in modulating contractile activity in the smooth muscle. Furthermore, telocytes can be viewed as somehow electrically active entities involved in bioelectrical signaling functions mediated by intracellular electrical messages. Their functional repertories began to be clarified by recently accumulating genomic and proteomic studies, which are as well pointing to telocytes' involvement in morphogenesis, tissue homeostasis, oxidative stress adaptation, and cancer suppression.

Acknowledgments This work was supported by the Romanian National Authority for Scientific Research CNCS-UEFISCDI project numbers 82/2012 and 194/2014.

References

1. Popescu LM, Hinescu ME, Ionescu N, et al. Interstitial cells of Cajal in pancreas. J Cell Mol Med. 2005;9:169–90.
2. Ciontea SM, Radu E, Regalia T, et al. C-kit immunopositive interstitial cells (Cajal-type) in human myometrium. J Cell Mol Med. 2005;9:407–20.
3. Popescu LM, Ciontea SM, Cretoiu D, et al. Novel type of interstitial cell (Cajal-like) in human fallopian tube. J Cell Mol Med. 2005;9:479–523.
4. Cretoiu D, Ciontea SM, Popescu LM, et al. Interstitial Cajal-like cells (ICLC) as steroid hormone sensors in human myometrium: immunocytochemical approach. J Cell Mol Med. 2006;10:789–95.
5. Popescu LM, Vidulescu C, Curici A, et al. Imatinib inhibits spontaneous rhythmic contractions of human uterus and intestine. Eur J Pharmacol. 2006;546:177–81.
6. Hinescu ME, Gherghiceanu M, Mandache E, et al. Interstitial Cajal-like cells (ICLC) in atrial myocardium: ultrastructural and immunohistochemical characterization. J Cell Mol Med. 2006;10:243–57.
7. Popescu LM, Ciontea SM, Cretoiu D. Interstitial Cajal-like cells in human uterus and fallopian tube. Ann N Y Acad Sci. 2007;1101:139–65.
8. Gherghiceanu M, Popescu LM. Interstitial Cajal-like cells (ICLC) in human resting mammary gland stroma. Transmission electron microscope (TEM) identification. J Cell Mol Med. 2005;9:893–910.
9. Radu E, Regalia T, Ceafalan L, et al. Cajal-type cells from human mammary gland stroma: phenotype characteristics in cell culture. J Cell Mol Med. 2005;9:748–52.
10. Suciu L, Popescu LM, Gherghiceanu M. Human placenta: de visu demonstration of interstitial Cajal-like cells. J Cell Mol Med. 2007;11:590–7.
11. Popescu LM, Faussone-Pellegrini MS. TELOCYTES – a case of serendipity: the winding way from Interstitial Cells of Cajal (ICC), via Interstitial Cajal-Like Cells (ICLC) to TELOCYTES. J Cell Mol Med. 2010;14:729–40.
12. Popescu LM, Manole CG, Gherghiceanu M, et al. Telocytes in human epicardium. J Cell Mol Med. 2010;14:2085–93.
13. Gherghiceanu M, Popescu LM. Cardiomyocyte precursors and telocytes in epicardial stem cell niche: electron microscope images. J Cell Mol Med. 2010;14:871–7.
14. Kostin S, Popescu LM. A distinct type of cell in myocardium: interstitial Cajal-like cells (ICLCs). J Cell Mol Med. 2009;13:295–308.

15. Suciu L, Nicolescu MI, Popescu LM. Cardiac telocytes: serial dynamic images in cell culture. J Cell Mol Med. 2010;14:2687–92.
16. Bani D, Formigli L, Gherghiceanu M, Faussone-Pellegrini MS. Telocytes as supporting cells for myocardial tissue organization in developing and adult heart. J Cell Mol Med. 2010;14:2531–8.
17. Kostin S. Myocardial telocytes: a specific new cellular entity. J Cell Mol Med. 2010;14:1917–21.
18. Faussone-Pellegrini MS, Bani D. Relationships between telocytes and cardiomyocytes during pre- and post-natal life. J Cell Mol Med. 2010;14:1061–3.
19. Manole CG, Cismasiu V, Gherghiceanu M, Popescu LM. Experimental acute myocardial infarction: telocytes involvement in neo-angiogenesis. J Cell Mol Med. 2011;15:2284–96.
20. Rusu MC, Pop F, Hostiuc S, et al. Telocytes form networks in normal cardiac tissues. Histol Histopathol. 2012;27:807–16.
21. Gherghiceanu M, Manole CG, Popescu LM. Telocytes in endocardium: electron microscope evidence. J Cell Mol Med. 2010;14:2330–4.
22. Gherghiceanu M, Hinescu ME, Andrei F, et al. Interstitial Cajal-like cells (ICLC) in myocardial sleeves of human pulmonary veins. J Cell Mol Med. 2008;12:1777–81.
23. Yang Y, Sun W, Wu SM, et al. Telocytes in human heart valves. J Cell Mol Med. 2014. doi:10.1111/jcmm.12285.
24. Zhou Q, Wei L, Zhong C, et al. Cardiac telocytes are double positive for CD34/PDGFR-alpha. J Cell Mol Med. 2015;19:2036–42.
25. Rusu MC, Nicolescu MI, Jianu AM, et al. Esophageal telocytes and hybrid morphologies. Cell Biol Int. 2012;36:1079–88.
26. Chen X, Zheng Y, Manole CG, et al. Telocytes in human oesophagus. J Cell Mol Med. 2013;17:1506–12.
27. Cantarero Carmona I, Luesma Bartolome MJ, Junquera Escribano C. Identification of telocytes in the lamina propria of rat duodenum: transmission electron microscopy. J Cell Mol Med. 2011;15:26–30.
28. Cretoiu D, Cretoiu SM, Simionescu AA, Popescu LM. Telocytes, a distinct type of cell among the stromal cells present in the lamina propria of jejunum. Histol Histopathol. 2012;27:1067–78.
29. Yang P, Liu Y, Ahmed N, et al. Ultrastructural identification of telocytes in the muscularis of chicken ileum. Exp Ther Med. 2015;10:2325–30.
30. Manetti M, Rosa I, Messerini L, Ibba-Manneschi L. Telocytes are reduced during fibrotic remodelling of the colonic wall in ulcerative colitis. J Cell Mol Med. 2015;19:62–73.
31. Xiao J, Wang F, Liu Z, Yang C. Telocytes in liver: electron microscopic and immunofluorescent evidence. J Cell Mol Med. 2013;17:1537–42.
32. Fu S, Wang F, Cao Y, et al. Telocytes in human liver fibrosis. J Cell Mol Med. 2015;19:676–83.
33. Hinescu ME, Ardeleanu C, Gherghiceanu M, Popescu LM. Interstitial Cajal-like cells in human gallbladder. J Mol Histol. 2007;38:275–84.
34. Matyja A, Gil K, Pasternak A, et al. Telocytes: new insight into the pathogenesis of gallstone disease. J Cell Mol Med. 2013;17:734–42.
35. Nicolescu MI, Bucur A, Dinca O, et al. Telocytes in parotid glands. Anat Rec (Hoboken). 2012;295:378–85.
36. Alunno A, Ibba-Manneschi L, Bistoni O, et al. Telocytes in minor salivary glands of primary Sjogren's syndrome: association with the extent of inflammation and ectopic lymphoid neogenesis. J Cell Mol Med. 2015;19:1689–96.
37. Nicolescu MI, Popescu LM. Telocytes in the interstitium of human exocrine pancreas: ultrastructural evidence. Pancreas. 2012;41:949–56.
38. Bosco C, Diaz E, Gutierrez R, et al. Ganglionar nervous cells and telocytes in the pancreas of Octodon degus: extra and intrapancreatic ganglionar cells and telocytes in the degus. Auton Neurosci. 2013;177:224–30.

39. Padhi S, Sarangi R, Mallick S. Pancreatic extragastrointestinal stromal tumors, interstitial Cajal like cells, and telocytes. JOP. 2013;14:1–14.
40. Zheng Y, Li H, Manole CG, et al. Telocytes in trachea and lungs. J Cell Mol Med. 2011;15: 2262–8.
41. Rusu MC, Jianu AM, Mirancea N, et al. Tracheal telocytes. J Cell Mol Med. 2012;16: 401–5.
42. Popescu LM, Gherghiceanu M, Suciu LC, et al. Telocytes and putative stem cells in the lungs: electron microscopy, electron tomography and laser scanning microscopy. Cell Tissue Res. 2011;345:391–403.
43. Qi G, Lin M, Xu M, et al. Telocytes in the human kidney cortex. J Cell Mol Med. 2012;16:3116–22.
44. Zheng Y, Zhu T, Lin M, et al. Telocytes in the urinary system. J Transl Med. 2012;10:188.
45. Gevaert T, De Vos R, Van Der Aa F, et al. Identification of telocytes in the upper lamina propria of the human urinary tract. J Cell Mol Med. 2012;16:2085–93.
46. Vannucchi MG, Traini C, Guasti D, et al. Telocytes subtypes in human urinary bladder. J Cell Mol Med. 2014. doi:10.1111/jcmm.12375.
47. Yang J, Chi C, Liu Z, et al. Ultrastructure damage of oviduct telocytes in rat model of acute salpingitis. J Cell Mol Med. 2015;19:1720–8.
48. Yang XJ, Yang J, Liu Z, et al. Telocytes damage in endometriosis-affected rat oviduct and potential impact on fertility. J Cell Mol Med. 2015;19:452–62.
49. Cretoiu SM, Simionescu AA, Caravia L, et al. Complex effects of imatinib on spontaneous and oxytocin-induced contractions in human non-pregnant myometrium. Acta Physiol Hung. 2011;98:329–38.
50. Hatta K, Huang ML, Weisel RD, Li RK. Culture of rat endometrial telocytes. J Cell Mol Med. 2012;16:1392–6.
51. Cretoiu SM, Cretoiu D, Marin A, et al. Telocytes: ultrastructural, immunohistochemical and electrophysiological characteristics in human myometrium. Reproduction. 2013;145: 357–70.
52. Hutchings G, Williams O, Cretoiu D, Ciontea SM. Myometrial interstitial cells and the coordination of myometrial contractility. J Cell Mol Med. 2009;13:4268–82.
53. Ullah S, Yang P, Zhang L, et al. Identification and characterization of telocytes in the uterus of the oviduct in the Chinese soft-shelled turtle, Pelodiscus sinensis: TEM evidence. J Cell Mol Med. 2015;18:2385–92.
54. Suciu L, Popescu LM, Gherghiceanu M, et al. Telocytes in human term placenta: morphology and phenotype. Cells Tissues Organs. 2010;192:325–39.
55. Bosco C, Diaz E, Gutierrez R, et al. Placental hypoxia developed during preeclampsia induces telocytes apoptosis in chorionic villi affecting the maternal-fetus metabolic exchange. Curr Stem Cell Res Ther. 2016;11:420–5.
56. Bosco C, Diaz E, Gutierrez R, et al. A putative role for telocytes in placental barrier impairment during preeclampsia. Med Hypotheses. 2015;84:72–7.
57. Popescu LM, Andrei F, Hinescu ME. Snapshots of mammary gland interstitial cells: methylene-blue vital staining and c-kit immunopositivity. J Cell Mol Med. 2005;9:476–7.
58. Mou Y, Wang Y, Li J, et al. Immunohistochemical characterization and functional identification of mammary gland telocytes in the self-assembly of reconstituted breast cancer tissue in vitro. J Cell Mol Med. 2013;17:65–75.
59. Cantarero I, Luesma MJ, Junquera C. The primary cilium of telocytes in the vasculature: electron microscope imaging. J Cell Mol Med. 2011;15:2594–600.
60. Li H, Lu S, Liu H, et al. Scanning electron microscope evidence of telocytes in vasculature. J Cell Mol Med. 2014;18:1486–9.
61. Zhang HQ, Lu SS, Xu T, et al. Morphological evidence of telocytes in mice aorta. Chin Med J (Engl). 2015;128:348–52.
62. Briggs Boedtkjer D, Rumessen J, Baandrup U, et al. Identification of interstitial Cajal-like cells in the human thoracic duct. Cells Tissues Organs. 2013;197:145–58.

63. Hinescu ME, Popescu LM, Gherghiceanu M, Faussone-Pellegrini MS. Interstitial Cajal-like cells in rat mesentery: an ultrastructural and immunohistochemical approach. J Cell Mol Med. 2008;12:260–70.
64. Hinescu ME, Gherghiceanu M, Suciu L, Popescu LM. Telocytes in pleura: two- and three-dimensional imaging by transmission electron microscopy. Cell Tissue Res. 2011;343:389–97.
65. Popescu LM, Manole E, Serboiu CS, et al. Identification of telocytes in skeletal muscle interstitium: implication for muscle regeneration. J Cell Mol Med. 2011;15:1379–92.
66. Suciu LC, Popescu BO, Kostin S, Popescu LM. Platelet-derived growth factor receptor-beta-positive telocytes in skeletal muscle interstitium. J Cell Mol Med. 2012;16:701–7.
67. Diaz-Flores L, Gutierrez R, Saez FJ, et al. Telocytes in neuromuscular spindles. J Cell Mol Med. 2013;17:457–65.
68. Rusu MC, Mirancea N, Manoiu VS, et al. Skin telocytes. Ann Anat. 2012;194:359–67.
69. Cretoiu D, Gherghiceanu M, Hummel E, et al. FIB-SEM tomography of human skin telocytes and their extracellular vesicles. J Cell Mol Med. 2015;19:714–22.
70. Manole CG, Gherghiceanu M, Simionescu O. Telocyte dynamics in psoriasis. J Cell Mol Med. 2015;19:1504–19.
71. Kang Y, Zhu Z, Zheng Y, et al. Skin telocytes versus fibroblasts: two distinct dermal cell populations. J Cell Mol Med. 2015;19:2530–9.
72. Luesma MJ, Gherghiceanu M, Popescu LM. Telocytes and stem cells in limbus and uvea of mouse eye. J Cell Mol Med. 2013;17:1016–24.
73. Corradi LS, Jesus MM, Fochi RA, et al. Structural and ultrastructural evidence for telocytes in prostate stroma. J Cell Mol Med. 2013;17:398–406.
74. Li H, Zhang H, Yang L, et al. Telocytes in mice bone marrow: electron microscope evidence. J Cell Mol Med. 2014;18:975–8.
75. Popescu BO, Gherghiceanu M, Kostin S, et al. Telocytes in meninges and choroid plexus. Neurosci Lett. 2012;516:265–9.
76. Xu T, Lu S, Zhang H. Transmission electron microscope evidence of telocytes in canine dura mater. J Cell Mol Med. 2016. doi:10.1111/jcmm.12726.
77. Dawidowicz J, Szotek S, Matysiak N, et al. Electron microscopy of human fascia lata: focus on telocytes. J Cell Mol Med. 2015;19:2500–6.
78. Yang P, Ahmad N, Hunag Y, et al. Telocytes: novel interstitial cells present in the testis parenchyma of the Chinese soft-shelled turtle Pelodiscus sinensis. J Cell Mol Med. 2015;19:2888–99.
79. Chang Y, Li C, Gan L, et al. Telocytes in the spleen. PLoS One. 2015;10:e0138851.
80. Cretoiu SM, Popescu LM. Telocytes revisited. Biomol Concepts. 2014;5:353–69.
81. Cretoiu SM, Cretoiu D, Popescu LM. Human myometrium – the ultrastructural 3D network of telocytes. J Cell Mol Med. 2012;16:2844–9.
82. Gherghiceanu M, Popescu LM. Cardiac telocytes – their junctions and functional implications. Cell Tissue Res. 2012;348:265–79.
83. Fertig ET, Gherghiceanu M, Popescu LM. Extracellular vesicles release by cardiac telocytes: electron microscopy and electron tomography. J Cell Mol Med. 2014;18:1938–43.
84. Cismasiu VB, Radu E, Popescu LM. miR-193 expression differentiates telocytes from other stromal cells. J Cell Mol Med. 2011;15:1071–4.
85. Zheng Y, Zhang M, Qian M, et al. Genetic comparison of mouse lung telocytes with mesenchymal stem cells and fibroblasts. J Cell Mol Med. 2013;17:567–77.
86. Cismasiu VB, Popescu LM. Telocytes transfer extracellular vesicles loaded with microRNAs to stem cells. J Cell Mol Med. 2015;19:351–8.
87. Zheng Y, Cretoiu D, Yan G, et al. Protein profiling of human lung telocytes and microvascular endothelial cells using iTRAQ quantitative proteomics. J Cell Mol Med. 2014;18:1035–59.
88. Zheng Y, Cretoiu D, Yan G, et al. Comparative proteomic analysis of human lung telocytes with fibroblasts. J Cell Mol Med. 2014;18:568–89.
89. Albulescu R, Tanase C, Codrici E, et al. The secretome of myocardial telocytes modulates the activity of cardiac stem cells. J Cell Mol Med. 2015;19:1783–94.

90. Chi C, Jiang XJ, Su L, et al. In vitro morphology, viability and cytokine secretion of uterine telocyte-activated mouse peritoneal macrophages. J Cell Mol Med. 2015. doi:10.1111/jcmm.12711.
91. Sun X, Zheng M, Zhang M, et al. Differences in the expression of chromosome 1 genes between lung telocytes and other cells: mesenchymal stem cells, fibroblasts, alveolar type II cells, airway epithelial cells and lymphocytes. J Cell Mol Med. 2014;18:801–10.
92. Zheng M, Sun X, Zhang M, et al. Variations of chromosomes 2 and 3 gene expression profiles among pulmonary telocytes, pneumocytes, airway cells, mesenchymal stem cells and lymphocytes. J Cell Mol Med. 2014;18:2044–60.
93. Wang J, Ye L, Jin M, Wang X. Global analyses of Chromosome 17 and 18 genes of lung telocytes compared with mesenchymal stem cells, fibroblasts, alveolar type II cells, airway epithelial cells, and lymphocytes. Biol Direct. 2015;10:9.
94. Zhu Y, Zheng M, Song D, et al. Global comparison of chromosome X genes of pulmonary telocytes with mesenchymal stem cells, fibroblasts, alveolar type II cells, airway epithelial cells, and lymphocytes. J Transl Med. 2015;13:318.
95. Song D, Cretoiu D, Zheng M, et al. Comparison of Chromosome 4 gene expression profile between lung telocytes and other local cell types. J Cell Mol Med. 2016. doi:10.1111/jcmm.12746.
96. Salama N. Immunohistochemical characterization of telocytes in rat uterus in different reproductive states. Egypt J Histol. 2013;36:185–94.
97. Rosenbaum ST, Svalo J, Nielsen K, et al. Immunolocalization and expression of small-conductance calcium-activated potassium channels in human myometrium. J Cell Mol Med. 2012;16:3001–8.
98. Campeanu RA, Radu BM, Cretoiu SM, et al. Near-infrared low-level laser stimulation of telocytes from human myometrium. Lasers Med Sci. 2014;29:1867–74.
99. Faussone-Pellegrini MS, Popescu LM. Telocytes. Biomol. Concepts. 2011;2:481–9.
100. Cretoiu SM, Radu B, Banciu A, et al. Isolated human uterine telocytes: immunocytochemistry and electrophysiology of T-type calcium channels. Histochem Cell Biol. 2014:1–12. doi:10.1007/s00418-014-1268-0.
101. Cretoiu SM, Cretoiu D, Suciu L, Popescu LM. Interstitial Cajal-like cells of human fallopian tube express estrogen and progesterone receptors. J Mol Histol. 2009;40:387–94.
102. Yang XJ, Wei W, Zhao J, Zheng FY. Inhibitory effects of methotrexate on spontaneous motility and Cajal-like type of tubal interstitial cells in rabbit oviduct. Fertil Steril. 2012;98: 215–21.
103. Yang XJ, Xu JY, Shen ZJ, Zhao J. Immunohistochemical alterations of Cajal-like type of tubal interstitial cells in women with endometriosis and tubal ectopic pregnancy. Arch Gynecol Obstet. 2013;288:1295–300.
104. Horn LC, Meinel A, Hentschel B. c-kit/CD 117 positive cells in the myometrium of pregnant women and those with uterine endometriosis. Arch Gynecol Obstet. 2012;286:105–7.
105. Terada T. Gastrointestinal stromal tumor of the uterus: a case report with genetic analyses of c-kit and PDGFRA genes. Int J Gynecol Pathol. 2009;28:29–34.
106. Ardeleanu C, Bussolati G. Telocytes are the common cell of origin of both PEComas and GISTs: an evidence-supported hypothesis. J Cell Mol Med. 2011;15:2569–74.
107. Roatesi I, Radu BM, Cretoiu D, Cretoiu SM. Uterine telocytes: a review of current knowledge. Biol Reprod. 2015;93:10.

Sanda Maria Cretoiu is currently the Associate Professor of Cell Biology and Histology at the Faculty of Medicine, "Carol Davila" University, Bucharest (Romania), and Principal Investigator and Senior Research Fellow at the "Victor Babes" National Institute of Pathology, Ultrastructural Pathology Laboratory. Her current research interest lies in uterine telocytes (e.g., electron microscopy, cell cultures, scanning confocal microscopy, time-lapse videomicroscopy, etc.). She has been working within telocyte research from the beginning in 2005, being a close collaborator of

Professor Laurentiu M. Popescu. For example, she participated in telocyte discovery and research involving the human female genital tract. She serves as Editorial Board Member of international journals, e.g., *Nature Scientific Reports* (IF 5.578) and *Romanian Biotechnological Letters* (IF 0.404). Moreover, she was Administrator and Assistant Editor for *Journal of Cellular and Molecular Medicine* (IF 4.125). Dr. Cretoiu has published 50 scientific papers in international peer-reviewed journals, having an h-index of 17 with a total of more than 1,300 independent citations.

Chapter 23
A Tale of Two Cells: Telocyte and Stem Cell Unique Relationship

Zeinab M. El Maadawi

Abstract Telocytes have been identified as a distinctive type of interstitial cells and have been recognized in most tissues and organs. Telocytes are characterized by having extraordinary long cytoplasmic processes, telopodes, that extend to form three-dimensional networks and commonly constitute specialized forms of cell-to-cell junctions with other neighboring cells. Telocytes have been localized in the stem cell niche of different organs such as the heart, lung, skeletal muscle, and skin. Electron microscopy and electron tomography revealed a specialized link between telocytes and stem cells that postulates a potential role for telocytes during tissue regeneration and repair. In this review, the distribution of telocytes in different stem cell niches will be explored, highlighting the intimate relationship between the two types of cells and their possible functional relationship.

Telocyte (TC) is a newly described type of interstitial cells which is characterized by having a small cell body and extremely long cytoplasmic prolongations that are called telopodes. Each telopode is composed of alternating thin segments (podomers) and dilated segments (podoms) [47].

Using electron microscopy (EM) is considered the ideal technique to identify TCs in most tissues. Meanwhile, double-positive immunohistochemistry for CD34/c-Kit (usually localized in TC cell body) or CD34/vimentin (localized in telopodes) is designated as a reliable method for TC visualization [48]. More recently, the emergence of cutting-edge technologies in subcellular three-dimensional (3D) imaging, such as FIB-SEM tomography, has provided more evidence-based approach for TC study with special emphasis on their relationships with neighboring cells [15].

Reviewing the literature establishes a growing body of evidence that TCs co-localize with stem and progenitor cells in different organs such as the heart [24], lungs [25], and skeletal muscles [53]. Furthermore, more emerging evidence has

Z.M. El Maadawi, MD
Histology & Cell Biology Department, Faculty of Medicine, Cairo University, Cairo, Egypt
e-mail: zeinab_elmaadawi@yahoo.com

X. Wang, D. Cretoiu (eds.), *Telocytes*, Advances in Experimental Medicine and Biology, 913, DOI 10.1007/978-981-10-1061-3_23

supported the notion of the potential role of TCs in tissue regeneration and/or repair after injury ([36, 41, 64, 69, 76]).

Being the fundamental cells for tissue regeneration, different categories of stem cells (SCs) have been identified in most adult human organs. However, resident SCs cannot survive only by their capacity of self-renewal. Therefore, SCs are strongly dependent on the microenvironment they live in each organ. Eventually, the term "stem cell niche" refers to the well-organized minute tissue areas housing SCs which ensure an appropriate microenvironment maintained with blood vessels, nerve endings, extracellular matrix, and supporting interstitial cells [19]. Stem cell niche commonly occurs at tissue intersections or transition zones, and it enables SCs to have spatial and dynamic interaction with neighboring cells as well as with remote cells through paracrine and endocrine signals [20].

Accordingly, several studies have confirmed the existence of TCs in SC niche microenvironment in a variety of tissues and organs such as the heart [2], lungs [51–53], skeletal muscle [6, 11, 64], skin [10], meninges and choroid plexus [54], liver [68, 71], eye [38], and aorta [74].

Moreover, TCs were found to express SC markers like c-kit, Sca-1, and Oct 4 [6, 23, 63]. Although the expression of particular SC markers by TCs varied among tissues, this finding suggests a major role of TCs in regeneration [14].

It seems prudent to indicate that no definite mechanism has been confirmed yet to explain the direct or indirect interaction between SCs and TCs. Meanwhile, several hypotheses have been proposed based on the morphological features of both types of cells in the microenvironment. For example, [57] suggested direct cell-to-cell signaling and close cross talk where TCs might provide antioxidant protective effect to SCs [78]. Another regenerative role of TCs was attributed to their expression of VEGF and PDGFR-b, both in situ and in vitro, speculating their potential effect in angiogenesis during tissue repair [51–53, 64].

Some investigators suggested a role of TCs in mesenchymal SC differentiation [10], while others [57] argued that TCs could be a subpopulation of mesenchymal SCs as they express some markers of stromal niche cells. In addition, TCs were proposed to play an important role in cell signaling, hence controlling the microenvironment in normal and malignant tissues [28].

In fact, increasing evidence reveals that TCs actively contribute to the coordination of cell signaling in SC microenvironment either through direct intercellular junctions or via shed vesicles that exert a paracrine effect [3, 26, 40].

In the following sections, tissue-specific localization of TCs in SC niche will be explored with special reference to the unique relationship between the two types of cells as evidenced by previous reports.

23.1 Telocytes in Cardiac Stem Cell Niche

TCs have been vigorously studied as a distinct type of cells in the heart interstitium [24, 27, 47, 60, 62]. More recently, 3D imaging using FIB-SEM tomography has added new evidence to the identification of cardiac TCs [15].

TCs were described as 3D network in the myocardium among cardiomyocytes, blood capillaries, nerve endings, and immune system cells [2, 3, 15]. In addition, telocytes were clearly identified under the pericardium in close association with cardiac SCs and progenitor cells (Fig. 23.1) [24].

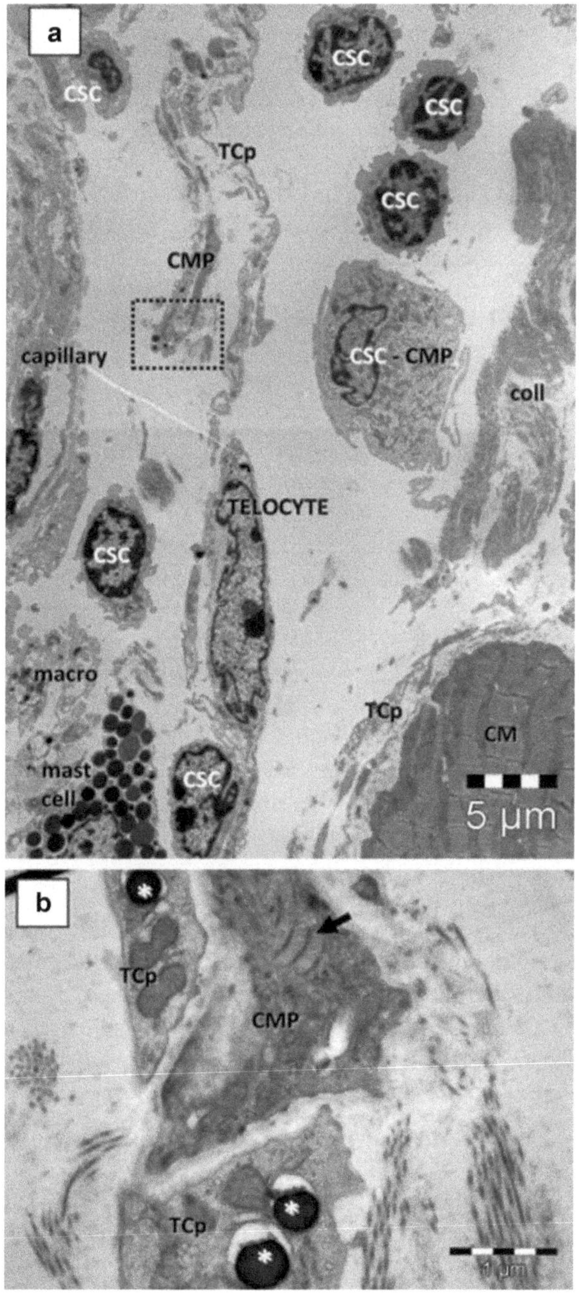

Fig. 23.1 (**a**) Electron microscopy of the subepicardial SC niche in the mouse showing the presence of putative cardiac stem cells (*CSC*), isolated or in small groups, cardiomyocyte progenitors (*CMP*), and cells with intermediate features (CSC–CMP). All these cells are placed in a loose extracellular matrix. (**b**) Higher magnification from area marked in (**a**) with a *dotted square*. The CMP have characteristic leptofibrils (*arrow*) and are closely assisted by telocyte processes (*TCp*) which contain few dense granules (*asterisk*) suggesting paracrine signaling. *CM* cardiomyocyte, *macro* macrophage, *coll* collagen fibers (Courtesy with permission from Gherghiceanu and Popescu [24])

The human myocardium is characterized by its extremely low ability to regenerate due to its limited number of cardiac SCs (0.01–1 %). Meanwhile, the number of cardiac TCs (0.5–1 %) exceeds that of cardiac SCs. Although TCs still represent a minute portion of human cardiac interstitial cells, their extremely long and extensive telopodes allow them to occupy more surface area that forms 3D reticulum probably for supporting other cells [55].

The exceptional relationship between cardiac TCs and SCs was best described as "tandem" where both types of cells integrate in a structural and functional manner [14, 24, 26]. Given such relationship, TCs were nominated to provide support for cardiac SCs [24, 26], guide cardiac progenitor cells [2, 50], and enhance neoangiogenesis [41, 76].

Different types of heterocellular junctions were described in adult murine and human heart tissue [25, 26] that have been localized between TCs in one side and cardiac SCs, cardiomyocytes, and other interstitial cells in the other side. The reported distance between junction cell membranes was 10–20 nm that goes well with transfer of macromolecules [26]. TCs and cardiomyocytes revealed electron-dense nanocontacts [25], while the junctions between TCs and cardiac SCs were described as stromal synapse or adherent junctions [26]. TCs were found also to have point and/or planar contacts with endothelial cells, pericytes, Schwann cells, and other interstitial cells such as fibroblasts, macrophages, and mast cells [26].

Similar types of junctions were described between cocultured cardiac TCs and SCs where cells exhibited heterocellular adherens junctions as well as nonclassical junctions such as the puncta adherentia (Fig. 23.2) and stromal synapses. The stromal synapse formed between TCs and SCs was frequently associated with small electron-dense structures (15 nm in length) that connect the two opposing cell membranes (Fig. 23.3) [56].

Such intimate cross-cellular relationships in situ and in vitro might help cardiac SCs to proliferate and differentiate refuting the hypothesis of a solitary role of SCs during cardiac regeneration [55]. Thus, the lack of such nursing role of TCs was proposed to explain the failure of solely engrafted cardiac SCs to survive in the microenvironment of injured myocardium [56].

Special attention has been drawn to the epicardium as a novel source of cardiac SCs [7]. Cardiac TCs and SC niche in the subepicardial region were shown to express common surface markers such as PDGFR-b and c-kit (Popescu et al. 2010, [8, 73, 79]). This finding argued for proposing cardiac TCs per se as a subpopulation of epicardium-derived progenitor cells [2]. Another postulation was attributed to the possibility that TCs might be a potential source of cardiac mesenchymal cells [17]. Accordingly, cardiac TCs were similar to bone marrow-derived SCs in being positive for mesenchymal marker CD29 and negative for hematopoietic marker CD45 [5].

Reviewing literature, in vitro and in vivo studies revealed the possible modulating effect of cardiac TCs on cardiac development and repair [2, 21, 24]. In vitro studies demonstrated that cardiac TCs embrace cell populations of growing cardiomyocytes supporting the proposed nursing role [2]. A tissue-engineered heart model demonstrated networks of TCs with their telopodes communicating with growing cardiomyocytes through junctions and shed vesicles [80]. Interestingly, studies on zebra fish revealed that, upon myocardial regeneration following amputation of the

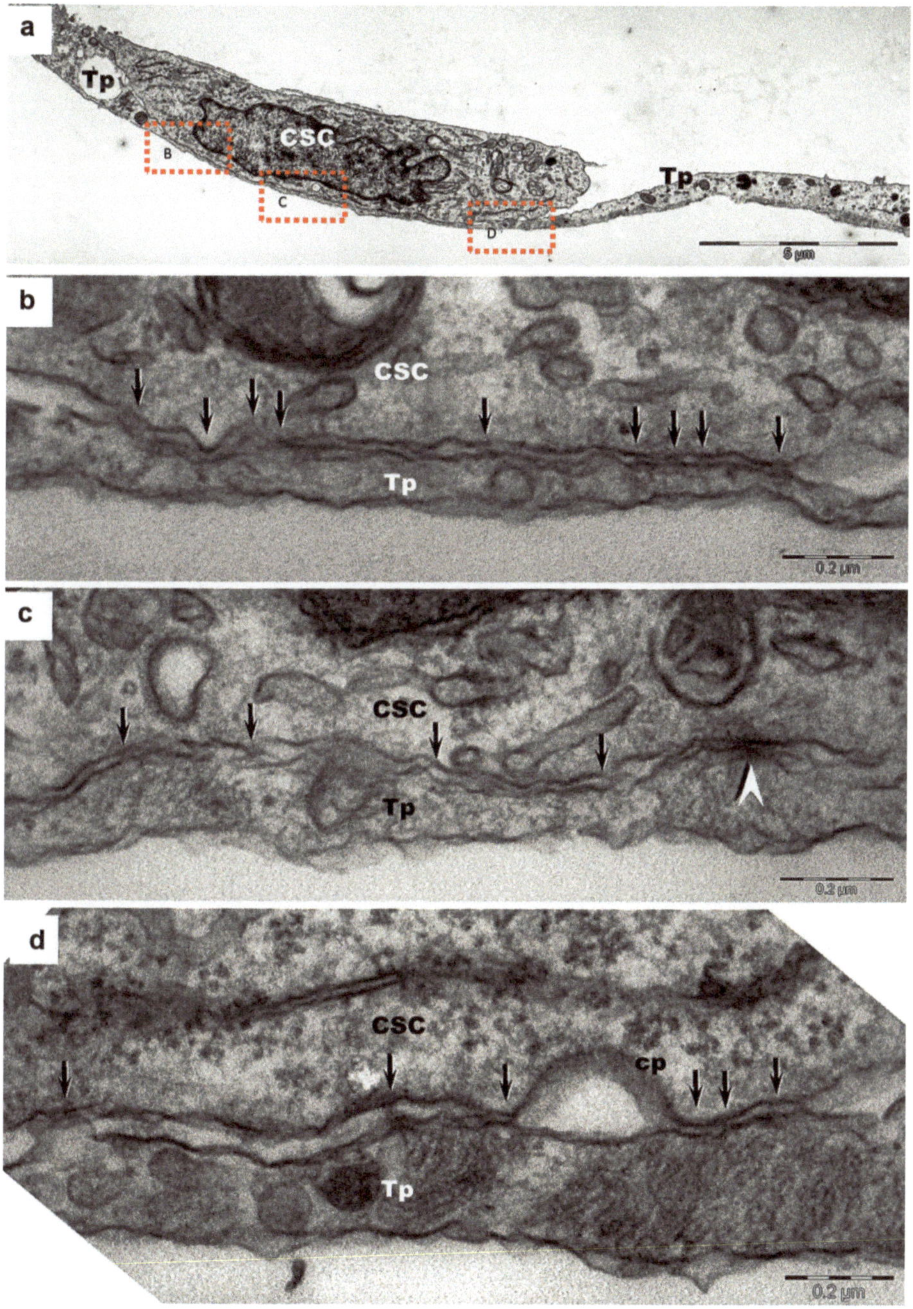

Fig. 23.2 (**a**) Transmission electron microscopy images of TC–CSC culture after 48 h shows a telopode (Tp) in close contact with a cardiac stem cell (CSC). (**b–d**) Marked areas from image A are shown at higher magnification in the corresponding panels. A planar contact (*stromal synapse*) between TC and CSC can be seen associated with a number of electron-dense structures (*arrows*). A *puncta adherentia* junction (*arrowhead*) is visible between TC and CSC in image (**c**). *Cp* coated pit (Courtesy with permission from Popescu et al. [56])

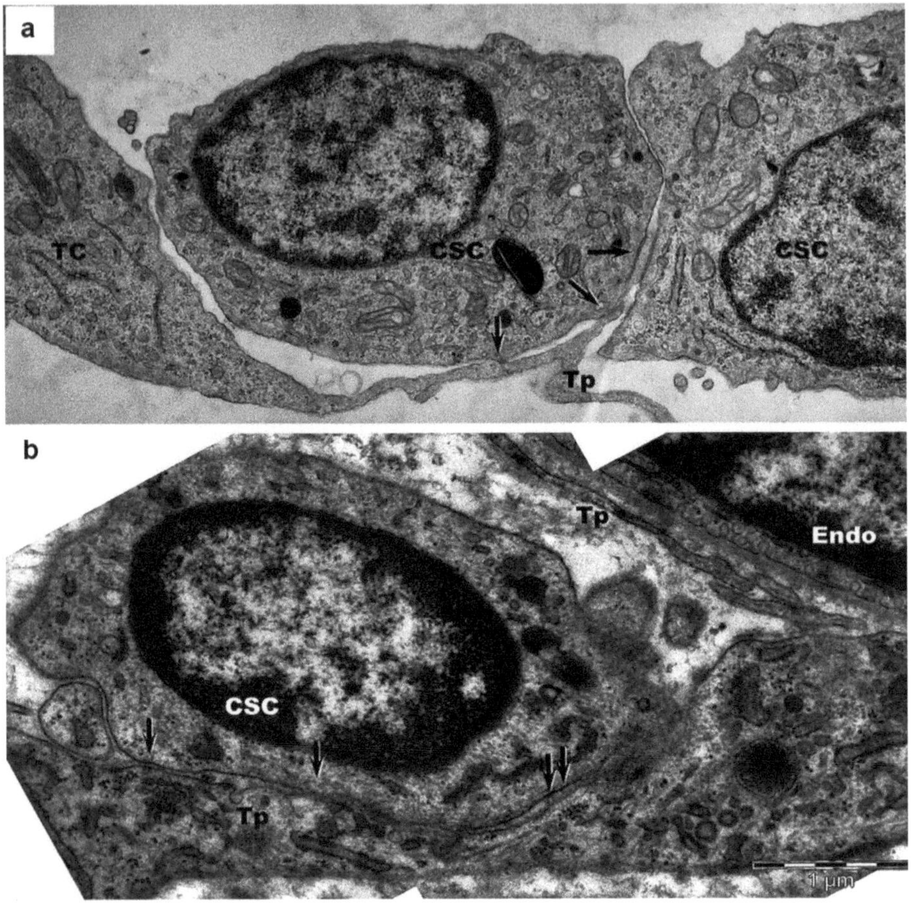

Fig. 23.3 Transmission electron microscopy shows similar intercellular connections (*plain stromal synapses*) between telocytes (TC) and cardiac stem cells (CSC) *in culture* (**a**) and *in tissue* (**b**). Telopodes (Tp) connect with a cardiac stem cell (CSC) through small electron-dense structures (*arrows*). *Endo* endothelial cell, *P* pericyte, *N* nerve ending, *CM* cardiomyocyte (Courtesy with permission from Popescu et al. [56])

ventricular apex, reorganized TCs with their telopodes were recognized in large number in close association with growing cardiomyocytes [32, 51].

In vivo transplantation of cardiac TCs in experimental rat model of myocardial infarction was found to decrease the size of infarction and to enhance myocardial function [75, 76]. Such improvement was attributed to the potential ability of TCs to promote angiogenesis and to reduce fibrosis following heart injury [4, 75, 76]. Another study revealed an improvement of myocardial infarction after transplantation of human iPS-derived mesenchymal SCs with specific networked arrangement of TCs in the interstitial space of infarction [42].

Thus, TCs might be critically involved in the integration of heterocellular communications that is crucial for proliferation, differentiation, and maturation of myocardial progenitor cells [4, 66]. Therefore, preconditioning of transplanted cardiac SCs with TCs should be considered as an alternative modality to regenerate cardiomyocyte instead of monocellular therapy protocols.

23.2 Telocytes in Skeletal Muscle Stem Cell Niche

The skeletal muscle is well known of its extraordinary capacity to regenerate after injury. Satellite cells are the primary cells involved in skeletal muscle regeneration and are found in specific satellite cell niche [33]. However, other non-satellite progenitor cell niche can be recognized in which other precursor cells exist such as bone marrow-derived cells and pericytes [16, 18, 34, 43].

It is worth mentioning that TCs have been well recognized in human adult skeletal muscle [51–53] where they closely communicate via telopodes with cells in both satellite cell niche and non-satellite progenitor cell niche suggesting a potential role for their proliferation and differentiation [6, 51–53].

TCs were found to constitute a 3D network in the interstitial tissue of skeletal muscle where they locate very close to blood capillaries, nerve fibers, myocytes, and satellite cells [6, 51–53]. Intimate contacts between TCs and both types of resident muscle SCs, which is satellite (Fig. 23.4) and non-satellite (Fig. 23.5), were clearly identified where telopodes were found in their vicinity [53]. Heterocellular junctions were also identified between TCs and the progenitor neighboring cells with shed vesicles and exosomes [4, 53].

TCs in the skeletal muscle were postulated to have proliferative and angiogenic capabilities based on their expression of proliferative marker Ki67, pluripotency marker Oct4, and vascular proliferation marker VEGF [6]. In vitro studies demonstrated better differentiation capacity of skeletal muscle-derived SCs into adipocytes, chondrocytes, and osteoblasts when cocultured with TCs, a finding that reinforces the potential role of TCs in tissue regeneration and repair [6].

23.3 Telocytes in Liver Growth and Regeneration

As many other solid organs, TCs in the liver have been clearly identified by both transmission electron microscopy and double immunofluorescent staining for CD34 with c-kit/CD117, vimentin, or PDGFR-a/b [71]. TCs were localized in the space of Disse, in close proximity with hepatocytes, putative stem/progenitor cells, and endothelial cells (Fig. 23.6) [71].

The liver is characterized by its remarkable ability to regenerate after exposure to traumatic or ischemic injury [31]. Such regenerative capacity is attributed to the proliferation of hepatocytes and, in severe injuries, the activation and differentiation of hepatic stem/progenitor cells [69].

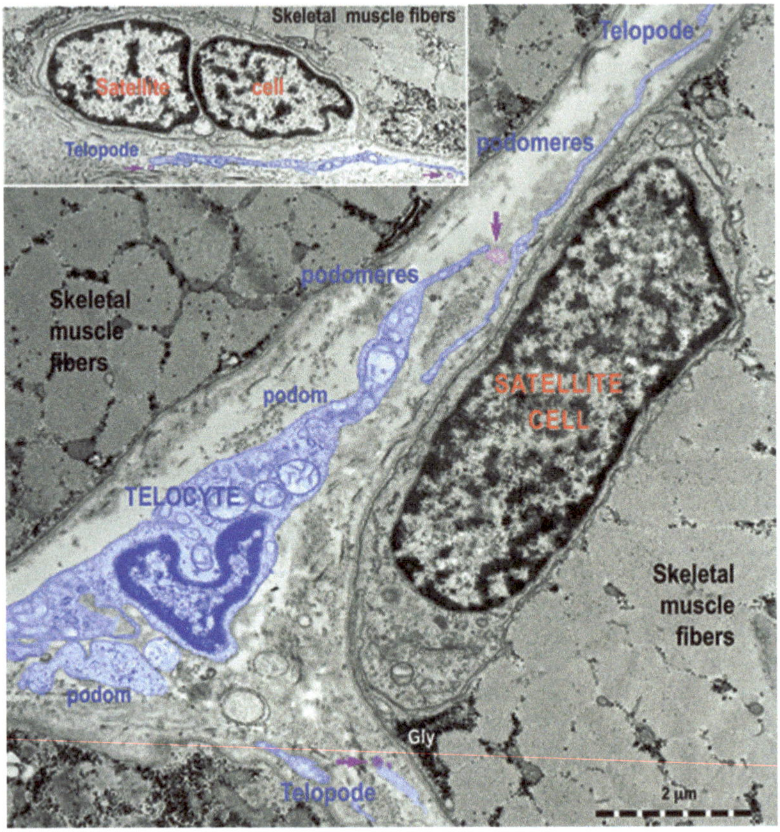

Fig. 23.4 Transmission electron micrographs show TC with Tps, podoms, and podomeres in between muscle fibers. Note the typical appearance of satellite cells. TCs (digitally *blue* colored) are positioned in the close vicinity of satellite cells. Two ultrastructural features are remarkable: the close spatial relationships of Tps with satellite cells and the fact that these Tps release shed vesicles (*purple arrows*). This may indicate that a transfer of chemical information flows from TC to satellite cells (Courtesy with permission from Popescu et al. [53])

Some studies have evaluated the involvement of TCs in posttraumatic regeneration and physiological growth of the liver. The possible role of TCs in liver regeneration has been investigated using a mouse model of partial hepatectomy [68]. It was reported that the peak activity of hepatocyte proliferation was recorded after the first 2 days of partial hepatectomy followed by another, yet lower, peak after the third day that was concurrently associated with an elevation of both TCs and CK19-positive hepatic SCs [68]. These findings suggested an exclusive relationship between TCs and other progenitor cells involved in liver regeneration. TCs might contribute to the control of proliferating hepatocytes and/or the differentiating stem cells [4].

Furthermore, an important insinuation of TCs was postulated in pregnancy-induced hepatic proliferation [69]. The two peaks of hepatocyte proliferation during pregnancy in mice correlated with an increase in CD34/PDGFR-α positive TCs

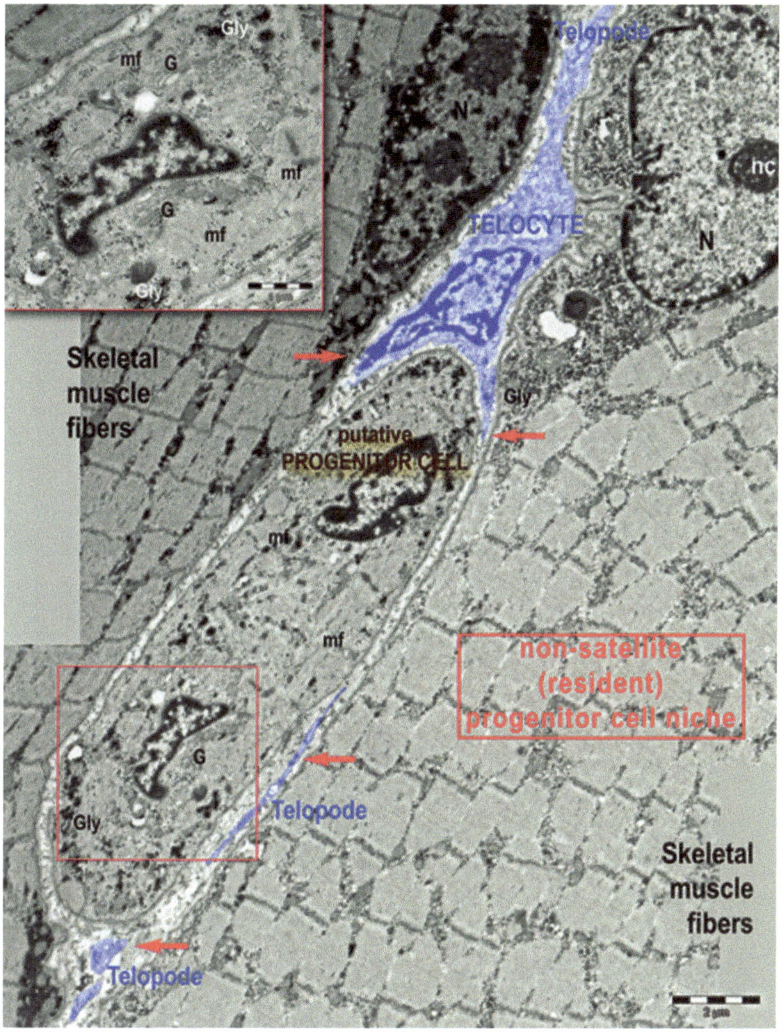

Fig. 23.5 Electron micrographs of human skeletal muscle show a TC (*blue colored*), which extends its Tps indicated by *red arrows* around a striated cell, in fact a (putative) progenitor cell. Note: the tandem TC–progenitor cell making a non-satellite (resident) progenitor stem cell niche. Inset: higher magnification of the progenitor cell shows incompletely differentiated features: unorganized myofilaments (mf), glycogen deposits (Gly), prominent Golgi complex (G). *N* nucleus, *nc* nucleolus (Courtesy with permission from Popescu et al. [53])

[69]. Further molecular mechanisms need to be investigated to explain the hepatic adaption in pregnancy with possible enrollment of TCs.

It is speculated that TCs contribute in liver regeneration through direct intercellular junctions or paracrine effect via shed vesicles [69]. However, further evidence is required to determine the functional relationship between TCs and other cells in the liver.

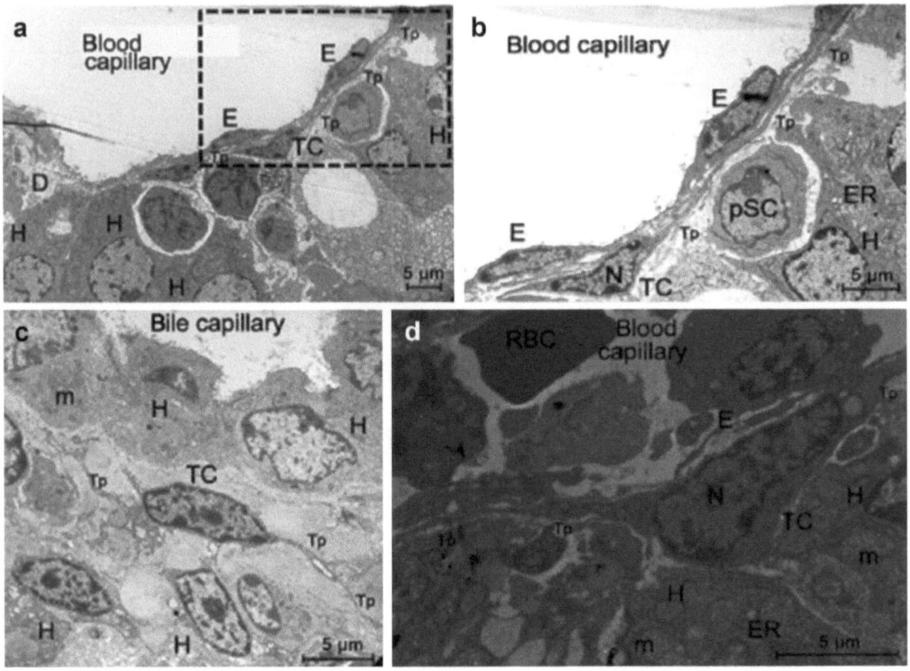

Fig. 23.6 Electron microscope images showing the ultrastructure of the liver (mice). (**a**) Telocytes (TCs) with telopodes (Tps) in the space of Disse (D) between endothelial cells (E) and hepatocytes (H). Note the upper telopode (Tp) which is more than 20 μm long. (**b**) Higher magnification of the field inside the *rectangle* in (**a**). Note in between the TC and hepatocytes (*H*) the presence of a putative stem cell (pSC) which has the features of a young cell (progenitor cell?); *ER* endoplasmic reticulum, *N* nucleus. (**c**) A TC with at least three Tps; *H* hepatocyte, *m* mitochondria. (**d**) A TC with a heterochromatic nucleus (*N*) at a higher magnification; (**e**) endothelial cell; *RBC* red blood cell; *H* hepatocyte, *m* mitochondria, *Tp* telopodes; scale bar = 5 μM (Courtesy with permission from Xiao et al. 2013)

23.4 Telocytes and Skin Stem Cells

The skin is characterized by its astonishing competency to regenerate featuring it as an ideal model for studying SCs [29]. Multiple locations for SC niche have been recognized in the skin including the bulge of the hair follicle, dermal papillae, and perivascular spaces [22, 70].

Bulge SCs express specific markers such as nestin [35]. However, they can proliferate and differentiate only after receiving signals from specialized neighboring cells in the stroma which is known as the “bulge activation hypothesis” [65].

As per other organs and tissues, electron microscopy and immunofluorescence confirmed the presence of TCs in human dermis [10, 58]. TCs were found to border a round cluster of bulge SCs with apparently two types of TCs/SCs contacts, point and planar contacts, forming atypical heterocellular junctions [10].

Interestingly, studying scleroderma skin samples revealed that TCs were found to decrease in areas of SC niches [39]. In addition, SCs could not be observed in the perivascular space in diffuse cutaneous scleroderma [39]. The depletion of TCs might be implicated to the impaired reparative and regenerative function of the skin in cases of scleroderma [39]. Nevertheless, TCs were also found to be reduced in organs other than the skin of systemic scleroderma patients such as the myocardium, lung, and stomach [40]. The pattern of TC distribution in the skin augments the hypothesis that TCs are "nursing cells" where they may interact with resident cells in SC niche through indirect (chemical) and/or direct (junctional) contacts [10].

23.5 Telocytes and Stem Cells in the Lung and Respiratory Passages

TCs have been described in the human and mouse respiratory tree including alveolar ducts as well as terminal and respiratory bronchioles [52]. The tandem of TC–SC was clearly demonstrated in subepithelial niches of the bronchioles particularly at the bronchoalveolar junction. Using electron tomography, telopodes were found to connect with SCs via bridging nanostructures [52].

The direct connection between TCs and resident lung SCs through nanocontacts, shed vesicles, and exosomes [52] postulates a major supporting, cell guiding, and communicating role of TCs [51].

In order to specifically identify the genes that regulate lung TCs, the genetic profile of murine lung TCs was vigorously studied and compared to mesenchymal SCs and fibroblasts [77]. At least 46 genes were found to be functionally connected in TCs, mesenchymal SCs, and fibroblasts [77]. Consequently, multiple functional roles have been implicated for lung TCs including tissue development and regeneration, remodeling of the extracellular matrix, and neovascularization with special emphasis on their potential effect on keeping the integrity of vascular basement membrane [77]. TCs usually locate in close vicinity of small blood vessels and express angiogenesis markers such as VEGF and NO [41].

In addition, lung TCs were found to express pluripotency marker Oct4, which is usually expressed in embryonic SCs (Fig. 23.7), a finding that urged the supposition that TCs could be a subpopulation of SCs [23].

23.6 Telocytes and Stem Cell Niche in the Eye

The limbus (corneoscleral junction) was described to host a rich SC niche [44, 59] that is essential for corneal regeneration [46]. TCs and SCs have been co-localized at the perivascular spaces of the limbus and the stroma of the iris (Fig. 23.8) among nerve endings. Heterocellular direct membranous junctions have been identified between TCs and SCs in the form of nanocontacts and planar contacts [38].

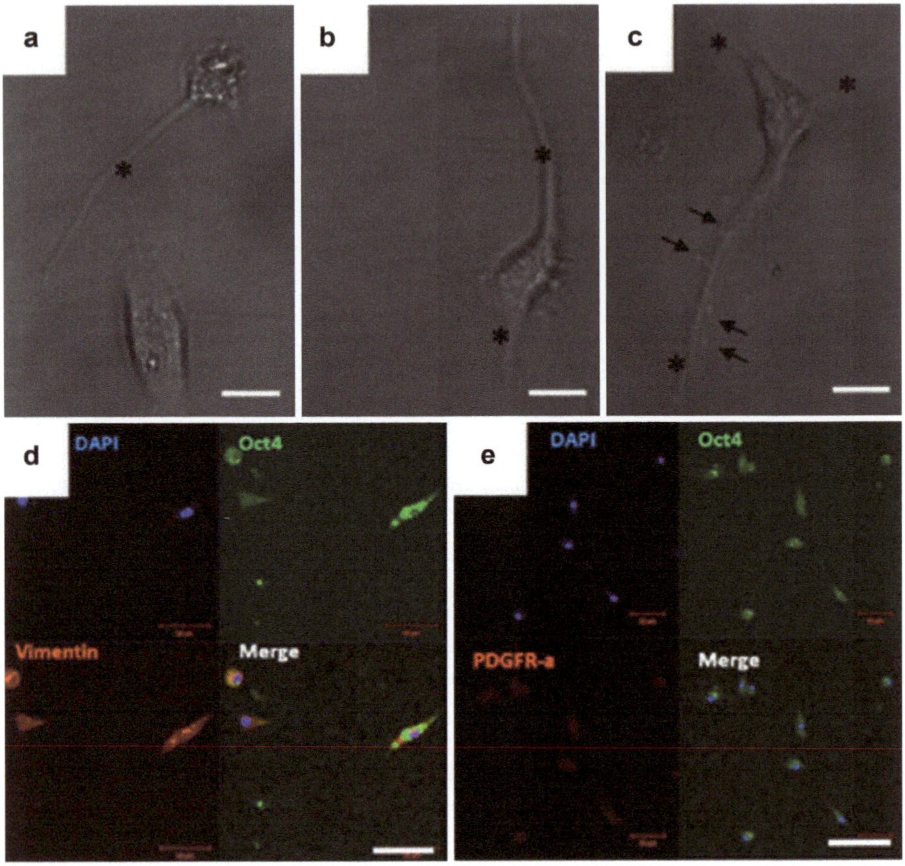

Fig. 23.7 Isolated lung Oct4-GFPpos cells that were maintained in culture for 5 days. Cells were sorted based on GFP fluorescence and kept in culture at 37 °C in MEM medium. (**a–c**) Bright-field images show cells with a fusiform telocyte-like shape with thin prolongations (*asterisks*). Examples are provided for cells having one (**a**), two (**b**), or three (**c**) cell prolongations (telopodes) with a particular shape resembling a string of beads (*arrows* in **c**). (**d**) Co-expression of Oct4 and vimentin in telocytes. (**e**) Co-staining of telocytes with Oct4 and PDGFR-α. Note that telocytes do not express PDGFR-α; bar = 50 μm (Courtesy with permission from Galiger et al. [23])

By securing a strategic 3D networked position in the eye among blood vessels, nerve endings, and other stromal cells such as SCs, melanocytes, and macrophages, TCs reinforce their highly proposed cell nursing role [38]. TCs might be able to influence other neighboring cells via short-distance signals through direct contacts, exosomes, and shed vesicles [1]. Meanwhile, their extremely long telopodes could exert a long-distance macromolecular signaling effect through which transfer of proteins, membrane receptors, and mRNAs occurs [37, 45]. Moreover, a specific modulatory effect [61] as well as guiding role of TCs [49] has been proposed in the immune system. Nevertheless, TCs per se could be an integral part of the mesenchymal SC niche [9, 38]. Eventually, TCs are candidates that pave the way for SC migration during tissue regeneration of the eye [38] and might be a promising new treatment modality for degenerative eye diseases.

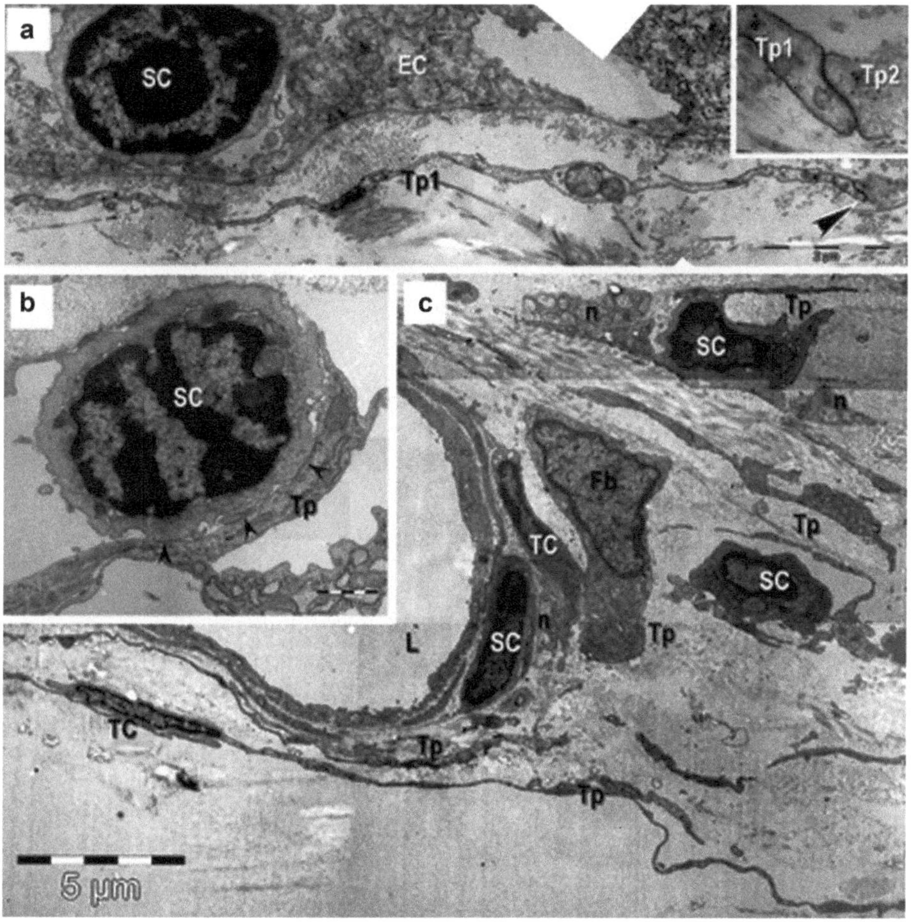

Fig. 23.8 Transmission electron microscopy images of epithelial (**a**) and stromal (**b**, **c**) stem cell (SC) niches in the mouse eye. (**a**) Basal SC is sited on the basement membrane of limbus epithelium. A telopode (Tp1) runs parallel with the basement membrane, and a gap junction (*arrowheads*) connects it with another one (Tp2; higher magnification in inset). (**b**, **c**) Stem cells in the stromal SC niches located in the corneoscleral junction. Direct contacts (*arrowheads*) between a Tp and the SC are visible in (**b**). *TC* telocytes, *Tp* telopodes, *Fb* fibroblast, *n* nerve endings, *L* lumen of an arteriole. Scale bars: (**a**) – 2 μm; inset – 0.1 μm; (**b**) – 1 μm; (**c**) – 5 μm (Courtesy with permission from Luesma et al. [38])

23.7 Telocytes and Hematopoietic Stem Cells in the Spleen

The spleen is one of the frequently reported sites for extramedullary hematopoiesis in adults [72]. Putative hematopoietic SCs have been identified adjacent to the sinusoidal endothelium of the spleen where they constitute SC niche that ensures a hypoxic environment sufficient for maintaining hematopoietic SC growth and proliferation [30].

The presence, characteristics, and distribution of TCs in the spleen have been recently reported [13]. TCs exhibited similar ultrastructural morphology to that described in other organs where they connected via their telopodes with leukocytes and red blood corpuscles in the red pulp of the spleen [13]. Moreover, splenic TCs formed a 3D network by their telopodes when cultured in vitro anchored by hetero- and homocellular junctions [13].

After 3 days of culture, splenic TCs formed circle-like structures in which their telopodes extended to support the circumference of these circles [13]. Double-labeling immunofluorescence revealed positive immuno-expression of vimentin with CD34, nanog, and Sca-1; however, splenic TCs were negative for c-kit [13].

Therefore, the expression of pluripotent SC marker nanog [12] and hematopoietic progenitor markers CD34 and Sca-1 [67] in splenic TCs suggests some role in regeneration [13]. Eventually, TCs were suggested to contribute in the control of splenic hematopoietic cell niche via signal transmission [13].

23.8 Summary and Conclusions

Morphological features and the characteristic distribution of TCs in SC niche remain the only current indicator that supports a structural relationship between the two types of cells. Therefore, more function-oriented studies need to be done in the future to prove a dual biological and structural liaison during migration, proliferation, and differentiation of SCs.

Moreover, genetic and proteomic analysis of TCs in different organs would clarify the functional relationship with other neighboring cells including SCs. The use of 3D culture techniques and lab-on-chip technology with microfluidic systems can help provide the ideal microenvironment in which TCs could effectively contribute to tissue repair and regeneration. In addition, coculture of SCs and TCs and the use of dual cell therapy protocols would enhance the proposed beneficiary effects in regenerative medicine.

Given the aforementioned techniques would make it possible to answer an open question about the exact nature of TC–SC tandem and whether TCs are a specific form of mesenchymal/progenitor SCs.

References

1. Bang C, Thum T. Exosomes: new players in cell-cell communication. Int J Biochem Cell Biol. 2012;44:2060–4.
2. Bani D, Formigli L, Gherghiceanu M, Faussone-Pellegrini M. Telocytes as supporting cells for myocardial tissue organization in developing and adult heart. J Cell Mol Med. 2010; 14:2531–8.
3. Bani D, Nistri S. New insights into the morphogenic role of stromal cells and their relevance for regenerative medicine, lessons from the heart. J Cell Mol Med. 2014;18:363–70.

4. Bei Y, Wang F, Yang C, Xiao J. Telocytes in regenerative medicine. J Cell Mol Med. 2015; 19(7):1441–54.
5. Bei Y, Zhou Q, Fu S, Lv D, Chen P, Chen Y, Wang F, Xiao J. Cardiac telocytes and fibroblasts in primary culture: different morphologies and immunophenotypes. PLoS One. 2015;10(2): e0115991.
6. Bojin FM, Gavriliuc OI, Cristea MI, Tanasie G, Tatu CS, Panaitescu C, Paunescu V. Telocytes within human skeletal muscle stem cell niche. J Cell Mol Med. 2011;15:2269–72.
7. Bollini S, Smart N, Riley PR. Resident cardiac progenitor cells: at the heart of regeneration. J Mol Cell Cardiol. 2011;50:296–303.
8. Bollini S, Vieira JMN, Howard S, Dubè KN, Balmer GM, Smart N, Riley PR. Re-activated adult epicardial progenitor cells are a heterogeneous population molecularly distinct from their embryonic counterparts. Stem Cells Dev. 2014;23:1719–30.
9. Cantarero I, Luesma MJ, Junquera C. The primary cilium of telocytes in the vasculature: electron microscope imaging. J Cell Mol Med. 2011;15:2594–600.
10. Ceafalan L, Gherghiceanu M, Popescu LM, Simionescu O. Telocytes in human skin – are they involved in skin regeneration? J Cell Mol Med. 2012;16:1405–20.
11. Ceafalan LC, Popescu BO, Hinescu ME. Cellular players in skeletal muscle regeneration. Biomed Res Int. 2014:957014.
12. Chambers I, Colby D, Robertson M, Nichols J, Lee S, Tweedie S, Smith A. Functional expression cloning of Nanog, a pluripotency sustaining embryonic stem cells. Cell. 2003;113:643–55.
13. Chang Y, Li C, Gan L, Li H, Guo Z. Telocytes in the Spleen. PLoS One. 2015; 23:10(9):e0138851.
14. Cretoiu SM, Popescu LM. Telocytes revisited. Biomol Concepts. 2014;5(5):353–69.
15. Cretoiu D, Hummel E, Zimmermann H, Gherghiceanu M, Popescu LM. Human cardiac telocytes: 3D imaging by FIB-SEM tomography. J Cell Mol Med. 2014;18(11):2157–64.
16. Dellavalle A, Sampaolesi M, Tonlorenzi R, Tagliafico E, Sacchetti B, Perani L, Innocenzi A, Galvez BG, Messina G, Morosetti R, Li S, Belicchi M, Peretti G, Chamberlain JS, Wright WE, Torrente Y, Ferrari S, Bianco P, Cossu G. Pericytes of human skeletal muscle are myogenic precursors distinct from satellite cells. Nat Cell Biol. 2007;9:255–67.
17. Díaz-Flores L, Gutiérrez R, García MP, Sáez FJ, Díaz-Flores Jr L, Valladares F, Madrid JF. CD34+ stromal cells/fibroblasts/ fibrocytes/telocytes as a tissue reserve and a principal source of mesenchymal cells. Location, morphology, function and role in pathology. Histol Histopathol. 2014;29:831–70.
18. Doyonnas R, LaBarge MA, Sacco A, Charlton C, Blau HM. Hematopoietic contribution to skeletal muscle regeneration by myelomonocytic precursors. Proc Natl Acad Sci. 2004;101:13507–12.
19. Drummond-Barbosa D. Stem cells, their niches and the systemic environment: an aging network. Genetics. 2008;180:1787–97.
20. Scadden DT. The stem-cell niche as an entity of action. Nature. 2006;441:1075–9.
21. Faussone-Pellegrini MS, Bani D. Relationships between telocytes and cardiomyocytes during pre- and post-natal life. J Cell Mol Med. 2010;14:1061–3.
22. Fuchs E. Skin stem cells: rising to the surface. J Cell Biol. 2008;180:273–84.
23. Galiger C, Kostin S, Golec A, Ahlbrecht K, Becker S, Gherghiceanu M, Popescu LM, Morty RE, Seeger W, Voswinckel R. Phenotypical and ultrastructural features of Oct4-positive cells in the adult mouse lung. J Cell Mol Med. 2014;18:1321–33.
24. Gherghiceanu M, Popescu LM. Cardiomyocyte precursors and telocytes in epicardial stem cell niche: electron microscope images. J Cell Mol Med. 2010;14:871–7.
25. Gherghiceanu M, Popescu LM. Heterocellular communication in the heart: electron tomography of telocyte-myocyte junctions. J Cell Mol Med. 2011;15(4):1005–11.
26. Gherghiceanu M, Popescu LM. Cardiac telocytes – their junctions and functional implications. Cell Tissue Res. 2012;348:265–79.
27. Gherghiceanu M, Manole CG, Popescu LM. Telocytes in endocardium: electron microscope evidence. J Cell Mol Med. 2010;14:2330–4.
28. Horch RE, Boos AM, Quan Y, Bleiziffer O, Detsch R, Boccaccini AR, Alexiou C, Sun J, Beier JP, Arkudas A. Cancer research by means of tissue engineering–is there a rationale? J Cell Mol Med. 2013;17:1197–206.

29. Hsu YC, Li L, Fuchs E. Emerging interactions between skin stem cells and their niches. Nat Med. 2014;20:847–56.
30. Johns JL, Christopher MM. Extramedullary hematopoiesis: a new look at the underlying stem cell niche, theories of development, and occurrence in animals. Vet Pathol. 2012;49(3):508–23.
31. Kandilis AN, Koskinas J, Tiniakos DG, Nikiteas N, Perrea DN. Liver regeneration: focus on cell types and topographic differences. Eur Surg Res. 2010;44:1–12.
32. Kostin S. Myocardial telocytes: a specific new cellular entity. J Cell Mol Med. 2010;14: 1917–21.
33. Kuang S, Gillespie MA, Rudnicki MA. Niche regulation of muscle satellite cell self-renewal and differentiation. Cell Stem Cell. 2008;2:22–31.
34. LaBarge MA, Blau HM. Biological progression from adult bone marrow to mononucleate muscle stem cell to multinucleate muscle fiber in response to injury. Cell. 2002;111:589–601.
35. Li L, Mignone J, Yang M, Matic M, Penman S, Enikolopov G, Hoffman RM. Nestin expression in hair follicle sheath progenitor cells. Proc Natl Acad Sci. 2003;100:9958–61.
36. Li L, Lin M, Li L, Wang R, Zhang C, Qi G, Xu M, Rong R, Zhu T. Renal telocytes contribute to the repair of ischemically injured renal tubules. J Cell Mol Med. 2014;18:1144–56.
37. Ludwig AK, Giebel B. Exosomes: small vesicles participating in intercellular communication. Int J Biochem Cell Biol. 2012;44(1):11–5.
38. Luesma MJ, Gherghiceanu M, Popescu LM. Telocytes and stem cells in limbus and uvea of mouse eye. J Cell Mol Med. 2013;17:1016–24.
39. Manetti M, Guiducci S, Ruffo M, Rosa I, Faussone-Pellegrini MS, Matucci-Cerinic M, Ibba-Manneschi L. Evidence for progressive reduction and loss of telocytes in the dermal cellular network of systemic sclerosis. J Cell Mol Med. 2013;17(4):482–96.
40. Manetti M, Rosa I, Messerini L, Guiducci S, Matucci-Cerinic M, Ibba-Manneschi L. A loss of telocytes accompanies fibrosis of multiple organs in systemic sclerosis. J Cell Mol Med. 2014;18:253–62.
41. Manole CG, Cismaşiu V, Gherghiceanu M, Popescu LM. Experimental acute myocardial infarction: telocytes involvement in neo-angiogenesis. J Cell Mol Med. 2011;15:2284–96.
42. Miao Q, Shim W, Tee N, Lim SY, Chung YY, Ja KP, Ooi TH, Tan G, Kong G, Wei H, Lim CH, Sin YK, Wong P. iPSC-derived human mesenchymal stem cells improve myocardial strain of infarcted myocardium. J Cell Mol Med. 2014;18:1644–54.
43. Mitchell KJ, Pann_erec A, Cadot B, Parlakian A, Besson V, Gomes ER, Marazzi G, Sassoon DA. Identification and characterization of a nonsatellite cell muscle resident progenitor during postnatal development. Nat Cell Biol. 2010;12:257–66.
44. Ordonez P, Di Girolamo N. Limbal epithelial stem cells: role of the niche microenvironment. Stem Cells. 2012;30:100–7.
45. Pant S, Hilton H, Burczynski ME. The multifaceted exosome: biogenesis, role in normal and aberrant cellular function, and frontiers for pharmacological and biomarker opportunities. Biochem Pharmacol. 2012;83(11):1484–94.
46. Pinnamaneni N, Funderburgh JL. Concise review: stem cells in the corneal stroma. Stem Cells. 2012;30:1059–63.
47. Popescu LM, Faussone-Pellegrini MS. TELOCYTES – a case of serendipity: the winding way from Interstitial Cells of Cajal (ICC), via Interstitial Cajal-Like Cells (ICLC) to TELOCYTES. J Cell Mol Med. 2010;14:729–40.
48. Popescu LM, Nicolescu MI (2013) Telocytes and stem cells. In: Goldenberg RC dos S, Carvalho ACC de, editors. Resid Stem Cells Regen Ther. Oxford:Academic Press; 205–231.
49. Popescu LM, Gherghiceanu M, Cretoiu D, Radu E. The connective connection: interstitial cells of Cajal (ICC) and ICC-like cells establish synapses with immunoreactive cells. Electron microscope study in situ. J Cell Mol Med. 2005;9:714–30.
50. Popescu LM, Gherghiceanu M, Manole CG, Faussone-Pellegrini MS. Cardiac renewing: interstitial Cajal- like cells nurse cardiomyocyte progenitors in epicardial stem cell niches. J Cell Mol Med. 2009;13(5):866–86.
51. Popescu LM, Gherghiceanu M, Kostin S. Telocytes and heart renewing. In: Wang P, Kuo CH, Takeda N, Singal PK, editors. Adaptation biology and medicine, vol 6. Cell adaptations and challenges. New Delhi: Narosa; 2011. p. 17–39.

52. Popescu LM, Gherghiceanu M, Suciu LC, Manole CG, Hinescu ME. Telocytes and putative stem cells in the lungs: electron microscopy, electron tomography and laser scanning microscopy. Cell Tissue Res. 2011;345:391–403.
53. Popescu LM, Manole E, Serboiu CS, Manole CG, Suciu LC, Gherghiceanu M, Popescu BO. Identification of telocytes in skeletal muscle interstitium: implication for muscle regeneration. J Cell Mol Med. 2011;15:1379–92.
54. Popescu BO, Gherghiceanu M, Kostin S, Ceafalan L, Popescu LM. Telocytes in meninges and choroid plexus. Neurosci Lett. 2012;516:265–9.
55. Popescu LM, Curici A, Wang E, Zhang H, Hu S, Gherghiceanu M. Telocytes and putative stem cells in ageing human heart. J Cell Mol Med. 2015;19(1):31–45.
56. Popescu LM, Fertig ET, Gherghiceanu M. Reaching out: junctions between cardiac telocytes and cardiac stem cells in culture. J Cell Mol Med. 2016;20(2):370–80.
57. Roatesi I, Radu BM, Cretoiu D, Cretoiu SM. Uterine telocytes: a review of current knowledge. Biol Reprod. 2015;93(1):10.
58. Rusu MC, Mirancea N, Manoiu VS, Vâlcu M, Nicolescu MI, Păduraru D. Skin telocytes. Ann Anat. 2012;194(4):359–67.
59. Schlotzer-Schrehardt U, Kruse FE. Identification and characterization of limbal stem cells. Exp Eye Res. 2005;81:247–64.
60. Sheng J, Shim W, Lu J, Lim SY, Ong BH, Lim TS, Liew R, Chua YL, Wong P. Electrophysiology of human cardiac atrial and ventricular telocytes. J Cell Mol Med. 2014;18:355–62.
61. Stoorvogel W, Kleijmeer MJ, Geuze HJ, Raposo G. The biogenesis and functions of exosomes. Traffic. 2002;3:321–30.
62. Suciu L, Nicolescu MI, Popescu LM. Cardiac telocytes: serial dynamic images in cell culture. J Cell Mol Med. 2010;14:2687–92.
63. Suciu L, Popescu LM, Gherghiceanu M, Regalia T, Nicolescu MI, Hinescu ME, Faussone-Pellegrini MS. Telocytes in human term placenta: morphology and phenotype. Cells Tissues Organs. 2010;192:325–39.
64. Suciu LC, Popescu BO, Kostin S, Popescu LM. Platelet-derived growth factor receptor-β-positive telocytes in skeletal muscle interstitium. J Cell Mol Med. 2012;16:701–7.
65. Sun TT, Cotsarelis G, Lavker RM. Hair follicular stem cells: the bulge-activation hypothesis. J Invest Dermatol. 1991;96:77S–8.
66. Vannucchi MG, Bani D, Faussone-Pellegrini MS. Telocytes contribute as cell progenitors and differentiation inductors in tissue regeneration. Curr Stem Cell Res Ther. 2016;11(5):383–9.
67. Wagner-Souza K, Diamond HR, Ornellas MH, Gomes BE, Almeida-Oliveira A, Abdelhay E, Bouzas LF, Rumjanek VM. Rhodamine 123 efflux in human subpopulation of hematopoietic stem cells: comparison between bone marrow, umbilical cord blood and mobilized peripheral blood CD34+ cells. Int J Mol Med. 2008;22:237–42.
68. Wang F, Song Y, Bei Y, Zhao Y, Xiao J, Yang C. Telocytes in liver regeneration: possible roles. J Cell Mol Med. 2014;18:1720–6.
69. Wang F, Bei Y, Zhao Y, Song Y, Xiao J, Yang C. Telocytes in pregnancy-induced physiological liver growth cell. Physiol Biochem. 2015;36:250–8.
70. Wong VW, Levi B, Rajadas J, Longaker MT, Gurtner GC. Stem cell niches for skin regeneration. Int J Biomater. 2012;2012:926059.
71. Xiao J, Wang F, Liu Z, Yang C. Telocytes in liver: electron microscopic and immunofluorescent evidence. J Cell Mol Med. 2013;17:1537–42.
72. Xue Y, Chen F, Zhang D, Lim S, Cao Y. Tumor-derived VEGF modulates hematopoiesis. J Angiogenes Res. 2009;1:9.
73. Yang Y, Sun W, Wu SM, Xiao J, Kong X. Telocytes in human heart valves. J Cell Mol Med. 2014;18:759–65.
74. Zhang HQ, Lu SS, Xu T, Feng YL, Li H, Ge JB. Morphological evidence of telocytes in mice aorta. Chin Med J (Engl). 2015;128(3):348–52.
75. Zhao B, Chen S, Liu J, Yuan Z, Qi X, Qin J, Zheng X, Shen X, Yu Y, Qnin TJ, Chan JY, Cai D. Cardiac telocytes were decreased during myocardial infarction and their therapeutic effects for ischaemic heart in rat. J Cell Mol Med. 2013;17:123–33.

76. Zhao B, Liao Z, Chen S, Yuan Z, Yilin C, Lee KK, Qi X, Shen X, Zheng X, Quinn T, Cai D. Intramyocardial transplantation of cardiac telocytes decreases myocardial infarction and improves post-infarcted cardiac function in rats. J Cell Mol Med. 2014;18:780–9.
77. Zheng Y, Zhang M, Qian M, Wang L, Cismasiu VB, Bai C, Popescu LM, Wang X. Genetic comparison of mouse lung telocytes with mesenchymal stem cells and fibroblasts. J Cell Mol Med. 2013;17(4):567–77.
78. Zheng Y, Cretoiu D, Yan G, Cretoiu SM, Popescu LM, Fang H, Wang X. Protein profiling of human lung telocytes and microvascular endothelial cells using iTRAQ quantitative proteomics. J Cell Mol Med. 2014;18:1035–59.
79. Zhou J, Zhang Y, Wen X, Cao J, Li D, Lin Q, Wang H, Liu Z, Duan C, Wu K, Wang C. Telocytes accompanying cardiomyocyte in primary culture: two- and three-dimensional culture environment. J Cell Mol Med. 2010;14:2641–5.
80. Zhou J, Wang Y, Zhu P, Sun H, Mou Y, Duan C, Yao A, Lv S, Wang C. Distribution and characteristics of telocytes as nurse cells in the architectural organization of engineered heart tissues. Sci China Life Sci. 2014;57(2):241–7.

Zeinab M. El Maadawi, MD is an associate professor of Histology & Cell Biology and e-learning expert at the Faculty of Medicine, Cairo University, Egypt, where she works as a medical educator, a biomedical researcher, and an e-learning expert with evident experience in international education management. Her biomedical research interest includes the study of Interstitial Cells of Cajal (ICC) that regulate gastrointestinal motility on which she pursued her master and PhD degree. She has got special interest in the field of stem cell research and regenerative medicine and has a list of publications in this field. She got postdoctoral training at the Institute of Cardiovascular Sciences, Faculty of Medical and Human Sciences, University of Manchester, UK, where she learned different stem cell culture techniques particularly induced pluripotent stem cells (iPSCs).

Zeinab M. El Maadawi

Chapter 24
Vascular Telocytes

Hongqi Zhang

24.1 Constitution of Circulatory System

Circulatory system is composed of cardiovascular system and lymph system, which is a transport system that carries blood and lymph to and from the tissue of the body. The cardiovascular system consists of a pump represented by the heart and blood vessels, which provide the route by which blood circulates to and from all parts of the body. The heart pumps the blood through the arterial system under significant pressure; blood is returned to the heart under low pressure with the assistance of negative in the thoracic cavity during inspiration and compression of the vein by the skeletal muscle. In our body, it is made up of arteries, veins, and capillaries between arteries and veins.

24.2 General Structure of Blood Vessels

There are three major types of blood vessels: the arteries, which carry the blood away from the heart; the capillaries, which enable the actual exchange of water and chemicals between the blood and the tissues; and the veins, which carry blood from the capillaries back toward the heart.

Except the capillaries, both the arteries and the veins have three layers, the tunica intima, tunica media, and tunica adventitia; the tunica intima and tunica media are separated by the internal elastic lamina, and the boundary between tunica media and tunica adventitia is the external elastic lamina.

H. Zhang, PhD
Department of Anatomy, Histology and Embryology, Shanghai Medical College of Fudan University, Shanghai 200032, China

Key Laboratory of Medical Imaging Computing and Computer Assisted Intervention of Shanghai, Shanghai, Room 308, Building 9, Yixueyuan Road 138, Shanghai 200032, China
e-mail: zhanghq58@126.com

X. Wang, D. Cretoiu (eds.), *Telocytes*, Advances in Experimental Medicine and Biology, 913, DOI 10.1007/978-981-10-1061-3_24

Tunica intima is the thinnest layer among three layers; it is made up of a single layer of continuous squamous endothelial cells (ECs) surrounded by a thin layer of subendothelial connective tissue interlaced with a number of circularly arranged elastic bands called the internal elastic lamina. Sometimes, within the subendothelial layer, a little of collagen fiber (CF), elastic fiber, and smooth muscle cells (SMCs) (the number of smooth muscle cells in it gradually increased with the age).

Tunica media locates between the internal elastic lamina and external elastic lamina. The thickness and elements of tunica media depend on the types and sizes of blood vessels. In large-sized arteries, the tunica media mainly consist of elastic membranes and a little quality of the smooth muscle; the tunica media of middle-sized arteries is composed mainly of smooth muscle cells, which can control the caliber of the vessel and produce collagen fiber, elastic fiber, and matrix. In some pathological condition, the smooth muscle cells in tunica media could migrate into the tunica intima and produce the connective tissues and result in the thickness increase of tunica intima, which is an important pathological basis of development of atherosclerosis.

Tunica adventitia is entirely made of connective tissue. It also contains circular or longitudinally arranged elastic fibers, collagen fibers, fibroblasts, and nerve endings that supply the vessel as well as nutrient capillaries (vasa vasorum) in the larger blood vessels.

24.3 A Novel of Interstitial Cell Telocytes

Tissues and organs where the telocytes have been found:

Telocyte (TC) is a novel type of interstitial cell, whose key features are their small cell body with very long prolongations of uneven caliber, termed telopodes (Tps) [1]. In fact, Tps are constituted by an alternation of dilated segments (podoms – which are harboring the mitochondria (M), endoplasmic reticulum, and caveolae) and thin segments (podomers). The telocytes have been recently described and termed in stromal connective tissue of many organs, such as the heart (endocardium [2], myocardium [3], epicardium [4]), skeletal muscles [5], gastrointestinal tract, and accessory annexes (esophagus [6], jejunum [7], salivary glands [8], gallbladder [9], [10, 11], liver [12]), respiratory system [13], (trachea [14], pleura [15]), urinary tract (kidney [16], renal pelvis [17], ureter [17], bladder [17], urethra [17]), meninges and choroid plexus, mammary glands [18], uterus [11], fallopian tube [19], placenta [20], skin [21], vasculature [22], and spleen [23]. Recently, telocytes are found in more and more tissues and organs.

Cantarero [24] and colleagues found telocytes in the loose connective tissue surrounding the arterioles, venules, and capillaries under electron microscope. These telocytes were morphologically similar to those previously found in other locations. In arterioles, the telocytes often send telopodes bordering the tunica adventitia. The adjacent telopodes may establish junction complexes and show small dilatations (podoms) that accommodate the rough endoplasmic reticulum (rER). In relation to venules and capillaries, the telocytes were located parallel with the longitudinal axis of the smooth muscle cells of the vessel wall. Also, the telopodes can form a wide network around the blood vessels. Vascular telocytes showed typical features: small cellular body; long, thin, and moniliform prolongations with a dichotomic branching pattern; and mitochondria, well-developed Golgi apparatus, and polyribosomes.

24.4 Blood Telocytes

As we know, the artificial blood vessel graft (more than 6 mm in diameter) is widely applied in the clinic or many operations; nevertheless, the use of artificial vascular graft (less than 6 mm in diameter) still encounters many problems such as; restenosis and is still in the experimental stage. The College of Textile of Donghua University and the Medical College of Fudan University in Shanghai cooperated to explore the possibility of small-caliber artificial blood vessels used in clinical operation.

In order to testify on the effect of artificial vascular graft (diameter is less than 6 mm), an animal experiment was done, that is, a 4-cm-long artificial vascular graft was intergrafted into the common carotid arteries in pigs for 1 month, 2 months, and 3 months. As a control group, normal morphological appearance of endothelial cells is observed under scanning electron microscope. Steps on the experimental animals to be treated are as follows: anesthetized pigs were fixed and the thoracic cavity was opened, through a puncture in the left ventricle, and 3500 ml of heparin physiological saline (for the removing of formed elements of the blood: erythrocytes, leukocytes, and thrombocytes or platelets on the endothelial surface) and 1000 ml 4 % paraformaldehyde solution were perfused at physiological pressure (for the fixation of endothelial cells and the tissues of the vascular wall in situ). After the perfusion, different arterial segments of large- and middle-sized arteries were removed and prepared for scanning electron microscopy according to routine treatment. Endothelial surface of various arteries was examined, and the images were captured by using Philips XL30E SEM.

Accidentally and luckily, a few of cells (Figs. 24.1, 24.2, 24.3, and 24.4) with long prolongation adhesive to the surface of endothelial cells were discovered mainly in middle-sized arteries, such as common carotid arteries, internal thoracic arteries, and coronary arteries [22]. The kind of cells with long prolongations is apparently different from the dendritic cell. Dendritic cells are present in those tissues that are in contact with the external environment, such as the skin (usually called the Langerhans cell) and the inner lining of the nose, lungs, stomach, and intestines. They can also be found in an immature state in the blood. At certain development stages, they grow branched projections, but their morphological features are with more and thinner dendrites than telocytes. Dendritic cells that circulate in blood do not have all the typical features of their counterparts in tissue, i.e., they are less mature and have no dendrites.

We used SEM to provide visual evidence for the presence of TCs in the blood. We studied both large arteries (different divisions of the aorta: the ascending aorta,

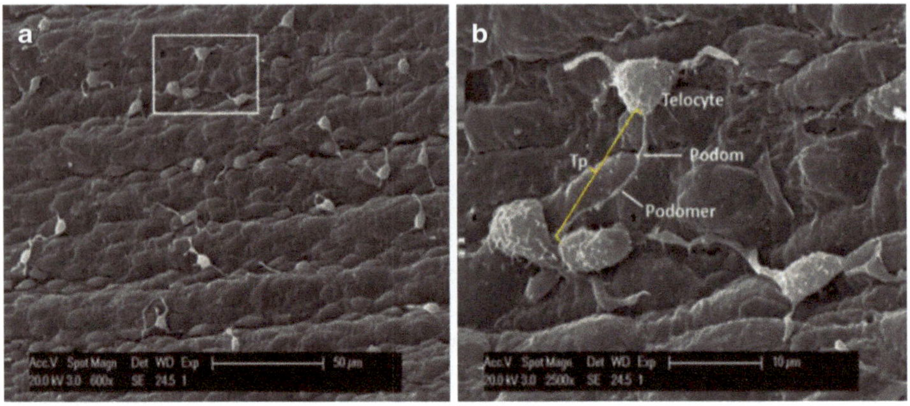

Fig. 24.1 Distribution of vascular telocytes (*TCs*) in pig, scanning electron microscope images. (**a**) Under lower power, several TCs are seen on the endothelial surface. (**b**) Local enlargement of white rectangle indicates a typical TC with triangular cell body and one long telopode (*Tp*) (Tp, 11.9 μm in length) with alternation of podomers and podoms

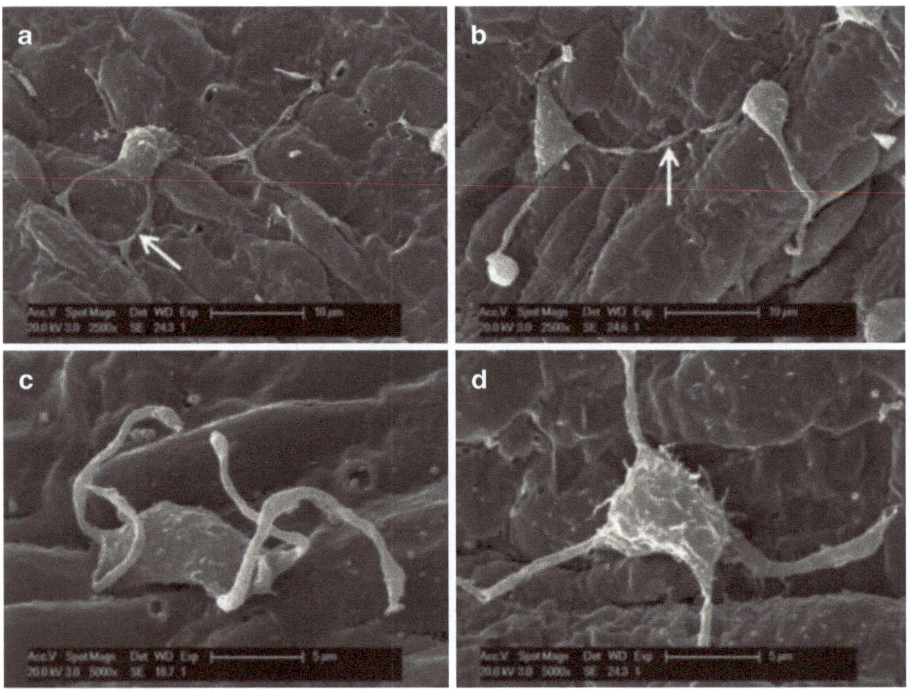

Fig. 24.2 The different types of telopodes (*Tps*) contacts and short prolongations, scanning electron microscope images of medium-sized arteries in pig. (**a**) Telocyte (*TC*) has four Tps, and two of them form direct contact and exist in the shape of ring (*arrow*). (**b**) Two Tps from different TCs form point contacts (*arrow*). (**c**) The Tps detach from the endothelial surface and the cell looks like a crab. (**d**) The shorter and thinner prolongations of cell body of a TC. They are apparently different from Tps and microvilli

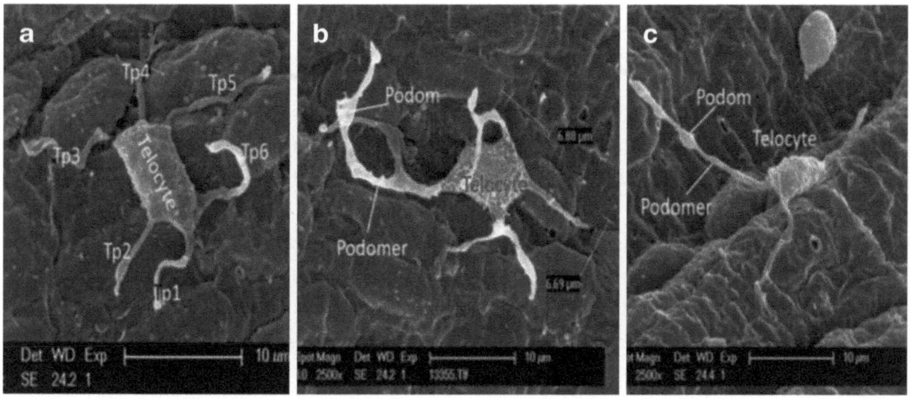

Fig. 24.3 Scanning electron microscope images of number and features of telopodes (*Tps*) adhering to endothelium of medium-sized arteries in pig. (**a**) The Telocyte (*TC*) with six Tps (Tp1–Tp6). (**b**) The TC with four Tps, three of them form bifurcations and present alternating podomers and podoms. The TC body is 6.88 μm and 6.69 μm in length and width, respectively. (**c**) The TC with three Tps, the distinctive Tps are observed with typical podomer and podom

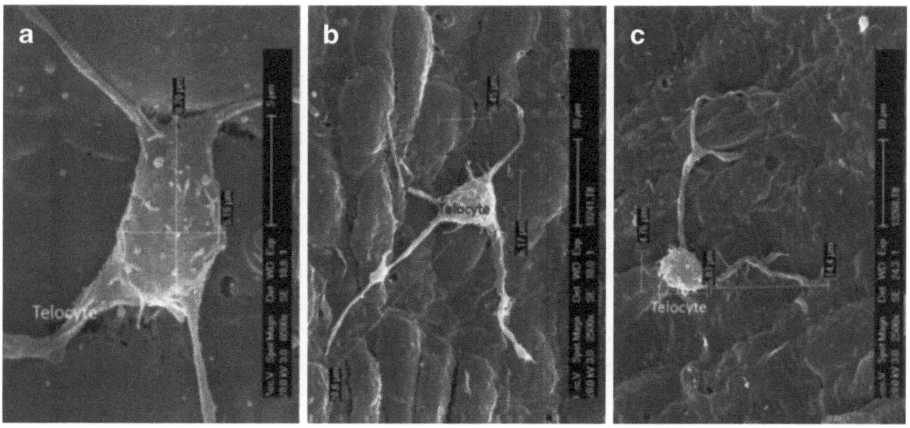

Fig. 24.4 The size of telocyte (*TC*) and its telopodes (*Tps*). Scanning electron microscope images of medium-sized arteries in pig. (**a**) The TC body looks rectangular and the long and short diameters of the cell body are 8.70 μm and 5.18 μm, respectively. (**b**) The TC with five Tps. The longest Tp is 20.8 μm. (**c**) Two Tps give rise to bifurcations. The cell body is 5.83 μm and 4.76 μm in length and width, respectively. The shorter Tps is 14.4 μm in length

aortic arch, thoracic part, and abdominal part of the aorta) and middle-sized arteries (internal thoracic arteries, common carotid arteries, and coronary arteries). The findings showed the presence of a new population of cells (TCs) on the pig endothelial surface with the same ultrastructural features as TCs, previously described by Popescu's group [1]. We found TCs only in the medium-sized arteries, rather than in large arteries. This might be because of the higher shear stress that occurred in these vascular regions and could be speculated as TCs are less tolerant for turbulent blood flow. On the other hand, it is well known that living cells, for self-assuring their survival, have their natural tendency for minimizing their surface in the liquid environment. Thus, TCs in vasculature could appear with slightly modified morphology, with more spherical, shorter, and thicker prolongations. In summary, we present visual evidence for the existence of TCs in the blood. The morphological specificity of blood TCs might be associated with the dynamic environment where they exist. Further morphological and functional correlations need to be established.

As to the origin of the blood telocytes, we speculate that they could be from the bone marrow; we all know that almost all the formed elements in the blood came from the bone marrow. Ten male C57BL/6 J mice aged 5 weeks (weight, 12–16 g) were used in the research. The femurs were harvested, and then the soft tissues attached to the femurs were removed. The femurs were then cut into two halves along their longitudinal axis by microsurgical scissors in preparation for sample treatment. The specimens were handled according to scanning electron microscope and transmission electron microscope (TEM) routine and observed under Philips XL30E SEM (Amsterdam, the Netherlands). Our research provide for the first time ultrastructural evidence for the presence of TCs within the bone marrow. Ultrastructurally, TCs found in mice bone marrow share the similar distinctive features with TCs described by Popescu's group within other organs [1]. TCs have the small cellular body with one to three suddenly emerging telopodes of uneven caliber. Similar to the previous reports, within the bone marrow, TCs appear interconnected by homocellular junctions, forming a network. In mice bone marrow, TCs are observed in close spatial relationships with small blood vessels and/or capillaries. Also, the presence of TCs in bone marrow tissue characterized by the generous presence of stem cells and progenitor cells somehow recapitulates the condition of TCs in stem cell niches. This aspect could offer new insights regarding the presumptive role(s) of TCs in cellular regeneration. Now we are not sure that the telocytes in the blood are surely from the bone marrow; the real origin and function of blood telocytes is an urgent problem to be solved in the future (Figs. 24.5 and 24.6).

24.5 The Telocytes Within the Arterial Wall

After the blood telocytes were found, we have been thinking whether the telocytes are present within the arterial wall. Usually, leukocytes in the blood could migrate into the vascular wall or tissues through the endothelial space and become the macrophage. Whether the telocytes in the blood also infiltrate the endothelial space into the vascular wall is still unclear. In order to explore this question, the aorta of mice is investigated under transmission electron microscope according to the specimen routine treatment.

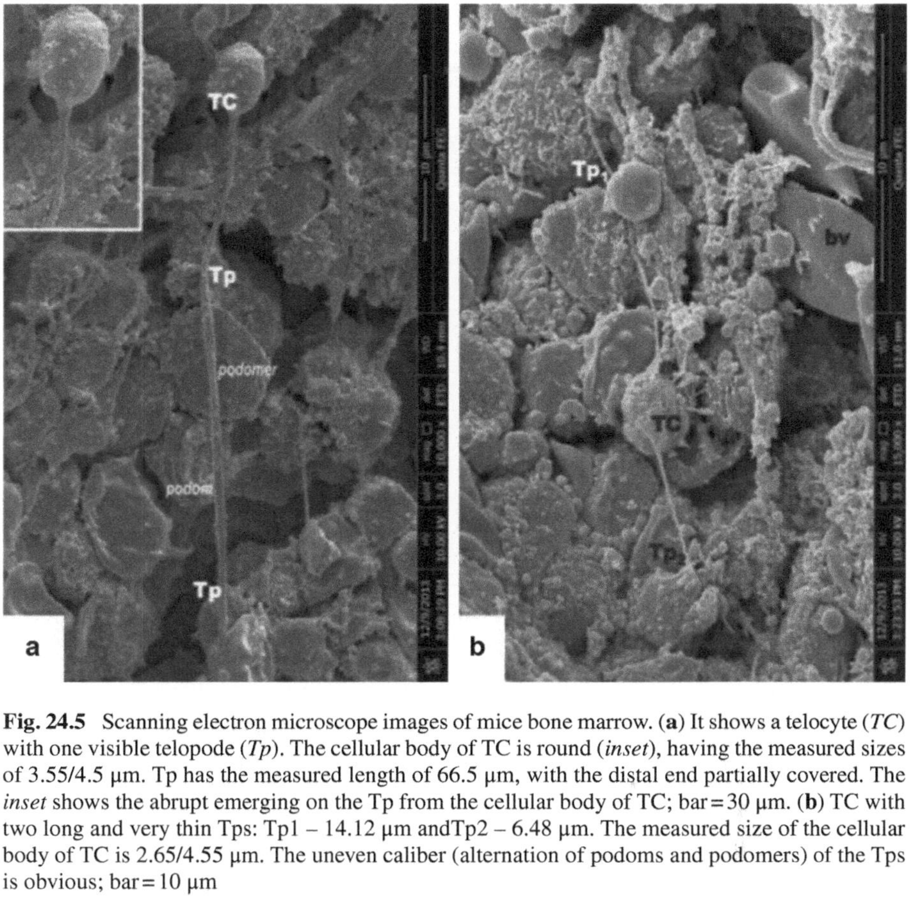

Fig. 24.5 Scanning electron microscope images of mice bone marrow. (**a**) It shows a telocyte (*TC*) with one visible telopode (*Tp*). The cellular body of TC is round (*inset*), having the measured sizes of 3.55/4.5 μm. Tp has the measured length of 66.5 μm, with the distal end partially covered. The *inset* shows the abrupt emerging on the Tp from the cellular body of TC; bar = 30 μm. (**b**) TC with two long and very thin Tps: Tp1 – 14.12 μm andTp2 – 6.48 μm. The measured size of the cellular body of TC is 2.65/4.55 μm. The uneven caliber (alternation of podoms and podomers) of the Tps is obvious; bar = 10 μm

24.5.1 Distinctive Features of Telocytes in Tunica Adventitia

Under transmission electron microscope, a novel type of interstitial cell with distinctive ultrastructural features defined as telocyte was observed in loose connective tissue of tunica adventitia of the mice aortic arch. TCs in the innermost layer of tunica adventitia, located at the juncture between tunica media and tunica adventitia, with their long axes are oriented parallel to the outer elastic membrane (Fig. 24.7). And no direct contacts between TCs and elastic membrane and no intercellular junctions between TCs and vascular SMCs (VSMCs) were observed (Fig. 24.7). TCs in the outer layer of tunica adventitia were intertwined with surrounding stromal CFs, which were organized into a highly complex three-dimensional meshwork among transverse-oriented fiber bundle and longitudinal array-oriented fiber bundle (Figs. 24.8 and 24.9). TCs in tunica adventitia were morphologically consistent with those previously reported in other tissues and organs. The characteristics of TCs in tunica adventitia demonstrated that the cell bodies were relatively small (range from 6.06 to 13.02 μm in length, from 1.05 to

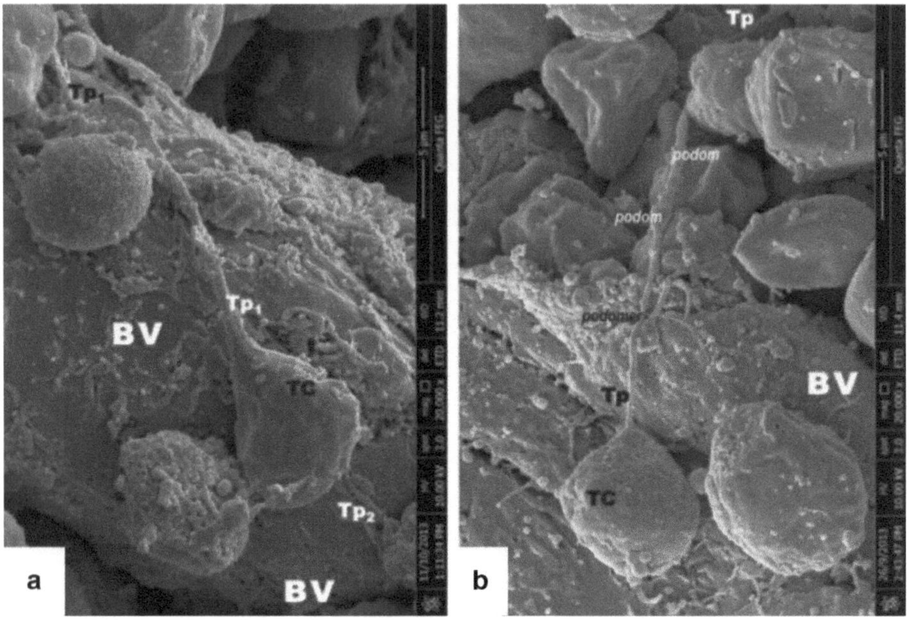

Fig. 24.6 Scanning electron microscope images of mice bone marrow. (**a**) Telocyte (*TC*) and its telopode (*Tp*) are located on the surface of an arteriole. TC cellular body size: 4.44/5.93 μm; Tp dimensions: length – 12.5 μm; diameter – 0.44 μm; bar = 5 μm. (**b**) TC with its Tp in close spatial relationship with surrounding interstitial cells. The moniliform aspect of the Tp is obvious: the alternation of dilated segments (**a**) with thin segments (**b**). TC cellular body size: 4.5 μm in diameter; Tp length – 15.12 μm (being which was partially sheltered by one cell); bar = 5 μm

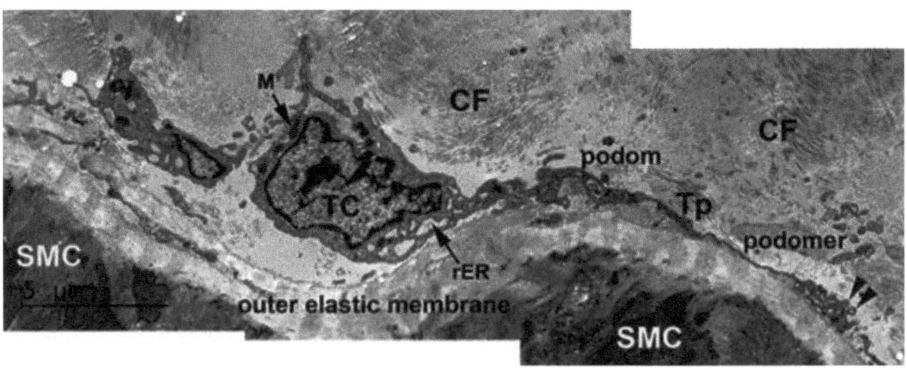

Fig. 24.7 Transmission electron microscope image of mouse aortic arch (merged image). Telocyte (*TC*) with a characteristic telopode (*Tp*) borders the outer elastic membrane of the aortic arch. The body of TC is 6.06 in length and 4.25 μm in width. The cytoplasm is scare and contains a moderate amount of mitochondria (*M*) and rough endoplasmic reticulum (*rER*). The visible Tp is 14.98 μm in length, presenting moniliform respect due to the alternation of podoms and podomers (average caliber 0.09 μm). Note the shed vesicles (*arrowheads*) in close proximity to the distal part of Tp. *CF* Collagen fibers, *SMC* Smooth muscle cell; Bar = 5 μm

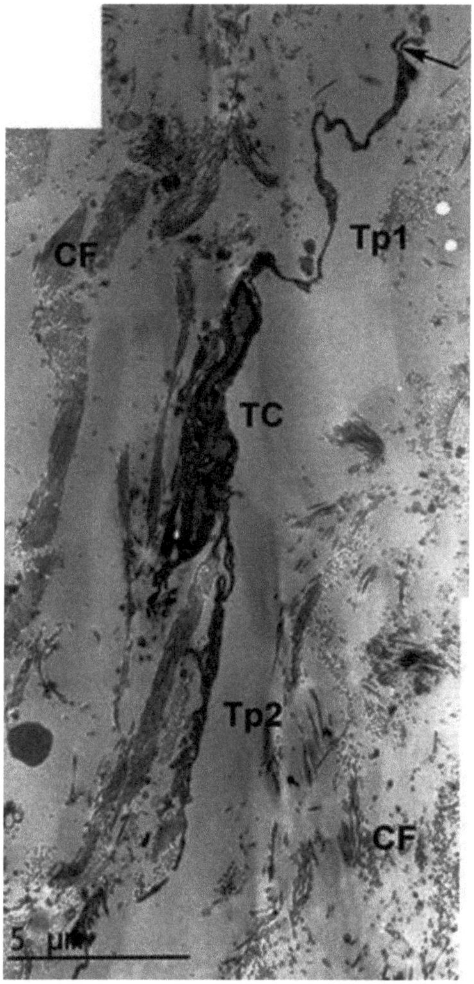

Fig. 24.8 Telocyte (*TC*) in adventitia of mouse aortic arch (merged image). TC with two telopodes (*Tps*) exists in tunica adventitia, embedded among collagen fibers (*CF*). The body of TC takes on the shape of irregular thin and long ellipse with 8.15 μm in length and the average width being 1.05 μm. The length of circuitous Tp1 and Tp2 is 12.90 μm and 13.07 μm, respectively. Note a dichotomous branch (*black arrow*) emerging from the end of Tp1. Bar = 5 μm

4.25 μm in width), with a high nuclear/cytoplasmic ratio (Figs. 24.7, 24.8, 24.9, 24.10, 24.11, and 24.12); the perinuclear cytoplasm contained some rER and mitochondria (Fig. 24.7); the thin and long (range from 7.74 to 39.05 μm) Tps were projecting from the cell body (Figs. 24.7, 24.8, 24.9, 24.10, 24.11, and 24.12), whose number per TC was variable, with one to three visible Tps in a single section, generally (Figs. 24.7, 24.8, 24.9, 24.10, 24.11, and 24.12); and the typical morphological features of convoluted and moniliform Tps (Fig. 24.4) occurred due to the alternation of podomers and podoms. The podomer was the thin segment whose caliber was about 0.09 μm (Fig. 24.7), and the podom was the dilated segment, which accommodated abundant organelles: rER, Golgi apparatus, lysosomes, and

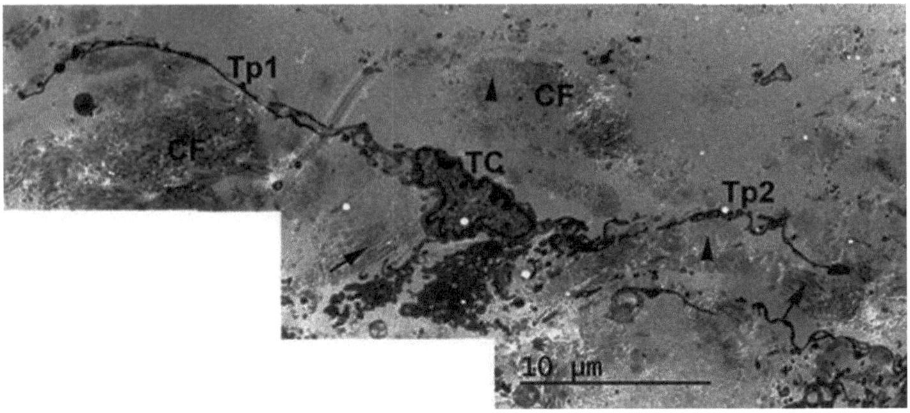

Fig. 24.9 Telocytes (*TCs*) and collagen fibers (*CFs*) in adventitia of mouse aortic arch (merged image). A TC (length: 7.17 µm, the average width: 2.18 µm) with two characteristic telopodes (*Tps*) (*Tp1* and *Tp2* are 22.31 µm and 17.37 µm in length, respectively) is intertwined with surrounding collagen fibers, which are organized into a highly complex three-dimensional meshwork among transverse-oriented fiber bundle (*arrowheads*) and longitudinal array-oriented fiber bundle (*black arrows*). Bar = 10 µm

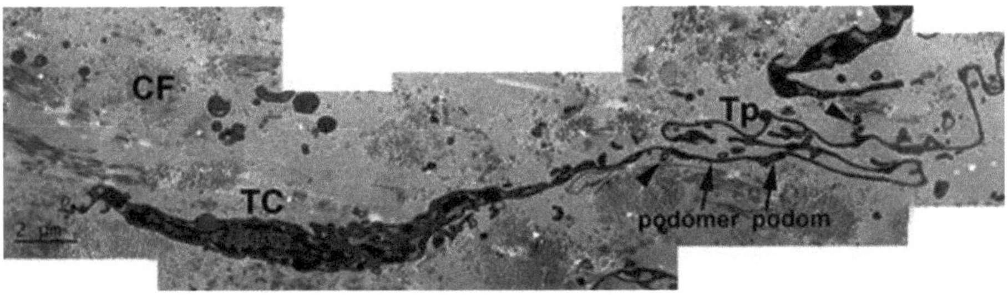

Fig. 24.10 Distinctive feature of telopode (Tp) of telocyte (TC) in adventitia of the mouse aortic arch (merged image). The Tp of a TC displays very thin, special long and convoluted aspect. The length of Tp is up to 39.05 µm, more than three times of the macro-axis length (13.01 µm in length) of the body of TC. The moniliform aspect of Tp consists of an alternation of podomers and podoms (*black arrows*). Plenteous shed vesicles (*arrowheads*) can be seen in the neighborhood of various regions of the Tp. *CF* Collagen fiber; Bar = 2 µm

caveolae (Fig. 24.11). In addition, the dichotomous branch emerged at various segments of Tps of TCs (Figs. 24.8 and 24.11), and vesicles shedding from TCs were present in the adjacent extracellular matrix (Figs. 24.7 and 24.10).

24.5.2 Cell Communication Between Telocytes and Other Cells in Tunica Adventitia

Homocellular junctions between TCs themselves were observed under TEM. The desmosomes (Fig. 24.12) were visible between Tps of different TCs, forming an intricate three-dimensional network in tunica adventitia. Moreover, macrophages

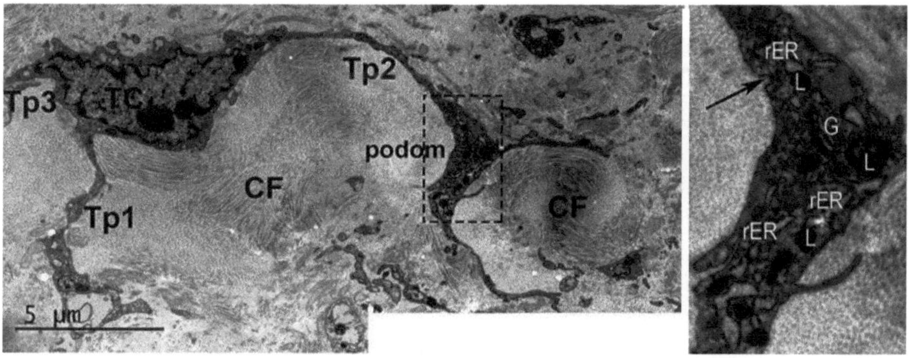

Fig. 24.11 Ultrastructure of podom of telopode (*Tp*) in adventitia of the mouse aortic arch (merged image). (**a**) A telocyte (*TC*) (length: 9.31 μm, the average width: 2.79 μm) with three Tps coexists with collagen fibers in adventitia of the aortic arch. Tp2 is the longest one of the three, reaching 16.43 μm, presenting the typical morphological feature of alternating podomer and podom and forming bifurcation. The length of Tp1 is 7.74 μm and Tp3 is only partially shown. (**b**) The higher magnification of dotted line rectangle area of (**a**) indicates that the dilated segment of Tp-podom has an irregular shape and accommodates rough endoplasmic reticulum (*rER*), Golgi apparatus (*G*), lysosomes (*L*), and caveolae (*black arrow*). Bar = 5 μm

and FB coexisted with TCs in the same region of loose connective tissue of tunica adventitia (Fig. 24.13), where a large amount of CFs survived, whereas no direct connections appeared among them.

Generally speaking, the tunica adventitia of the large artery is relatively a thin connective tissue layer containing cells (fibroblasts and macrophages mainly) and collagen fibers and a few elastic fibers. The morphological features of telocytes in the aorta are also in accord with previous research that TCs are located on the connective tissue of rat duodenal blood vessels including arterioles, venules, and capillaries [18, 24]. Although TEM alone allowed identification of TCs, caution should be taken to differentiate TCs from other interstitial cells in tunica adventitia. Macrophages should not give rise to diagnostic problems due to their appearance of a large Golgi apparatus, abundant lysosomes, and irregular cytoplasmic projections. But attentions should be taken when in the face of fibroblasts and TCs, which could be confused by their similar structural features. According to the results of this study, TCs were different from other interstitial cells, including fibroblasts, by presence of Tps, which are extremely long, thin, and moniliform prolongations. The main function of fibroblast is to synthesize collagen fibers, elastic fibers, and some extracellular constituents. The fibroblast body is large and pleomorphic; the Golgi apparatus is prominent; the rough endoplasmic reticulum is well developed; and the cell processes are few, short, and of large caliber, thus being easily appreciable under a light microscope [19, 20, 25, 26]. Based on these, these cells are markedly different from TCs, which present an irregular cell body, a large nucleus, and a scarce cytoplasm, with a small quantity of Golgi apparatus, some mitochondria, few endoplasmic reticula, and, the most striking feature of special long, thin Tps [19, 25]. Numerous studies have documented the possibility of present interstitial Cajal-like cells (ICLCs) in the blood vessels.

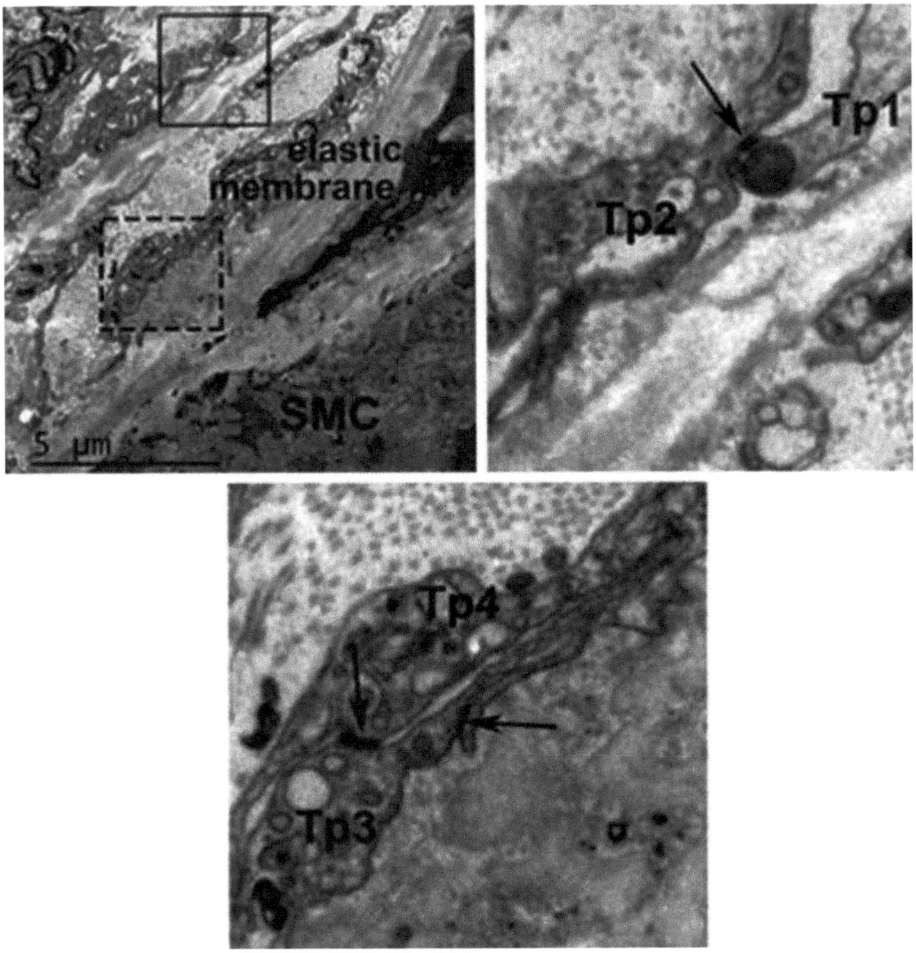

Fig. 24.12 Cell communication between different telopodes (*Tps*). (**a**) Direct contacts come into existence between Tps of different telocytes (TCs) in adventitia of the aortic arch. (**b**) Local higher magnification of the line rectangular area of (**a**) shows that the end of Tp1 forms desmosome (*black arrow*) with Tp2 of another TC. (**c**) Local higher magnification of dotted line rectangular area of (**a**) demonstrates that Tp3 inserts into a cupped space formed by Tp4 of another TC, and desmosomes (*black arrows*) connecting Tp3 and Tp4 are visible. *SMC* Smooth muscle cell. Bar = 5 μm

The studies by Pucovský et al. [21, 27] and Harhun et al. [22, 28] described a novel cell type, which had similar morphological features of ICC in mesenteric arteries of the guinea pig and rabbit portal vein, respectively. It seemed that there were some differences in both morphology and physiological roles of ICCs of rabbit portal veins and mesenteric arteries of guinea pig. Harhun et al. [23, 29] utilized ICs (interstitial cells) for all subtypes of interstitial cells found in blood vessels. A study performed by Bobryshev [2, 30] firstly showed the presence of arterial cells

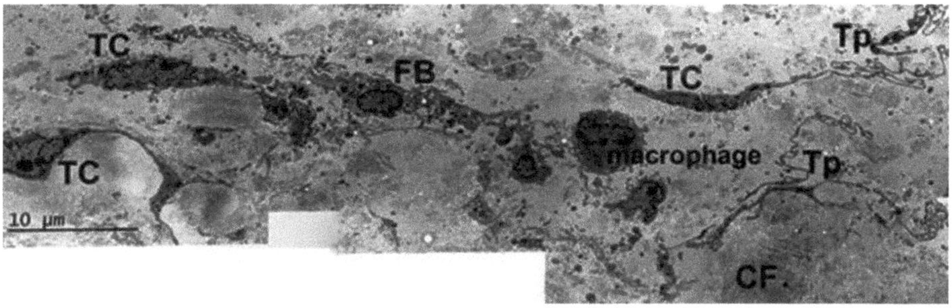

Fig. 24.13 Relationship between telocytes (TCs) and other cells (merged image). In adventitia of mouse aortic arch, three TCs are observed to be present with fibroblast and macrophage in the same region of extracellular matrix of adventitia where a large amount of collagen fibers (CF) exist, but no direct contact among them appears. Bar = 10 μm

with typical structural characteristics of ICC in situ, which was known as arterial ICC, and perhaps represents a distinct subtype within the ICC family. These cells were in direct contact with both SMCs and nerve endings at the juncture of media and adventitia of human large arteries. According to the review of interstitial cells of blood vessels, [24, 31] a new cell type termed interstitial cell was indicated in the tunica media of both veins and arteries. These cells possessed the characteristic of irregularly shaped, thin processes and noncontractile, which were totally different from VSMCs. The main role of portal vein ICs may tend to be a pacemaker in the wall of blood vessels, while the physiological role of interstitial cells in arteries is still unclear, and according to their phenotypes, the arterial interstitial cells may belong to the SMC lineage. From the research so far, ICCs or ICs of blood vessels are more likely a subset of SMC subpopulation, which shares most of the features with the ICCs of the gastrointestinal tract [25, 32]. In the present study, a subset of interstitial cells with ultrastructure characteristics enabling these cells to be regarded as TCs (see in detail: http://www.telocytes.com/) was usually located in tunica adventitia of the aorta, not in tunica media, and no direct intercellular junctions were found between them and SMCs. It seems to be clear that there are plenty of differences between ICCs and TCs in blood vessels, such as ultrastructure and function. Previous study by Corselli et al. [26, 33] demonstrated that cells expressed mesenchymal stem cells markers resided in the outmost layer of blood vessels. Besides pericytes, which encircle capillaries and microvessels, the tunica adventitia might be another source of mesenchymal stem cells. The new finding in our current study showed the presence of TCs in the tunica adventitia of the mice aorta. Since TCs had been reported in close relationship with several stem cells, such as cardiac progenitors [27, 34] subepithelial lung stem cells [5], skeletal muscle stem cells [28, 35], skin stem cell clusters [7, 21]. There are grounds for believing that TCs play an important role in increasing the efficiency and efficacy of resident local stem cells in the process of repair/regeneration through cell-to-cell communication or shed vesicles. In conclusion, our study provided TEM evidence for the presence of TCs with representative features in tunica adventitia of the mice aortic arch. And their biologically functional significance in vasculature is needed to be further explored.

24.6 Telocytes Within the Wall of the Inferior Vena Cava

Telocytes (TCs), distinctive population of stromal cells extending special long prolongations with thin segments (podomers) and dilations (podoms), have been identified in various organs and tissues including the blood, the large-sized artery, and the connective tissue around the venule, arteriole, and capillaries [36]. However, whether TCs exist in the large-sized vein is still not clear.

Three male beagles (aged 3 months weighted 23–26 g) were anesthetized with diazepam and ketamine hydrochloride by intramuscular injection. Then, their thoracic cavities were opened, through a puncture into the left ventricle; 2500 ml of heparin physiological saline and 1000 ml 4 % formaldehyde solution for each animal were perfused at physiological pressure, respectively. After the perfusion, the inferior vena cava was removed and prepared according to routine treatment and observed under transmission electron microscope (FEI TECAI SPIRITTEM) [34].

A new type of interstitial cell with typical ultrastructural features defined as TC was observed in the canine inferior vena cava under TEM. TCs in the inferior vena cava were morphologically consistent with those previously reported in other tissues and organs. TCs with high nuclear/cytoplasmic ratio had relatively small cell bodies (range from 4.49 to 7.44 μm in length, from 1.12 to 1.78 μm in width) (Figs. 24.14, 24.15, and 24.16). Their thin and long Tps (range from 5.71 to 31.75 μm) were projecting from the cell body (Figs. 24.14, 24.15, and 24.16). The morphological features of TPs were identified: very long and thin cellular prolongations, with uneven caliber (moniliform aspect) (Fig. 24.14). Inferior vena cava TCs were surrounded by a great deal of collagen fibers and adjacent to the capillary (Figs. 24.15 and 24.16). Figure 24.3 displayed the typical morphology of cells with very long, thin, and moniliform Tps.

Conventionally, the inferior vena cava is formed by the joining of the left and right common iliac veins and carries deoxygenated blood from the lower half of the body into the right atrium. The inferior vena cava belongs to the large-sized vein, in which the tunica media is relatively thin and the tunica adventitia relatively thick. The tunica intima of the large vein consists of an endothelial lining with its basal lamina, a small amount of subendothelial connective tissue and some smooth muscle cells. The tunica media is relatively thin and contains circumferentially arranged smooth muscle cells, collagen fibers. The tunica adventitia of large veins is the thickest layer of the vessel wall, along with the usual collagen and elastic fiber and fibroblasts. The tunica adventitia also contains longitudinally disposed smooth muscle cells. Our research indicated that telocyte located in subendothelial layer is

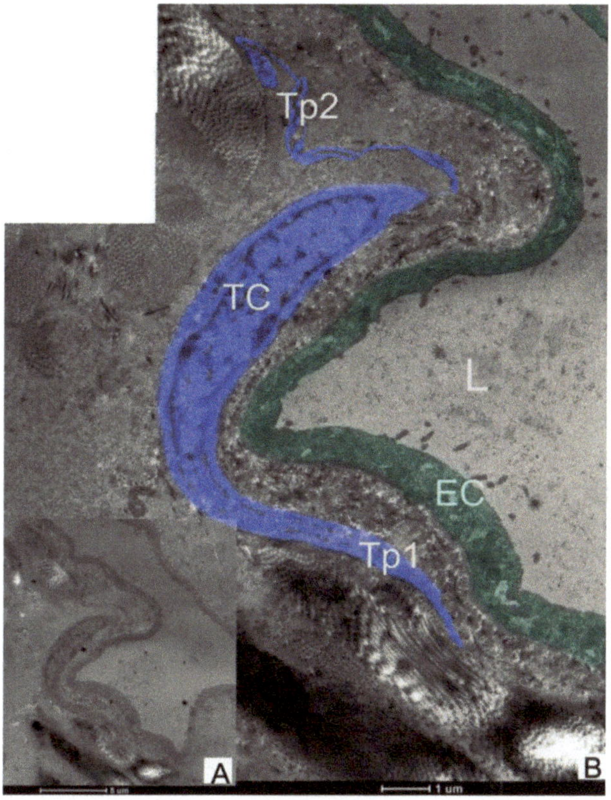

Fig. 24.14 TEM image shows the spatial relationship between telocytes (TCs) and collagen fibers of the inferior vena cava. (**a**) The strategic position of TCs, adjacent to the capillary, is visible; bar = 5 μm. (**b**) The higher magnification of Fig. a shows that TCs coexisted with collagen fibers in the inferior vena cava and were in close vicinity with capillary *EC* endothelial cell; bar = 1 μm

regarded as a kind of new cell element. The morphology of telocytes in the aorta is similar to that of the inferior vena cava, venule, arteriole, and capillary. But the location of telocytes in both the aorta and inferior vena cava is apparently different. The telocytes in the aorta are in the tunica adventitia and close to the external elastic lamina; nevertheless, the telocyte in the inferior vena cava is located in the subendothelial layer in tunica intima. In conclusion, inferior vena cava TCs, as a population of presumed accessory cells for stem cells [13, 34] and capillaries, might play an important role in maintaining the homeostasis including angiogenesis, regeneration, and reparation programming during inferior vena cava injury. Although the accurate functional association of inferior vena cava TCs needs further exploration, when all these would be confirmed, the corresponding clinical application prospect would be fully broad.

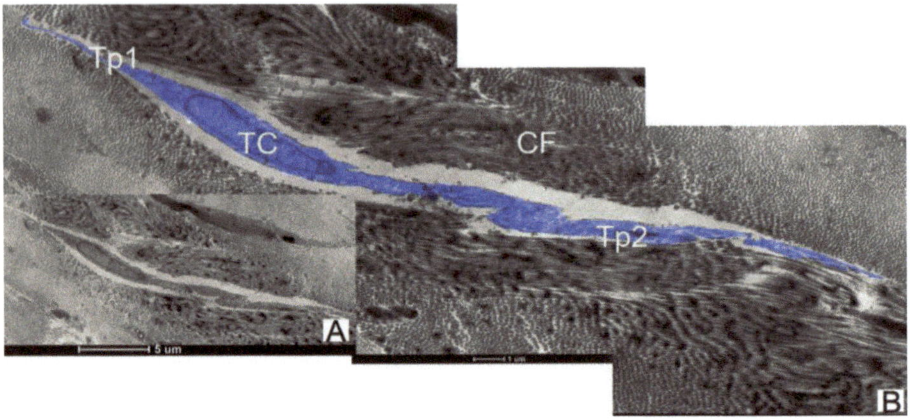

Fig. 24.15 Transmission electron microscope (TEM) images of the canine inferior vena cava. (**a**) One typical telocyte (*TC*) in the inferior vena cava is visible; bar = 5 μm. (**b**) It is the image of high power of figure (**a**). The TC with small spindly body extended specially long and thin telopodes (*Tps*). The TC body was 6.71 μm in length and 1.29 μm in the average width in size. And the measured length of Tp1 and Tp2 is 5.71 μm and 19.71 μm, respectively. *CF* collagen fiber, bar = 1 μm

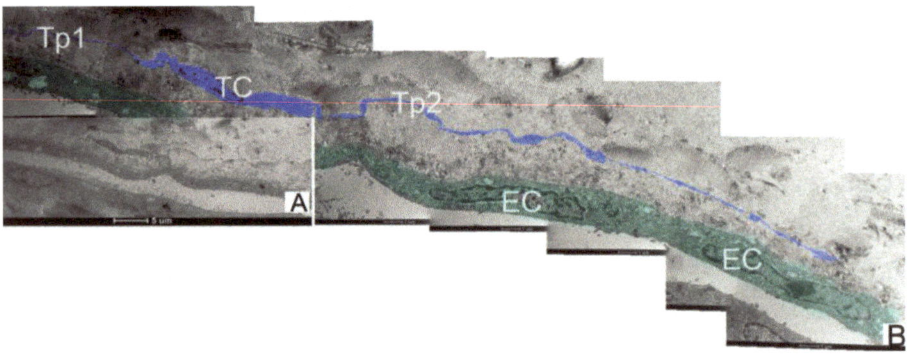

Fig. 24.16 TEM image shows the topographic localization between TC and the capillary of the inferior vena cava. (**a**) Telocytes (TCs) are located in the inferior vena cava, surrounded by collagen fibers, adjacent to the capillary; bar = 5 μm. (**b**) The morphological feature of the Tps with thin segments (podomers) and dilations (podoms) is shown. Tps, characteristic cellular prolongations of inferior vena cava TCs, were special long (about 31.75 nm) and moniliform forms, accommodating mitochondria, endoplasmic reticulum, and vesicles in podoms. *CF* collagen fiber, bar = 1 μm

24.7 Summary

Telocytes are a distinct type of interstitial cells characterized by a small cell body and extremely long and thin telopodes (Tps). They have been discovered to be present in many tissues and organs (http://www.telocytes.com). In the cardiovascular system, they have been found in the epicardium, myocardium, endocardium, large-sized artery (aorta), large-sized vein (inferior vena cava), and connective tissue around the venule, arteriole, and capillaries. We are sure that it will be found in more and more organs and tissues in the coming time. Functionally, telocytes form a three-dimensional (3D) interstitial network by homocellular and heterocellular communication (between telocytes and capillaries, between telocytes and telocytes, between telocytes and nerve endings, and between telocytes and host cells) and are involved in the maintenance of tissue homeostasis. As important interstitial cells to guide or nurse putative stem and progenitor cells in stem cell niches in a spectrum of tissues and organs, TCs contribute to tissue repair and regeneration. The coming research should focus on searching for the specific markers of telocytes of different tissues and organs, exploring the method to isolate or separate telocytes from cardiovascular tissue or organs such as the heart, artery, or vein, through culturing telocytes independently and clarifying their biological features.

References

1. Popescu LM, Faussone-Pellegrini MS. TELOCYTES – a case of serendipity: the winding way from interstitial cells of Cajal (ICC), via interstitial Cajal-like cells (ICLC) to TELOCYTES. J Cell Mol Med. 2010;14(4):729–40.
2. Gherghiceanu M, Manole CG, Popescu LM. Telocytes in endocardium: electron microscope evidence. J Cell Mol Med. 2010;14(9):2330–4.
3. Popescu LM, Fertig ET, Gherghiceanu M. Reaching out: junctions between cardiac telocytes and cardiac stem cells in culture. J Cell Mol Med. 2015.
4. Suciu L, Popescu LM, Regalia T, Ardelean A, Manole CG. Epicardium: interstitial Cajal-like cells (ICLC) highlighted by immunofluorescence. J Cell Mol Med. 2009;13(4):771–7.
5. Popescu LM, Manole E, Serboiu CS, Manole CG, Suciu LC, Gherghiceanu M, Popescu BO. Identification of telocytes in skeletal muscle interstitium: implication for muscle regeneration. J Cell Mol Med. 2011;15(6):1379–92.
6. Vannucchi MG, Traini C, Manetti M, Ibba-Manneschi L, Faussone-Pellegrini MS. Telocytes express PDGFRalpha in the human gastrointestinal tract. J Cell Mol Med. 2013;17(9):1099–108.
7. Cretoiu D, Cretoiu SM, Simionescu AA, Popescu LM. Telocytes, a distinct type of cell among the stromal cells present in the lamina propria of jejunum. Histol Histopathol. 2012;27(8):1067–78.
8. Alunno A, Ibba-Manneschi L, Bistoni O, Rosa I, Caterbi S, Gerli R, Manetti M. Telocytes in minor salivary glands of primary Sjogren's syndrome: association with the extent of inflammation and ectopic lymphoid neogenesis. J Cell Mol Med. 2015;19(7):1689–96.
9. Hinescu ME, Ardeleanu C, Gherghiceanu M, Popescu LM. Interstitial Cajal-like cells in human gallbladder. J Mol Histol. 2007;38(4):275–84.
10. Nicolescu MI, Popescu LM. Telocytes in the interstitium of human exocrine pancreas: ultrastructural evidence. Pancreas. 2012;41(6):949–56.
11. Popescu LM, Ciontea SM, Cretoiu D. Interstitial Cajal-like cells in human uterus and fallopian tube. Ann N Y Acad Sci. 2007;1101:139–65.
12. Liu J, Cao Y, Song Y, Huang Q, Wang F, Yang W, Yang C. Telocytes in liver. Curr Stem Cell Res Ther. 2015.
13. Popescu LM, Gherghiceanu M, Suciu LC, Manole CG, Hinescu ME. Telocytes and putative stem cells in the lungs: electron microscopy, electron tomography and laser scanning microscopy. Cell Tissue Res. 2011;345(3):391–403.

14. Zheng Y, Li H, Manole CG, Sun A, Ge J, Wang X. Telocytes in trachea and lungs. J Cell Mol Med. 2011;15(10):2262–8.
15. Hinescu ME, Gherghiceanu M, Suciu L, Popescu LM. Telocytes in pleura: two- and three-dimensional imaging by transmission electron microscopy. Cell Tissue Res. 2011;343(2):389–97.
16. Qi G, Lin M, Xu M, Manole CG, Wang X, Zhu T. Telocytes in the human kidney cortex. J Cell Mol Med. 2012;16(12):3116–22.
17. Gevaert T, De Vos R, Van Der Aa F, Joniau S, van den Oord J, Roskams T, De Ridder D. Identification of telocytes in the upper lamina propria of the human urinary tract. J Cell Mol Med. 2012;16(9):2085–93.
18. Gherghiceanu M, Popescu LM. Interstitial Cajal-like cells (ICLC) in human resting mammary gland stroma. Transmission electron microscope (TEM) identification. J Cell Mol Med. 2005;9(4):893–910.
19. Popescu LM, Ciontea SM, Cretoiu D, Hinescu ME, Radu E, Ionescu N, Ceausu M, Gherghiceanu M, Braga RI, Vasilescu F, et al. Novel type of interstitial cell (Cajal-like) in human fallopian tube. J Cell Mol Med. 2005;9(2):479–523.
20. Suciu L, Popescu LM, Gherghiceanu M, Regalia T, Nicolescu MI, Hinescu ME, Faussone-Pellegrini MS. Telocytes in human term placenta: morphology and phenotype. Cells Tissues Organs. 2010;192(5):325–39.
21. Ceafalan L, Gherghiceanu M, Popescu LM, Simionescu O. Telocytes in human skin – are they involved in skin regeneration? J Cell Mol Med. 2012;16(7):1405–20.
22. Li H, Lu S, Liu H, Ge J, Zhang H. Scanning electron microscope evidence of telocytes in vasculature. J Cell Mol Med. 2014;18(7):1486–9.
23. Chang Y, Li C, Gan L, Li H, Guo Z. Telocytes in the Spleen. PLoS One. 2015;10(9):e0138851.
24. Cantarero I, Luesma MJ, Junquera C. The primary cilium of telocytes in the vasculature: electron microscope imaging. J Cell Mol Med. 2011;15(12):2594–600.
25. Faussone Pellegrini MS, Popescu LM. Telocytes. Biomol Concepts. 2011;2(6):481–9.
26. Rusu MC, Nicolescu MI, Jianu AM, Lighezan R, Manoiu VS, Paduraru D. Esophageal telocytes and hybrid morphologies. Cell Biol Int. 2012;36(12):1079–88.
27. Pucovsky V, Moss RF, Bolton TB. Non-contractile cells with thin processes resembling interstitial cells of Cajal found in the wall of guinea-pig mesenteric arteries. J Physiol. 2003;552(Pt 1):119–33.
28. Harhun MI, Gordienko DV, Povstyan OV, Moss RF, Bolton TB. Function of interstitial cells of Cajal in the rabbit portal vein. Circ Res. 2004;95(6):619–26.
29. Harhun MI, Pucovsky V, Povstyan OV, Gordienko DV, Bolton TB. Interstitial cells in the vasculature. J Cell Mol Med. 2005;9(2):232–43.
30. Bobryshev YV. Subset of cells immunopositive for neurokinin-1 receptor identified as arterial interstitial cells of Cajal in human large arteries. Cell Tissue Res. 2005;321(1):45–55.
31. Pucovsky V. Interstitial cells of blood vessels. Sci World J. 2010;10:1152–68.
32. Al-Shboul OA. The importance of interstitial cells of cajal in the gastrointestinal tract. Saudi J Gastroenterol: Off J Saudi Gastroenterol Assoc. 2013;19(1):3–15.
33. Corselli M, Chen CW, Sun B, Yap S, Rubin JP, Peault B. The tunica adventitia of human arteries and veins as a source of mesenchymal stem cells. Stem Cells Dev. 2012;21(8):1299–308.
34. Gherghiceanu M, Popescu LM. Cardiomyocyte precursors and telocytes in epicardial stem cell niche: electron microscope images. J Cell Mol Med. 2010;14(4):871–7.
35. Bojin FM, Gavriliuc OI, Cristea MI, Tanasie G, Tatu CS, Panaitescu C, Paunescu V. Telocytes within human skeletal muscle stem cell niche. J Cell Mol Med. 2011;15(10):2269–72.
36. Zhang HQ, Lu SS, Xu T, Feng YL, Li H, Ge JB. Morphological evidence of telocytes in mice aorta. Chin Med J (Engl). 2015;128(3):348–52.

Hongqi Zhang PhD, born in 1959, postgraduate supervisor, got his Medical Doctor Degree in Fudan University in 2004. He worked as a professor at Xinxiang Medical University since 2001 and enrolled in the Department of Anatomy, Histology, and Embryology, School of Basic Medical Sciences, Fudan University, in 2006. He went to the Downstate Medical Center of State University of New York for his postdoctoral research from January of 2008 to May of 2009. Research interests mainly focus on mechanism of atherosclerosis and cardiovascular remodeling and recently major in exploring the origin and function of telocytes (TCs).

HONGQI ZHANG

Chapter 25
Juxtacerebral Tissue Regeneration Potential: Telocytes Contribution

Laura Cristina Ceafalan and Bogdan Ovidiu Popescu

Abstract It is well proved already that neurogenesis does take place in mammals' brain, including human brain. However, neurogenesis by itself is not able to compensate for brain tissue loss in serious neurological diseases, such as stroke, brain trauma or neurodegenerative disorders. Recent evidences show that neural stem cell niches are present not only in classical locations, such as subventricular or subgranular zones, but in other areas as well, including tissues contiguous to the brain (meninges and choroid plexus). In this chapter we revise the relationship of neural stem cells with interstitial cells (mainly telocytes), which we think is significant, and we describe what is known about the juxtacerebral tissue neurogenesis potential.

25.1 Adult Neurogenesis: Biological and Therapeutical Perspectives

Discovered by Joseph Altman more than half a century ago [1], adult neurogenesis represents the formation of new neurons and glial cells from stem cells (SCs) in brains of mature mammals. Basic features of stem cells are their capacity to both self-renew and differentiate [2]. The classical neural stem cell (NCS) populations seem to be embedded in the walls of cerebral ventricles (in the so-called

---

L.C. Ceafalan
Department of Cellular and Molecular Biology and Histology,
'Carol Davila' University of Medicine and Pharmacy, Bucharest, Romania

Department of Molecular Medicine and Neuroscience,
'Victor Babeş' National Institute of Pathology, Bucharest, Romania

B.O. Popescu (✉)
Department of Neurology, Colentina Clinical Hospital,
'Carol Davila' University of Medicine and Pharmacy, Bucharest, Romania

Department of Molecular Medicine and Neuroscience,
'Victor Babeş' National Institute of Pathology, Bucharest, Romania
e-mail: bogdan_ovidiu_popescu@yahoo.com

X. Wang, D. Cretoiu (eds.), *Telocytes*, Advances in Experimental Medicine and Biology, 913, DOI 10.1007/978-981-10-1061-3_25

subventricular zone (SVZ) and subgranular zone (SGZ)), and in small brains, these cells are able to proliferate and migrate through considerable distances [3]. Recently, a different neurogenic zone in the dorsal hippocampus was described – the subhippocampal zone (SHZ) [4]. A variety of experimental studies, performed mainly in rodents, suggested that different types of injuries, such as trauma or stroke, are able to trigger the proliferation of NCS progenitors [5]. However, whether secondary differentiation to specialized nervous cells occurs and how all this process is able to compensate for cell loss in brains are still to be answered [6]. Instead of literally replacing the lost cells in the central nervous system, it seems that NCS might favor neurological recovery by synthesis and release of trophic factors [7]. The brain exerts an important endogenous defense activity (EDA), and generation of new brain cells is part of EDA [8]. Nonetheless, neurogenesis is not efficacious in neurodegenerative diseases, such as Alzheimer's disease [9] or amyotrophic lateral sclerosis [10], apparently due to an alteration of cell cycle reentry and failure of cell differentiation [11, 12].

Starting from the considerations above, NCS could be interesting from at least two angles, namely, stimulation of endogenous neurogenesis and transplantation of autologous/allogeneic stem cells [13]. In order either to stimulate NCS or to facilitate integration of transplanted stem cells in the peculiar brain environment, understanding of regulation of stem cell niche activation and of maturation and migration of newly born neurons is essential. The NCS niches are governed by cell-cell interactions between NCS (A, B, and C types), ependymal cells (in SVZ), endothelial cells, microglial cells, and extracellular matrix molecules [14–16]. Growth factors, cytokines, neurotransmitters, and other signaling molecules such as NO regulate the niche activation or inhibition in SVZ, SGZ, and SHZ [4, 17–19].

25.2 Juxtacerebral Tissue as Neural Stem Cell Host

Besides classical location of NSC in mammalian adult brain ventricular walls, recently, progenitor cells were identified in meninges [20] and choroid plexus [21], which we called here and elsewhere *juxtacerebral tissues*, since they are nonneuronal structures contiguous to brain tissue. According to different experimental paradigms, mainly models of stroke and traumatic brain injury (TBI) in rodents, the stem cell niches hosted by these juxtacerebral tissues respond to brain injuries by intense activation, with cell multiplication, differentiation, and migration to the location of lesion [20]. Interestingly, similar NSC niches were identified in spinal cord meninges as well, even though its embryonic origin is different from telencephalic meninges, and these niches also react to injury [22]. Markers of NCS used usually in immunofluorescence or immunohistochemical methods are nestin [20], doublecortin [22], or PAX 3/7 [23, 24]. Moreover, cell explants from meningeal biopsies gave rise in culture to neurospheres [25]. Interestingly enough, pia mater folds within the brain, going along with every penetrating artery and creating the perivascular space. Therefore, the pial NSC niches are spread out in cerebral tissue much more than the ones from SVZ, SGZ, and SHZ. Further on, to be noted is the proximity of the choroid plexus to the SVZ, actually a subependymal zone. Signaling through CSF might modulate the activity of stem cell niches from choroid plexus and meninges.

25.3 Telocytes: Peculiar Interstitial Cells in Juxtacerebral Stem Cell Niches

Telocytes (TCs) are peculiar interstitial cells which are found in virtually all mammalian tissues, characterized by a small cell body and a few very long, moniliform cell processes, called telopodes [26]. We demonstrated in previous reports that in different tissues, telocytes are in contact with many other cell partners, at the level of stem cell niches [27–29]. We also showed, for the first time, that telocytes are present both in meninges and choroid plexus in rat and mouse brain [23]. By electron microscopy (EM) and immunofluorescence, we demonstrated that telocytes are in close contact with putative stem cells [23]. In Fig. 25.1 we show such an example of telocytes at the junction between arachnoid and pia mater, in the proximity of

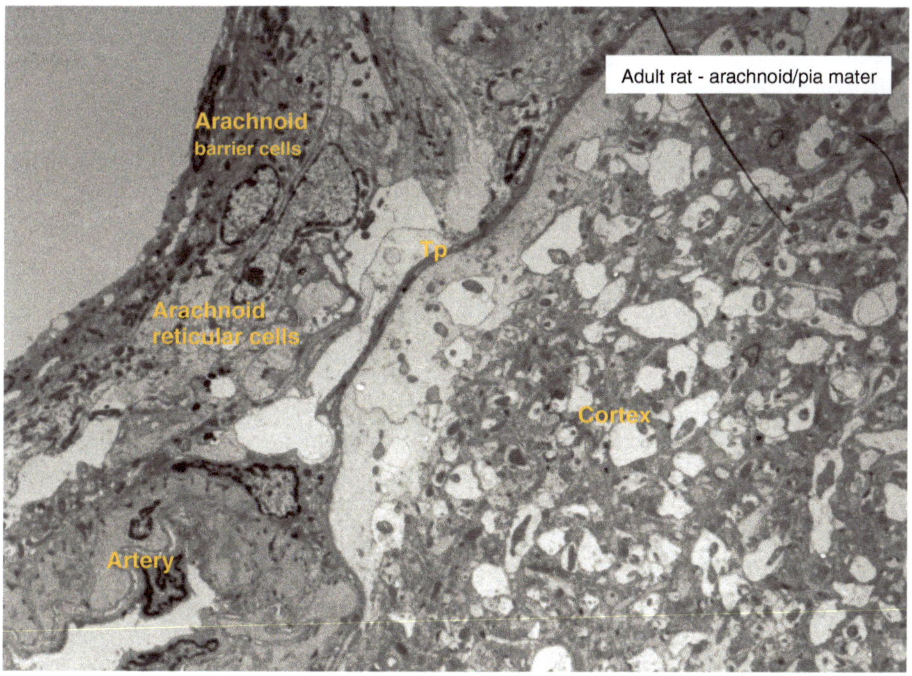

Fig. 25.1 Adult rat meninges – transmission electron microscopy. A telocyte with a long telopode (Tp) is identified at the arachnoid/pia mater border, in the vicinity of an artery. Arachnoid cells and the cortex are marked as well (Courtesy of Dr. Mihaela Gherghiceanu, Laboratory of Ultrastructural Pathology, 'Victor Babeş' National institute of Pathology)

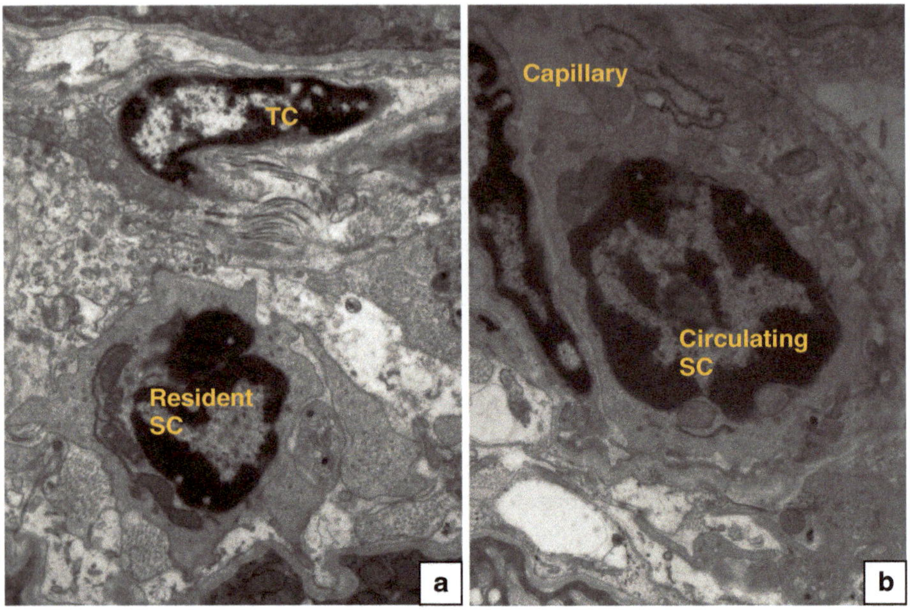

Fig. 25.2 Adult rat choroid plexus – transmission electron microscopy. (**a**) A telocyte (TC) is in contact proximity to a resident stem cell (SC), in a niche. (**b**) A circulating stem cell (SC) is identified at the level of a capillary (Courtesy of Dr. Mihaela Gherghiceanu, Laboratory of Ultrastructural Pathology, 'Victor Babeş' National institute of Pathology)

cortex and cortical penetrating vessels (EM). In Fig. 25.2a please note a telocyte in contact with a resident stem cell and in Fig. 25.2b a circulating stem cell in a capillary, at the level of adult rat choroid plexus. In this light, telocytes might participate in regulation of NSC niche activation in juxtacerebral reservoirs.

25.4 Future Perspectives

The recent identification of NSC niches in juxtacerebral structures opens an interesting new treatment target both for acute conditions (such as stroke and TBI) and neurodegenerative disorders (such as Alzheimer's and Parkinson's diseases). Both choroid plexus and meninges are in contact with cerebrospinal fluid, so one therapeutic possibility would be to deliver endogenous neurogenesis stimulants, such as growth factors, intrathecally. A different approach could base on meningeal biopsy, multiplication of stem cells and interstitial cells in cell cultures, and injection at the affected sites.

Acknowledgments We do thank Dr. Mihaela Gherghiceanu for providing us the EM figures and for the fruitful discussions and advice.

References

1. Altman J. Are new neurons formed in the brains of adult mammals? Science. 1962;135:1127–8.
2. Lajtha LG, Schofield R. Regulation of stem cell renewal and differentiation: possible significance in aging. Adv Gerontol Res. 1971;3:131–46.
3. Paredes MF, Sorrells SF, Garcia-Verdugo JM, Alvarez-Buylla A. Brain size and limits to adult neurogenesis. J Comp Neurol. 2016;524:646–64.
4. Belmadani A, Ren D, Bhattacharyya BJ, Rothwangl KB, Hope TJ, Perlman H, Miller RJ. Identification of a sustained neurogenic zone at the dorsal surface of the adult mouse hippocampus and its regulation by the chemokine SDF-1. Hippocampus. 2015;25:1224–41.
5. Yu TS, Washington PM, Kernie SG. Injury-induced neurogenesis: mechanisms and relevance. Neuroscientist. 2016;22:61–71.
6. Kernie SG, Parent JM. Forebrain neurogenesis after focal Ischemic and traumatic brain injury. Neurobiol Dis. 2010;37:267–74.
7. Wang L, Chopp M, Gregg SR, Zhang RL, Teng H, Jiang A, Feng Y, Zhang ZG. Neural progenitor cells treated with EPO induce angiogenesis through the production of VEGF. J Cereb Blood Flow Metab. 2008;28:1361–8.
8. Muresanu DF, Popa-Wagner A, Stan A, Buga AM, Popescu BO. The vascular component of Alzheimer's disease. Curr Neurovasc Res. 2014;11:168–76.
9. Winner B, Winkler J. Adult neurogenesis in neurodegenerative diseases. Cold Spring Harb Perspect Biol. 2015;7:a021287.
10. Hugnot JP, Franzen R. The spinal cord ependymal region: a stem cell niche in the caudal central nervous system. Front Biosci (Landmark Ed). 2011;16:1044–59.
11. Enciu AM, Nicolescu MI, Manole CG, Mureşanu DF, Popescu LM, Popescu BO. Neuroregeneration in neurodegenerative disorders. BMC Neurol. 2011;11:75.
12. Brocardo PS, Patten A, Gil-Mohapel J. Altered adult neurogenesis in neurodegenerative diseases. Int J Med Biol Front. 2012;18:469–520.
13. Haidet-Phillips AM, Maragakis NJ. Neural and glial progenitor transplantation as a neuroprotective strategy for Amyotrophic Lateral Sclerosis (ALS). Brain Res. 2015;1628(Pt B):343–50.
14. Alvarez-Buylla A, Garcia-Verdugo JM. Neurogenesis in adult subventricular zone. J Neurosci. 2002;22:629–34.
15. Mothe AJ, Tator CH. Proliferation, migration, and differentiation of endogenous ependymal region stem/progenitor cells following minimal spinal cord injury in the adult rat. Neuroscience. 2005;131:177–87.
16. Lin R, Iacovitti L. Classic and novel stem cell niches in brain homeostasis and repair. Brain Res. 2015;1628(Pt B):327–42.
17. Sun W, Sun C, Zhao H, Lin H, Han Q, Wang J, Ma H, Chen B, Xiao Z, Dai J. Improvement of sciatic nerve regeneration using laminin-binding human NGF-beta. PLoS One. 2009;4:e6180.
18. Ma DK, Kim WR, Ming G, Song H. Activity-dependent extrinsic regulation of adult olfactory bulb and hippocampal neurogenesis. Ann N Y Acad Sci. 2009;1170:664–73.
19. Decimo I, Fumagalli G, Berton V, Krampera M, Bifari F. Meninges: from protective membrane to stem cell niche. Am J Stem Cells. 2012;1:92–105.
20. Nakagomi T, Molnár Z, Nakano-Doi A, Taguchi A, Saino O, Kubo S, Clausen M, Yoshikawa H, Nakagomi N, Matsuyama T. Ischemia-induced neural stem/progenitor cells in the pia mater following cortical infarction. Stem Cells Dev. 2011;20:2037–51.
21. Falcão AM, Marques F, Novais A, Sousa N, Palha JA, Sousa JC. The path from the choroid plexus to the subventricular zone: go with the flow! Front Cell Neurosci. 2012;6:34.
22. Decimo I, Bifari F, Rodriguez FJ, Malpeli G, Dolci S, Lavarini V, Pretto S, Vasquez S, Sciancalepore M, Montalbano A, Berton V, Krampera M, Fumagalli G. Nestin- and doublecortin-positive cells reside in adult spinal cord meninges and participate in injury-induced parenchymal reaction. Stem Cells. 2011;29:2062–76.

23. Popescu BO, Gherghiceanu M, Kostin S, Ceafalan L, Popescu LM. Telocytes in meninges and choroid plexus. Neurosci Lett. 2012;516:265–9.
24. Buckingham M, Relaix F. PAX3 and PAX7 as upstream regulators of myogenesis. Semin Cell Dev Biol. 2015;44:115–25.
25. Bifari Fl, Decimo I, Chiamulera C, Bersan E, Malpeli G, Johansson J, Lisi V, Bonetti B, Fumagalli G, Pizzolo G, Krampera M. Novel stem/progenitor cells with neuronal differentiation potential reside in the leptomeningeal niche. J Cell Mol Med. 2009;13(9B):3195–208.
26. Hinescu ME, Gherghiceanu M, Suciu L, Popescu LM. Telocytes in pleura: two- and three-dimensional imaging by transmission electron microscopy. Cell Tissue Res. 2011;343: 389–97.
27. Popescu LM, Gherghiceanu M, Suciu LC, Manole CG, Hinescu ME. Telocytes and putative stem cells in the lungs: electron microscopy, electron tomography and laser scanning microscopy. Cell Tissue Res. 2011;345:391–403.
28. Popescu LM, Manole E, Serboiu CS, Manole CG, Suciu LC, Gherghiceanu M, Popescu BO. Identification of telocytes in skeletal muscle interstitium: implication for muscle regeneration. J Cell Mol Med. 2011;15:1379–92.
29. Suciu LC, Popescu BO, Kostin S, Popescu LM. Platelet-derived growth factor receptor-β-positive telocytes in skeletal muscle interstitium. J Cell Mol Med. 2012;16:701–7.

Chapter 26
Telocytes of Fascial Structures

Joanna Dawidowicz, Natalia Matysiak, Sylwia Szotek, and Krzysztof Maksymowicz

Abstract Currently, the exact role of telocytes within fascial structures is unknown. The morphology, distribution and behaviour of fascial telocytes as well as the mutual relationship between telocytes and other cellular fascia constituents should be definitely a subject of further studies. It will contribute to better understanding of the role of the fascial system in health and diseases, may shed light on the regeneration potential of these tissues and may help to find targets for future treatments for locomotor disorders, including fascial diseases. Last but not least, confirmation of the presence of telocytes within fascia may contribute to optimise the use of fascia as a graft material.

J. Dawidowicz (✉)
Veterinary Clinic "Brynów", Brynowska 25c, Kolibrów 15/5, Katowice 40-585, Poland
e-mail: joanna@dawidowicz.pl

N. Matysiak
Department of Histology and Embryology, School of Medicine with the Division of Dentistry in Zabrze, Medical University of Silesia, Zabrze 41-800, Poland
e-mail: natalia.matysiak@o2.pl

S. Szotek
Department of Biomedical Engineering, Mechatronics and Theory of Mechanisms, Faculty of Mechanical Engineering, Wrocław University of Science and Technology, Łukasiewicza 7/9, Wrocław 50-371, Poland
e-mail: sylwia.szotek@pwr.edu.pl

K. Maksymowicz
Department of Forensic Medicine, Faculty of Medicine, Wroclaw Medical University, J. Mikulicza-Radeckiego 4, Wrocław 50-368, Poland
e-mail: krzysztof.maksymowicz@umed.wroc.pl

X. Wang, D. Cretoiu (eds.), *Telocytes*, Advances in Experimental Medicine and Biology, 913, DOI 10.1007/978-981-10-1061-3_26

26.1 Introduction

It has only been several months since the presence of telocytes was described in fascial structures (human fascia lata of the thigh) [14]. Let us take a closer look at the fascial system of the body to try to understand the potential role telocytes might play in the physiology and pathology of these interesting and ubiquitous tissues.

26.1.1 Fascial Structures: What They Are and What They Do

The term "fascia" comes from Latin and means "band of material" [28]. However, it should be emphasised that fasciae are not just simple bands connecting joints and muscles. As a matter of fact, these structures form a sophisticated, anatomical and functional three-dimensional network surrounding all structures of the body, protecting them and enhancing their function [5, 18, 29, 64]. One might even be tempted to conclude that normal movement of the body is only possible because of fascial tissues. It would not be too much of an exaggeration to say that fascial tissues are omnipresent, extending literally "from head to toe" [31], and serial anatomical continuity of different fasciae was demonstrated (for instance, gluteal fascia continues with the fascia lata, the crural fascia and finally the plantar fascia). This continuity provides not only anatomical but also functional integration of all fascial structures in the body [62].

Under pathological conditions, fasciae undergo both functional and structural changes that manifest themselves as any of a wide range of fascial diseases. This group of diseases is often difficult to diagnose and treat. The main reason of diagnostic difficulties is the fact that clinical symptoms of these disorders are to a great extent nonspecific and subjectively experienced by the patient, with persistent pain of varying severity and a marked reduction in the normal range of movement usually being the main complaints.

Some fascial structures are commonly used as valuable graft material. They are used in many different fields of reconstructive and plastic surgery [16, 72]. Fascial grafts, for instance, are applied in tracheal reconstruction, surgical treatment of aortic aneurysm, glaucoma surgery [36], urological reconstructive surgery [68] and repair of different kinds of hernia.

Despite the crucial role of fasciae, these structures are relatively poorly investigated, especially compared with other elements of the broadly defined locomotor system. This might be a consequence of a general misconception that fasciae are just uncomplicated and non-specialised structures, whose only role is to passively assist muscle and bone activity [55]. Fortunately, recently fascial structures have become an increasingly important subject of investigation among scientists and clinicians [69].

26.1.2 Fascial Terminology and Heterogeneity

Despite a rise in the number of studies concerning fascial structures, their official defining terminology, and especially the general classification system, is still being debated [9, 30, 32, 57]. According to the Federative Committee on Anatomical Terminology (FCAT), fasciae can be classified based on their location. Following this classification, one can distinguish fascia of the head and neck, fascia of the trunk and fascia of limbs. Another slightly different approach is to classify fasciae based on their relation to the surrounding structures into subcutaneous fascia, fascia of muscles, visceral fascia, parietal fascia and fascia extraserosalis [1]. Fasciae are also named based on their given topographic location (thoracolumbar fascia, crural fascia and plantar fascia) and sometimes even on their morphological character (fascia lata).

The aforementioned approaches, despite being wildly used in practice, do not take into account microscopic fascial features or the diversity of its functional characteristics [31]. As an answer to this problem, Kumka proposed a functional classification system, which distinguishes four main categories of fascial structures: linking, fascicular, compression and separating fasciae. Compression fasciae are represented, e.g. by fascia lata, crural fascia, brachial fascia and antebrachial fascia, while thoracolumbar fascia belongs to the group of linking fasciae and plantar fascia belongs to passive fasciae [31]. Each region of the body has a complex mixture of these different fascial types.

Standard anatomical approach distinguishes superficial, deep and visceral fasciae. The superficial fascia (*tela subcutanea*) is located directly beneath the skin. The deep fascia (*fascia profunda*) typically envelops muscles, adhering to them more or less tightly. However, it should be stressed that superficial or deep fascia may be locally absent. From the histological point of view, fascial structure constitutes loose connective tissue and dense connective tissue. Most often, however, they are a combination of these two tissue types (Fig. 26.1).

The superficial fascia is generally considered loose connective tissue while the deep fascia is classified as mostly dense, irregular connective tissue. The histological distinction between the superficial and deep fascia is not always possible and clear because the superficial fascia can occasionally contain sheets of dense connective tissue (and even muscle tissue, e.g. muscles of facial expression). Also, the deep fascia is not always synonymous with dense connective tissue because it may contain varying (sometimes significant) amounts of loose connective tissue [33]. Moreover, under some conditions, when the deep fascia (e.g. plantar fascia) becomes a subject of marked compression, its cells can undergo chondrocytic differentiation (fibrocartilaginous tissue).

It could be concluded that, at present, no unified and clear fascia classification systems exist. Below we have described the telocyte populations of four examples of fascial structures: fascia lata, thoracolumbar fascia, crural fascia and plantar fascia. The above choice was dictated by the fact that selected fasciae are often affected

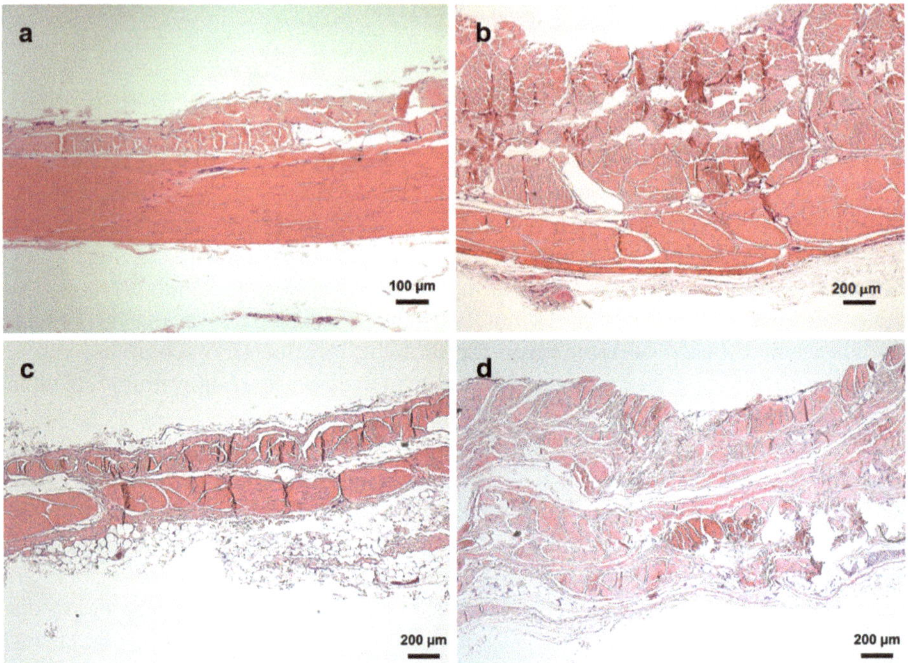

Fig. 26.1 Histological images showing layered organisation of collagen bundles of human fasciae: (**a**) fascia lata, (**b**) plantar fascia, (**c**) thoracolumbar fascia and (**d**) crural fascia; note the alternating arrangement of dense and loose connective tissue layers. H&E staining. Scale bars: (**a**) 100 μm; (**b**–**d**) 200 μm

in the course of painful, troublesome and sometimes severe fascial disorders. Moreover, the fascia lata, due to its durability, elasticity and relative ease of harvesting, is most often used as a comprehensive graft material.

26.1.3 Common Fascial Disorders

Fascial disorders are a heterogeneous group of diseases, whose pathogenesis remains largely unknown. Overuse, inflammation, autoimmune processes as well as genetic factors are listed among possible etiologies. As was mentioned before, the most common features of fascial diseases are pain and a reduction of the range of motion. Many patients may also report localised or generalised stiffness and discomfort. It should be stressed that no specific diagnostic markers of fascial disorders exist and that most often they are "diagnoses of exclusion". The best known examples of fascial disorders are plantar fasciitis and fibromatosis (*Morbus Ledderhose*), proximal iliotibial band (ITB) syndrome, Dupuytren's contracture, congenital fascial dystrophy, compartment syndromes and fibromyalgia [4, 57].

26.2 Morphology and Characterisation of Fascial Telocytes: Are Fascial Telocytes Somehow Different?

Telocytes are recently discovered cells involved in a number of essential biological processes [13, 19, 45]. So far, the presence of telocytes has been confirmed within different organs and tissues [13]. However, it has only been several months since the presence of telocytes was described in human fascial structures [14]. As it was mentioned above, fascial structures are forming a ubiquitous, anatomical and functional network within the body. Taking into consideration that at the same time telocytes are forming a unique, three-dimensional network in the stromal compartment of many different tissues [19, 45, 70], it may be assumed that these two networks must be strictly interwoven and interdependent, creating a complex, sophisticated and integrative "network in network" system (Fig. 26.2).

Fig. 26.2 Fascia–telocytes "network in network" system (Photo courtesy of Cedron M. Sterling, L.M.P. http://www.CedronSterling.com, modified by the authors)

Even though telocyte presence within a given tissue could be confirmed using several techniques, such as immunofluorescence, immunostaining or proteomics and genetic methods, transmission electron microscopy (TEM) remains a valuable diagnostic tool in identifying this distinct cell population [7, 10, 12, 74]. Characteristic morphological features of telocytes are clearly visible under a transmission electron microscope, which allows them to be distinguished from other kinds of cells, especially fascial fibroblasts and fascial nerve cells [47]. In general, fascial telocytes morphologically resemble telocytes of other tissues (Fig. 26.3). As observed in TEM images, fascial telocytes exhibit a small, oval-shaped or sometimes triangular cellular body of an average size 2.4±0.5 μm/5±1.7 μm, with the nucleus occupying about 25-35 % of the cellular volume.

The cytoplasm is generally sparse and the perinuclear region is usually rich in mitochondria. The cytoplasm also contains a Golgi complex, elements of rough and smooth endoplasmic reticulum and cytoskeletal elements. However, obviously the most characteristic morphological feature of these cells is the presence of a different number (up to four) of very long processes – telopodes (Fig. 26.3). It is interesting to note that, except for the axons of some types of neurons, telopodes are probably the longest cellular prolongations in the human body. The observed fascial telopodes achieved a maximum length of 22 μm. Fascial telocytes exhibit typical telopodes – with the presence of dilated podoms harbouring the mitochondria, endoplasmic reticulum as well as caveolae and thin podomers [45]. Most often, telopodes of fascial tissues form a complex network due to their branching and overlapping.

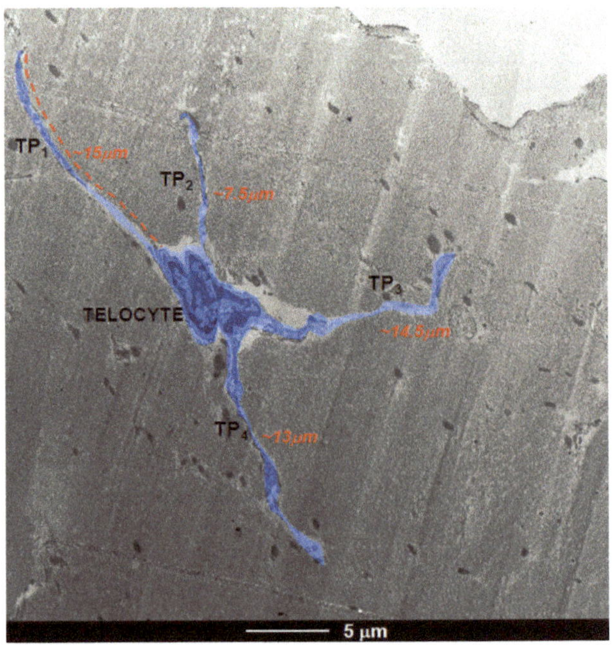

Fig. 26.3 Overview of a fascia lata telocyte; note the relatively small cell body and 4 long characteristic processes – telopodes (TP_{1-4}); a digitally coloured (*blue*) transmission electron microscope image. Scale bar: 5 μm (Reproduced with permission from Dawidowicz et al. [14])

26.2.1 Telocytes of the Fascia Lata

The fascia lata is an integral part of the fascial system of the body. It envelops the entire thigh and hip area and its main tasks are to reduce tissue friction, protect underlying structures from injury, encourage venous return and, most importantly, transmit mechanical forces generated by the musculoskeletal system of the lower extremity [5, 18, 29, 64]. The fascia lata is also important because this structure relatively often becomes a target for pathological processes resulting from repetitive traumas or overuse. The best known example of a disease affecting the fascia lata is the ITB syndrome [5, 27, 60]. It is most common in athletes or older overweight females and its main clinical symptoms are pain and tenderness at the iliac tubercle area [60]. The proximal ITB syndrome is often underdiagnosed and, consequently, difficult to treat.

From the histological perspective, the fascia lata is a poorly cellular structure (Fig. 26.1a) with the cellular components represented mainly by fibroblasts, fibrocytes and lower numbers of mast cells, adipocytes, single cells showing features of myofibroblastic differentiation and, as recently observed, telocytes. Telocytes populating the fascia lata (compared to telocytes of other fascial structures) have been examined most intensively. Fascia lata telocytes are rather scarce (0–2 cells per FOV) when the number of cells is expressed as cells per field of view – FOV (FOV = 7.065 mm^2 – the surface of the TEM grid of 3 mm diameter). As far as morphological features are concerned, they exhibit the presence of typical telopodes with podomers alternating with much thicker podoms (Figs. 26.4 and 26.5).

Each fascia lata telocyte had several (up to four) telopodes (Fig. 26.3) of the maximum length of about 22 µm in the place of the section (Fig. 26.6).

The size of the podomers was slightly thicker compared to the thickness of adjacent collagen fibres (Fig. 26.6). Within some of the podoms, significant accumulations of mitochondria were present (Figs. 26.4 and 26.5). Telopodes form a kind of semi-circular arrangement (Fig. 26.7), dichotomous branching (Fig.26.5a) or strongly convoluted three-dimensional structures (Fig. 26.4).

Within the fascia lata, both homocellular (telocyte-to-telocyte) junctions (Fig. 26.4) and heterocellular (telocyte to other cells) junctions were observed (Fig. 26.8).

This finding is consistent with previous studies concerning other tissues and organs, showing that telocytes frequently establish close contacts not only with each other but also with other kinds of cells [11, 26, 42, 43, 47, 48, 52, 53, 65].

26.2.2 Telocytes of the Plantar Fascia

The plantar fascia originates on the bottom surface of the calcaneus and extends along the sole of the foot towards the toes. Its main roles are to transmit forces from the hind to the forefoot and maintain the medial longitudinal arch. The presence of

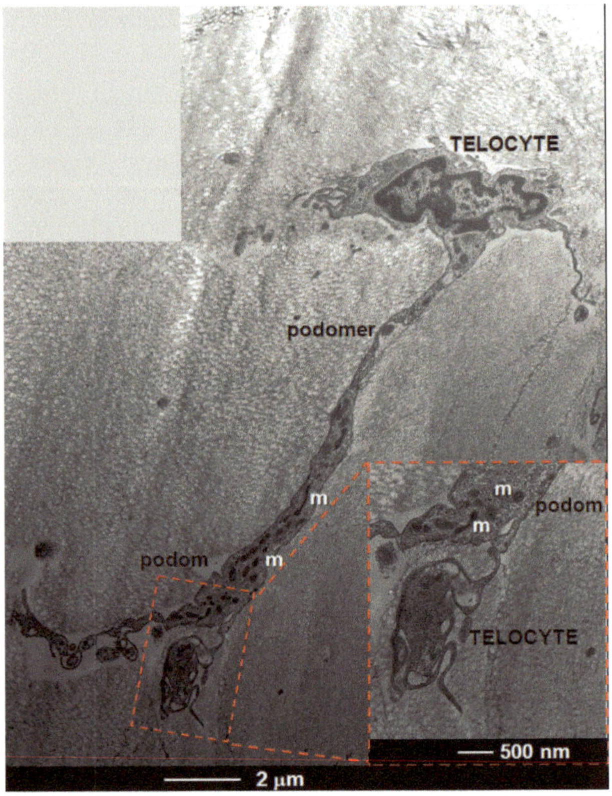

Fig. 26.4 Transmission electron microscopy images of telocytes of the fascia lata; note the large telopode extending from the cell body of a telocyte with podomers alternating with much thicker podoms containing abundant mitochondria (*m*); the inset shows a part of the telocyte with a fragment of the telopode forming a circular, convoluted appearance. Scale bars: 2 μm; 0.5 μm (*inset*) (Reproduced with permission from Dawidowicz et al. [14])

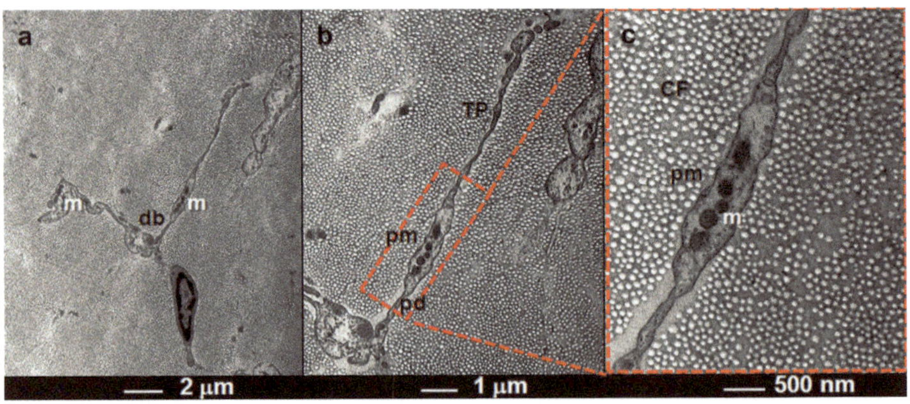

Fig. 26.5 (**a**) Electron micrographs of a telocyte in the fascia lata with the presence of typical telopodes (TP), forming dichotomous branching (*db*) and consisting of podomers (*pd*) alternating with podoms (*pd*); (**b**–**c**) the insets show the podom at higher magnification, with clearly visible focal accumulation of mitochondria (*m*). Scale bars: (**a**) 2 μm; (**b**) 1 μm; (**c**) 0.5 μm (Reproduced with permission from Dawidowicz et al. [14])

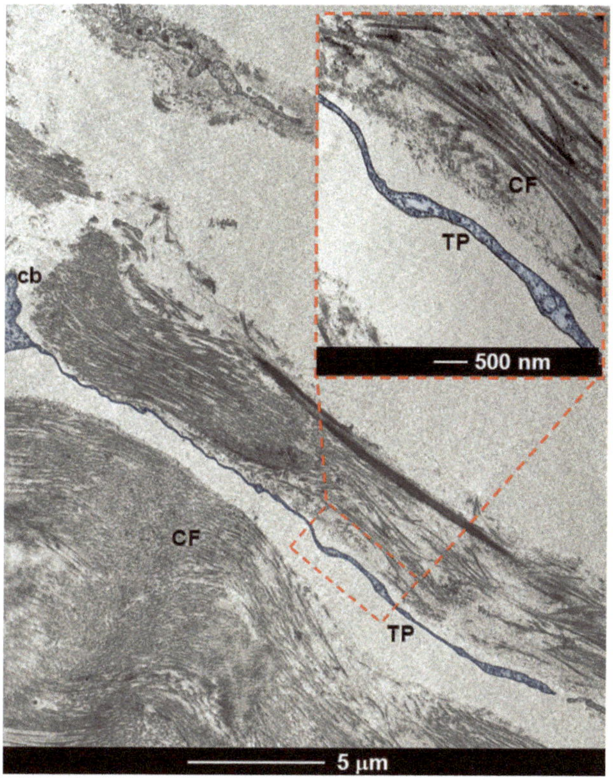

Fig. 26.6 Telocyte of the fascia lata with a small part of the cell body (*cb*) and the very long telopode (*TP*) situated between collagen fibres (*CF*); the inset shows higher magnification of a fragment of a telopode that is slightly thicker compared to the thickness of neighbouring collagen fibres (longitudinal sections); digitally coloured (*blue*) electron micrographs. Scale bars: 5 μm; 0.5 μm (inset) (Reproduced with permission from Dawidowicz et al. [14])

Pacinian and Ruffini corpuscles suggests that plantar fascia has a role in proprioception, stability and control of foot movement. It carries 14 % of the total load of the foot [62]. The plantar fascia is investigated more frequently than the other fascial structures not only because it plays a crucial role in foot biomechanics but mainly because it is involved in several severe and difficult-to-treat pathologies. One of the main diseases affecting the plantar fascia is plantar fasciitis. This disorder is particularly common in runners and is considered an overuse syndrome caused by repetitive stress and excessive pronation of the foot. An especially interesting plantar fascia disease is plantar fibromatosis, also known as Ledderhose's disease. Similarly to palmar fibromatosis (Dupuytren's disease) and penile fibromatosis (Peyronie's disease), it belongs to superficial fibromatoses. It is characterised by local proliferation of abnormal fibrotic tissue in the plantar fascia. Clinically, in patients suffering from Ledderhose's disease, one can observe slow-growing nodules most often found in the central and medial portions of the plantar fascia.

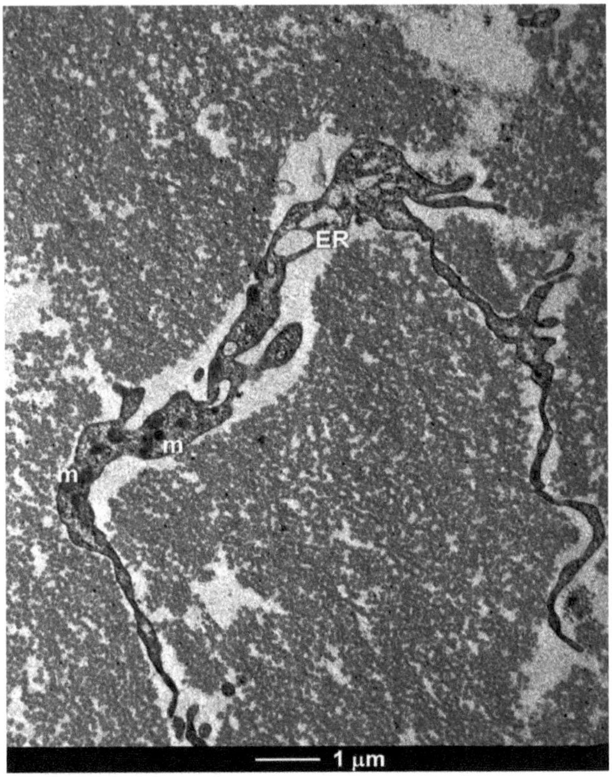

Fig. 26.7 Electron micrograph of a section of a fascia lata telocyte; note the telopode forming a semi-circular arrangement with focal accumulations of mitochondria (*m*) and endoplasmic reticulum elements (*ER*). Scale bar: 1 μm (Reproduced with permission from Dawidowicz et al. [14])

The plantar fascia is composed mainly of dense connective tissue (Fig. 26.1b). Telocytes populating the plantar fascia morphologically resemble those of the fascia lata (Fig. 26.9) except they exhibit abundant glycogen granule (Fig. 26.10).

It could explain at least to some extent the fact why the plantar fascia is very susceptible to glycation and oxidation, which results in its changes and thickening in the course of diabetes mellitus.

Moreover, telocytes of the plantar fascia form clearly visible homocellular junctions more frequently compared to telocytes populating other fascia types (Figs. 26.10 and 26.11). These cells junctions exhibited mainly the morphology of *macula adherens* (spot desmosomes) and sometimes they were present in a close proximity to each other.

Telocytes communicate with other cells not only by forming junctions (e.g. spot desmosomes, atypical junctions or nano-contacts) but also through extracellular vesicles. Extracellular vesicles are membrane-covered spheres that transport various biomolecules and proteins. They are comparable to intercellular cargos allowing telocytes to fulfil their roles in many crucial processes such as intercellular signalling, immune response and angiogenesis [23]. These distinct structures were observed in the close vicinity of some plantar fascia telocytes (Figs. 26.11 and 26.12).

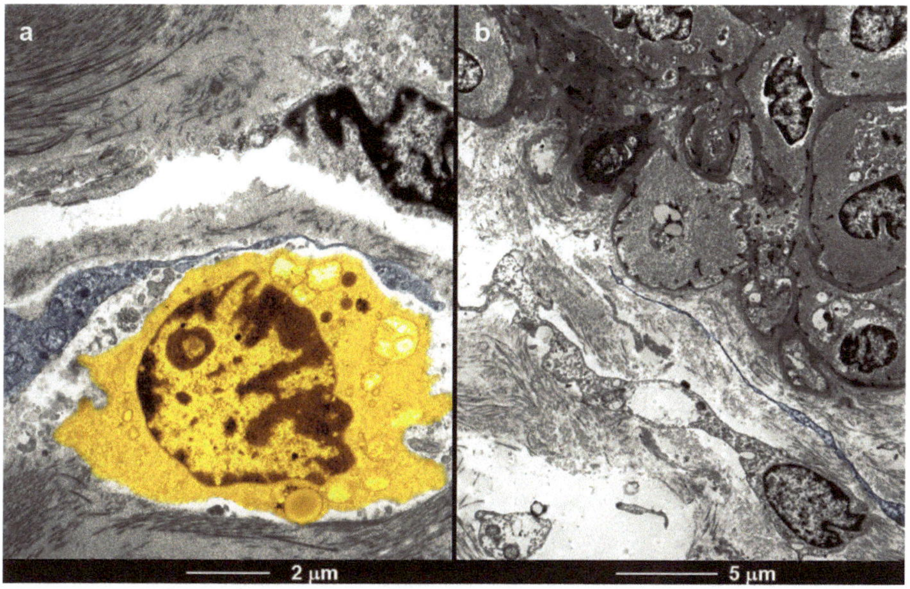

Fig. 26.8 Digitally coloured electronogram of a fragment of telopode (*blue*) (**a**) in close proximity with a fascia lata macrophage (*yellow*) and (**b**) a fragments of the telopode (*blue*) in close proximity with cells of the vascular wall. Scale bars: (**a**) 2 μm; (**b**) 5 μm

26.2.3 Telocytes of the Thoracolumbar Fascia

The thoracolumbar fascia (TLF) plays a very important role in sustaining the postural stability of the body and stabilises the lumbar vertebrae on the sacral base. The TLF also takes part in respiration [22, 58]. The TLF envelops the back muscles from the sacral region through the thoracic region and could be described as a complex arrangement of multilayered fascial planes and aponeurotic sheets [3]. There are several anatomical models for the TLF, with the two-layer model and the three-layer model being the most common [58]. Abnormalities affecting the thoracolumbar fascia can be one of the reasons of progressive spinal instability and chronic low back pain (LBP) [35]. In humans suffering from chronic or recurrent LBP, increased thickness and disorganisation of collagen layers of the thoracolumbar fascia were observed and could be considered important factors in pathophysiology of this entity [34]. Long-term consequences of the abnormalities affecting the TLF include its degeneration and degeneration of the surrounding anatomical structures. The thoracolumbar fascia, as other deep fasciae, consists mainly of dense irregular connective tissue (Fig. 26.1c), so by definition it contains a low number of cells. Among those cells, telocytes deserve special attention. The TLF telocytes are distributed among tightly packed collagen bundles and elastic fibres. Their long telopods (the longest observed telopod measured about 6.5 μm) are often interdigitated among numerous collagen fibres, being oriented both horizontally and longitudinally towards them (Fig. 26.13).

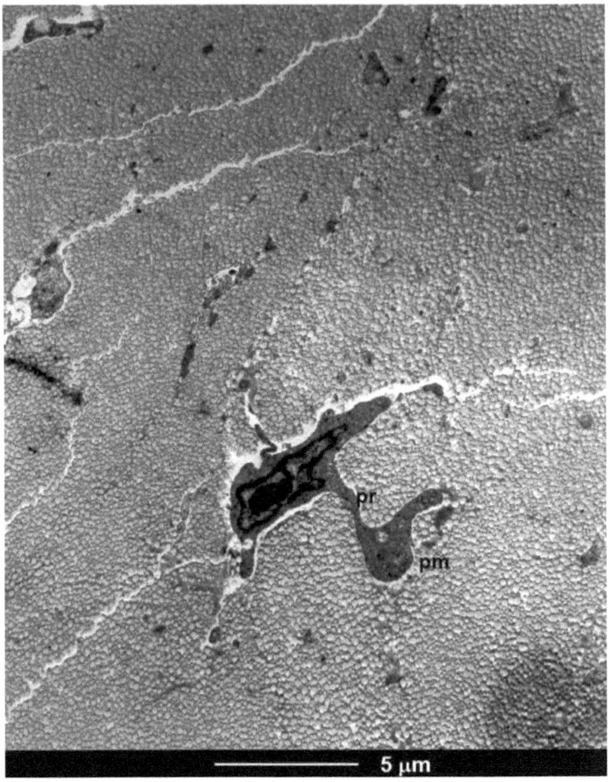

Fig. 26.9 Electron micrograph showing low magnification of a cross section of a plantar fascia telocyte with a podomer (*pr*) and a podom (*pm*). Scale bar: 5 μm

26.2.4 Telocytes of the Crural Fascia

The crural fascia ensheathes the muscles of the leg forming an anatomical and functional continuum with the fascia lata and the deep fascia of the foot. The crural fascia is known to be involved in etiopathogenesis of the medial tibial stress syndrome (MTSS), commonly known as "shin splints". It is a frequent and often disabling disorder in athletes such as runners and jumpers. The patients suffering from MTSS feel exertional pain (usually vague and diffuse, rather than focal) of the lower extremity centred on the posteromedial tibial border [21]. Reported incidence rate of MTSS varies from 4 % to 35 % in military and athletic populations [2]. Because the pathogenesis of this disorder is not fully understood, MTSS is not always easy to treat in everyday practice. It was indicated that the crural fascia, similarly to the fascia lata, has a layered structure (Fig. 26.1d) consisting of three, or at some points two, layers of parallelly aligned main collagen bundles separated by a thin layer of loose connective tissue [3]. Among those abundant fibrous structural proteins, telocytes are present together with fibroblasts, fibrocytes and some mast cells. Relatively often telopodes of the crural fascia telocytes were forming convoluted forms and running in close proximity to nerves and vessels. The higher observed length of the crural fascia telopode was about 13.5 μm in the place of the section (Fig. 26.14).

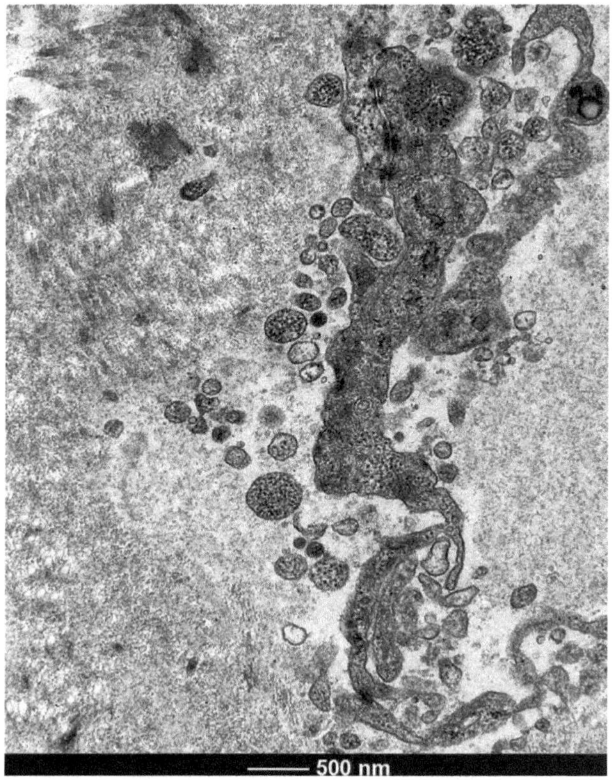

Fig. 26.10 Electron microscopy image of portions of plantar fascia telocytes with multiple homocellular junctions (*circle*) between adjutant fragments of telopodes; note the presence of abundant glycogen granules and abundant cytoskeletal elements within those cells. Scale bar: 0.5 μm

26.3 Potential Telocyte Crosstalks with Other Cell Constituents of Fascial Tissues

As has already been mentioned, telocytes of different tissues are known to form homo- and heterocellular junctions with, respectively, each other and with other cells (e.g. with endothelial cells, pericytes, Schwann cells, immune cells and fibroblasts) [6, 13, 23, 39, 46, 71]. These junctions allow cell–cell interactions that are essential to maintain tissue architecture, for cell growth, renewal and repair [8, 24, 37, 38, 44, 49, 59]. As it was shown by other researchers, different homocellular junction types between telocytes may exist (e.g. *puncta adhaerentia minima, processus adhaerentes* and *manubria adhaerentia)*. As far as heterocellular contacts are concerned, they generally form minute junctions (*point contacts*, *nano-contacts* and *planar contacts*) [23]. The close relationship of telocytes with other cells, especially with immune cells, indicates that they may play a potentially important role in physiological immunoregulation as well as in the pathogenesis of many different diseases. The presence of telocytes, recently confirmed within the fascial tissue, makes us look at fascia from a new perspective. It also encouraged us to seek possible relationships between telocytes and the remaining cells populating fascial

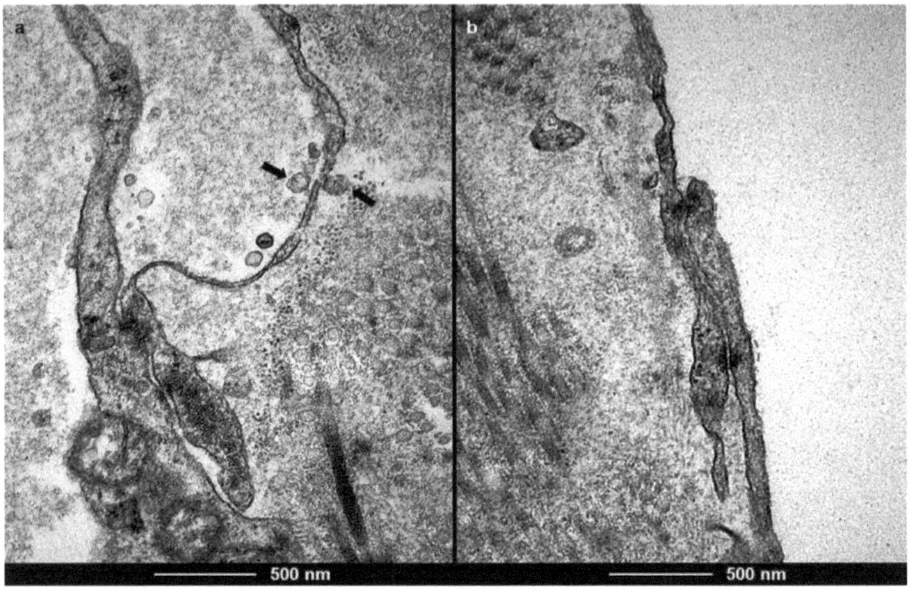

Fig. 26.11 Electron microscopy images of fragments of plantar fascia telocytes: (**a**) homocellular junctions (*circle*) evident between telopodes; note the extracellular vesicles released by telocyte (*arrows*) in close proximity with its plasma membrane; (**b**) the presence of homocellular junctions (*circles*) between telopodes. Scale bars: 0.5 μm

structures, such as fibroblasts and fibrocytes, cells showing myofibroblastic differentiation or mast cells [14, 63] (Fig. 26.15).

26.3.1 Telocytes and Fibroblasts

Fibroblasts and fibrocytes constitute the highest number of cells populating fascial structures. Fibroblasts are known for their high adaptability to their environmental conditions and the capacity to remodel in response to the direction of various mechanical stimuli [17]. Fibroblasts play a crucial role in the formation and maintenance of the fascial structure because they produce main fibrous proteins (collagens, elastin) and ground substance components. It is important to note that a considerable part of fascial disorders is accompanied by excessive proliferation of cells of fibroblastic differentiation. From this perspective, it is interesting that the number of telocytes turned out to be reduced in fibrotic areas of different organs, such as the heart muscle, intestinal wall, skin as well as other organs in patients suffering from systemic sclerosis [20, 40]. This suggested a potential role of those cells in the process of fibrotic remodelling and fibrosis. In examined fascia samples, telocytes were often surrounded by abundant and tightly packed collagen bundles (Fig. 26.13). Both the telocytes cell body as well as their telopodes

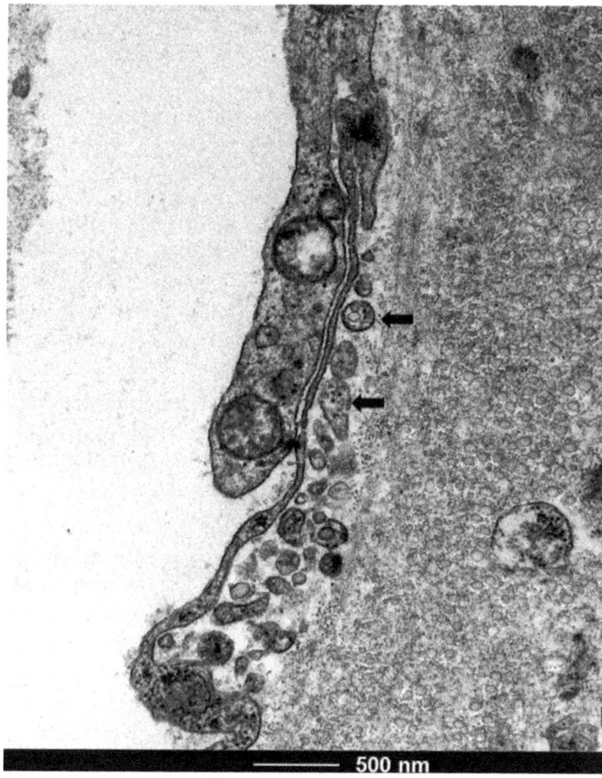

Fig. 26.12 Electron microscopy image of fragments of plantar fascia telocytes with visible homocellular junctions (*circles*) and abundant extracellular vesicles (*arrows*) in close proximity with the plasma membrane. Scale bar: 0.5 μm

were often closely associated with, or even in contact with, fascia fibroblasts (Fig. 26.16).

These telocytes–fibroblasts junctions may be of special interest with regard to research into fibrotic fascial diseases, such as Dupuytren's contracture or plantar fibromatosis, and should be definitely a subject of future investigations.

26.3.2 *Telocytes and Myofibroblasts*

Well-defined myofibroblasts are known to play an important role in establishing tension during wound healing and pathological contracture [66]. Even at rest, healthy connective tissues seem to be under the influence of a kind of mechanical tension [15, 66] and myofibroblasts are suspected of being involved in this tensional homeostasis. Moreover, it has been shown that myofibroblasts play a role in the synthesis and reorganisation of the connective tissue extracellular matrix, which, in the case of fascial structures, might significantly influence biomechanical properties of those tissues [55, 66]. Increased concentration of myofibroblasts in pathologically changed fascia is suspected of creating tissue contractures in clinical

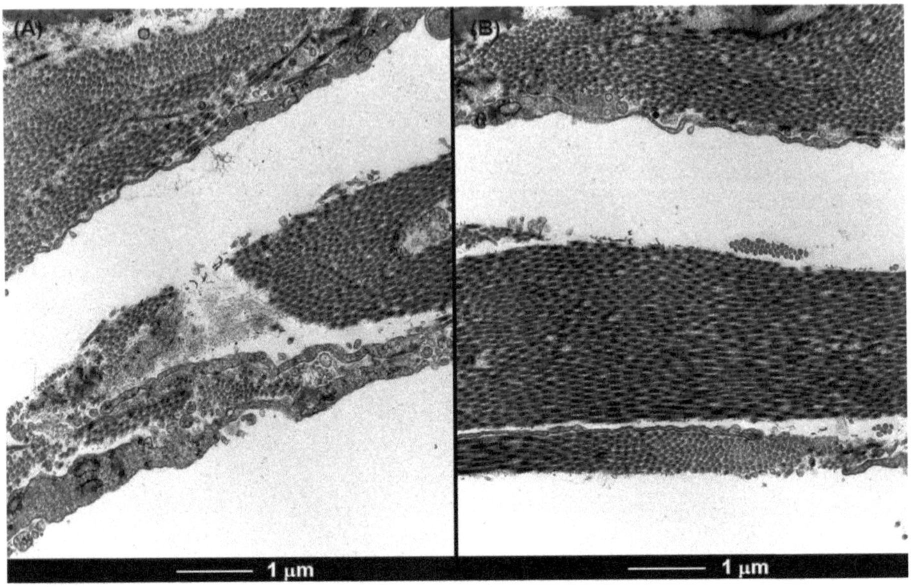

Fig. 26.13 Digitally coloured electron micrographs of thoracolumbar fascia (**a**–**b**); long telopodes (*blue*) penetrating tightly packed collagen fibres. Scale bars:1 μm

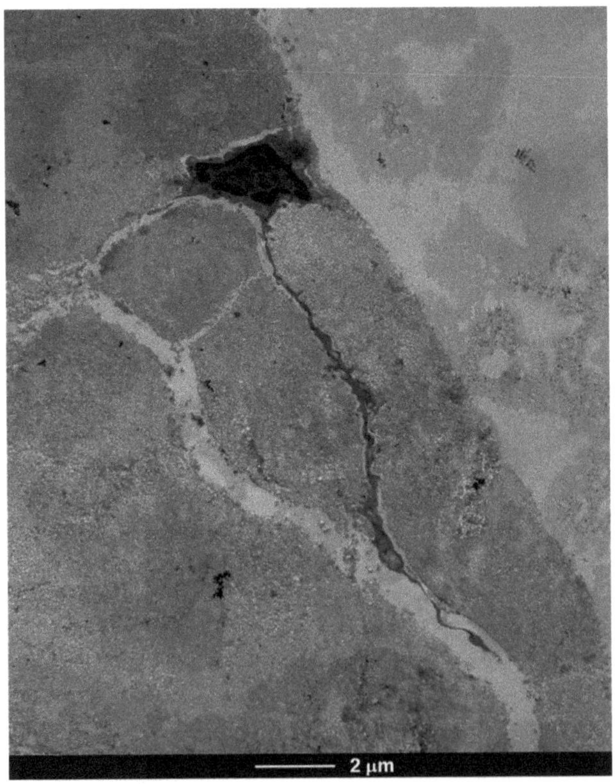

Fig. 26.14 Electron microscopy of a telocyte coming from the crural fascia demonstrates one long telopode and other, much shorter, prolongations; the longest telopode exhibits characteristic thinner (podomers) and thicker (podoms) fragments. Scale bar: 2 μm

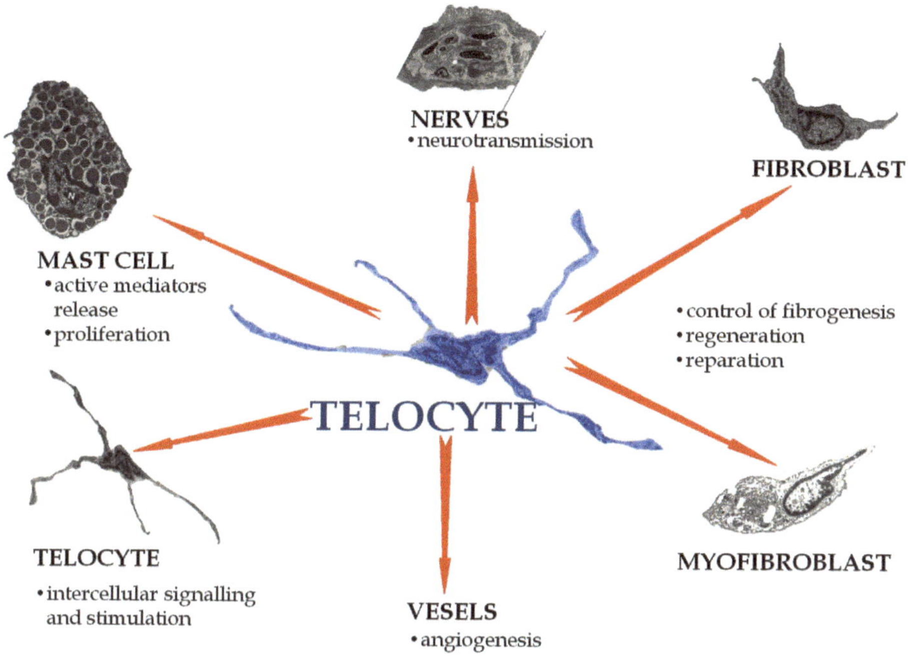

Fig. 26.15 Diagram showing the possible crosstalks between telocytes and other fascial cell populations (cells of the nervous system, fibroblasts, myofibroblasts, cells of the vessel structures, telocytes and mast cells)

conditions like palmar fascial fibromatosis (Dupuytren's disease), plantar fascial fibromatosis (Ledderhose's disease) and adhesive capsulitis (frozen shoulder) [25]. As far as possible telocyte–myofibroblast crosstalk is concerned, a hypothesis exists that telocytes might be involved in controlling fibroblast activation and that, consequently, a loss of telocytes (confirmed in fibrotic tissues) might promote the differentiation of fibroblasts into profibrotic myofibroblasts [40].

26.3.3 Telocytes: Mast Cells

Mast cells are bone marrow-derived cells found in many different tissues of the body. These cells are best known as potent effector cells mediating allergic diseases. They are also involved in a number of other important processes, such as fibrogenesis and wound healing [50]. Mast cells fulfil their biological roles thanks to the presence of their active mediators, some of which are currently being investigated as potential targets for therapeutic interventions [41]. Mast cell mediators are able to alter significantly the behaviour of various cell types they have contact with. Most probably it is also true in the case of mast cell–telocytes interaction. Telocytes are known to form heterocellular junctions with mast cells. They were found, among others, in close vicinity of mast cells in human pregnant and nonpregnant uteri [51].

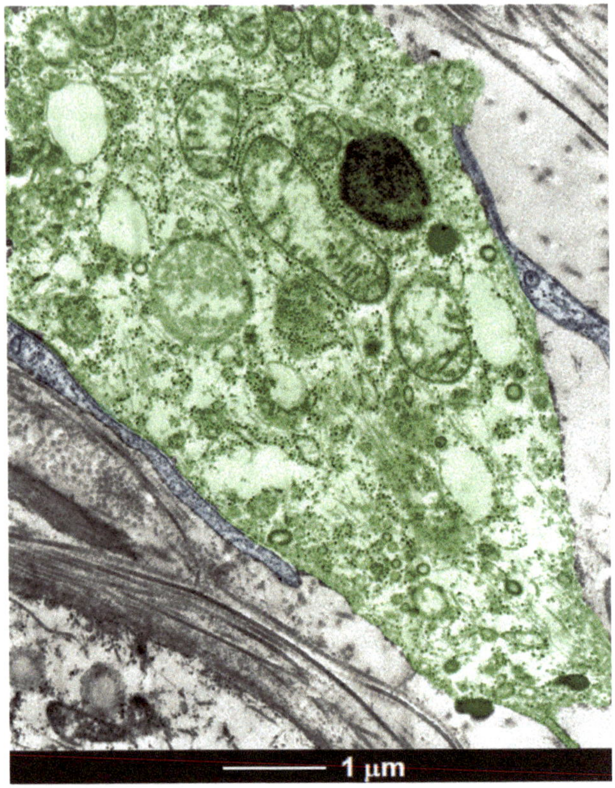

Fig. 26.16 Digitally coloured electronogram of the fascia lata; note the telopodes (*blue*) surrounding the fibroblast (*green*). Scale bar: 1 μm

Mast cell–telocyte tandems were also observed in mouse trachea, where they were suspected to be involved in the tracheal functional regulation (e.g. secretion, contractility) [52]. It was also indicated that telocytes established stromal synapses with mast cells (and eosinophils) in the course of acute salpingitis in rats and so play a role in local immunoregulation [71]. The presence of mast cells neighbouring fascial telopodes was especially pronounced in case of fascia lata samples (Fig. 26.17).

Consequently, it could be speculated that not only mast cell–fibroblast interactions but also mast cell–telocyte interactions might be of crucial importance for the mechanical properties of fascial tissues as well as fascial fibrotic diseases. Mast cell–telocyte tandem presence within fascial structure may also be taken to indicate that fasciae are immunologically active tissues with protective and regulatory potential.

26.3.4 Elements of the Nervous System

Our understanding of fascial innervation still remains incomplete. So far, the presence of abundant free and encapsulated nerve endings (including Ruffini and Pacinian corpuscles) has been described in some types of fascial tissues [4, 54]. In

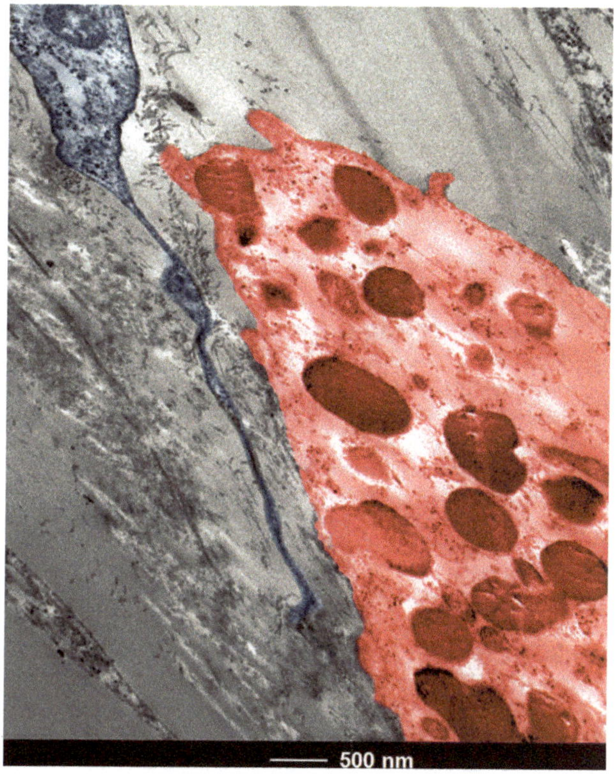

Fig. 26.17 Electron microscopy image of a fascia lata telocyte (*blue*) with a telopode in close proximity with a fragment of a mast cell (*red*). Scale bar: 0.5 μm

the fascia lata, the presence of mainly nonmyelinated nerve fibres and only single nerve endings has been confirmed. Observed fascial nerve endings were closely connected to the surrounding collagen fibres, suggesting that they may be easily stretched, and consequently activated, every time the surrounding fascia is stretched. In this way, the neural supply of fascia may influence its contractility both in health and disease. At the cellular level, the telocyte network running through the body sometimes is considered a very primitive nervous system [61]. Moreover telocytes of other (then fasciae) tissues are known to create links with cell of nervous system [6]. It was found that in the human heart, telocytes make contact with Schwann cells. In the urinary system and the uterus, they establish numerous contacts with nerve endings. Also, in the small intestine they come into contact with nerve structures [23, 67, 73]. Likewise, telocytes of fascial tissues may also create heterocellular junctions with neural cells and might be involved in the integration of the neural processes.

Acknowledgement The work was financed by grants from Wroclaw Medical University, Poland - National Science Centre grant no. N N518 286540. We would like to thank MSc Łukasz Mielańczyk for his technical support. TEM analysis was performed in the School of Medicine with the Division of Dentistry in Zabrze, Department of Histology and Embryology, Medical University of Silesia in Katowice, using equipment financed by the "Silesian Bio-farma" program with infrastructure supported by POIG.02.03.01-24-099/13.

References

1. Allen WE. Terminologia anatomica: international anatomical terminology and terminologia histologica: international terms for human cytology and histology. J Anat. 2009;215(2):221.
2. Batt ME. Medial tibial stress syndrome. Br J Sports Med. 2011;45:e2.
3. Benetazzo L, Bizzego A, De Caro R, Frigo G, Guidolin D, Stecco C. 3D reconstruction of the crural and thoracolumbar fasciae. Surg Radiol Anat. 2011;33(10):855–62.
4. Benjamin M. The fascia of the limbs and back-a review. J Anat. 2009;214(1):1–18.
5. Bordoni B, Zanier E. Clinical and symptomatological reflections: the fascial system. J Multidiscip Healthc. 2014;18:401–11.
6. Bosco C, Díaz E, Gutiérrez R, González J, Pérez J. Ganglionar nervous cells and telocytes in the pancreas of Octodon degus: extra and intrapancreatic ganglionar cells and telocytes in the degus. Auton Neurosci. 2013;177(2):224–30.
7. Cantarero I, Luesma MJ, Alvarez-Dotu JM, Muñoz E, Junquera C. Transmission electron microscopy as key technique for the characterization of telocytes. Curr Stem Cell Res Ther. 2016;11(5):410–4.
8. Cavey M, Lecuit T. Molecular bases of cell-cell junctions stability and dynamics. Cold Spring Harb Perspect Biol. 2009;1(5):a002998.
9. Chaitow L. The fascia debate. J Bodyw Mov Ther. 2014;18(3):443.
10. Cismasiu VB, Radu E, Popescu LM. miR-193 expression differentiates telocytes from other stromal cells. J Cell Mol Med. 2011;15(5):1071–4.
11. Cretoiu D, Cretoiu SM, Simionescu AA, Popescu LM. Telocytes, a distinct type of cell among the stromal cells present in the lamina propria of jejunum. Histol Histopathol. 2012;27(8):1067–78.
12. Cretoiu SM, Cretoiu D, Marin A, Radu BM, Popescu LM. Telocytes: ultrastructural, immunohistochemical and electrophysiological characteristics in human myometrium. Reproduction. 2013;145(4):357–70.
13. Cretoiu SM, Popescu LM. Telocytes revisited. Biomol Concepts. 2014;5(5):353–69.
14. Dawidowicz J, Szotek S, Matysiak N, Mielańczyk Ł, Maksymowicz K. Electron microscopy of human fascia lata; focus on telocytes. J Cell Mol Med. 2015;19(10):2500–6.
15. Demirci N, Tönük E. Non-integer viscoelastic constitutive law to model soft biological tissues to in-vivo indentation. Acta Bioeng Biomech. 2014;16(4):13–20.
16. Derwin KA, Baker AR, Spragg RK, Leigh DR, Farhat W, Iannotti JP. Regional variability, processing methods, and biophysical properties of human fascia lata extracellular matrix. J Biomed Mater Res A. 2008;84(2):500–7.
17. Eagan TS, Meltzer KR, Standley PR. Importance of strain direction in regulating human fibroblast proliferation and cytokine secretion: a useful in vitro model for soft tissue injury and manual medicine treatments. J Manipulative Physiol Ther. 2007;30(8):584–92.
18. Eng CM, Pancheri FQ, Lieberman DE, Biewener AA, Dorfmann L. Directional differences in the biaxial material properties of fascia lata and the implications for fascia function. Ann Biomed Eng. 2014;42(6):1224–37.
19. Faussone-Pellegrini MS, Popescu LM. Telocytes. Biomol Concepts. 2011;2(6):481–9.
20. Fu S, Wang F, Cao Y, Huang Q, Xiao J, Yang C, Popescu LM. Telocytes in human liver fibrosis. J Cell Mol Med. 2015;19(3):676–83.
21. Galbraith RM, Lavallee ME. Medial tibial stress syndrome: conservative treatment options. Curr Rev Musculoskelet Med. 2009;2(3):127–33.
22. Gatton ML, Pearcy MJ, Pettet GJ, Evans JH. A three-dimensional mathematical model of the thoracolumbar fascia and an estimate of its biomechanical effect. J Biomech. 2010;43(14):2792–7.
23. Gherghiceanu M, Popescu LM. Cardiac telocytes-their junctions and functional implications. Cell Tissue Res. 2012;348(2):265–79.
24. Green KJ, Getsios S, Troyanovsky S, Godsel LM. Intercellular junction assembly, dynamics, and homeostasis. Cold Spring Harb Perspect Biol. 2010;2(2):a000125.
25. Hedley G. Notes on visceral adhesions as fascial pathology. J Bodyw Mov Ther. 2010;14(3):255–61.
26. Hinescu ME, Gherghiceanu M, Suciu L, Popescu LM. Telocytes in pleura: two- and three-dimensional imaging by transmission electron microscopy. Cell Tissue Res. 2011;343(2):389–97.

27. Huang BK, Campos JC, Ghobrial P, Peschka M, Pretterklieber ML, Skaf AY, Chung CB, Pathria MN. Injury of the gluteal aponeurotic fascia and proximal iliotibial band: anatomy, pathologic conditions, and MR maging. Radio Graph. 2013;33(5):1437–52.
28. Hyrtl J. Onomatologia Anatomica: Geschichte und Kritik der Anatomichen Sprache der Gegenwart. Vienna: Wilhelm Braumüller; 1880.
29. Klingler W, Velders M, Hoppe K, Pedro M, Schleip R. Clinical relevance of fascial tissue and dysfunctions. Curr Pain Headache Rep. 2014;18(8):439.
30. Kumka M. Kumka's response to Stecco's fascial nomenclature editorial. J Bodyw Mov Ther. 2014;18(4):591–8.
31. Kumka M, Bonar J. Fascia: a morphological description and classification system based on a literature review. J Can Chiropr Assoc. 2012;56(3):179–91.
32. Langevin HM. Langevin's response to Stecco's fascial nomenclature editorial. J Bodyw Mov Ther. 2014;18(3):444.
33. Langevin HM, Huijing PA. Communicating about fascia: history, pitfalls, and recommendations. Int J Ther Massage Bodyw. 2009;2(4):3–8.
34. Langevin HM, Stevens-Tuttle D, Fox JR, Badger GJ, Bouffard NA, Krag MH, Wu J, Henry SM. Ultrasound evidence of altered lumbar connective tissue structure in human subjects with chronic low back pain. BMC Musculoskelet Disord. 2009;10:151.
35. Langevin HM, Fox JR, Koptiuch C, Badger GJ, Greenan- Naumann AC, Bouffard NA, Konofagou EE, Lee WN, Triano JJ, Henry SM. Reduced thoracolumbar fascia shear strain in human chronic low back pain. BMC Musculoskelet Disord. 2011;12:203.
36. Lemound J, Stoetzer M, Kokemüller H, Schumann P, Gellrich NC. Modified technique for rehabilitation of facial paralysis using autogenous fascia lata grafts. J Oral Maxillofac Surg. 2015;73(1):176–83.
37. Li J, Radice GL. A new perspective on intercalated disc organization: implications for heart disease. Dermatol Res Pract. 2010;2010:207835.
38. Li J, Swope D, Raess N, Cheng L, Muller EJ, Radice GL. Cardiac tissue-restricted deletion of plakoglobin results in progressive cardiomyopathy and activation of {beta}-catenin signaling. Mol Cell Biol. 2011;31(6):1134–44.
39. Li H, Lu S, Liu H, Ge J, Zhang H. Scanning electron microscope evidence of telocytes in vasculature. J Cell Mol Med. 2014;18(7):1486–9.
40. Manetti M, Rosa I, Messerini L, Guiducci S, Matucci-Cerinic M, Ibba-Manneschi L. A loss of telocytes accompanies fibrosis of multiple organs in systemic sclerosis. J Cell Mol Med. 2014;18(2):253–62.
41. McEuen AR, Walls AF. Purification and characterization of mast cell tryptase and chymase from human tissues. Methods Mol Med. 2008;138:299–317.
42. Nicolescu MI, Popescu LM. Telocytes in the interstitium of human exocrine pancreas: ultrastructural evidence. Pancreas. 2012;41(6):949–56.
43. Nicolescu MI, Bucur A, Dinca O, Rusu MC, Popescu LM. Telocytes in parotid glands. Anat Rec (Hoboken). 2012;295(3):378–85.
44. Palatinus JA, Rhett JM, Gourdie RG. Translational lessons from scarless healing of cutaneous wounds and regenerative repair of the myocardium. J Mol Cell Cardiol. 2010;48(3):550–7.
45. Popescu LM, Faussone-Pellegrini MS. Telocytes – a case of serendipity: the winding way from Interstitial Cells of Cajal (ICC), via Interstitial Cajal-Like Cells (ICLC) to telocytes. J Cell Mol Med. 2010;14:729–40.
46. Popescu LM. The tandem: telocytes – stem cells. Int J Biol Biomed Eng. 2011;2:83–92.
47. Popescu LM, Gherghiceanu M, Suciu LC, Manole CG, Hinescu ME. Telocytes and putative stem cells in the lungs: electron microscopy, electron tomography and laser scanning microscopy. Cell Tissue Res. 2011;345(3):391–403.
48. Popescu LM, Manole E, Serboiu CS, Manole CG, Suciu LC, Gherghiceanu M, Popescu BO. Identification of telocytes in skeletal muscle interstitium: implication for muscle regeneration. J Cell Mol Med. 2011;15(6):1379–13792.
49. Raju H, Alberg C, Sagoo GS, Burton H, Behr ER. Inherited cardiomyopathies. BMJ. 2011;343:d6966.
50. Rao KN, Brown MA. Mast cells: multifaceted immune cells with diverse roles in health and disease. Ann N Y Acad Sci. 2008;1143:83–104.

51. Roatesi I, Radu BM, Cretoiu D, Cretoiu SM. Uterine telocytes: a review of current knowledge. Biol Reprod. 2015;93(1):10.
52. Rusu MC, Jianu AM, Mirancea N, Didilescu AC, Mănoiu VS, Păduraru D. Tracheal telocytes. J Cell Mol Med. 2012;16(2):401–5.
53. Rusu MC, Pop F, Hostiuc S, Curca GC, Jianu AM, Paduraru D. Telocytes form networks in normal cardiac tissues. Histol Histopathol. 2012;27(6):807–16.
54. Schleip R. Fascial plasticity-a new neurobiological explanation part 1. J Bodyw Mov Ther. 2003;7(1):11–9.
55. Schleip R, Klingler W, Lehmann-Horn F. Active fascial contractility: fascia may be able to contract in a smooth muscle-like manner and thereby influence musculoskeletal dynamics. Med Hypotheses. 2005;65(2):273–7.
56. Schleip R, Findley TW, Chaitow L, Huijing P. Fascia – the tensional network of the human body: the science and clinical applications in manual and movement therapy. Edinburgh: Churchill Livingstone; 2012.
57. Schleip R, Jäger H, Klingler W. What is 'fascia'? A review of different nomenclatures. J Bodyw Mov Ther. 2012;16(4):496–502.
58. Schuenke MD, Vleeming A, Van Hoof T, Willard FH. A description of the lumbar interfascial triangle and its relation with the lateral raphe: anatomical constituents of load transfer through the lateral margin of the thoracolumbar fascia. J Anat. 2012;221(6):568–76.
59. Sheikh F, Ross RS, Chen J. Cell-cell connection to cardiac disease. Trends Cardiovasc Med. 2009;19(6):182–90.
60. Sher I, Umans H, Downie SA, Tobin K, Arora R, Olson TR. Proximal iliotibial band syndrome: what is it and where is it? Skeletal Radiol. 2011;40(12):1553–6.
61. Smythies J, Edelstein L. Telocytes, exosomes, gap junctions and the cytoskeleton: the makings of a primitive nervous system? Front Cell Neurosci. 2013;7:278–81.
62. Stecco C, Hammert WC. Functional atlas of the human fascial system. Edinburgh: Churchill Livingstone, Elsevier; 2015.
63. Stecco C, Pavan PG, Porzionato A, Macchi V, Lancerotto L, Carniel EL, Natali AN, De Caro R. Mechanics of crural fascia: from anatomy to constitutive modelling. Surg Radiol Anat. 2009;31(7):523–9.
64. Stecco C, Pavan P, Pachera P, De Caro R, Natali A. Investigation of the mechanical properties of the human crural fascia and their possible clinical implications. Surg Radiol Anat. 2014;36(1):25–32.
65. Suciu L, Popescu LM, Gherghiceanu M, Regalia T, Nicolescu MI, Hinescu ME, Faussone-Pellegrini MS. Telocytes in human term placenta: morphology and phenotype. Cells Tissues Organs. 2010;192(5):325–39.
66. Tomasek JJ, Gabbiani G, Hinz B, Chaponnier C, Brown RA. Myofibroblasts and mechano-regulation of connective tissue remodelling. Nat Rev Mol Cell Biol. 2002;3(5):349–63.
67. Vannucchi MG, Traini C, Manetti M, Ibba-Manneschi L, Faussone-Pellegrini MS. Telocytes express PDGFRα in the human gastrointestinal tract. J Cell Mol Med. 2013;17(9):1099–108.
68. Warwick RM, Brubacer SA. Tissue and cell clinical use: an essential guide. Chichester: Wiley-Blackwell; 2012.
69. Wearing SC, Schleip R, Chaitow L, Klingler W, Findley T editors. Fascia research IV basic science and implications for conventional and complementary health care. Kiener Press, Munich. (In Press) 2015.
70. Yang Y, Sun W, Wu SM, Xiao J, Kong X. Telocytes in human heart valves. J Cell Mol Med. 2014;18(5):759–65.
71. Yang J, Chi C, Liu Z, Yang G, Shen ZJ, Yang XJ. Ultrastructure damage of oviduct telocytes in rat model of acute salpingitis. J Cell Mol Med. 2015;19(7):1720–8.
72. Zandi AR. Harvesting fascia lata for brow suspension procedure via two small skin incisions. JRMS. 2006;11(6):396–9.
73. Zheng Y, Zhu T, Lin M, Wu D, Wang X. Telocytes in the urinary system. J Transl Med. 2012;10:188.
74. Zheng Y, Cretoiu D, Yan G, Cretoiu SM, Popescu LM, Fang H, Wang X. Protein profiling of human lung telocytes and microvascular endothelial cells using iTRAQ quantitative proteomics. J Cell Mol Med. 2014;18(6):1035–59.

Chapter 27
Hepatic Telocytes

Yingying Zhao, Xiaoyu Chen, Fei Wang, and Changqing Yang

Abstract Telocytes (TCs), a novel peculiar interstitial cell found in many tissues and organs, play pivotal roles in maintaining tissue homeostasis and regulating tissue and organ development and immune surveillance. In recent years, the existence of TCs in liver has been confirmed. In this chapter, we evaluate the role of TCs on promoting liver regeneration and the therapeutic effects on liver fibrosis.

27.1 Introduction

A century ago, a Spanish histologist, Ramon y Cajal, pioneered a microscopy-based study of neuroanatomy and first discovered the existence of a peculiar interstitial cell in the intestines with similar morphological characters to neurons [1]. Based on the specific staining profile (methylene blue and argentaffin) and its location in the intercellular space between smooth muscle cells (SMCs) and nerve endings, they named cells *interstitial neurons* and hypothesized its function in neural activity. Through the development of transmission electron microscopes, Faussone-Pellegrini et al. revealed the ultrastructure of this cell type in the human esophagus and stomach [2, 3]. Subsequently, Thuneberg et al. verified the function of the cells as a pacemaker and conductor of intestinal impulse [4]. As there was consistent and sufficient evidence that these cells were not neurons, the name interstitial cells of Cajal (ICC) was adopted. In 1992, Christensen J summarized the general features of the ultrastructure of ICC: a round or oval nucleus, scattering chromatins, scant cytoplasm with marked smooth endoplasmic reticula (SER), mitochondria, and well-developed Golgi apparatus and several vacuoles in the membrane; incomplete basement membranes; and closer-to-nerve fiber and gap junctions between the processes and SMCs [5]. Over the following decades, researchers found similar ICC-like cells in many organs, with a fundamental morphological difference in the

Y. Zhao • X. Chen • F. Wang • C. Yang (✉)
Division of Gastroenterology and Hepatology, Digestive Disease Institute, Shanghai Tongji Hospital, Tongji University School of Medicine, 389 Xin Cun Road, Shanghai 200065, China
e-mail: changqingyang_tj@hotmail.com

X. Wang, D. Cretoiu (eds.), *Telocytes*, Advances in Experimental Medicine and Biology, 913, DOI 10.1007/978-981-10-1061-3_27

extending antennae. As such, these cells were named interstitial Cajal-like cells (ICLC). In 2010, Popescu LM et al. demonstrated that the ICLC phenotype was quite different from ICC. To distinguish the two types of cells, they officially named ICLC as telocytes (TCs) and extending antennae telopodes (TPs) [6].

27.2 Ultrastructure and Immunophenotype of TCs

To date, TCs have been found widely in numerous human organs, such as the pancreas [7], mammary gland [8–10], urinary tract [11], reproductive organs [12–14], placenta [15, 16], skeletal muscle [17], myocardium [18–24], and lung [25, 26], among others. The general morphology of TCs is described as relatively small cell body compared with other interstitial cells, with a nucleus that accounts for one quarter of the total cell volume, surrounded by the mitochondria, Golgi apparatus, and rough and smooth endoplasmic reticula. There are 1–5 triangular or spindly TPs in a TC, of which the length ranges from tens to hundreds of micrometers, with inhomogeneous thickness of less than 0.2 μm. There are narrow segments (podomers) on the TPs alternating with dilations (podoms). TPs could form a complex set of network through three-dimensional crimps and overlaps. Due to the nearly invisible structure under light microscopy, TCs were only officially discovered in recent decades. In addition, TCs can make connections with other cells by hetero- and homocellular junctions or extracellular secretion of vesicles [6, 27].

Notably, TCs display different immunophenotypes according to the organ and/or the animal species, and while they share the same ultrastructural features, the identification of TC can be made only by transmission electron microscope (TEM) [28]. TCs have been reportedly positive for c-kit [29], CD34 [15, 26, 30, 31], S-100 [30], α-smooth muscle actin (α-SMA) [26], VEGF [15, 17, 32, 33], vimentin [29], PDGFR-α [28], and PDGFR-β [34].

27.3 Discovery of TCs in Liver

In the study by Xiao et al., the authors provide the first verification of the existence of TCs in the liver of mice using TEM (Fig. 27.1) [35]. In addition, they extracted liver tissues from frozen sections and identified immunofluorescence markers for TC based on double labeling for CD34 combined with c-kit, vimentin, PDGFR-α, or PDGFR-β.

The isolation and primary culture of hepatic TCs could be potentially modified according to the cardiac TCs isolation procedures [36]. Ten- to 12-week-old C57/BL6 male mice were anesthetized using 1 % pentobarbital sodium, and liver tissues were subsequently obtained. The tissues were washed with Hank's balanced salt solution (HBSS), minced, and then washed again. Afterward, the proper amount of 0.25 % trypsin in DMEM was added at 37 °C for digestion. The tissues were oscillated with a shaker for 5 min, and equal volumes of pre-chilled Dulbecco's modified eagle

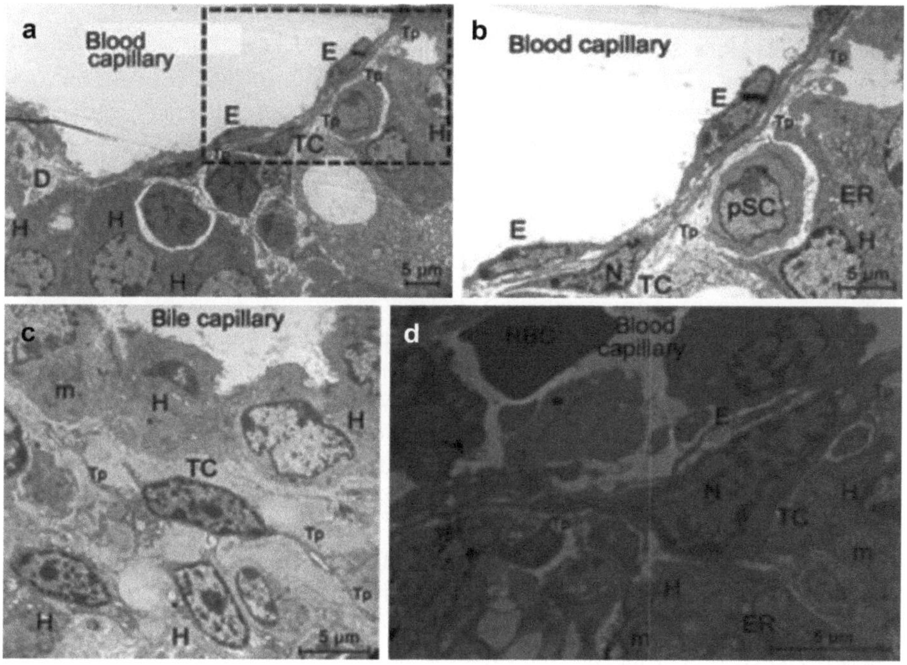

Fig. 27.1 Electron microscopy images of the ultrastructure of the liver (from mice). (**a**) TCs with a TP located in the Disse space (*D*) between hepatocytes (*H*) and endothelial cells (*E*). (**b**) A higher magnification of the field inside the rectangle in A. A putative stem cell (*pSC*) is located between the hepatocytes (*H*) and TC; *N* nucleus. *ER* endoplasmic reticulum. (**c**) A TC with three Tps; *H* hepatocyte, *m* mitochondria. (**d**) A higher magnification of a TC with a nucleus (*N*); *RBC* red blood cell, *H* hepatocyte, *E* endothelial cell, *m* mitochondria; scale bar = 5 μm (Reproduced with permission from [35])

medium (DMEM)/F12 were added to terminate digestion. The tissues were filtered through a 40 μm pore membrane and centrifuged at 1,500 rpm for 5 min. The cells were then washed again and resuspended with DMEM/F12 containing 10% fetal bovine serum (FBS). The above procedures (including digestion, termination of digestion, and cell isolation) were repeated five times. The collected cell suspension was mixed, resuspended, and centrifuged at 1,000 rpm for 5 min and cultured with 10 ml DMEM/F12 containing 10% FBS in a constant temperature incubator. After that, we would suggest using flow cytometric sorting with antibodies to CD34 and PDGFR-α. After that, immunofluorescence staining could be used to identify hepatic TCs.

27.4 TCs and Liver Regeneration

The liver has tremendous capacity for regeneration following various injuries. By example, it can restore to its original weight in 2 weeks following partial hepatectomy (PH) [37, 38]. Hepatocyte proliferation represents a vital role during the regeneration process [37, 39]. In the study by Wang et al., a murine model of liver regeneration was induced by PH [40]. They calculated the residual liver lobes/body weight of mice at 0, 12, 24, 48, 72, 96, 120, and 168 h following PH to evaluate the regeneration of liver mass. The regenerated liver tissues were made into frozen sections for immunofluorescence double labeling (CD34/PDGFR-α, CD34/PDGFR-β, or CD34/vimentin) and EdU (5-ethynyl-20-deoxyuridine) staining. The remaining tissues were collected for Western blotting analysis of proliferating cell nuclear antigen (PCNA). The results of EdU staining and Western blotting demonstrated that cell proliferation reached its peak at 48 and 72 h following PH. The results of immunofluorescence staining demonstrated that the number of double-labeled TCs significantly increased at 72 h, consistent with the point of maximum cell proliferation post-PH. In addition, Wang et al. also demonstrated that the number of CK-19-positive hepatic stem cells most significantly increased 72 h post-PH. Taken together, this study proposed a potential role for TCs in liver regeneration and a close relationship between TCs and hepatocytes or hepatic stem cells. There were three possibilities regarding the relationship between TCs and liver regeneration: (i) it was merely coincidental that the numbers of TCs and proliferating cells during liver regeneration had concurrently increased, (ii) the increased TCs were associated with increased hepatocytes and hepatic stem cells in liver regeneration, and (iii) the increased TCs induced the increased hepatocytes and hepatic stem cells in liver regeneration. Considering that the increase in TCs following PH occurs concurrently with an increase in hepatocytes and hepatic stem cells suggests that TCs participate in liver regeneration through intercellular junctions or a paracrine effect via ectovesicles. Improved understanding of the impact of TCs on hepatocytes and/or hepatic stem cells and liver regeneration is crucial and will aid in the development of novel therapeutics.

27.5 TCs and Liver Fibrosis

Liver fibrosis is a pathophysiologic process characterized by intrahepatic extracellular matrix (ECM) dysplasia, following chronic damages, which can progress to cirrhosis [41, 42]. As the ECM is mainly secreted by hepatic stellate cell (HSC), it is widely believed that activation of HSC is central to the pathobiology of liver fibrosis. In recent decades, the reversible nature of liver fibrosis or even early-stage cirrhosis has been demonstrated. Unfortunately, no effective or specific medication is currently available [43]. As it has been reported that the number of TCs was significantly decreased in fibrotic areas of myocardial infarction [44], Fu et al. collected human liver fibrosis samples obtained upon liver biopsy to determine the relation between TCs and liver fibrosis [45]. Paraffin sections from liver tissues of fibrosis or healthy controls were prepared for hematoxylin-eosin and Masson's trichrome staining and frozen sections for immunofluorescence studies (CD34/PDGFR-α, CD34/PDGFR-β, CD34/vimentin, or CD34/c-kit). In the fibrosis group, there were an obvious inflammatory response, collagen deposition, and cell necrosis. While

TEM examination is the golden standard for TC identifications, the authors used four double immunolabeling methods to identify and semiquantitatively analyze these cell populations. Double labeling of CD34 and PDGFR-α noted a significant reduction of TCs in the fibrosis samples. This study highlighted a role for TCs in liver fibrosis. As the number of HSCs is increased in the context of liver fibrosis, the study also supports the conclusion that TCs are distinct entities from HSCs. However, it remains to be determined whether TCs reduction is a cause or consequence of this pathological process. Taken together, this study proposed a potential role of TCs in liver fibrosis: (i) TCs reduction in liver fibrosis may impact ECM deposition [46] and regulate HSCs activation and (ii) as the numbers of TCs is increased in liver regeneration [40], the disappearance of TCs may impair the regenerative actions mediated by hepatocytes and stem cells in liver fibrosis. In addition, it has been reported that intramyocardial transplantation of cardiac TCs ameliorated MI and improved postinfarcted cardiac function [47]. As such, TCs transplantation may provide a new treatment strategy for liver fibrosis.

27.6 Conclusion

The existence of TCs has been well established in many hollow and solid viscera and even in skin dermis [6]. Unfortunately, studies on the physiological functions and impact of TCs remain very superficial; thus the mechanisms by which TCs affect their function remain largely uncharacterized. As a type of interstitial cell, connection between TCs and vessels, nerve fiber tracts, and immune cells is established via its hetero- and homocellular junctions with other cell types, which indicates a mechanical support effect of TCs [48]. It is notable that some microRNAs (miRNAs, miRs) display proangiogenic potential, including let-7 family [49, 50], miR-10 [51], miR-126 [52], miR-155 [53], and miR-503 [54], while other miRNAs, such as miR-130a, may downregulate the expression of antiangiogenic factors [55]. Studies on the role of miRNAs in TCs will enhance our understanding of the functions of TCs. Meanwhile, studies on the pathological effects of TCs in various disease states have only just begun. By example, reduction of TCs was observed in the ileum of patients with systemic sclerosis and Crohn's disease, when compared with healthy controls [56]. Specifically, Manole et al. noted that numerous TCs were found in the marginal area of infarcted myocardium [32]. In addition, TCs have an inhibitory effect on cell apoptosis when co-cultured with breast cancer EMT-6 cells [57]. As studies have increasingly demonstrated the pathological effects of TCs, this improved understanding will help usher in new prospect for liver diseases treatment.

Acknowledgments This work was supported by the grants from National Natural Science Foundation of China (81070343 and 81370559 to C. Yang; 81400635 to F. Wang), Joint Projects in Major Diseases funding from Shanghai Municipal Commission of Health and Family Planning (2014ZYJB0201 to C. Yang), Joint Projects for Novel Frontier Technology in Shanghai Municipal Hospital from Shanghai Municipal Commission of Health and Family Planning (SHDC1204122 to C. Yang), Shanghai Medical Guide Project from Shanghai Science and Technology Committee (14411971500 to F. Wang), grants from the Chinese Foundation for Hepatitis Prevention and Control (TQGB20140141 to F. Wang), and funds from the Shanghai Innovation Program (12431901002 to C. Yang).

Conflict of Interest The authors declare there are no conflicts of interest.

References

1. Llinas RR. The contribution of santiago ramon y cajal to functional neuroscience. Nat Rev Neurosci. 2003;4:77–80.
2. Faussone Pellegrini MS, Cortesini C, Romagnoli P. Nature and significance of the so-called cajal's "interstitial cells" in the human esophagus and stomach. Boll Soc Ital Biol Sper. 1976;52:1764–7.
3. Faussone Pellegrini MS, Cortesini C, Romagnoli P. Ultrastructure of the tunica muscularis of the cardial portion of the human esophagus and stomach, with special reference to the so-called cajal's interstitial cells. Arch Ital Anat Embriol Ital J Anat Embryol. 1977;82:157–77.
4. Thuneberg L, Rumessen JJ, Mikkelsen HB. Interstitial cells of cajal – an intestinal impulse generation and conduction system? Scand J Gastroenterol Suppl. 1982;71:143–4.
5. Christensen J. A commentary on the morphological identification of interstitial cells of cajal in the gut. J Auton Nerv Syst. 1992;37:75–88.
6. Popescu LM, Faussone-Pellegrini MS. Telocytes – a case of serendipity: the winding way from interstitial cells of cajal (icc), via interstitial cajal-like cells (iclc) to telocytes. J Cell Mol Med. 2010;14:729–40.
7. Popescu LM, Hinescu ME, Ionescu N, Ciontea SM, Cretoiu D, Ardelean C. Interstitial cells of cajal in pancreas. J Cell Mol Med. 2005;9:169–90.
8. Gherghiceanu M, Popescu LM. Interstitial cajal-like cells (iclc) in human resting mammary gland stroma. Transmission electron microscope (tem) identification. J Cell Mol Med. 2005;9:893–910.
9. Radu E, Regalia T, Ceafalan L, Andrei F, Cretoiu D, Popescu LM. Cajal-type cells from human mammary gland stroma: phenotype characteristics in cell culture. J Cell Mol Med. 2005; 9:748–52.
10. Popescu LM, Andrei F, Hinescu ME. Snapshots of mammary gland interstitial cells: methylene-blue vital staining and c-kit immunopositivity. J Cell Mol Med. 2005;9:476–7.
11. Hinescu ME, Ardeleanu C, Gherghiceanu M, Popescu LM. Interstitial cajal-like cells in human gallbladder. J Mol Histol. 2007;38:275–84.
12. Popescu LM, Ciontea SM, Cretoiu D, Hinescu ME, Radu E, Ionescu N, Ceausu M, Gherghiceanu M, Braga RI, Vasilescu F, Zagrean L, Ardeleanu C. Novel type of interstitial cell (cajal-like) in human fallopian tube. J Cell Mol Med. 2005;9:479–523.
13. Popescu LM, Ciontea SM, Cretoiu D. Interstitial cajal-like cells in human uterus and fallopian tube. Ann N Y Acad Sci. 2007;1101:139–65.
14. Cretoiu SM, Cretoiu D, Suciu L, Popescu LM. Interstitial cajal-like cells of human fallopian tube express estrogen and progesterone receptors. J Mol Histol. 2009;40:387–94.
15. Suciu L, Popescu LM, Gherghiceanu M, Regalia T, Nicolescu MI, Hinescu ME, Faussone-Pellegrini MS. Telocytes in human term placenta: morphology and phenotype. Cells Tissues Organs. 2010;192:325–39.
16. Suciu L, Popescu LM, Gherghiceanu M. Human placenta: De visu demonstration of interstitial cajal-like cells. J Cell Mol Med. 2007;11:590–7.
17. Popescu LM, Manole E, Serboiu CS, Manole CG, Suciu LC, Gherghiceanu M, Popescu BO. Identification of telocytes in skeletal muscle interstitium: implication for muscle regeneration. J Cell Mol Med. 2011;15:1379–92.

18. Gherghiceanu M, Manole CG, Popescu LM. Telocytes in endocardium: electron microscope evidence. J Cell Mol Med. 2010;14:2330–4.
19. Hinescu ME, Popescu LM. Interstitial cajal-like cells (iclc) in human atrial myocardium. J Cell Mol Med. 2005;9:972–5.
20. Hinescu ME, Gherghiceanu M, Mandache E, Ciontea SM, Popescu LM. Interstitial cajal-like cells (iclc) in atrial myocardium: ultrastructural and immunohistochemical characterization. J Cell Mol Med. 2006;10:243–57.
21. Popescu LM, Gherghiceanu M, Hinescu ME, Cretoiu D, Ceafalan L, Regalia T, Popescu AC, Ardeleanu C, Mandache E. Insights into the interstitium of ventricular myocardium: interstitial cajal-like cells (iclc). J Cell Mol Med. 2006;10:429–58.
22. Kostin S, Popescu LM. A distinct type of cell in myocardium: interstitial cajal-like cells (iclcs). J Cell Mol Med. 2009;13:295–308.
23. Suciu L, Popescu LM, Regalia T, Ardelean A, Manole CG. Epicardium: interstitial cajal-like cells (iclc) highlighted by immunofluorescence. J Cell Mol Med. 2009;13:771–7.
24. Gherghiceanu M, Popescu LM. Human epicardium: ultrastructural ancestry of mesothelium and mesenchymal cells. J Cell Mol Med. 2009;13:2949–51.
25. Popescu LM, Gherghiceanu M, Suciu LC, Manole CG, Hinescu ME. Telocytes and putative stem cells in the lungs: electron microscopy, electron tomography and laser scanning microscopy. Cell Tissue Res. 2011;345:391–403.
26. Zheng Y, Li H, Manole CG, Sun A, Ge J, Wang X. Telocytes in trachea and lungs. J Cell Mol Med. 2011;15:2262–8.
27. Cretoiu SM, Popescu LM. Telocytes revisited. Biomol Concepts. 2014;5:353–69.
28. Vannucchi MG, Traini C, Manetti M, Ibba-Manneschi L, Faussone-Pellegrini MS. Telocytes express pdgfralpha in the human gastrointestinal tract. J Cell Mol Med. 2013;17:1099–108.
29. Zheng Y, Bai C, Wang X. Telocyte morphologies and potential roles in diseases. J Cell Physiol. 2012;227:2311–7.
30. Popescu LM, Manole CG, Gherghiceanu M, Ardelean A, Nicolescu MI, Hinescu ME, Kostin S. Telocytes in human epicardium. J Cell Mol Med. 2010;14:2085–93.
31. Bani D, Formigli L, Gherghiceanu M, Faussone-Pellegrini MS. Telocytes as supporting cells for myocardial tissue organization in developing and adult heart. J Cell Mol Med. 2010;14:2531–8.
32. Manole CG, Cismasiu V, Gherghiceanu M, Popescu LM. Experimental acute myocardial infarction: telocytes involvement in neo-angiogenesis. J Cell Mol Med. 2011;15:2284–96.
33. Bojin FM, Gavriliuc OI, Cristea MI, Tanasie G, Tatu CS, Panaitescu C, Paunescu V. Telocytes within human skeletal muscle stem cell niche. J Cell Mol Med. 2011;15:2269–72.
34. Sheng J, Shim W, Lu J, Lim SY, Ong BH, Lim TS, Liew R, Chua YL, Wong P. Electrophysiology of human cardiac atrial and ventricular telocytes. J Cell Mol Med. 2014;18:355–62.
35. Xiao J, Wang F, Liu Z, Yang C. Telocytes in liver: electron microscopic and immunofluorescent evidence. J Cell Mol Med. 2013;17:1537–42.
36. Bei Y, Zhou Q, Fu S, Lv D, Chen P, Chen Y, Wang F, Xiao J. Cardiac telocytes and fibroblasts in primary culture: different morphologies and immunophenotypes. PLoS One. 2015; 10:e0115991.
37. Zimmermann A. Liver regeneration: the emergence of new pathways. Med Sci Monit Int Med J Exp Clin Res. 2002;8:RA53–63.
38. Fausto N. Liver regeneration. J Hepatol. 2000;32:19–31.
39. Gilgenkrantz H. Collin de l'Hortet A: new insights into liver regeneration. Clin Res Hepatol Gastroenterol. 2011;35:623–9.
40. Wang F, Song Y, Bei Y, Zhao Y, Xiao J, Yang C. Telocytes in liver regeneration: possible roles. J Cell Mol Med. 2014;18:1720–6.
41. Li LC, Gao J, Li J. Emerging role of hmgb1 in fibrotic diseases. J Cell Mol Med. 2014;18:2331–9.
42. Wynn TA, Ramalingam TR. Mechanisms of fibrosis: therapeutic translation for fibrotic disease. Nat Med. 2012;18:1028–40.

43. Iwaisako K, Jiang C, Zhang M, Cong M, Moore-Morris TJ, Park TJ, Liu X, Xu J, Wang P, Paik YH, Meng F, Asagiri M, Murray LA, Hofmann AF, Iida T, Glass CK, Brenner DA, Kisseleva T. Origin of myofibroblasts in the fibrotic liver in mice. Proc Natl Acad Sci U S A. 2014;111: E3297–305.
44. Zhao B, Chen S, Liu J, Yuan Z, Qi X, Qin J, Zheng X, Shen X, Yu Y, Qnin TJ, Chan JY, Cai D. Cardiac telocytes were decreased during myocardial infarction and their therapeutic effects for ischaemic heart in rat. J Cell Mol Med. 2013;17:123–33.
45. Fu S, Wang F, Cao Y, Huang Q, Xiao J, Yang C, Popescu LM. Telocytes in human liver fibrosis. J Cell Mol Med. 2015;19:676–83.
46. Cismasiu VB, Popescu LM. Telocytes transfer extracellular vesicles loaded with micrornas to stem cells. J Cell Mol Med. 2015;19:351–8.
47. Zhao B, Liao Z, Chen S, Yuan Z, Yilin C, Lee KK, Qi X, Shen X, Zheng X, Quinn T, Cai D. Intramyocardial transplantation of cardiac telocytes decreases myocardial infarction and improves post-infarcted cardiac function in rats. J Cell Mol Med. 2014;18:780–9.
48. Diaz-Flores L, Gutierrez R, Saez FJ, Diaz-Flores Jr L, Madrid JF. Telocytes in neuromuscular spindles. J Cell Mol Med. 2013;17:457–65.
49. Kuehbacher A, Urbich C, Zeiher AM, Dimmeler S. Role of dicer and drosha for endothelial microrna expression and angiogenesis. Circ Res. 2007;101:59–68.
50. Otsuka M, Zheng M, Hayashi M, Lee JD, Yoshino O, Lin S, Han J. Impaired microrna processing causes corpus luteum insufficiency and infertility in mice. J Clin Invest. 2008;118: 1944–54.
51. Shen X, Fang J, Lv X, Pei Z, Wang Y, Jiang S, Ding K. Heparin impairs angiogenesis through inhibition of microrna-10b. J Biol Chem. 2011;286:26616–27.
52. Wang S, Aurora AB, Johnson BA, Qi X, McAnally J, Hill JA, Richardson JA, Bassel-Duby R, Olson EN. The endothelial-specific microrna mir-126 governs vascular integrity and angiogenesis. Dev Cell. 2008;15:261–71.
53. Urbich C, Kuehbacher A, Dimmeler S. Role of micrornas in vascular diseases, inflammation, and angiogenesis. Cardiovasc Res. 2008;79:581–8.
54. Caporali A, Meloni M, Vollenkle C, Bonci D, Sala-Newby GB, Addis R, Spinetti G, Losa S, Masson R, Baker AH, Agami R, le Sage C, Condorelli G, Madeddu P, Martelli F, Emanueli C. Deregulation of microrna-503 contributes to diabetes mellitus-induced impairment of endothelial function and reparative angiogenesis after limb ischemia. Circulation. 2011;123:282–91.
55. Chen Y, Gorski DH. Regulation of angiogenesis through a microrna (mir-130a) that down-regulates antiangiogenic homeobox genes gax and hoxa5. Blood. 2008;111:1217–26.
56. Manetti M, Guiducci S, Ruffo M, Rosa I, Faussone-Pellegrini MS, Matucci-Cerinic M, Ibba-Manneschi L. Evidence for progressive reduction and loss of telocytes in the dermal cellular network of systemic sclerosis. J Cell Mol Med. 2013;17:482–96.
57. Mou Y, Wang Y, Li J, Lu S, Duan C, Du Z, Yang G, Chen W, Zhao S, Zhou J, Wang C. Immunohistochemical characterization and functional identification of mammary gland telocytes in the self-assembly of reconstituted breast cancer tissue in vitro. J Cell Mol Med. 2013;17:65–75.

Index

X. Wang, D. Cretoiu (eds.), *Telocytes*, Advances in Experimental Medicine and Biology, 913, DOI 10.1007/978-981-10-1061-3

R

S

T

U

FSC
www.fsc.org
MIX
Papier | Fördert
gute Waldnutzung
FSC® C083411

Kolibri 360 GmbH
Ferdinand-Jühlke-Straße 7
99095 Erfurt, Deutschland
produktsicherheit@kolibri360.de